NUTRITION SUPPORT

THEORY AND THERAPEUTICS

Chapman & Hall Series in Clinical Nutrition

Ronni Chernoff, Ph.D., R.D.
Series Editor

Adolescent Nutrition:
Assessment and Management
Edited by:
Vaughn I. Rickert, Psy.D., F.S.A.M.

Enteral Nutrition
Edited by:
Bradley C. Borlase, M.D., M.S.
Stacey J. Bell, M.S., R.D., C.N.S.D.
George L. Blackburn, M.D., Ph.D.
R. Armour Forse, M.D., Ph.D.

Nutrition Support:
Theory and Therapeutics
Edited by:
Scott A. Shikora, M.D.
George L. Blackburn, M.D., Ph.D.

Obesity:
Pathophysiology, Psychology, and Treatment
Edited by:
George L. Blackburn, M.D., Ph.D.
Beatrice S. Kanders, Ed.D., M.P.H., R.D.

Obesity Assessment:
Tools, Methods, Interpretations
Edited by:
Sachiko T. St. Jeor, Ph.D., R.D.

Pediatric Enteral Nutrition
Edited by:
Susan S. Baker, M.D., Ph.D.
Robert D. Baker, Jr., M.D., Ph.D.
Anne Davis, M.S., R.D., C.N.S.D.

Pediatric Parenteral Nutrition
Edited by:
Robert D. Baker Jr., M.D., Ph.D.
Susan S. Baker., M.D., Ph.D.
Anne Davis, M.S., R.D., C.N.S.D.

NUTRITION SUPPORT

THEORY AND THERAPEUTICS

Chapman & Hall Series in Clinical Nutrition
Series Editor: Ronni Chernoff, Ph.D., R.D.

Edited by
Scott A. Shikora, M.D.
Tufts University School of Medicine, Faulkner Hospital
George L. Blackburn, M.D., Ph.D.
Harvard Medical School, New England Deaconess Hospital

CHAPMAN & HALL

ITP® International Thomson Publishing

New York • Albany • Bonn • Boston • Cincinnati • Detroit • London • Madrid • Melbourne
Mexico City • Pacific Grove • Paris • San Francisco • Singapore • Tokyo • Toronto • Washington

Cover Design: Andrea Meyer, emDASH inc.

Printed in the United States of America

For more information, contact:

Chapman & Hall
115 Fifth Avenue
New York, NY 10003

Chapman & Hall
2-6 Boundary Row
London SE1 8HN
England

Thomas Nelson Australia
102 Dodds Street
South Melbourne, 3205
Victoria, Australia

Chapman & Hall GmbH
Postfach 100 263
D-69442 Weinheim
Germany

International Thomson Editores
Campos Eliseos 385, Piso 7
Col. Polanco
11560 Mexico D.F.
Mexico

International Thomson Publishing-Japan
Hirakawacho-cho Kyowa Building, 3F
1-2-1 Hirakawacho-cho
Chiyoda-ku, 102 Tokyo
Japan

International Thomson Publishing Asia
221 Henderson Road #05-10
Henderson Building
Singapore 0315

1 2 3 4 5 6 7 8 9 10 XXX 01 00 99 98 97

Library of Congress Cataloging-in-Publication Data

Nutrition support : theory and therapeutics / edited by Scott A. Shikora and George L. Blackburn.
p. cm. — (Chapman & Hall series in clinical nutrition)
Includes bibliographical references and index.
ISBN 0-412-06681-5 (alk. paper)
1. Dietetics. 2. Diet therapy. I. Shikora, Scott A., 1959– . II. Blackburn, George L., 1936– . III. Series.
[DNLM: 1. Nutritional Support. 2. Nutrition Assesment. 3.Nutritional Requirements. WB 400 N97535 1996]
RM216.N8624 1996
613.2—dc20
DNLM/DLC 96-28658
for Library of Congress CIP

British Library Cataloguing in Publication Data available

To order this or any other Chapman & Hall book, please contact **International Thomson Publishing, 7625 Empire Drive, Florence, KY 41042.** Phone: (606) 525-6600 or 1-800-842-3636.
Fax: (606) 525-7778, e-mail: order@chaphall.com.

For a complete listing of Chapman & Hall's titles, send your requests to
Chapman & Hall, Dept. BC, 115 Fifth Avenue, New York, NY 10003.

Contents

PART III ENTERAL NUTRITION

PART IV SPECIAL SITUATIONS IN NUTRITIONAL SUPPORT

Series Editor's Foreword

It has been several years since a new approach to nutrition has been captured in a reference book that provides the most up-to-date information. Since its beginning, nutrition support has been a specialty field that has evolved rapidly, with advances occurring in all its aspects: parenteral and enteral nutrition formulations; delivery systems; equipment, including tubes, catheters, and pumps; care protocols; and nutrient profile of substrate solutions, including nutrient supplements and additives. This is a compendium of the why, what, when and how of nutrition support. A new book that captures the state of the art as well as the science of nutrition support is welcome!

Nutrition Support: Theory and Therapeutics, edited by Scott A. Shikora, M.D. and George L. Blackburn, M.D., is divided into four sections: rationale for nutrition support; parenteral nutrition; enteral nutrition; and special situations in nutrition support. The first part, Rationale for Nutritional Support, provides the base for nutrition support by presenting the foundation of nutrition information needed for competent practice. These first six chapters address the identification of need for nutritional support and what the practitioner has to know to assure nutrient adequacy of nutrition support protocols. Also included in this section are chapters on assessing nutritional status and on basic human nutritional requirements including macro- and micronutrients. These discussions lay the foundation for the rest of this volume on nutrition support theory and practice.

The second part of this book focuses on parenteral nutrition, its indications and the various aspects of its provision, including access, formulation, delivery systems, and potential complications. There are topics covered in this section that are generally not included in other similar texts, such as a chapter on peripheral parenteral nutrition. Each chapter looks at the theory and therapeutics of parenteral nutrition in some depth.

Topics associated with enteral nutrition are addressed in the third section of the book. It is not common to find both parenteral and enteral nutrition discussed in the same book, but in the approach used by the editors, Drs. Shikora and Blackburn, the inclusion of all aspects of nutrition support adds to the value of this volume. This part of the book examines the state of the art and science of enteral nutrition access routes, infusion system issues, formula selection with a chapter on special nutrients for gut feeding, and the prevention and management of enteral feeding complications.

The last part of this book addresses special issues in nutritional support. These include problems unique to individuals with diabetes; liver, pulmonary, or renal disease; pancreatitis, HIV; cancer; inflammatory bowel disease; organ transplantation; morbid obesity; critical illnesses; and trauma and burns. There are chapters on nutrition support in pregnancy, for pediatric patients, and for patients on home nutrition support. This section is very important for all clinicians as it deals with the application of theory into practice.

This book is a very thorough compendium of issues encountered by nutrition support professionals. What is unique about this volume is that it gives equal treatment to both parenteral and enteral nutrition. It also, in the choice of authors, demonstrates a valuable perspective on multidisciplinary health care which is very valuable in nutrition support. This book brings a dimension to the topic of nutrition support that is often overlooked; it is multifaceted, multidisciplinary and and thorough. It is a wonderful addition to the Chapman & Hall Series in Clinical Nutrition.

Ronni Chernoff, Ph.D., R.D.
Series Editor, Chapman & Hall Series in Clinical Nutrition

Foreword

As defined primarily by anthropometric and serum albumin determinations, malnutrition has been estimated to be manifested in up to fifty percent of hospitalized patients, occurring especially in the indigent population and surgical patients in intensive care units. Although the vast majority of patients are malnourished upon admission to the hospital as a result of inadequate nutrient intake induced by, or associated with, their diseases or disorders, others become malnourished subsequently during hypercatabolism associated with trauma, major surgical interventions, infection, sepsis, and/or other forms of stress. By and large, malnutrition occurs insidiously, and to some extent permissibly, secondary to a myriad of complex interacting biological and social factors. Moreover, malnutrition encompasses a wide variety of situations in different individuals ranging from morbid obesity to single-nutrient deficiency states to protein deficiency, to combined protein-calorie malnutrition to cachexia with alterations in host resistance and defense, wound healing, and organ function which may be manifested under each of these circumstances. Although a clear demonstration of cause and effect in malnutrition and diminished host resistance to sepsis is lacking, it is most likely that malnutrition and sepsis predispose to each other, and/or co-exist in a vicious cycle with other factors. In this "chicken or egg" situation, it does not make much difference from the therapeutic point of view, because the ultimate objective is to improve patient survival overall with combined nutritional repletion and immunologic restoration. Evidence of malnutrition may not be easily detected during a casual clinical examination. However, a reasonably accurate evaluation of the nutritional status of most patients can be readily obtained from a carefully directed history, physical examination, and rational combination of clinical and laboratory studies. The editors and their carefully selected authors address these and other related issues comprehensively as they present their rationale for nutritional support in Part I of this volume.

Historically, most clinicians have directed their interests and efforts primarily toward achieving or maintaining optimal nutrition of the body as a whole. Traditionally, it has been of highest priority to achieve fundamental nutritional goals, such as growth and development of infants; weight gain to ideal body weight in adults; nitrogen equilibrium or positive nitrogen balance; increased strength and a sense of well-being; improved capacity for function and activity; decreased morbidity and mortality; improved risk:benefit ratio; improved recovery, convalescence and rehabilitation; decreased hospital stay; and improved cost:benefit ratio. These have been the general, but very important, goals of the early endeavors to provide and improve nutritional support.

Interest in these broad clinical objectives continues among today's enlightened clinicians. However, a major advance in the potential for nutritional support was made almost thirty years ago when laboratory and clinical investigations demonstrated unequivocally for the first time that all essential nutrient substrates could be administered intravenously for prolonged periods of time to support normal growth and development in infants and to achieve or maintain a state of anabolism in nutritionally depleted adults. Subsequent studies of the technique of total parenteral nutrition by several basic investigators and clinical scientists demonstrated improved or positive nitrogen balance, improved wound healing, and accelerated convalescence and rehabilitation in various groups of malnourished patients, but most importantly, demonstrated the obvious positive clinical effects of nutritional support. During the past quarter century, the methodology and technology for providing safe and effective long-term supplemental or total parenteral feeding have been modified and improved continuously as a result of numerous scientific developments and clinical applications in the rapidly advancing field of nutritional support. Moreover, the dramatic results of the technique, together with acquired expertise in catheter insertion, solution formulation and preparation, and safe and effective delivery to the patient, have resulted in the widespread acceptance and application of central venous nutrient infusion. Indeed, it is possible for patients with inadequate or compromised gastrointestinal function, who might otherwise succumb directly to starvation or malnutrition, to be maintained practically and effectively for prolonged periods of time with various parenteral feeding regimens and techniques even on an ambulatory basis at home. However, the same considerations and concerns that have existed through the years continue to be addressed, as efforts are increasingly directly toward improving the materials and methods for providing nutritional support and assessing its safety and effectiveness. As a result, the substrates for parenteral nutrition and enteral nutrition have been greatly improved and expanded, leading to a prodigious increase in the use of parenteral and enteral feedings in a wide variety of patients with inadequate oral intake to support normal nutritional status.

Basic and clinical investigations with fat emulsions; essential and nonessential fatty acid; short-, medium-, and long-chain triglycerides; structured lipids; and lipoproteins and other lipid substrates continue to progress and will demand more investigative effort and evaluation in the future. Energy substrates in addition to the fats include glucose, fructose, sorbitol, xylitol, carbohydrate polymers, glycerol, ethanol and many other biochemical compounds and intermediary metabolites that might be considered as fuel sources in future formulations. Major electrolyte and mineral investigations must continue, especially those concerned with trace element imbalances and deficiencies. The precise requirements for both fat- and water-soluble vitamins must also be determined under various conditions, even while integrating these micronutrients into the major components of enteral and parenteral nutrition regimens. Amino acids have become, and will remain, a vital component of the nutritional and metabolic management of critically ill patients. As current knowledge of the altered regulation of amino acid metabolism in injured or critically ill patients increases, the formulation and administration of more effective parenteral and enteral therapeutic feeding regimens will inevitably evolve. The use of specific amino acids in pharmacologic doses and in special combinations and rations is likely to be beneficial as experience and technology improve. A working knowledge of the multiple and complex functions of the fundamental molecules known as amino acids will be essential to the competent and efficacious practice of medicine in the 21st century. Because derangements in amino acid metabolism are common in pathologic states and are very detrimental to optimal metabolic function, reversal of these pathophysiologic alterations by optimal nutritional support will be obligatory if the outcomes of critically ill patients are to be significantly improved.

Metabolic interests of clinicians have been increasingly directed toward disease-specific nutritional support. A primary concern and responsibility of all physicians, regardless of their specialty, must be the orientation and dedication of their efforts to maintain optimal function of the maximal number of cells in the total body cell mass of the patient at all times in order to ensure optimal organ and system function. These important aspects of, and considerations for, parenteral and enteral nutrition support are presented and discussed practically and comprehensively by an elite group of expert authors and organized in a rational manner by the editors in Part II and Part III of this volume.

Special situations in nutritional support are presented by additional experts assembled by the editors in Part IV. The pancreas has been a subject of focal or organ concern and attention, and many special formulas have been studied in an attempt to reduce the ravages of acute and chronic pancreatitis, cystic fibrosis, and diabetes, or to increase the efficacy of both the endocrine and exocrine cells of the pancreas selectively by manipulating specific substrates enterally or parenterally. Much investigation is required in this most stimulating area of endeavor especially considering the array and variety of endocrine, exocrine and paracrine cells crowded into this organ and responsible for its many important and highly specialized nutritional and metabolic functions. In managing patients with respiratory failure, problems have been encountered at times with high-carbohydrate diets. Great progress has been made in manipulating the substrates in such patients in order to maximize the effects of ventilator support and to minimize the problems of weaning patients from ventilators while accomplishing normal ventilatory and respiratory function. The role of fat substrates as a major source of non-protein calories in patients requiring lung support is a timely topic requiring additional study. The use of branched-chain amino acids has also been advocated in the management of these patients and may prove to be beneficial.

The kidney has been studied from the standpoint of the advantage of low-protein oral diets for many decades and from the inception of TPN with multiple renal failure formulas based on the varied etiology and severity of renal decompensation. Because many types and causes of renal failure exist, the therapy of each individual patient with impaired renal function secondary to a number of pathological processes will probably require some modification in formula composition in order to provide optimal nutritional support. Donor kidney perfusion with special nutrient formulations designed to prolong the quality and duration of preservation prior to transplantation continues to be a fruitful area for investigation. Refinement of the keto acid concept for recycling urea and the development of other nitrogenous substrate support for renal transplant patients that might act as adjuncts to both the nutritional and immunosuppressant requirements for success in this vital area will continue to stimulate the ingenuity of investigators.

In the area of specific organ nutrition, the liver presents the most difficult challenge because of its multiple and complex functions. From a clinical standpoint, much interest has been focused on the general biochemical activities of the liver which include protein synthesis, especially of albumin, globulins and other specialized proteins; enzyme synthesis and function in the various metabolic pathways; energy metabolism; cholesterol synthesis, breakdown and excretion; bile composition, secretion and stasis; and gallstone formation. Clinical investigation is likely to continue in the area of hepatic encephalopathy which has been shown to respond to specific nutritional manipulation by increasing the branched-chain amino acids and decreasing the aromatic amino acids in the ration. Judicious use of intermediary metabolite nutrient substrates to achieve results more efficacious than those now possible with the use of the common amino acid, fatty acid and dextrose substrates will continue to test the mettle of surgical and medical hepatologists.

Optimal diets remain to be perfected to ensure absorption of all required nutrients in order to maximize adaptation of the short bowel, to support bowel transplantation, and to maintain the mucosal barrier and integrity against invasion by pathological microorganisms. Much recent progress has been accomplished in this area with the demonstration of the importance of glutamine as a major intestinal nutrient, but much more study and work are indicated before this deceptively simple, but actually sophisticated, tubular organ responsible for many complex functions is to be truly understood. Furthermore, the role of various types of fiber and other substrates provided enterally or by mouth in the genesis and treatment of adenocarcinoma and other malignancies, atherosclerosis, bowel motility disorders, inflammatory bowel diseases, severe obesity, and other disorders must be identified and addressed. Finally, attention must be paid to the special needs of AIDS patients, other immunologically compromised patients, cancer patients, trauma and burn patients, pediatric patients and pregnant women. As nutritional and investigative work progresses to the cellular and subcellular level, the potential for developing nutrient regimens specifically tailored to inhibit the growth and division of cancer cells or to render the malignant cells more susceptible to antineoplastic agents will be more readily exploited. Currently, nutritional support of cancer patients has reduced the ravages and morbidity of starvation but has not uniformly been of obvious benefit in lowering mortality rates. Studies are underway to determine the general effects of feeding cancer patients with enteral and parenteral arginine-enriched diets and the potential for improving the capacity of the immune system to contain, or even kill, the malignant cells. The reversal and treatment with specific nutrients of such conditions as multiple systems organ failure, hematopoietic disorders, and brain and central nervous system disorders remain great challenges for future clinicians and investigators. The current strategies of nutritional support of the whole organism and of the key organ systems must proceed and advance to the cellular level of nutritional support if the ultimate goal of providing optimal nutrition for all patients under all conditions at all times is to be realized. The relationships between nutrient substrates, cellular biology, molecular biology, and the human genome are myriad, and their identification, classification and beneficial exploitation in the management of patients will, in all likelihood, open new frontiers for clinical nutrition support in the next century.

Dr. Blackburn and Dr. Shikora have commendably achieved their goal of producing a manual which is a scholarly, state of the art, user friendly and utilitarian instrument for clinicians of all specialties caring for nutritionally compromised patients. It is yet another testament to their extraordinary expertise and contributions to the vital area of nutritional support. I am honored to have the opportunity to introduce this valuable addition to our clinical armamentarium, and I salute the authors, editors, and publishers for their impressive accomplishment.

Stanley J. Dudrick, MD
Program Director and Associate Chairman
Department of Surgery
St. Mary's Hospital
Waterbury, Connecticut

Preface

Since the mid-1970s, the discipline of nutritional support has expanded greatly in both knowledge and application. Malnutrition and related morbidities are now increasingly recognized and less readily tolerated. Nutritional intervention for patients unable to consume adequate protein and calories is in most cases initiated earlier, and with greater frequency. The options available for nutritional support has also expanded. Improved technology has led to the development of better catheters, tubes, pumps, and solutions. Feeding strategies now consider not only the most appropriate route of administration but also the best formula prescription for a patient's specific condition and disease state. This notion of disease and patient-specific nutritional support requires a greater understanding of the discipline by the prescribing clinician who has traditionally only been able to order from a menu that had few choices and assumed that "one size fits all."

Further complicating the use of nutritional support is the ever increasing interest in considering the pharmacologic characteristics of certain nutrients and the potential to modulate metabolic processes. The concept of "nutritional immunomodulation" or "neutraceuticals" as it is currently termed, is an exciting new application for nutritional support, one that offers the field a more aggressive role in the overall management of patients, particularly those who are critically ill or seriously injured.

It is somewhat ironic that, as the discipline of nutritional support expands into new directions for the betterment of patient care, its increased sophistication may actually restrict its use. More and more, clinicians who are not specifically trained in nutritional support are finding it too difficult to prescribe. There lies the incentive for this book: to compile from the experts in the field a comprehensive and scholarly, yet practical "how to feed" manual that covers most of the major issues and disease states.

Nutritional Support: Theory and Therapeutics is divided into four sections. Part I covers the general concepts of the field of nutritional support. It describes the growing importance of nutritional support in medicine, the complications associated with malnutrition, the components of the comprehensive nutritional assessment and the methods used to calculate both the macronutrient and micronutrient requirements. Also included in this section is a chapter on the controversial topic of preoperative nutritional intervention for the severely malnourished patient who requires surgery.

Part II of this book is devoted exclusively to the application of parenteral nutrition. It includes chapters that describe the indications for the parenteral route of feeding, access issues, admixture considerations, nutrient–drug interactions, and the complications associated with intravenous feeding. Also included in this section of the book are chapters that describe the proper use of the peripheral vein for feeding and the prescription of disease- and patient-specific parenteral solutions.

Part III covers the salient issues pertaining to enteral nutritional support. Over the last decade, enteral feeding has been revisited. Recent investigations have suggested that it may be a superior route of feeding than parenteral. Chapters in this section describe the numerous access options and formula choices to aid the clinician in selecting the most appropriate access and formula from the multitude of available products. The chapter on complications not only discusses the risks associated with tube feeding, but also outlines steps to prevent and manage these problems. Also included in this section are two chapters on the benefits of early enteral feeding and special nutrients for gut feeding that review the recent literature concerning the relationship of novel enteral feeding techniques with patient outcome.

The final section of the book, part IV, provides a description of feeding strategies for most of the major types of patients who might require nutritional intervention, both while in the hospital and after discharge. Each of these chapters is written by authors who have extensive experience with the nutritional support of the patient's they have written about.

Our hope is that this book will help the clinician understand the rationale and mechanics for safe, effective nutritional support. By educating the medical community and alleviating some of the mystery of parenteral and enteral nutrition, we hope to expand the use of nutritional support and thereby improve the total care package for the hospitalized patient.

Scott A. Shikora, MD
George L. Blackburn, MD, PhD

Contributors

Roger Anderson, M.D.
Clinical Faculty
Department of Obstetrics and Gynecology
University of Washington
Staff Obstetrician
Providence Hospital and Swedish Medical Center
Seattle, Washington

Caroline Apovian, M.D.
Associate Physician
Department of Gastroenterology and Nutrition
Geisinger Medical Center
Danville, Pennsylvania

Timothy J Babineau, M.D.
Assistant Professor of Surgery
Harvard Medical School
Department of Surgery
New England Deaconess Hospital
Boston, Massachusetts

Robert D. Baker, M.D., Ph.D.
Associate Professor of Pediatrics
Division of Pediatric GI/Nutrition
Medical University of South Carolina
Charleston, South Carolina

Susan S. Baker, M.D., Ph.D.
Associate Professor of Pediatrics
Division of Pediatric GI/Nutrition
Medical University of South Carolina

Department of Pediatrics
Children's Hospital
Charleston, South Carolina

Tom Baumgartner, Pharm.D., M.Ed., B.C.N.S.P.
Clinical Pharmacy Specialist
Nutritional Support Consult Services
Clinical Professor
Colleges of Pharmacy/Medicine/Nursing/Dentistry
University of Florida
Gainesville, Florida

Susan L. Baumgartner, Pharm.D., M.B.A.
Consultant Pharmacists of America, Inc.
Gainesville, FL

Mary Ellen Beindorff, R.D.
Division of Nutrition
Barnes Hospital
St. Louis, Missouri

Stacey J. Bell, D.Sc., R.D.
Instructor in Surgery
Harvard Medical School
Research Dietitian
Surgical Metabolism Laboratory
New England Deaconess Hospital
Boston, Massachusetts

Peter N. Benotti, M.D.
Professor of Surgery
Mt Sinai School of Medicine
New York
Chief of Surgery
Englewood Hospital and Medical Center
Englewood, New Jersey

Carl T. Bergren, M.D.
Staff Attending
Oakwood Hospital and Medical Center
Dearborn, Michigan

George L. Blackburn, M.D., Ph.D.
Associate Professor of Surgery
Harvard Medical School
Chief, Nutrition/Metabolism Laboratory
Cancer Research Institute
New England Deaconess Hospital
Boston, Massachusetts

Wendy Swails Bollinger, R.D.
Research Dietitian
Nutrition Support Service
New England Deaconess Hospital
Boston, Massachusetts

Gordon P. Buzby, M.D.
Associate Professor of Surgery
University of Penn. School of Medicine
Philadelphia, Pennsylvania

Yoshimi L. Clark, Pharm. D., B.C.N.S.P.
(*Formerly*), Chief O.R. Pharmacy
Nutrition Support Pharmacist
Wilford Hall Medical Center
Lackland AFB, Texas

Linda Codina, R.D., C.N.S.D.
Nutrition Support Service
Wilford Hall Medical Center
Lackland AFB, Texas

Anne Davis, M.S., R.D., C.N.S.D.
Assistant Professor of Pediatrics
Division of Pediatric GI/Nutrition
Medical University of South Carolina
Charleston, South Carolina

Donald R. Duerksen, M.D.
Assistant Professor of Medicine
Section of Gastroenterology
University of Manitoba
Winnipeg, Manitoba Canada

Raed Dweik, M.D.
Fellow,
Department of Pulmonary and Critical Care Medicine
Cleveland Clinic Foundation
Cleveland, Ohio

Laura Ellis, Ph.D.
Executive Director of the Oley Foundation
Albany Medical Center
Albany, New York

John Fitzpatrick, M.D.
Denver Childrens Hospital
Denver, Colorado

R. Armour Forse, M.D., Ph.D.
Associate Professor of Surgery
Harvard Medical School
Chief, Division of General Surgery
New England Deaconess Hospital
Boston, Massachusetts

M. Patricia Fuhrman, R.D., C.N.S.D.
Director, Nutrition Support Service
St. Louis University Medical Center
St. Louis, Missouri

E. Kerry Gallivan, M.D., M.P.H.
Surgical Resident
New England Medical Center
Boston, Massachusetts

Robert C. Gorman, M.D.
Instructor of Surgery
University of Penn. School of Medicine
Philadelphia, Pennsylvania

Bharat Gupta, M.D.
Resident,
Department of General Internal Medicine
Cleveland Clinic Foundation
Cleveland, Ohio

Jane Heetderks-Cox, R.D., C.N.S.D.
Clinical Dietitian/Dietetic Internship Instructor
Malcolm Grow Medical Center
Andrews AFB, Maryland

George Henderson, Ph.D.
Assistant Research Scientist
Division of Endocrinology
Department of Medicine
College of Medicine
University of Florida
Gainesville, FL

Virginia Herrmann, M.D.
Professor of Surgery
Medical Director, Nutrition Support Service
St. Louis University Medical Center
St. Louis, Missouri

Lyn Howard, M.D.
Professor of Medicine
Associate Professor of Pediatrics
Medical and Research Director of the Oley Foundation
Albany Medical College
Albany, New York

Peter G. Janu, M.D.
Research Fellow in Nutrition and Trauma
University of Tennessee
Memphis, Tennessee

Gordon L. Jensen, M.D., Ph.D.
Director, Section of Nutrition Support
Department of Gastroenterology and Nutrition
Geisinger Medical Center
Danville, Pennsylvania

Mitchell V. Kaminski, Jr., M.D.
Clinical Professor of Surgery
FUHS/Chicago Medical School and
Thorek Hospital and Medical Center
Chicago, Illinois

Richard E. Karulf, M.D.
Associate Professor of Surgery
Uniformed Services University of the Health Sciences
Bethesda, Maryland
Chief, Colorectal Surgery
Wilford Hall Medical Center
Lackland AFB, Texas

Kenneth A. Kudsk, M.D.
Professor of Surgery
Director of Surgical Research
University of Tennessee
Memphis, TN

Vernon T. Lew, R.Ph., B.C.N.S.P.
Chief, Decentralized Pharmacy Services
Nutrition Support Pharmacist
Wilford Hall Medical Center
Lackland AFB, Texas

Jeffrey A. Lowell, M.D.
Assistant Professor of Surgery
Section of Transplantation
Washington University School of Medicine
St. Louis, Missouri

Margaret Malone Ph.D.
Professor of Pharmacy Practice
Albany College of Pharmacy
Albany, New York

Edward A. Mascioli, M.D.
Department of Nutrition Support
New England Deaconess Hospital
Boston, Massachusetts

George Melnik, Pharm.D., F.A.C.N., B.C.N.S.P.
Clinical Pharmacy Specialist
Audie L. Murphy Memorial Veterans' Hospital
Clinical Assistant Professor Department of Pharmacology
University of Texas Health Science Center San Antonio
Clinical Assistant Professor of Pharmacy University of Texas, Austin

Gayle Minard, M.D.
Assistant Professor of Surgery
University of Tennessee
Memphis, Tennessee

Michael Moore, M.D.
Instructor in Medicine
State University of New York at Buffalo
Chief of Gastroenterology
Chairman of Nutritional Care
Our Lady of Victory Hospital
Lackawana, New York
Chairman of Nutritional Care
Mercy Hospital
Buffalo, New York

Jon B. Morris, M.D.
Assistant Professor of Surgery
University of Pennsylvania School of Medicine
Philadelphia, Pennsylvania

James Mullen, M.D.
Associate Professor of Surgery
University of Pennsylvania School of Medicine
Philadelphia, Pennsylvania

Suzanne Murray, R.N., M.S.N.
Director of the Patient Learning Center
Albany Medical Center
Albany, New York

Peter C. Muskat, M.D.
Major, USAF, MC
Department of Surgery
Wilford Hall Medical Center
Lackland AFB, Texas

Angela M. Ogawa, R.D., C.N.S.D.
Nutrition Support Service
Wilford Hall Medical Center
Lackland AFB, Texas

Faith D. Ottery, M.D., Ph.D.
Assistant Professor of Surgery
Temple University School of Medicine
Medical Director,
Cancer Recovery Center
Philadelphia, Pennsylvania

Stephen Phinney, M.D., Ph.D.
Professor of Medicine
University of California at Davis
Clinial Nutrition
University of California
Davis, California

Basil A. Pruitt, Jr., M.D.
Col USA MC
Commander and Director
US Army Institute of Surgical Research
Fort Sam Houston, Texas

Patricia Queen Samour, M.M.Sc., R.D.
Director, Department of Dietetics
New England Deaconess Hospital
Boston, Massachusetts

Douglas L. Seidner, M.D.
Staff, Department of Gastroenterology
Co-director, Nutrition Support Team
Cleveland Clinic Foundation
Cleveland, Ohio

Scott A. Shikora, M.D.
Assistant Professor of Surgery
Tufts University School of Medicine
Assistant Chief of Surgery
Director, Nutrition Support Service
Director, Surgical Intensive Care Units
Faulkner Hospital
Boston, Massachusetts

John Siepler, Pharm.D., B.C.N.S.P.
Associate Clinical Professor
Division of Clinical Pharmacy
University of California at San Francisco
Nutrition Support Pharmacist
University of California at Davis Medical Center
Davis, California

Christopher Still, M.S., D.O.
Nutrition Fellow
Department of Gastroenterology and Nutrition
Geisinger Medical Center
Danville, Pennsylvania

Elaine B. Trujillo, M.S., R.D.
Nutrition Support Service
Brigham & Women's Hospital
Boston, Massachusetts

J. Elizabeth Tuttle-Newhall, M.D.
Fellow, Surgical Critical Care
University of North Carolina
Chapel Hill, North Carolina

Laurie Mello Udine, R.D., C.N.S.D., M.P.A, L.D.N.
Nutrition Support Consultant
Private Practice
Boca Raton, Florida

David W. Voigt, M.D.
Major USA MC
US Army Institute of Surgical Research
Fort Sam Houston, Texas

H. David Willcutts, Jr.
President and CEO of Health Integration Strategies
Pasadena, California

Harrison D. Willcutts, M.D.
Vice President of Nutrition for Coram Healthcare, Denver
Colorado; Private Practice - Surgery, Metabolism, Nutrition,
West Springfield, Massachusetts

RATIONALE FOR NUTRITION SUPPORT

The Importance of Nutrition in Medical Practice

CHAPTER 1

Patricia Queen Samour, M.M.Sc., R.D. and
George L. Blackburn, M.D., Ph.D.

INTRODUCTION

Nutrition medicine refers to the interdisciplinary knowledge in the fields of human nutrition and medicine for the purpose of diagnosis, prevention, and treatment of disease.[1–2] It is a very good fit for the new health care delivery system where the goal is to extend health, fitness, and quality of life. It involves many allied health professions together with physicians to prescribe and implement nutritional therapy, particularly to prevent malnutrition of the hospitalized patient.[3] Thus, the diagnosis and treatment of the nutritionally challenged hospitalized patient is the keystone of clinical expertise in this speciality.[4]

Recent advances in our understanding of the pathophysiology of chronic disease associated with catabolic illness and malnutrition requires a new paradigm that includes metabolic support and metabolic fitness through nutritional medicine therapy. The knowledge required includes bionutrition, nutritional biochemistry, physiology, metabolism, and genetics. In this new paradigm we must rediscover Claude Bernard rather than Louis Pasteur, because the idea of balance as *milieu interieur* best fits the challenges of treating chronic disease.[5] Also, we need to overcome the limits of medicine that exist with regard to infectious disease, which underlie medicine today.[6] In addition to returning to the physiology of medicine, the development of nutritional medicine requires better measurement tools to supervise and monitor changes in the metabolic response to injury and nutrient intake.

Health care reform is, in part, responsible for the development of these new challenges to all health care providers and payors. There has been an evolution from a patient provider focus to one that defines the entire health maintenance organization (HMO) enrolled population. Compensation has gone from a fee-for-service basis to capitated care; and case-by-case quality control is replaced by an enrolled-population approach to benefit/cost analysis and health care outcome. In the early 1980s, there were 1217 inpatient days per 1000; in 1991 it was down to 795 inpatient days per 1000 and the future trend is for 180 inpatient days per 1000. Many critical issues face us, such as regulatory restrictions, budget reductions, reduction in labor, and equipment. Preferred Provider Organizations (PPOs), HMOs and other insurance providers are the gatekeepers of the future, creating restraints on what health insurance is available to whom. Critical pathways require proven therapy with measurable outcome.

This chapter provides a case study from our institution of our evolving response to these conditions, which are mostly outside our control. We have had to develop new skills and new knowledge to sustain this new clinical specialty. Although critical pathways that have been tested by prospective clinical control trials would be desirable, we feel that valuable knowledge exists that needs to be accessible to the nutrition support practitioners now.

Fifteen years ago, nutrition support of the hospitalized patient was directed toward the intensive care unit (ICU). Inadequate knowledge existed about the hypermetabolic state and its response to overfeeding calories in treatment called "Hyperalimentation". Today ICU therapy is metabolic support and immunomodulating therapy. The need and rationale for parenteral and enteral nutrient therapy has changed with the 50% reduction in ICU stays (10 days vs 20 days).

These dramatic changes in health care delivery require practitioners involved in nutrition and metabolic support to have an updated manual to support the critical pathways for most of the illnesses that lead to hospitalization and the risk of malnutrition. Practitioners involved in the nutritional care of patients requiring parenteral and/or enteral nutrition must maintain a state-of-the-art knowledge base, not only of the treatment of various conditions or diseases but also of the various specialized medical nutritional products and delivery methods. This manual is designed to provide a practical knowledge base of generally accepted standards of practice for nutrition support. Implementation of these evidence-based scientific opinions favors the most appropriate, safe, and effective nutritional management for hospitalized patients.

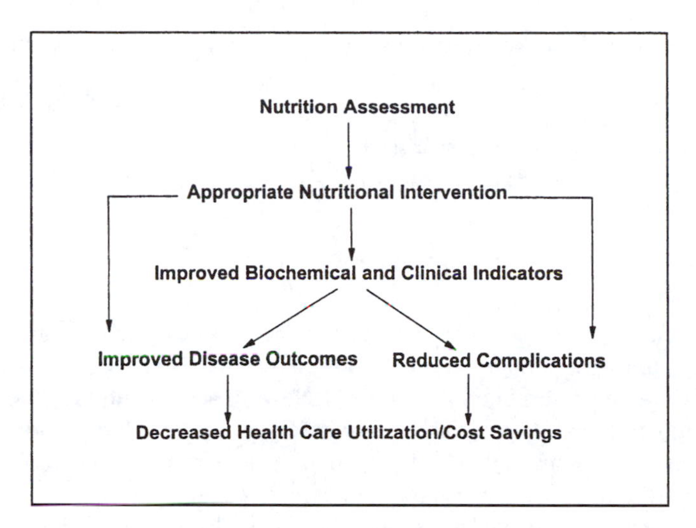

Figure 1.1. Model of the effect of nutrition intervention on outcomes and cost savings. Adapted from Mason M., ed. *Costs and Benefits of Nutritional Care. Phase I.* Chicago: American Dietetic Association, 1979. From: Gallagher-Allred, et al.[8]

HOSPITALIZED PATIENTS

Feeding patients with adequate nutrition is an important component of medical care. The present focus on nutrition and metabolic support of the hospitalized patient (a) promotes disease healing/recovery, (b) avoids complications, fatigue, and/or disability, and (c) minimizes length of stay. Proactive nutritional intervention, such as providing nutritional support to nutritionally high-risk patients before surgery or initiating enteral feedings within 24 to 48 hours of hospital admission can contribute to good clinical outcomes and can reduce the need for longer hospitalizations and other associated increased resource utilization.[7]

New skills are needed to properly feed the right patient at the right time (just in time) and in the right location. This requires cross training among physicians, nurses, dietitians, and other workers to provide screening, assessment, and treatment that is efficient and cost effective (see Figure 1.1).

NUTRITIONAL SCREENING AND ASSESSMENT

Obtaining data to assess a patient's nutritional status is essential for optimal patient care, especially for patients at high risk for malnutrition. Nutritional screening and assessment can be done with readily available and relatively inexpensive methods. Four key questions should be addressed for a nutritional assessment:

1. Was the patient normally nourished?
2. Is the patient at morbidity or mortality risk?
3. What is the cause of under/overnutrition?
4. Is the patient responding to nutritional treatment?

Nutritional assessment is only useful if there is an effective treatment; it is not enough to assess and identify malnutrition. Outcomes are improved and cost saved *only* when appropriate intervention follows. There is substantial evidence that nutritional therapy including nutritional assessment and appropriate nutritional interventions improves health outcomes, lowers costs, saves lives, and reduces morbidity.[9] Such early nutritional assessment and appropriate nutritional intervention *must* be accepted as essential for quality health care delivery. Appropriately selected nutritional support can address the problems of malnutrition, improve clinical outcomes, and help reduce the costs of health care.

There is no *one* universally accepted nutrition status index. The selection of the initial screening and follow-up assessment parameters depends on the available resources and data, such as serum albumin, and the amount of recent weight loss as a percent of usual weight. Some nutritional markers include food intake, weight stature, immune tests, serum protein, anthropometric measurements, and body composition. Serum albumin is a good predictor of outcome and is associated with physiologic stress.

Many tools are available for evaluating nutritional status such as the Nutritional Screening Initiative (NSI), the Subjective Global Assessment (SGA), the Nutrition Risk Index (NRI).[10–13] The successful role of nutrition in medicine is the early detection of malnutrition in nutritionally high-risk patients. At the Deaconess Hospital, the SGA was used as the foundation for developing our own nutritional risk evaluation (NRE) form, shown in Figure 1.2, according to a defined algorithm as shown on Figure 1.3.[11] This form is used to determine whether a patient is well nourished, at risk for developing malnutrition, or is moderately or severely malnourished. Coding for malnutrition is a tool used to assist the dietitian and physician in determining whether the patient is at level A, B, or C. Such a coding system is shown in Figure 1.4. Proper documentation of malnutrition by the physician in the discharge summary may also improve reimbursement.[14]

A list of high-risk diagnoses developed collaboratively by the Department of Dietetics and the Nutrition Support Service is used during baseline screening prior to completing the NRE form. See Table 1.1 for a list of these nutritionally high-risk diagnoses.

Implementing this system of nutrition evaluation improved the efficiency and

Nutritional risk evaluation form, Deaconess Hospital, Boston, Massachusetts[1]

History

1. Admission Dx:

2. Height:____ Weight:____(Kg)/____(Lbs) UBW:____(Kg)/____(Lbs) Date:______ IBW:____(Kg)/____(Lbs)

3. Weight change:

   a. Overall loss in past ___ months: ____(Kg)/____(Lb) _____% weight loss: ____ %IBW ____
   b. Change in past 2 weeks ❑ No change ❑ Increase ❑ Decrease

4. GI symptoms (circle if >3 days):

Nausea Vomiting Constipation Diarrhea Anorexia Dysphagia Dental problems

Physical examination

Loss of subcutaneous fat (triceps): ____mm TSF ____percentile ____cm AC ____cm AMC ____percentile

Calorie needs:

Protein needs:

ICD 9 code: __________

Subjective global assessment:

____A (Well nourished)
____B (Moderately, or suspected of being, malnourished)
____C (Severely malnourished)

❑ Transfer from:_____

Date	Wt	Na	K	CO2	Cl	BS	BUN	Cr	Alb	Ca	Mg	P	WBC	Other

Appetite:
Intolerances:
SPCL diets:
Previous diet education:
Previous supplements:

Comments:

PMH:

RX:

Meds:

Care Plan:

Date:	Diet order:	Date:	Diet Order:

RD	DT	Level:	Date:	Chart:	Next:

Patient seen: __________

Figure 1.2. Nutritional risk evaluation. From *Hospital Food & Nutrition Focus.* Rockville, MD: Aspen Publications, September 1995.

Screening for nutritional risk at Deaconess Hospital, Boston, Massachusetts[1]

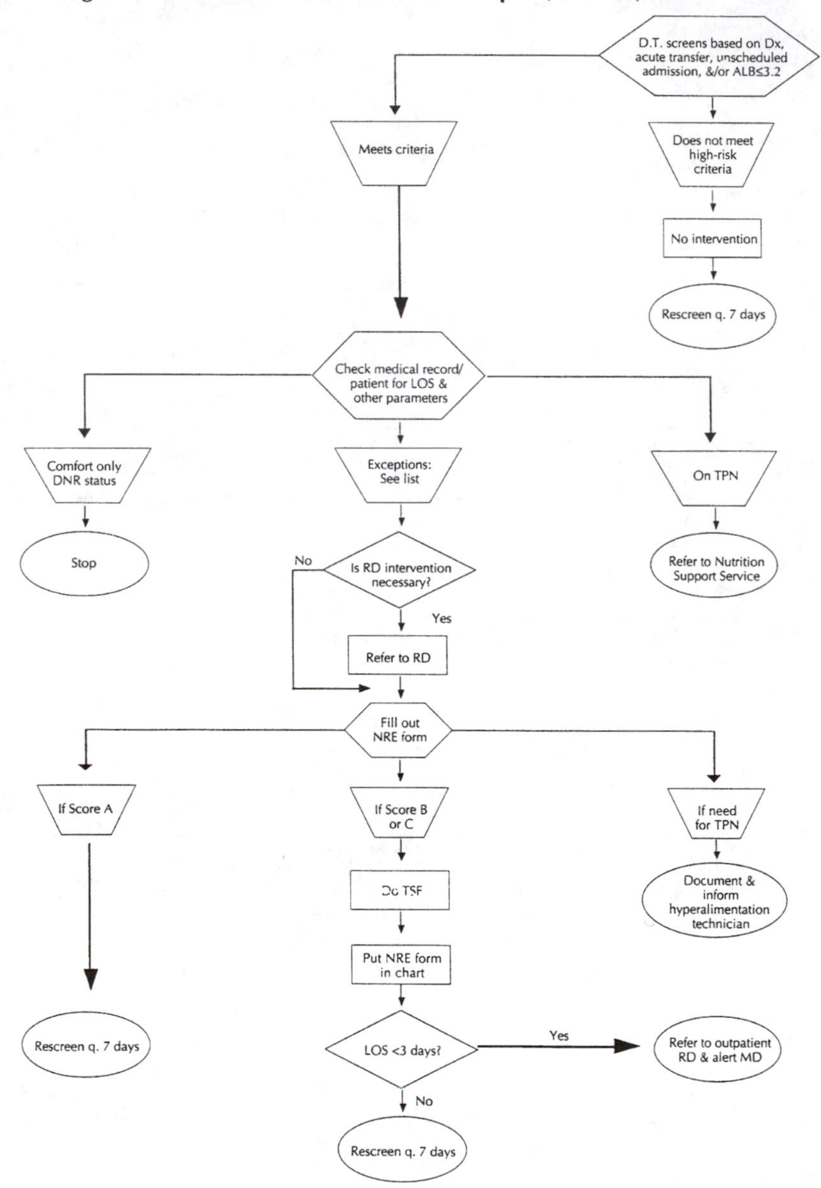

Figure 1.3. Screening for nutritional risk at Deaconess Hospital, Boston, Massachusetts. From: *Hospital Food & Nutrition Focus.* Rockville, MD: Aspen Publications. September 1995.

Kwashiorkor
ICD-9-CM code 260
Weight loss ≤10%
Albumin level <25 g/L[a]

Malnutrition of moderate degree
Mixed marasmus-hypoalbuminemia
ICD-9-CM code 263.0
Weight loss >15%
Albumin level ≤32 g/L

Nutritional marasmus
ICD-9-CM code 261
Weight loss >20% and <80% ideal body weight or <70% ideal body weight alone
Albumin level ≥25 g/L[b]

Malnutrition of mild degree
Mixed marasmus-hypoalbuminemia
ICD-9-CM code 263.1
Weight loss 10% to 15%
Albumin level ≤32 g/L

Other severe protein-energy malnutrition
ICD-9-CM code 262
Weight loss >10%
Albumin level <25 g/L

Other protein-energy malnutrition
ICD-9-CM code 263.8
Anticipate prolonged length of stay

1. Not depleted but stressed or septic
   Weight loss <5%
   Albumin level ≤32 g/L
2. Moderate weight loss with planned major surgery
   Weight loss >10%
   Albumin level >32 g/L[b]
3. Moderate depletion with mild weight loss
   Weight loss >5%
   Albumin level ≤32 g/L
4. Inability to eat ≥7 days
   (actual or predicted)

Figure 1.4. Proposed schema for defining adult protein-energy malnutrition (PEM) based on the malnutrition codes of the *International Classification of Diseases,* ninth revision. *Clinical Modification* (ICD-9-CM) (8). If a patient can be classified in a major code (260, 261, or 262) and a minor code (263.0, 263.1, or 263.8), always pick the major code; this choice may affect reimbursement. "To convert g/L albumin to g/dL, multiply g/L by 0.1. To convert g/dL albumin to g/L, multiply g/dL by 10." Albumin level not necessary for diagnosis of malnutrition. From: Swails et al.[14]

Table 1.1. Nutritionally High-Risk Diagnoses

Irritable bowel disease, ulcerative colitis
Crohn's disease
Pressure ulcer
Pancreatitis
Upper gastrointestinal cancer
Hepatic failure, ascites, cirrhosis
Congestive heart failure and >70 years of age
Head-neck, lung cancer
Renal failure, dialysis
Intractable diarrhea, vomiting, dehydration
Dysphagia
HIV, AIDS, pneumocystis carinii pneumonia, fever of unknown origin, cytomegalovirus
Failure to thrive, malnutrition
CVA, transient ischemic attack
TPN and tube feedings

Source: Hospital Food & Nutrition Focus. Rockville, MD: Aspen Publishers, Sept. 1995.

effectiveness of the nutrition staff and allowed them to focus on nutritionally high-risk patients rather than on low-risk patients or those hospitalized for a short stay.

FEEDING THE MALNOURISHED PATIENT—NUTRITIONAL INTERVENTION

Numerous studies document the high incidence of hospitalized adult patients who are malnourished or are at risk for malnutrition.[15–22] Malnutrition is a significant health problem among all ages and across all diagnosis groups.[15–22] Malnutrition is associated with a longer hospital stay and higher costs; malnourished patients experience slower healing, more complications, and higher mortality rate.[11,16,23–29] Malnourished surgical patients are more likely to have a greater incidence of complications and excess mortality.[30–31]

Hospital charges may be up to 75% higher for malnourished versus well-nourished patients due to longer stays and increased utilization of resources for treating the complications associated with malnutrition.[23,25,27,28,32]

It is not enough to assess patients and identify malnutrition. Many studies support the findings that health outcomes of malnourished patients can be improved and overall resources (such as cost) reduced when appropriate nutrition intervention is given. These interventions include parenteral and enteral nutrition, oral diet and supplements, and nutrition counseling. Appropriately selected nutrition support can address the problem of malnutrition, improve clinical outcomes, and reduce the costs of health care both in hospital and home settings. Patients with diseases such as head and neck cancer, AIDS, liver disease, Crohns' disease and others can have an improved nutritional status and quality of life with specialized nutritional management and/or supplementation.[33–36] Providing nutritional supplements with an oral diet is associated with a desirable clinical outcome across the continuum of care.[37,38]

The use of cost and outcome of enteral nutrition support and total parenteral nutrition (TPN) has been studied extensively. The advantages of the enteral route far outweighs the costs and complications associated with parenteral nutrition, assuming the gut can be used.[37,39,40] Home care enteral or parenteral nutrition support can be a cost saving alternative for patients requiring preoperative or long-term nutrition support.[41,42]

A team approach, such as the "seamless" department between the Nutrition Support Service and Department of Dietetics at Deaconess Hospital is an example of a collaborative, efficient, and effective system for managing the nutritional care of patients (see Figure 1.5). This approach handles the entire nutrition care process from assessment to monitoring as further detailed in Figure 1.6.

The Nutrition Support Service (NSS) team of attending physicians, fellows, and dietetic technicians handle the complex set of skills required in the assessment and management of patients requiring parenteral nutrition. Patients on hyperalimentation are followed on a daily basis with formal rounds held three times each week.

Several of the physicians from the Nutrition Support Service also serve on the Enteral Nutrition Service (ENS) where patients requiring an enteral tube feeding are assessed and monitored by the dietitians, Department of Dietetics.

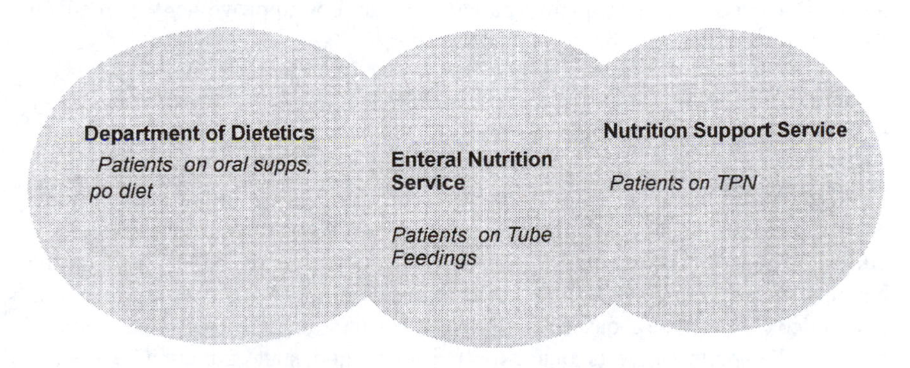

Figure 1.5. Seamless Nutrition Care. *Source:* Deaconess Hospital, Boston, Massachusetts.

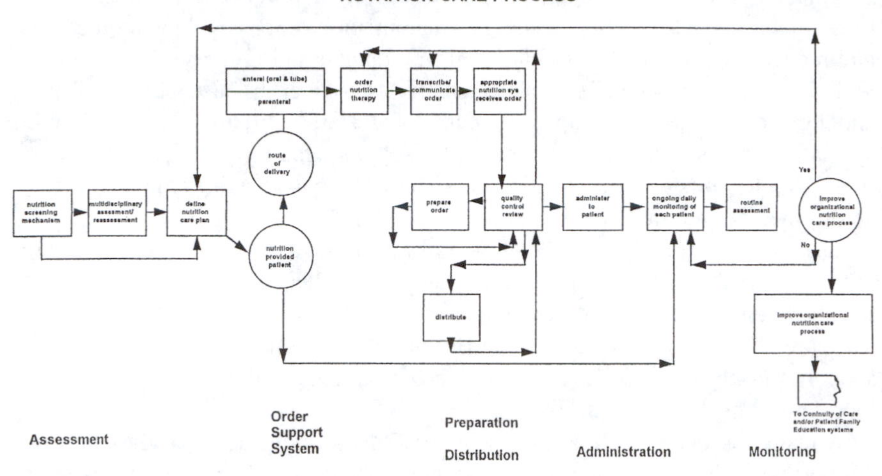

Figure 1.6. Nutritional care process. Nutrition care involves a series of steps: *Step 1* gives nutritional screening to triage the patients to nutritional assessment and/or nutritional care. *Step 2* (a) utilizes the Nutrition Support Service for nonvolitional parenteral and enteral feeding or (b) provides routine dietary services. *Step 3* provides appropriate monitoring and supervision to nutritional care particularly in high-risk patients. *Step 4* provides quality control review, appropriate reassessment, and continuity of care on discharge. Nutritional prescriptions are a critical step given the short length of stays (LOS). Reprinted from *The 1995 Dietary Guidelines.*[60]

ENS rounds are held three times weekly as a team to review each patient's status and make recommendations or changes in their nutrition care plan.

Patients may be discharged by the NSS on parenteral nutrition or tube feeding or gradually transitioned to an oral diet prior to discharge. Qualified professionals are then needed to insure quality patient care and/or improvement in medical and nutritional status in the home care setting.

EVIDENCE-BASED MEDICINE

Evidence-based medicine emphasizes the need to move beyond clinical experience and physiological principles to rigorous evaluations of the consequences of clinical actions.[43] It is essential to provide patients with care that is based on the best evidence currently available.[44] Variations in clinical practice, comparisons of practice with evidence-based standards, and evaluation of the recommendations of clinical experts suggests that expert opinion and standard practice do not provide adequate mechanisms for the transfer of scientific information into clinical decisions.[45,46]

VALUE OF NUTRITIONAL MEDICINE

Public policy makers, administrators of health systems, third-party payers, and patients all require documentation of health care quality and cost effectiveness. As health care expenditures consume ever larger portions of budgets, there are mounting pressures to eliminate services that do not have clearly demonstrable benefits. Health plan "report cards" or published summaries of health plan performance are a new way to help consumers select their health plan based on cost and quality. The Health Plan Employer Data and Information Set (HEDIS) includes a set of health plan performance measures and methods for data collection. The current list of HEDIS performance measures do not include medical nutrition therapy. Although measures related to medical nutrition therapy have been proposed for managed care report cards (such as medical nutrition therapy for high cholesterol levels and cardiovascular disease), none are included to date. Nutrition experts need to influence representatives of health plans and employers to include these measures.[47]

The knowledge and use of published guidelines, standards of care, and patient treatment protocols are critical for providing high-quality nutritional care that results in effective and efficient outcomes. More and more emphasis is being placed on measuring performance and improving the quality of care provided. Organizations such as the Joint Commission on Accreditation of Healthcare Organizations (JCAHO) evaluate an institution's performance based on an outcome-oriented monitoring and an evaluation of what is done to improve care. Additionally, hospital functions that have the greatest impact on patient outcomes are the focus of the JCAHO.

In response to the JCAHO's new initiative called "Agenda for Change," the American Society for Parenteral and Enteral Nutrition (ASPEN) "1995 Standards for Nutrition Support: Hospitalized Patients" was revised to coincide with the 1995 Accreditation Manual for Hospitals (AMH).[48] The standards aim to assure sound and efficient nutrition care, including enteral and parenteral support, for those hospitalized patients in need of specialized nutrition support. "Suggested Guidelines for Nutrition and Metabolic Management of Adult Patients Receiving Nutrition Support" is a valuable resource that includes guidelines for enteral and parenteral nutrition.[49] Through validation by field testing, they allow the user to have a sound, scientific starting point for their quality assessment and outcome monitors. Numerous other examples of practice guidelines are available, such as the nationally produced guidelines developed by the Agency for Health Care Policy and Research (AHCPR). Their publications on prediction and prevention and treatment of pressure ulcers are widely used by nutrition and other health care professionals.[9,50] Other guidelines are available for the use of parenteral and enteral nutrition from ASPEN.[51] These guidelines are intended to provide practical advice to clinicians who administer oral, tube enteral, or TPN support to patients.

Indicator-based monitoring systems, such as the one developed by the JCAHO (Indicator Measurement System), are tools intended to provide information for health care organizations and providers to use internally to improve performance and externally to meet the demands of patients, payors, and others for health care quality data. Clinical indicators have been developed by the American Dietetic Association specifically for surgical, cardiovascular, and oncology patients.[52] Other lists of indicators are available in the literature or can be developed with interdisciplinary quality-improvement teams.

The best practice guidelines synthesize information in a given field according to explicit rules (evidence-based medicine) and give recommendations, or guidelines, on what to do. Some key practice guidelines dealing with prevention are available.[53–55] Outcomes describe the results of (nutritional) interventions; for example, what happens in terms of identifying patients at risk of nutritional deficiencies, the results of providing nutritional support to a select group of patients or complications that may result from such treatment. Outcomes include intermediate and cumulative events and may be quantitative and qualitative.

Well-designed studies report the outcomes of nutrition modalities. Malnutrition is associated with negative health outcomes and increased utilization of resources. Nutrition support provided to malnourished patients across a range of medical conditions reduces complications, morbidity, length of stay, and mortality; and nutrition support of malnourished patients reduces overall resource utilization. Although the costs and other health problems associated with malnutrition are known, data is lacking to causally link malnutrition to increased morbidity and mortality in all patient populations. More data are needed on the interaction of nutrition risk with age and the type and severity of the underlying disease state. Outcome studies are essential to better quantify the benefits and cost effectiveness of nutrition interventions and services.[56]

Managed care is our future. The nutrition experts need to find their niche. A recent monograph by Ross Laboratories listed several strategies for dietetic professionals to achieve this objective:

- Make sure the inpatient nutrition services department is included in the list of services their hospital used to negotiate for membership in a PPO. Development of nutrition screening and assessment programs add value to the patient hospital services when negotiating provider contracts with managed care organizations.
- Expand the role of the outpatient dietitian counselors to include providing programs such as wellness and smoking cessation in addition to nutrition counseling.
- Develop referral relationships with physician members of network HMOs and PPOs.
- Include nutrition services provided by dietitians in private practice as a component of the physicians practice.
- Promote the employment of dietitians by nursing agencies, home care agencies, or home-infusion companies that have managed care contracts.[57]

Regardless of how nutrition professionals work in managed care, provision of outcome data will be essential to demonstrate cost effectiveness and quality. As acute care patients are discharged earlier, their risk of malnutrition is greater. Dietitians can become more involved in discharge planning by providing recommendations and guidance that minimize the risk of rehospitalization. Nutrition professionals need to look for opportunities and be attuned to how managed care models change as health care reform takes shape.

In this chapter we wish to briefly address other important aspects of nutritional medicine needed to fully serve the needs of the community and the public. Important new advances in preventive medicine and nutrition need to be appreciated by a clinical nutritionist regardless of their primary graduate-level training.

WELLNESS AND NUTRITION

Nutrition medicine is focused on comprehensive lifestyle change in which individuals take ownership of their preventive healthy behavior. It is multidisciplinary, involving many allied health professions together with physicians to prescribe and implement the skills and cognition needed for individual responsibility.

Preventative and proactive nutrition intervention is an essential component of health care. Steps must be taken to *improve* the nutritional status of Americans, *prevent* nutrition problems, and *prevent* the health consequences of poor nutritional status among the public. Many people have poor eating habits or conditions associated with nutritional problems that put them at nutritional risk.[58]

The Healthy Eating Index (HEI), together with the 1995 U.S. Dietary Guidelines (Fig. 1-7), food labeling and food guide pyramid have helped this effort in a very important way.[59–61] Since the early part of the 20th century, the major areas of concern in public health nutrition have changed from problems of nutritional deficiency to problems of excesses and imbalances. Problems of over consumption are, on the average, more prevalent than problems of underconsumption.[62] The casual links between diet and chronic diseases have also become clearer.[63,64]

An extensive body of scientific evidence suggest that diets high in total fat, saturated fat, and cholesterol and low in fiber and complex carbohydrates are linked to coronary artery disease, stroke, diabetes, and certain forms of cancer.[65] Improving dietary patterns and, in turn, improving nutritional status, is viewed as a key way to improve public health.

The Healthy Eating Index (HEI) is a single, summary measure of diet quality that can be used to monitor changes in consumption patterns as well as serve as a useful tool for nutrition education and health promotion. Additionally, the HEI provides important evidence on the types of dietary improvements that need

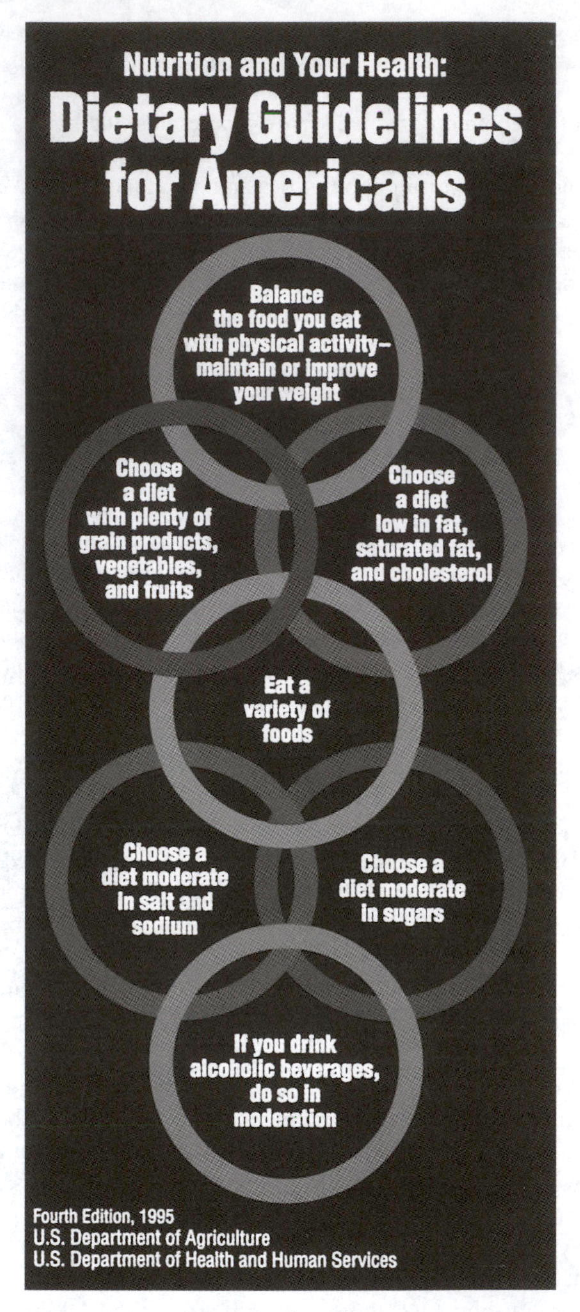

Figure 1.7. *The 1995 Dietary Guidelines.*[60]

to be made to bring U.S. food consumption patterns more in line with the recommendations of the food guide pyramid and the dietary guidelines.[59,61,66]

REFERENCES

1. Blackburn GL. Nutritional assessment and support during infection. *Am J Clin Nutr* 1977;30:1493–1497.
2. Blackburn GL. The interaction of the science of nutrition with the science of medicine. *J Parent Ent Nutr* 1979;3:131–136.
3. Blackburn GL. Protein metabolism and nutritional support. *J Trauma* 1981;21:707–711.
4. Blackburn GL, Ahmad A. Skeleton in hospital closet—then and now. *Nutrition* 1995;11:193–195.
5. Donald Stokes' *Pasteur's Quadrant,* Woodrow Wilson School of Princeton University, 1995.
6. Golub ES. *The limits of medicine: how science shapes our hope for the cure.* New York: Times Books, 1994.
7. Moore FA, Feliciano DV, Andrassy RJ, et al. Early enteral feeding, compared with parenteral, reduces postoperative septic complications: The results of a meta-analysis. *Ann Surg* 1992;216:172–183.
8. Gallagher-Allred GR, Voss AC, Finn SC, McCamish MA. Malnutrition and clinical outcomes: the case for medical nutrition therapy. *J Am Diet Assoc* 1996;96:361–366.
9. Bergstrom N, Allman RM, Alvarez OM, et al. *Treatment of Pressure Ulcers.* Clinical Practice Guideline #15, Rockville, MD; United States Department of Health and Human Services, Public Health Service, Agency for Health Care Policy and Research, Publication #950652, December, 1994.
10. Greer, Marglis, Mitchell, Grunwald and Associates, Inc. *Nutrition Interventions Manual for Professionals Caring for Older Americans.* Washington, DC: Nutrition Screening Initiative, 1991.
11. Detsky AS, Smalley PS, Chang J. Is this patient malnourished? *JAMA* 1994;27(1):54–58.
12. Baker JP, Detsky As, Wesson DE, et al. Nutritional assessment: A comparison of clinical judgment and objective measurements. *N Engl J Med* 1982;306:969–972.
13. Veterans Affairs TPN Cooperative Study group. Perioperative total parenteral nutrition in surgical patients. *N Engl J Med* 1991;325:525–532.
14. Swails SW, Samour PQ, Babineau TJ, Bistrian BR. A proposed revision of current ICD-9-CM malnutrition code definitions. *J Am Diet Assoc* 1996;96:370–373.
15. Agradi E, Messina V, Campanella G, et al. Hospital malnutrition: incidence and prospective evaluation of general medical patients during hospitalization. *Acta Vitaminol Enzmol* 1984;6:235–242.

16. Weinsier RL, Hunker EM, Krumdieck CL, Butterworth CE. Hospital malnutrition: a prospective evaluation of general medical patients during the course of hospitalization. *Am J Clin Nutr* 1979;32:418–426.
17. Hill GL, Pickford GA, Schorah CJ, et al. Malnutrition in surgical patients: an unrecognized problem. *Lancet* 1977;1:689–692.
18. Bistrian BR, Blackburn GL, Hallowell E, Heddle R. Protein status of general surgical patients. *JAMA* 1974;230:858–860.
19. Bistrian BR, Blackburn GL, Vitale J, et al. Prevalence of malnutrition in general medical patients. *JAMA* 1976;235:1567–1570.
20. Sullivan DH, Moriarty MS, Chernoff R, Lipschitz DA. Patterns of care: an analysis of the quality of nutritional care routinely provided to elderly hospitalized veterans. *J Parent Ent Nutr* 1989;13:249–254.
21. Messner RL, Stephens N, Wheeler WE, Hawes MC. Effect of admission nutritional status on length of hospital stay. *Gastroenterol Nurs* 1991;Spring:202–205.
22. Detsky AS, Baker JP, O'Rourke K, Goel V. Perioperative parenteral nutrition: a meta-analysis. *Ann Intern Med* 1987;107:195–203.
23. Christensen KS. Hospital wide screening increases revenue under prospective payment system. *J Am Diet Assoc* 1986;86:1234–1235.
24. Reilly JJ, Hull SF, Albert N, et al. Economic impact of malnutrition: A model system for hospitalized patients. *J Parent Ent Nutr* 1988;12:371–376.
25. Robinson G, Goldstein M, Levine GM. Impact of nutrition status on DRG length of stay. *J Parent Ent Nutr* 1987;11:49–51.
26. Walesby RK, Goode AW, Spinks TJ, et al. Nutritional status of patients requiring cardiac surgery. *J Thorac Cardiovas Surg* 1979;77:570–576.
27. Epstein AM, Read JL, Hoefer M. The relation of body weight to length of stay and charges for hospital services for patients undergoing elective surgery: a study of two procedures. *Am J Public Health* 1987;77:993–997.
28. Christensen KS, Gstundtner KM. Hospital-wide screening improves basis for nutrition intervention. *J Am Diet Assoc* 1985;85:7005–706.
29. Riffer J. Malnourished patients feed rising costs: study. *Hospitals.* 1986; Vol 60 86.
30. Hickman DM, Miller RA, Bonbeau JL, et al. Serum albumin and body weight as predictors of postoperative course in colorectal cancer. *J Parent Ent Nutr* 1980;4:314–316.
31. Klidjian AM, Archer TJ, Foster KJ, Careen SO. Detection of dangerous malnutrition. *J Parent Ent Nutr* 1982;6:119–121.
32. Maoris SA, Takiguchi S, Slavich S, Rose CL. Consistent wound care and nutritional support in treatment. *Decubitus* 1990;3(3):16–28.
33. Chlebowski RT, Gildon B, Grosvenor M, et al. Long term effects of early nutrition support with new enterotrophic peptide-based formula vs. standard enteral formula in HIV-infected patients: randomized prospective trial. *Nutrition* 1993;9:1–6.
34. Hirsch S, Bunout D, DeLaMaza P, et al. Controlled trial on nutrition supplementation in outpatients with symptomatic alcoholic cirrhosis. *J Parent Ent Nutr* 1993;17:119–124.

35. Harries AD, Jones LA, Danis V, et al. Controlled trial of supplemented oral nutrition in Crohn's disease. *Lancet* 1983;1:887–890.
36. Lipschitz DA, Mitchell CO, Steele RW, Milton KY. Nutritional evaluation and supplementation of elderly subjects participating in "Meals on Wheels" program. *J Parent Ent Nutr* 1985;9:343–347.
37. Gray-Donald K, Payette H, Boutier V, Page S. Evaluation of the dietary intake of homebound elderly and the feasibility of dietary supplementation. *J Am Coll Nutr* 1994:13:277–284.
38. Larsson J, Unosson M, Ek A-C. Effect of dietary supplement on nutritional status and clinical outcome in 501 geriatric patients: a randomized study. *Clin Nutr* 1990;9(4):179–184.
39. Kudsk KA, Croce MA, Fabian TC, et al. Enteral versus parenteral feeding: effects on septic morbidity after blunt and penetrating abdominal trauma. *Ann Surg* 1992;215:503–513.
40. Moore EE, Jones TN. Benefits of immediate jejunostomy feeding after major abdominal trauma: a prospective, randomized study. *J Trauma* 1986;26:875–879.
41. National Alliance for Infusion Therapy. *Cost effectiveness of home infusion therapy: a review of the literature.* Washington, DC: National Alliance for Infusion Therapy, December 17, 1993.
42. Kishibuchi M, Tsujinaka T, Iijima S, et al. Interrelation of intercellular proteases with total parenteral nutrition-induced gut mucosal atrophy and increase of mucosal macromolecular transmission in rats. *J Parent Ent Nutr* 1995;19:187–192.
43. Oxman AD, Sackett DL, Guyatt GH. Users' guide to the medical literature I. How to get started. *JAMA* 1993;270:2093–2095.
44. Evidence-Based Medicine Working Group. Evidence-based medicine: a new approach to teaching the practice of medicine. *JAMA* 1992;268:2420–2425.
45. Stross JK, Harlan WR. The dissemination of new medical information. *JAMA* 1979;241:2622–2624.
46. Williamson JW, German PS, Weiss R, et al. Health science information management and continuing education of physicians: a survey of US primary care practitioners and their opinion leaders. *Ann Intern Med* 1989;110:151–160.
47. Turner MA, Dwyer JT. Nutrition measure for managed care report cards. *J Am Diet Assoc* 1996:96:374–380.
48. *1995 standards for nutrition support: hospitalized patients.* ASPEN, 1995.
49. Winkler MF, Lysen LK (ed). *Suggested guidelines for nutrition and metabolic management of adult patients receiving nutrition support.* Chicago: American Dietetic Association, 1993.
50. *Pressure ulcers in adults: prediction and prevention.* Clinical Practice Guidelines #3, Rockville, MD; United States Department of Health and Human Services, Public Health Service, Agency for Health Care Policy and Research, Publication #92-0047, 1992.
51. Guidelines for the use of parenteral and enteral nutrition in adult and pediatric patients. *J Parent Ent Nutr* 1993;17:1–52SA.

52. Queen PM, Caldwell M, Balogun L. Clinical indicators for oncology, cardiovascular, and surgical patients: report of the ADA Council on Practice Quality Assurance Committee. *J Am Diet Assoc* 1993;93(3):338–344.
53. *Report of the U.S. Preventive Services Task Force. Guide to preventive services. An assessment of the effectiveness of 169 interventions.* Baltimore: Williams and Wilkins, 1989.
54. Eddy DM (ed). *Common Screening Tests.* Philadelphia: American College of Physicians, 1991.
55. *Guidelines for health supervision II.* Elk Grove Village, IL: American Academy of Pediatrics, 1988.
56. Derelian D, Gallagher A, Snetselaar L. President's page: letting the outcomes justify the reimbursement. *J Am Diet Assoc* 1995;95:371.
57. Mathieu-Harris M, Foltz MB, Calvert Finn S presented at the American Dietetics Annual Meeting and Exhibition. Anheim, CA. *Finding your niche in the managed-care market.* Ross Laboratories, 1993.
58. Ryan AS, Craig LD, Finn SC. Nutrient intakes and dietary patterns of older Americans. A national study. *J Gerontol* 1992;47(5):M145–150.
59. Kennedy ET, Ohls J, Carlson S, Fleming K. The Health Eating Index: design and applications. *J Am Diet Assoc* 1995;95:1103–1108.
60. U.S. Department of Agriculture, Agriculture Research Service, Dietary Guidelines Advisory Committee. *Report of the Dietary Guidelines Advisory Committee on the Dietary Guidelines for Americans.* 1995.
61. *The Food Guide Pyramid.* Home and Garden Bulletin No 252. Washington, DC: US Department of Agriculture, Human Nutrition Information Service; 1992.
62. *Healthy People: The Surgeon General's report on health promotion and disease prevention.* DHEW(PHS) Publ. 79-55071. Washington, D.C.: US Dept of Health, Education, and Welfare, 1979:177.
63. *The Surgeon General's report on nutrition and health.* DHHS(PHS) Publ. 88-50210. Washington, D.C.: U.S. Dept of Health and Human Services, 1988.
64. Food and Nutrition Board. *Diet and health: implications for reducing chronic disease risk.* Washington, D.C.: National Academy Press, 1989.
65. *Healthy People 2000: National Health Promotion and Disease Prevention Objectives.* DHHS(PHS) Publ. 91-50213. Washington, D.C.: U.S. Dept. of Health and Human Services, 1991.
66. *Nutrition and Your Health: Dietary Guidelines for Americans.* 3rd ed. Home and Garden Bulletin No. 232. Washington, D.C.: U.S. Dept. of Agriculture/Dept. of Health and Human Services, 1990.

Malnutrition and Related Complications

CHAPTER 2

Christopher Still, M.S., D.O.,
Caroline Apovian, M.D., and
Gordon L. Jensen, M.D., Ph.D.

INTRODUCTION

Protein energy malnutrition results when the body's needs for proteins, energy fuels, or both cannot be satisfied by the diet. It includes a wide spectrum of clinical manifestations determined by the relative intensity of protein or energy deficit, the duration of the deficiency, the age of the host, the causes of the deficiency, and the association with other nutritional and infectious diseases and stressors.[1]

Protein energy malnutrition may be differentiated into two clinically distinct syndromes: marasmus and hypoalbuminemic malnutrition. This chapter defines each, describes its pathophysiology, and examines its consequences on physiologic function.

MALNUTRITION SYNDROMES

Marasmus

Marasmus, or semistarvation, is a syndrome that develops gradually over months to years with insufficient energy intake. Marasmus will be observed in individuals with chronic disease processes that adversely affect energy intake, such as anorexia nervosa, cardiac cachexia, malabsorption, or esophageal carcinoma. Patients appear cachectic with generalized muscle wasting and absence of subcutaneous fat, which leads to a "skin and bones" appearance. Peripheral edema is usually not a component of this disease. Visceral proteins are often normal despite depressed anthropometric measures.

During periods of adequate energy balance, the body first provides fuel for ongoing metabolic processes, then refills glycogen and protein reserves, and finally stores excess calories as fat.[2] During starvation, endogenous energy stores in the form of free glucose, stored glycogen, fat, and protein are used as energy sources to preserve visceral proteins. Free glucose is the initial fuel for hematopoietic cells and the brain, but beyond 24 hours, hepatic gluconeogenesis and glycogenolysis lead to the depletion of hepatic glycogen and release of hepatic glucose. The liver removes amino acids (primarily alanine and glutamine), glycerol, lactate, pyruvate, and free fatty acids (FFA) from the blood to provide for newly synthesized glucose and ketone bodies.[3] Plasma insulin level falls, which, in turn, stimulates lipolysis, ketogenesis, amino acid metabolism, gluconeogenesis, and a decrease in protein synthesis. The brain shifts to the metabolism of ketone bodies and free fatty acids to minimize the use of visceral proteins for fuel.[4]

The decrease in energy intake in early starvation is followed quickly by a decrease in basal energy expenditure.[5] This decrease in basal metabolic rate is primarily regulated through a decrease in thyroid and sympathetic nervous system activity. The result is a syndrome of generalized wasting, with prominent weight loss, usually normal visceral protein stores, and a clinical course lasting months to years.

Hypoalbuminemic Malnutrition

Hypoalbuminemic malnutrition (albumin < 3.5 gm/dL) is a manifestation of the body's response to infection or inflammation.[2] In contrast to marasmus, hypoalbuminemic malnutrition occurs more quickly and is modulated by hormones and cytokines that work to deplete visceral protein stores (albumin). It usually follows a major stressor (sepsis, head injury, burns, trauma, etc.) and has multiple sequellae affecting metabolism and immune functions. Hypoalbuminemic malnutrition is a term popularized by Blackburn and colleagues to emphasize the presence of hypoalbuminemia in patients who appear well-nourished but have defects in visceral protein stores and immune function.[6]

Stress-induced hormonal changes mediated by catecholamines can increase metabolic rate and sympathoadrenal axis stimulation, as well as raise levels of antidiuretic hormone (ADH) and aldosterone.[2] Stress also stimulates counter-regulatory hormones—glucagon, epinephrine, cortisol, and growth hormone—which act to cause hyperglycemia and the catabolism of skeletal muscle. These changes provide for the synthesis of glucose and protein as well as the stimulation of lipolysis.[7]

Cytokines are protein hormones that mediate host immune responses during times of stress. Interleukin-1 and tumor necrosis factor (TNF) are examples of cytokines that are synthesized primarily by cells lining the liver and spleen in response to infection and inflammation. Interleukin-1 uniquely activates lymphocytes and can replicate many of the acute phase responses observed with inflammation including fever, anorexia, increase in blood leukocyte count with release of immature forms from the marrow, changes in concentrations of acute phase proteins, altered intermediary metabolism, and trace mineral redistribution.[8]

It is important to understand the pathophysiology of albumin synthesis and the affect cytokines have on the serum albumin levels during stress.[9] The daily hepatic synthesis rate is 120–270 mg/kg or 15 g in a 70-kg individual.[10] It is usually distributed between the intravascular and extravascular spaces; however, during injury, the liver increases its production of acute phase proteins rather than albumin.[2] Interleukin-1 shares with TNF the ability to down-regulate the albumin gene and to decrease the rate of albumin messenger ribonucleic acid (mRNA) translation.[11] It is this decrease in production of albumin coupled with increased catabolism and extravascular extravasation that culminates in the hypoalbuminemia that results from injury.[12] Therefore, during the injury response, the serum albumin concentration may be an unreliable marker for nutritional status but does serve as a marker for injury or stress. Hypoalbuminemia has prognostic importance and has been associated with increased morbidity and mortality among hospitalized patients.[9] It is important to note that many hospitalized patients will manifest components of both marasmus and hypoalbuminemia, because they are subject to both semistarvation and stress response.

PHYSIOLOGIC CONSEQUENCES OF MALNUTRITION

Weight Loss

Weight loss is one of the most obvious consequences of poor nutrition. Practitioners should carefully review prior weights and weight trends with patients. Unexplained weight loss often signifies an occult disease process. Most patients can tolerate a loss of 5% to 10% of body weight without significant consequences, but losses greater than 40% of usual body weight are often fatal.[13] Nonvolitional weight loss of 10% or greater of usual body weight over a 6-month period is significant and warrants evaluation. Survival during starvation correlates with the quantity of fat stores that exist at the onset of the fast.[14] Overall changes in body composition that occur in simple starvation include a relative increase in extravascular water component, loss of adipose stores, and, to a lesser degree, a loss of lean body mass.[3] It is important to note that weight may actually increase during a critical illness, not due to an increase in adipose or protein stores, but due to retention of third space fluids.

Respiratory System

Starvation and stress can affect respiratory muscle structure and function. Malnutrition is associated with a decrease in diaphragmatic muscle mass, maximum voluntary ventilation (MVV), and respiratory muscle strength.[15] Malnutrition impairs ventilatory drive, which may affect the ability to clear secretions and manifest adequate tidal volumes to prevent atelectasis. In addition, parenchymal changes occur secondary to nutrient depletion. Specific abnormalities include emphysematous changes, decreased lipogenesis,[16] protein loss,[17] and biochemical changes in connective tissue components (protein, hydroxyproline, and elastin).[18] Refeeding of excess carbohydrates can result in increased carbon dioxide production and precipitate respiratory failure in those patients with poor pulmonary reserve.

Cardiovascular System

The effects of protein calorie malnutrition include loss of cardiac muscle and decrease in absolute cardiac output.[19] In autopsy specimens, gross examination revealed a decreased myocardial weight, atrophy of subepicardial fat, and interstitial edema.[20] Echocardiographic findings indicate that 60% of the loss of cardiac volume is due to an increase in internal chamber volume and 40% to a decrease in cardiac mass, especially the left ventricle.[21] Electrocardiographic abnormalities are nonspecific and may include sinus bradycardia, low voltage of the *QRS* complex, reduction in *T*-wave amplitude, and prolongation of the Q-T_c interval.[22] Thiamine deficiency can result in acute high-output cardiac failure, which is often fatal. Subjects often present with mixed manifestations of thiamine, alcohol, and protein calorie cardiomyopathies.[23]

Gastrointestinal System

The rapid turnover of the enterocytes and colonocytes that line the gastrointestinal tract causes malnutrition can have a profound effect on both intestinal mass and function. Without enteral stimulus, as in total starvation or use of parenteral alimentation alone, intestinal epithelial cells atrophy with a resultant decrease in intestinal mass, villus size, cellular size and number, mitotic index, and disaccharidase activity.[24,25] Nutrient malabsorption may ensue from the architectural flattening of villi and infiltration of lymphocytes. Translocation of gut flora and associated endotoxin may precipitate occult bacteremia and the injury response.[26–28] It appears that the provision of enteral nutrition is paramount in maintaining structural integrity and function.

Renal System

Malnutrition has little affect on kidney function, although renal mass is decreased. Urinalysis is usually unremarkable, without evidence of protein, casts, white cells, or abnormal sediment.[29] The specific gravity is often low, and a defect in concentrating ability is sometimes present.[30]

Wound Healing

Hypoalbuminemic malnutrition commonly affects wound healing. Research dating back to the 1930s revealed that malnutrition in animals delayed or retarded wound healing.[31] Neovascularization, fibroblast proliferation, collagen synthesis, and wound remodeling are delayed. In addition, local factors, such as edema associated with hypoalbuminemia and micronutrient deficiencies, may contribute to poor wound healing in undernourished and stressed patients.[1,32,33]

Vitamin C is required for hydroxylation of proline and lysine and is, therefore, necessary for collagen synthesis.[34] Neutrophils also require vitamin C for superoxide production and bacterial killing. Collagen lysis will continue during Vitamin C deficiency, so one may see reopening of old wounds, impairment of tensile strength, and compromise of local antibacterial activity. Zinc functions in wound healing in its role as a cofactor in a variety of enzyme systems, including nucleic acid, messenger rRNA, and protein synthesis. Zinc deficiency is observed in stressed and/or malnourished patients, with elevated renal losses being a contributing factor. Intestinal fluids may contain appreciable zinc, with diarrheal or ileostomy losses precipitating deficiency.[35] Chronic steroid administration is also associated with decreased serum zinc levels. Magnesium is also necessary for wound healing, serving as a cofactor for enzymes of protein synthesis. Magnesium may be depleted in patients with chronic diarrhea, intestinal fistulas, or renal tubular dysfunction.

Immune Status

The effects of protein calorie malnutrition on the immune system are myriad.[36] Humoral immunity can be affected by protein calorie malnutrition, although specific antibody responses are variable.

Cellular immune function is commonly affected in hypoalbuminemic or severely marasmic patients. Skin test anergy to recall antigens can occur within one week of nutritional depletion, however, with adequate nutritional repletion, positive skin test results can be restored within two to three weeks.[37]

A total lymphocyte count is readily available and can be correlated with cellular immune function. Low lymphocyte counts reflect changes in nutritional status, especially protein depletion.[38] Although there are many nonnutritional factors that may affect this count, counts less than 1,200 lymphocytes/mm^3 may reflect a mild deficiency and counts less than 800 cells/mm^3, a severe deficiency. As a screening test, lymphopenia has correlated with increased morbidity and mortality in hospitalized patients.[38]

Neutropenia to a varying degree may occur in individuals with protein calorie malnutrition. Neutrophils are normal morphologically, but some measures of neutrophil function, including chemotaxis and bacterial killing, are abnormal.[38] Phagocytosis, however, is usually normal.

Complement components, as well as other nonspecific host defense mechanisms, are commonly depressed. Interferon production, opsonization, and plasma lysosome production may be adversely affected as well as acute phase reactants such as C-reactive proteins and alpha$_1$-antitrypsin. Changes in the body's anatomic barriers to infection, including atrophy of the skin and gastrointestinal mucous membranes may also increase the risk of infection.

CONCLUSION

The study of the pathophysiology of malnutrition has evolved to our present day understanding of protein-energy malnutrition as a spectrum spanning the clinical syndromes of marasmus and hypoalbuminemia (Figure 2.1).

A thorough nutrition assessment is a crucial part of the work-up of any patient as nutritional status will affect prognosis. The early recognition and treatment of an underlying nutritional disorder is an adjunct to other medical therapies and can promote favorable outcomes.

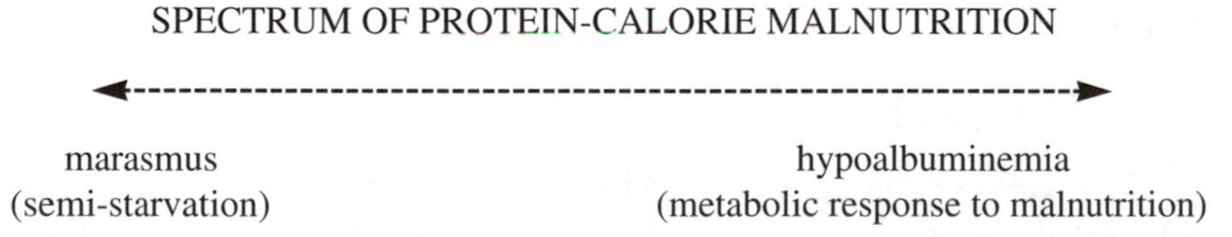

Figure 2.1. Spectrum of protein-calorie malnutrition. Where one falls on this continuum is determined by the degree and duration of semistarvation and the severity of the stressor.

REFERENCES

1. Torun B, Viteri FE. Protein energy malnutrition. In Shils ME, Young VR. *Modern Nutrition in Health and Disease,* 7th ed. Philadelphia: Lea & Febiger, 1988: 1183–1194.
2. McMahon M, Bistrian BR. The physiology of nutritional assessment and therapy in protein calorie malnutrition. *Disease Month* 1990;36:373–417.
3. Foxx-Orenstein, Kirby D. Understanding malnutrition and refeeding syndrome. In Kirby DF, Dundrick SJ, eds. *Practical Handbook of Nutrition in Clinical Practice.* Orlando: CRC Press, 1994:19–30.
4. Randle PJ, Garlan PB, Hales CN, et al. The glucose fatty-acid cycle: its role in insulin sensitivity and metabolic disturbances of diabetes mellitus. *Lancet* 1963;1:785–789.
5. Love A. Metabolic response to malnutrition: its relevance to enteral feeding. *Gut* 1986;27 (Suppl. 1):9–13.
6. Blackburn G, Bistrian B, Maini B, et al. Nutritional and metabolic assessment of the hospitalized patient. *J Parent Ent Nutr* 1977;1:11–22.
7. Gore DC, Jahoor F, Wolfe PR, et al. Acute response of human muscle protein to catabolic hormones. *Ann Surg* 1993;679–684.
8. Dinarello CA. Interleukin-1 and the pathogenesis of the acute-phase response. *N Engl J Med* 1984 311:1413–1418.
9. Doweiko JP, Nompleggi DJ. The role of albumin in hormone physiology, Part III: albumin and disease sttus. *J Parent Ent Nutr* 1991;15:476–483.
10. Gersovitz M, Munro HN, Udall J, et al. Albumin synthesis in young and elderly subjects using new stable isotope methodology: response to level of protein intake. *Metabolism* 1980;29:1075–1086.
11. Perlmutter DH, Dinarello CA, Punsal PI et al. Tumor necrosis factor regulating hepatic active phase gene expression. *J Clin Invest* 1986;78:1349–1354.
12. Rothschild MA, Oratz M, Schreiber SS et al. Albumin synthesis. *N Engl J Med* 1972;286:748–757.
13. Bistrian BR. Nutritional assessment in the hospitalized patient: a practical approach. In Wright RA, Heymsfield S, eds. *Nutritional Assessment.* Boston: Blackwell Scientific, 1984:183–194.
14. Bistrian BR, Blackburn GL, Vitole J, et al. Prevalence of malnutrition in general medical patients. *JAMA* 1976;235:1567–1570.
15. Arora NS, Rochester DF. Respiratory muscle strength and maximal voluntary ventilation in undernourished patients. *Am Rev Respir Dis* 1982;126:5–8.
16. Scholz RW. Lipid metabolism by rat lung *in vitro:* utilization of citrate by normal and starved rats. *Biochem J* 1972;126:1219–1224.
17. Jacad G, Dickie K, Massaro D. Protein synthesis in lung: influence on starvation on immunoacid incorporation into protein. *J Appl Physiol* 1972;33:381–384.

18. Sahebjami H, MacGee J. Changes in connective tissue composition of the lung in starvation and refeeding. *Am Rev Respir Dis* 1983;128:644–651.
19. Keys A, Henchel A, Taylor HL. The size and function of the human heart at rest in semistarvation and in subsequent rehabilitation. *Am J Physiol* 1947;150:153–169.
20. Chauhan S, Nayack N, Ramalingaswami V et al. The heart and skeletal muscle in experimental protein malnutrition in rhesus monkeys. *J Pathol Bacteriol* 1965;90:301–309.
21. Heymsfield SB, Bethel RA. Cardiac abnormalities in cachectic patients before and during nutritional repletion. *Am Heart J* 1978;95:584–594.
22. Sonis HE, Fratelli VP. Sudden death associated with very low calorie weight reduction regimes. *Am J Clin Nutr* 1981;34:453–461.
23. Foxx-Orenstein A, Jensen GL, McMahon MM. Over zealous resuscitation of an extremely malnourished patient with nutritional cardiomyopathy. *Nutr Rev* 1990;48:406–411.
24. Levine GM, Deren JJ, Steiger E et al. Role of oral intake in maintenance of gut mass and disaccharide activity. *Gastroenterology* 1974;67:975–982.
25. Goodlad RA, Wright NA. The effects of starvation and refeeding on intestinal cell proliferation in the mouth. *Virchows Arch* 1984;45:63–73.
26. Alverdy JC, Aoy SE, Moss GS. Total parenteral nutrition promotes bacterial translocation from the gut. Surgery 1988;104:185–190.
27. Fink MP. Gastrointestinal mucosal injury in experimental models of shock, trauma, and sepsis. *Crit Care Med* 1991;19:627–641.
28. Kudsk KA, Croce MA, Fabian TC et al. Enteral versus parenteral feeding: effects on septic morbidity after blunt and penetrating abdominal trauma. *Ann Surg* 1992;251:503–511.
29. Keys A, Brozek J, Henschel A et al. *The biology of human starvation,* Vol. 1 and 2. Minneapolis: University of Minnesota Press, 1950.
30. Levenson SM, Crowley LM, Seifter E. Starvation. In *Manual of Surgical Nutrition,* ed. Ballinger WF, Collins JA, and Drucker WR, et al. Philadelphia: W. B. Saunders, 1975:236–264.
31. Thompson WD, Ravdin JS, Rhoads JE, et al. Use of lyophile plasma in correction of hypoproteinemia and preservation of wound disruption. *Arch Surg* 1938;36:509–518.
32. Ward MWN, Danzi M, Lewin MR, et al. The effects on subclinical malnutrition and refeeding in the healing of experimental colonic anastomoses. *Br J Surg* 1982;69:308–310.
33. Haydock DA, Hill GL. Improved wound healing response in surgical patients receiving intravenous nutrition. *Br J Surg* 1987;74:320–323.
34. Ross R, Benditt EP. Wound healing and collagen formation. V: quantitative electron microscope radioautographic observation of proline-H3 utilization by fibroblasts. *J Cell Biol* 1965;27:83–106.
35. Wolman SL, Anderson GH, Marliss EB, Jeejeebhoy KN. Zinc in total parenteral nutrition—requirements and metabolic effects. *Gastroenterology* 1979;76:458–467.

36. Bower RH. Nutrition and immune function. *Nutr Clin Prac* 1990;5:189–195.
37. Christou NV, MeaKing JL, MacLean LD. The predictive role of delayed hypersensitivity in preoperative patients. *Surg Gynecol Obstet* 1981;152:197–301.
38. Bistrian BR, Blackburn GL, Scrimshaw NS, et al. Cellular immunity in semi-starved states in hospitalized patients. *Am J Clin Nutr* 1975;28:1148–1155.

CHAPTER

3 The Comprehensive Nutritional Assessment

Jane E. Heetderks-Cox, R.D., C.N.S.D.

OVERVIEW AND PURPOSE

Nutrition assessment may be considered an art as well as a science. A plethora of tools are available to aid the clinician in evaluating the nutritional status of and identifying those at risk for undernutrition. However, one must be careful when interpreting a single value as indicative of a patient's nutritional status. This chapter describes the most common methods available at most institutions along with some of their applications, strengths, and weaknesses.

CLINICAL EVALUATION

A detailed history and physical exam is integral to the nutrition assessment process. Specific areas to address should include medical diseases and surgeries that could impair nutrient intake or metabolism. Changes in weight, appetite, or food intake should be noted. The patient's medical records are useful for verifying weight status, especially if the patient cannot recall usual weight. One should inquire about the patient's general eating habits to include the number of meals consumed per day, size of meals and snacks, and use of nutritional supplements. Other areas that should be addressed are the presence of problems that could potentially impair oral nutrient intake such as problems chewing or swallowing, poor appetite, or early satiety. Presence of gastrointestinal problems such as nausea, vomiting, diarrhea, or extreme constipation should be identified. If the patient is experiencing vomiting, what is the frequency; if diarrhea, how many bowel movements per day? Is the stool malodorous and does it float (indicative of fat malabsorption)? Other areas to address include alcohol use, socioeconomic or living conditions, mental status, and functional capacity. Finally, medications should be reviewed for drug nutrient interactions and the patient examined for physical signs of vitamin or trace element deficiencies (Table 3-1). Iron deficiency anemia is often difficult to diagnose. However, laboratory tests may be ordered

Table 3.1 Clinical Findings Associated With Specific Nutrient Deficiencies

	Clinical Findings	Deficiency
Hair	Alopecia	Protein-energy malnutrition (PEM)
	Dry, easily pluckable, brittle, sparse	
	Dyspigmentation	Biotin, PEM
	Flag sign	Protein
Skin	Xerosis	Vitamin A, essential fatty acids
	Follicular hyperkeratosis	Vitamin A, essential fatty acids
	Perifollicular petechiae	Vitamin C, vitamin K
	Dermatitis	Niacin
	Nasolabial seborrhea	Niacin, riboflavin, vitamin B_6
Eyes	Xerophthalmia	Vitamin A
	Bitot's spots	Vitamin A
	Night blindness	Vitamin A, riboflavin, niacin
	Angular stomatitis	Riboflavin, vitamin B_6, iron
	Angular palpebritis	Riboflavin
Lips	Cheilosis	Vitamin B_6
	Angular stomatitis	Riboflavin, vitamin B_6, iron
Gums	Bleeding/spongy	Vitamin C
Tongue	Magenta tongue	Riboflavin
	Atrophic papillae	Iron, niacin, folate, vitamin B_{12}
	Glossitis	Niacin, folate, iron, vitamin B_6, vitamin B_{12}
Nails	Koilonychia	Iron
Subcutaneous tissue	Edema	PEM, thiamine
Musculoskeletal system	Muscle Wasting	PEM
	Bowlegs	Vitamin D, calcium
	Beading of ribs	Vitamin D, PEM

Source: Adapted with permission from Bernard, Jacobs, and Rombeau.[1]

Table 3.2 Examples Of Drugs That May Adversely Affect Nutritional Status

Drug/Drug Group	Effect
Digoxin, Calopril, Nonsteroidal anti-inflammatory drugs	Decreased appetite
Anti-Parkinson drugs, tricyclics, antihistamines	Dry mouth
Antihistamines	Dry mouth, loss of appetite
Captopril, penicillamine	Decreased taste perception
Methotrexate	Decreased ability to swallow
Laxatives	Potassium deficiency, malabsorption
Antacids	Phosphate depletion; muscle weakness; osteomalacia
Aspirin and ibuprofen	Iron deficiency anemia
Phenytoin, trimethoprim-sulfamethoxazole	Folate deficiency with anemia
Furosemide	Thiamin deficiency, potassium depletion

Source: Adapted with permission from Roe.[2]

to rule out other forms of anemia including B_{12} or folate deficiency. The elderly, in particular, are at risk for polypharmacy, resulting in adverse drug reactions which affect nutrient intake and metabolism (Table 3.2).

NUTRITION SCREENING

Timely detection of patients at risk for malnutrition can allow for earlier nutrition intervention and reversal of malnutrition. Screening all patients upon admission or in the outpatient clinic can more quickly identify patients at risk for undernutrition. A simple assessment form can be administered on admission or in the waiting room for evaluation by members of nursing or nutritional medicine staff. Various forms, ranging from basic to very detailed in content, have been developed.[3,4] Screening forms are also available for use in the outpatient setting.[4] If the patient is determined to be at risk for undernutrition, a more detailed assessment may be performed by a registered dietitian.

METHODS OF ASSESSING FOOD INTAKE

There are a variety of methods available for assessing both food intake of individuals as well as populations. In the hospital setting, clinicians are usually interested in the intake of individuals. Observation of dietary intake can validate a diet suspected to be inadequate or adequate, demonstrating the need for initiation of or discontinuation of nutrition support, respectively. The 24-hr recall or "usual intake" (retrospective methods), food records, and calorie count (prospective methods) are the most common and practical methods for gathering information about an individual's food consumption. Using the 24-hr recall or "usual intake," individuals recall foods and beverages consumed over the previous 24 hours or during a usual day. Memory loss (elderly, Alzheimer's, dementia) limits the accuracy of this method. Using food records, individuals record all foods and beverages consumed shortly after consumption for, generally, 3 to 7 days. Food records and calorie counts tend to be more accurate if they do not rely on memory (especially for the elderly), data are quantifiable, and daily variation is reduced.[5–7] However, both recall and food records have been associated with individuals under- or overreporting food intake.

The calorie count is one of the most widely used assessment tools for institutionalized individuals. However, some serious limitations include the labor intensity required of involved staff and the difficulty of obtaining complete results. Moreover, calorie counts are generally performed for 3 to 7 days. These recommendations are based on studies to assess the intake of healthy, free living populations. When calorie counts are performed in the hospital or in an institutional setting, it is usually to confirm or validate suspected inadequate dietary intake. Breslow and Sorkin[8] conducted a pilot study using prospective, nonconcurrent review of medical records to determine whether a 1-day calorie count could replace the more labor intensive 3-day calorie count. The inclusion criteria included adult patients with calorie counts ordered who were not receiving nutrition support (TPN or enteral feedings). In this study, documentation was lacking (one or more meals not recorded) for 62% of patients (65 out of 105). Of the 30 patients with complete documentation, the 1-day caloric count correctly identified 28 patients with inadequate intake and 2 patients with adequate intake compared to individually determined estimated needs. Thus, it appears that a shorter calorie-count period is a valid method with results similar to a longer collection period. However, it is questionable whether calorie counts ought to be performed in those with documented malnutrition, as this may further delay initiation of nutrition support.

MEASUREMENTS OF BODY COMPOSITION

Body Weight and Stature

The oldest and most common way of obtaining information about a person's body composition is to measure body weight, which is a very rough measure of total body energy stores. Total body weight is comprised of the sum of various body compartments to include adipose, lean body mass, skeletal and visceral mass, and intra- and extracellular fluid. Body weight and height are easy and inexpensive to obtain and can be compared to desirable weights or ideal body weight (IBW) using reference tables. The Metropolitan Life Insurance Company tables express ideal body weights as a range and by frame size for individuals greater than 25 years of age.[9] One limitation is that the tables were based on measurements of people under 59 years of age. Thus, they may not be applicable to the elderly, who often experience changes in height and weight with age. Tables to evaluate appropriate weights for the elderly are not available at this time. Furthermore, weight tables do not always accurately describe body composition. For instance, athletes may appear overweight or obese because they have a greater amount of lean body mass. Finally, tables do not account for ethnic populations who display variations in lean body and skeletal mass.

The Hamwi method[11] or "rules of 5s and 6s" is a quick way to determine IBW. For females, IBW is calculated as 100 lb for the first 5 feet and 5 lb for each additional inch. For males, the calculation is 106 lb for the first 5 feet and 6 lb for each individual foot. For example: if a male is 6 feet tall, IBW would be $106 + (12 \times 6) = 178$ lb. Using either a standard weight table or the Hamwi method, the individual's actual weight can be compared to the reference weight and expressed as a percentage. An acceptable weight range for IBW is considered between 90% and 110%. The minimal survivable weight for humans is between 48% and 55% IBW.[9] Table 3.3 defines various degrees of malnutrition or obesity based on percentage IBW and other commonly available nutrition assessment tools.[10,11]

Body mass index (BMI)[11] assesses the degree of lean body mass and body fat and is performed by dividing weight in kilograms by height in meters squared. Guidelines have been developed based on extensive reviews and functional measurements and health outcomes at different levels of BMI.[9,12,13] The minimal survivable weight for humans is a BMI of approximately 13 to 15.[9] Numerical values less than 18.5 indicate presence of malnutrition, whereas values above 27 correlate with obesity. This method also is subject to inaccuracy as it does not account for age, sex, lean body mass, or fluid status. Males and athletes tend to have more lean body mass at any given BMI value. Excess body fluid or edema can falsely elevate values.[9]

Usual weight is likely the most clinically useful parameter for use in the nutrition assessment process, that is, because many individuals may be normal to overweight compared to a reference population yet lose a significant amount of body mass resulting in functional impairment. Using this method, the patient's current weight is compared to his or her "usual" or premorbid weight using stated information or, preferably, a previously measured weight. In some instances, such as chronic disease, weight loss may occur over several years. The severity of weight loss is determined by both the rate of weight change over time and the total reduction in weight. Greater than 10% of weight loss in 6 months is

Table 3.3 Common Indicators Of Nutritional Status

Indicator	Degree of Malnutrition		
	Mild	Moderate	Severe
Weight			
IBW (%)	80–90	70–79	<69
BMI	17.0–18.4	16.0–16.9	<16.0
Usual weight (%)	85–90	75–84	<74
Recent weight loss (%)	5–10	10–20	>20
Anthropometrics			
Arm muscle circumference (percentiles)	—	5th–10th	<5th
Triceps skinfold (percentiles)	—	5th–10th	<5th
Laboratory			
Albumin (g/dL)	2.8–3.5	2.1–2.7	<2.1
Transferrin (mg/dL)	150–200	100–150	<200
Prealbumin (mg/dL)	10–15	5–10	<5
Immunologic			
Total lymphocyte count (mm^3)	1200–1500	800–1200	<800
Delayed hypersensitivity skin testing (DH)	Reactive	+/–	Unreactive
Creatinine height index (CHI) %	80–90	60–80	<60

Source: Adapted from Shikora, Blackburn, and Forse[10] and from Hopkins.[11]

considered clinically significant. Studley was the first in 1936 to associate weight loss with increased mortality in patients undergoing surgery for peptic ulcer disease.[14] Loss of more than 20% of body weight was associated with 33% mortality compared to 4% mortality in patients who lost less than 20% of their body weight. Seltzer et al. observed that weight loss of more than 10 lb in patients undergoing elective surgery was associated with increase in mortality of greater than 19-fold.[15] In another study, weight loss was associated with an increased rate of complications (44% incidence) with loss of greater than 6% of premorbid weight compared to those with less than 6% weight loss (0% incidence).[16] However, others have not found weight loss to be predictive of postoperative complications.[17,18]

Measurement Techniques and Interpretation

Height can normally be measured by a sliding bar attached to a standing scale. However, in patients who are bedridden or have postural changes, such as in elderly patients, knee height or arm span length may be used.[11] These indices are based on the fact that the long bones of the arms and legs do not change with age. Formulas are also available for calculating ideal weight and BMI for patients with amputations.[11,19,20] Weight should be taken on admission, monitored daily for patients receiving nutrition support, and be obtained at each clinic visit for outpatients. Standard weight varies less than +/– 0.1 kg in healthy adults. If weight loss is greater than 0.5 kg/day, this implies negative energy or water balance or a combination of the two.[9] Weight is more often than not a poor indicator of nutritional status in the hospitalized patient and must be interpreted cautiously. Patients may often experience false elevations from, for example, edema, ascites, or massive tumor growth, organomegaly, or intravenous fluids, which mask the loss of lean body mass. Large fluid shifts are often associated with certain disease states and are usually seen in the postsurgical or critically ill patient. Conversely, weight may be falsely depressed in patients admitted with severe dehydration or who become dehydrated on admission. Large changes in energy intake over several days affects glycogen mass and bound water over several days.[9] Large changes in sodium intake also result in weight changes.[9] For these reasons, weight is often criticized as a nutrition assessment tool. Nevertheless, it remains one of the quickest, easiest measures and costs nothing. A trend downward in absence of fluid shifts may indicate inadequate nutrient intake and loss of body mass, therefore, it should be monitored closely.

ANTHROPOMETRICS

Anthropometric data are designed to further describe the total body compartment specifically related to the amount of adipose tissue and lean body or fat free mass. Losses of either compartment can result in malnutrition and functional impairment. Anthropometrics are simple to perform, practical, and inexpensive. Methods for determining the degree of body fat are the single skinfold method, multiple skinfold measurements, and limb fat area. Skinfold caliper and tape measure are used and measurement sites for skinfolds include biceps, subscapular, suprailiac, thigh, and calf. The triceps skinfold (TSF) is the most common measurement site. Limb circumference measurement sites include the upper arm, thigh, and calf. Values obtained are compared to standard values calculated as a percentage of the standard or compared with percentile values using reference tables. Those interested in more information regarding measurement techniques and interpretation of values may seek additional references.[9]

Including anthropometrics as a part of the initial nutrition assessment may give a more accurate assessment of energy stores than using weight alone. Lansey et al. compared the use of percent IBW and anthropometry in 47 hospitalized frail elderly patients.[21] Approximately 45% of the patients had two anthropometric measurements below the 5th percentile indicating severe malnutrition. However, only 28% of patients were identified as malnourished using percent IBW (<90%) alone. When biochemical indicators (such as albumin and total lymphocyte count, which are discussed in detail later in this chapter) were used in combination with %IBW, 19% of patients were identified as malnourished. However, when anthropometrics were used with serum indicators, 36.2% were identified for malnutrition. This suggests that anthropometrics are more sensitive than weight alone or biochemical markers of nutritional status. Others have experienced similar findings. Fernandez et al. studied 45 male patients ranging in age from 20 to 60 years old who were undergoing elective abdominal surgery.[22] When patients were grouped by weight or BMI (<52kg or BMI <18.5 vs weight >52kg or BMI>18.5), there were no differences in biochemical indices (hemoglobin, albumin, total protein) between the undernourished or normally nourished. However, there were marked differences in anthropometric values between the nourished and the undernourished groups. Midarm circumference (MAC) and TSF were significantly decreased in those who were undernourished by body weight.

Data are conflicting as to whether anthropometrics are more accurate in short term (<1 month) or in long term assessment.[23] Special problems in their use in the hospitalized patients include fluid retention or fluid fluctuation, which is often experienced by such patients. Acute weight loss in surgical or critically ill patients may be secondary to changes in lean body mass and fluid status. In the aforementioned study,[22] 10 of the 45 patients were followed postoperatively, and the same anthropometric and biochemical data was collected on postoperative days 1, 4, and 8. All experienced net negative nitrogen balance with a 2-kg net weight loss with loss of lean and fat tissue. Body water changes were variable with retention on postoperation day 1, followed by loss mid study, and regain in the latter part of the study. However, anthropometric indices showed no significant difference and, thus, did not reflect these losses in body fat and nitrogen.

In healthy individuals, multiple skinfold sites are thought to be more accurate than measuring a single site. Limitations of a single skinfold measurement include large interindividual differences in fat distribution. The elderly tend to experience intramuscular deposition of fat rather than subcutaneous deposition as with younger people, and standard reference tables are not available for this population. Also, it is not possible to perform skinfold measurements in the massively obese. Finally, interobserver differences and day-to-day variability are limitations of this assessment tool.[17] Many prediction equations have been developed for calculating total body fat from measured skinfold thicknesses, circumferences, and body weight. However, these may be better used in the treatment of obesity and may not be accurate in patients with severe weight loss or fluid accumulation.[9]

Bioelectrical Impedance

Bioelectric impedance (BIA) measures electrical conductance (the inverse of resistance) through the body and is an indirect measurement of body composition. Because BIA is relatively inexpensive, portable, and simple to use, many institutions have been utilizing this nutritional assessment technique. Bioelectrical impedance involves attaching electrodes to the wrist and ankle while applying an alternating current through the body mass. Tissues containing water and salt such as blood, visceral organs, and muscle are good conductors of electricity, whereas fat is a poor conductor of electricity resulting in high resistance. Because conductance is not a direct measure of total body water, regression equations or nomograms are used to predict total body water. These equations were validated in heterogeneous groups of healthy subjects. The equations include height and weight to adjust for the length and width of the current. A taller, thinner, person will encounter a longer path for the electrical current to pass from hand to foot. Thus, resistance will be greater and conductance will be lower.

Most studies using BIA have been conducted in healthy populations. These studies have found strong correlations with BIA compared to other methods of body composition.[24–27] However, others have demonstrated lack of precision.[28] When comparing BIA to skinfold anthropometry, skinfold measurements do not appear to be comparable to BIA, especially in extreme ranges of body fat.[29] Few studies using BIA have been conducted using acutely ill patients. To use BIA in the ill, regression equations specific to that population are necessary as equations using healthy individuals may not accurately predict total body water (TBW). Using end-stage cancer patients, Simones et al. found that TBW was overestimated in the underweight (87% IBW) group using prediction equations for the normal, healthy population.[30] The same was found in underweight patients suffering from Crohns disease.[31] Other uses for BIA include monitoring adequacy of nutrient intake.[32] A recent study found BIA useful in monitoring nutritional status in the critically ill patient with a correlation between body-cell-mass (BCM) changes and energy and protein intake.[33] Bioelectrical impedance is currently being investigated for determining the distribution of intracellular versus extracellular water, especially in states such as trauma or malnutrition, thus better defining body compartments.[34] Its usefulness for use in longitudinal assessment of nutrition assessment and application to patients with large fluid fluctuations is not known.[35]

URINARY MEASURES OF NUTRITIONAL STATUS

Urinary measurements can also be used as an index of protein nurtriture. Nitrogen balance studies are commonly conducted to measure changes in the protein compartment of the body to maintain nitrogen equilibrium. Nitrogen balance will be discussed in detail in the following chapter. Two other urinary measures, urinary creatinine and 3-methylhistidine, have been validated and used by investigators to estimate lean body mass.

Creatinine Height Index

Urinary creatinine excretion is proportional to the amount of muscle mass. The creatinine height index (CHI) estimates the amount of muscle mass by measuring urinary creatinine excretion compared to normal age- and height-matched controls of the same sex and is expressed as a percentage[11]:

$$\% \text{ CHI} = \frac{\text{actual 24-hour creatinine excretion}}{\text{expected 24-hour creatinine excretion}} \times 100.$$

Expected creatinine excretion values are derived by tables. A total of 100% indicates normal muscle mass, whereas a value less than 80% indicates skeletal muscle depletion. A value less than 60% represents severe muscle depletion.[36] Studies have shown strong correlation between lean body mass and 24-hour excretion of creatinine.[37,38] Bistrian et al. found CHI to be more sensitive than weight for height ratio in distinguishing between the malnourished, especially in the presence of edema.[39] Patients with catabolic or potentially catabolic diseases (e.g., trauma or diabetes) may experience increased muscle turnover; however, creatinine excretion may or may not increase.[40,41] Rosenfalck et al. followed 147 adults with type I diabetes mellitus for 7 years and found that individuals obtained whole-body protein mass between 80% and 120% of expected normal values.[42] Although both CHI and lean body mass can decrease during starvation, stress, and prolonged immobility, CHI does not appear to be predictive of complications or negative clinical outcome.[43] Valid results for creatinine excretion require accurate urine collection, normal renal function, and a normocatabolic state. Limitations include daily variations in excretion and the effect of meat ingestion. Tables used to predict creatinine excretion are based medium frame only, thus results are not as accurate in the thin or very obese. Finally, creatinine excretion decreases with age; thus, the tool may not prove accurate in the elderly.[11]

3-Methylhistidine

3-Methylhistidine (3-MeH) is a nonrecyclable component of muscle derived from the catabolism of myofibril protein, and excretion approximates muscle turnover. It is normally used in research and requires an amino acid analyzer for measurement. Several criticisms of the tool include reports that other nonskeletal muscle sources may falsely influence the measurement and that levels may not correlate well with nitrogen balance during starvation, trauma, stress, and burns.[44,45] Finally, a major limitation of urinary measures of nutritional status is obtaining an accurate 24-hr urine collection.

BIOCHEMICAL INDICATORS OF NUTRITIONAL STATUS

One of the main goals of nutrition support is to replete or preserve the protein compartment of the body. Several serum proteins have been identified that give a picture of the patient's overall protein compartment. If interpreted correctly, these visceral protein measurements can be easy and reliable indicators of nutritional status. Proteins, which have a shorter half life and smaller total body pool, are generally better as early measures of malnutrition as well as monitoring the effectiveness of nutrition support therapy. In addition, certain indicators or combinations of indicators have been able to predict morbidity and mortality. However, it must be emphasized that no single indicator alone accurately assesses nutritional status; rather, each indicator must be viewed as only one piece of the "puzzle." The hydration status of the patient as well as a number of illnesses and metabolic sequelae can affect the ability of these markers to predict nutritional status. Because most of these proteins are produced in the liver, hepatic disease affects their synthesis.

Albumin

Albumin is the oldest and most common laboratory measurement of visceral protein status. Low levels have been correlated with increased morbidity, mortality, infection, and length of hospital stay.[46,47] The total body pool of albumin is large, with greater than 60% present in the extravascular space. The total amount of albumin depends on the rate of synthesis, rate of catabolism, and characteristics of the distribution space. During chronic protein energy malnutrition, albumin shifts from extravascular to intravascular space. Concurrently, albumin production by the hepatocytes increases with a decrease in the rate of degradation. The combination of these events can result in a preservation in serum albumin making the patient appear nourished. This form of malnutrition is called marasmus, which is primarily caused by inadequate calorie intake. In the earlier-cited study by Fernandez et al., preoperative albumin was not significantly different between those malnourished by low body weight or BMI compared to healthy subjects.[22] Because of its long half life (20 days), albumin is not a good indicator of short-term energy and protein deprivation nor does it respond quickly to nutritional repletion. However, as time progresses, total body albumin becomes depleted helping to identify various levels of protein-energy malnutrition. Levels corresponding to different degrees of malnutrition have been developed (Table 3.3). Low albumin can also be observed when calorie intake is adequate but protein intake is inadequate. This form of malnutrition is known as kwashiorkor. Marasmic kwashiorkor occurs when both energy and protein are deficient. This form of malnutrition can also present when stress is superimposed on a chronically starved patient. On the other hand, a low albumin can be misinterpreted as a variety of medical diseases and conditions result in depression of albumin. Examples include those involving fluid retention (ascites, edema, CHF), chronic liver disease, renal disease, cancer, pancreatitis, and sepsis.[9] Albumin has also been found to decrease with age. However, population studies concluded that this decrease was less than previously believed and mean levels in healthy individuals were within normal limits.[48] Although, albumin has several limitations in the ill or hospitalized patient, it remains a fairly inexpensive, readily accessible, and useful assessment tool.

Transferrin

Transferrin is best known for its function of binding and transporting the ferric portion of iron. Approximately 99% of the iron in the serum is bound to about one third of the transferrin pool. It's biologic half life is shorter than albumin, being only about 10 days. Transferrin is synthesized by the liver and its synthesis is regulated by iron stores. An elevated transferrin level indicates early iron deficiency as transferrin levels rise to transport available iron. Transferrin has a smaller body pool and a shorter half life than albumin deficiency. Therefore, compared to albumin, it tends to be more sensitive to concentration changes and is quicker to identify protein depletion in malnutrition and quicker to rise following nutritional repletion. In some studies, transferrin has been shown to be a sensitive indicator of nutritional status as well as a prognostic tool predicting morbidity and mortality in hospitalized and long-term care patients.[49,50] However, transferrin may not be sufficiently sensitive in detecting changes following initiation of nutrition support. In malnourished patients receiving parenteral nutrition for 2 weeks, transferrin levels did not change despite improvement in nutritional status.[51] In addition, certain conditions alter transferrin levels independently of nutritional status. For example, transferrin levels increase with pregnancy, iron deficiency, and use of birth control pills. Levels decrease with end-stage liver disease, nephrotic syndrome, anemia, and neoplastic disease. High-dose antibiotics may also result in falling levels.[52]

Transthyretin

Transthyretin is most commonly called prealbumin and is a transport protein for thyroxine and vitamin A. Prealbumin also has a high content of the amino acid tryptophan, which is involved in the initiation of protein synthesis. Prealbumin concentrations depend on the presence and adequacy of calories and protein. Because of its shorter half life (2 to 3 days), small body pool, and early response to nutritional deficits and repletion it has become a more popular indicator than albumin or transferin. Thus, the ability to monitor prealbumin daily provides for earlier assessment of the adequacy of a given nutrition support regimen. Bernstein et al. have chosen a response time of greater than 20 mg/L in 7 days as an outcome measure indicating adequate response or adequate nutritional support.[53] Both prealbumin and transferrin have been positively correlated with adequate calorie and protein intake.[54] Prealbumin appears to be more sensitive than albumin or transferrin in determining the adequacy of oral nutrient intake in patients in transition from parenteral to enteral or oral intake. In malnourished patients who had 1 week of adequate oral or enteral feeding, 98% of the patients had normal values for prealbumin compared with only 80% for transferrin and 38% for albumin.[55] Prealbumin has also been used to approximate nitrogen balance. A level of 180 mg/L correlated with a positive nitrogen balance.[53] Limitations in interpreting the prealbumin level include patients with chronic renal failure on dialysis secondary to decreased renal catabolism. Prealbumin is also decreased in metabolic stress, postoperatively, in hyperthyroidism, and in protein-losing enteropathy. Levels may increase in some cases of nephrotic syndrome.[9]

Retinol Binding Protein

A less commonly used visceral protein, retinol binding protein (RBP) has a small body pool size and a half life of 24 hr. Retinol binding protein is a single polypeptide chain that interacts with plasma prealbumin. It usually circulates in plasma as a 1:1 mol/L REP–prealbumin complex; thus, levels parallel each other. Because of its short half-life, RBP has been reported to respond quickly to energy and protein deprivation.[56] In patients with kidney disease, RBP poses limitations for use in determining visceral protein status even more so than prealbumin. Plasma concentrations of RBP rise from decreased catabolism by the kidney. In contrast, catabolism of prealbumin occurs only to a small extent in the kidney, resulting in less elevated serum concentrations. Because RBP binds to the alcohol form of vitamin A, levels are low in vitamin A deficiency. As with prealbumin, metabolic stress, surgery, and hyperthryroidism result in decreased levels.[52]

Other Visceral Proteins

More recently investigated biochemical markers include somatomedin C (also known as insulinlike growth factor) and fibronectin. Further studies are needed before these indicators become routine assessment tools. Like insulin, SMC helps stimulate growth. It is similar in molecular size and structure to proinsulin, which is strongly anabolic, and is thought to mediate many effects of growth hormone. Advantages over previously mentioned serum proteins include its sensitivity, shorter half life (2 to 4 hr), and that it is not subject to acute stress.[52] Concentrations rise and fall rapidly during starvation and refeeding.[57] Somatomedin C has also been positively correlated with nitrogen balance and has been used to monitor the effectiveness of nutritional therapy.[52,58,59] Limitations of its use include hypothyroidism, with estrogen therapy, and possibly, obesity.[9] Levels may also vary in individuals with liver or renal disease.[52] Fibronectin is unique in that it is produced in the liver and is also synthesized by endothelial cells, peritoneal macrophages, and fibroblasts. Fibronectin has a half-life ranging from 4 to 24 hr and studies have shown sensitivity to both caloric deprivation and repletion.[60,61] In one study, fibronectin levels rose within 5 days in response to refeeding, whereas transferrin was slower to respond and albumin remained the same.[60]

FUNCTIONAL ASSESSMENTS

Suppressed markers of nutritional status in and of themselves do not lead to the demise of a patient's clinical course. Rather, it is the impairment and failure in physiologic and protein-requiring functions in response to these events. Thus, several "functional" measures of a patient's nutritional status have been developed.

Immune System

The most important complication of malnutrition is the development of infection. The purpose of the immune system is to protect against foreign invaders known as antigens (e.g., bacteria, viruses, fungi, parasites). An intact immune system is required for a patient to mount the appropriate response to antigens by cellular and humoral response. Malnutrition has been shown to affect many components of the immune system, including lower numbers of circulating T lymphocytes and reduced ability to mount an immune response. Effects of immune dysfunction in malnourished hospitalized patients include development of infection, sepsis, organ failure, poor wound healing, and increased morbidity and mortality.[62] Recent studies have associated the use of an enteral diet containing known immunostimulatory components with significant reductions in the incidence of infection, wound complications, and decreased length of stay.[63]

Delayed hypersensitive skin test. Delayed hypersensitivity (DH) skin testing is a common assessment tool in which 1 or more antigens (including tetanus, diphtheria, Streptococcus, tuberculin, Candida, and mumps) are injected subcutaneously into the skin of the forearm. If there is no response, the person is classified as anergic. A recent study found bulemic patients to have a reduced response as measured by DH and lower lymphocyte subsets compared to controls, although anthropometric and most other clinical parameters remained normal.[64] Besides protein and energy malnutrition, specific vitamin, trace element (zinc, iron, folate, and B6), and fatty acid deficiencies result in a lowered response.[65] Limitations of this test are that studies have shown that mild malnutrition does not consistently result in anergy[66] and that a multitude of factors lead to anergy in the absence of malnutrition. This partial list includes infections, immunosuppressive diseases [acquired immunosuppressive deficiency syndrome (AIDS) and cancer], liver disease, immunosuppressive drugs (chemotherapy, steroids), radiation therapy, surgery and metabolic stress.[35] Thus, it may not predict nutritional status in the ill patient.

Total lymphocyte count. Total lymphocyte count (TLC)[11] is a simple, inexpensive tool that can help define nutritional status as well as immunocompetency. It is calculated according to the following formula:

$$\text{TLC} = \frac{\%\ \text{lymphocytes} \times \text{white blood cells (cells/mm}^3\text{)}}{100}$$

Values less than 1500 are considered to be suggestive of nutritional depletion.[36] Circulating T lymphocytes are reduced in malnutrition. A depressed TLC has also found to correlate with increased morbidity and mortality in surgical patients as well as those with anorexia nervosa and AIDS.[65,35] Limitations of TLC include alterations by metabolic stress, sepsis, cancer, and steroid use.[35]

Muscle Function

Tools have been developed to measure skeletal muscle function for use as another indicator of nutritional status. In fact, these measurements have been shown to identify early malnutrition even when anthropometric and biochemical markers have not been sensitive.[67,68,69] Hand grip dynometry (grip strength) has been identified as a good indicator of total body protein, was found to be proportional to protein loss in surgical patients,[70] and is an indicator of postoperative risk.[71,72] For unconscious patients involuntary muscle function test using electrical stimulation of the ulnar nerve has been developed.[73] Skeletal and respiratory muscle function decreases rapidly during starvation and improves readily with refeeding.[67,68,69,74] Others have shown that nutritional support improves skeletal and diaphragmatic function in only a few days before increases in lean body mass or body protein occur.[75] Ventilator patients are also at risk for malnutrition and, during critical illness, are at an increased risk of morbidity and mortality. Driver[76] found ventilator patients received only 73% of estimated nutritional needs. Thus, health care providers must take care to prevent iatrogenic malnutrition resulting from underfeeding these critically ill patients. An advantage of using muscle function tests is that changes in muscle are related specifically to changes in nutritional status and are not influenced by sepsis, trauma, renal failure, or steroid administration.[77,78]

Subjective Global Assessment

Because there are several indicators of malnutrition and no one indicator or group of indicators in itself are reliable, it is imperative to look at a variety clinical parameters. The purpose of a subjective global assessment (SGA) is to determine whether there has been a true restriction in food intake or presence of malabsorption or abnormal digestion and, secondly, to see if there has been a change in functional status as a result. Thus, SGA aims at identification of patients who are malnourished and at risk for developing nutrition-related complications and, thus, poor outcomes. It is based on the hypothesis that restoring food intake can rapidly reduce the risk of malnutrition and that restoring nutrient intake to optimal levels will result in lower risk of complications, although the patient may remain wasted and underweight.[79] Also, SGA considers restrictions in food intake, the presence of abnormal digestion or absorption, and the patient's functional status. After obtaining all the relevant information from the medical history, physical exam, and patient interview, the examiner relies on clinical judgment to determine overall nutritional status.[80,81] The sample form[80] in Table 3.4 contains features based on the history, physical, and finally, a rating for nutritional status. Ratings are based on the clinician's subjective impressions. An alternative approach is to assign a numerical weighting factor to each indicator. Total amount and rate of weight loss are considered. However, if weight is lost but regained, the patient is considered better nourished than one who continues to lose weight. Gastrointestinal symptoms (anorexia, nausea, vomiting, diarrhea) may be considered significant if they are experienced daily or have persisted for more than 2 weeks. A high-stress disease might include trauma, whereas a low-stress disease could include a nonhealing ulcer or a primary tumor. Observation of subcutaneous fat can be determined by observation but also by palpating a number of skinfolds between the finder and thumb. If the dermis can be felt between the finder and thumb when pinching the triceps and biceps skin folds, considerable losses of body fat have occurred. Inspection and palpatation of muscle groups can be observed for protein depletion. Long muscles are considered profoundly protein depleted when the tendons are prominent to palpatation.[82] In a recent study, Lupo et al. found clinical judgment reliable in assessing the nutritional status of surgical patients.[83] In this study, three independent observers categorized 64 surgical patients to one of four nutritional states—normal nourishment, and mild, moderate, and severe malnutrition—using a questionnaire and clinical exam. Agreement was found in 77% of the assessments and partial agreement in the remaining patients. Clinical judgment was significantly correlated with biochemical indicators (including total protein, albumin, transferrin, hemoglobin, immunological skin tests, and weight loss). Also, SGA was recently validated as an assessment tool in dialysis patients.[84] An advantage of SGA has been its ability to predict negative clinical outcome or the development of nutritionally associated complications as opposed to a single objective parameter. Those identified as severely malnourished in a study by Baker et al. had an increased incidence of infection, more antibiotics prescribed, and increased length of stay.[83] Although SGA appears to have reasonable interobserver agreement, it

Table 3.4 Subjective Global Assessment Form

Select Appropriate Category with A Check mark, or Enter Numerical Value Where Indicated By A "#"

A. History

1. Weight change and height
   Overall loss in past 6 months: Amt. +/– ___ #/kg; % loss = ___ Ht(cm) ___
   Change in the past 2 weeks: ___ increase ___ , no change ___ , decrease ___ .
2. Dietary intake change (compared to normal)
   ___ No change ___ Change duration = ________________ (number of weeks)
   If yes, type: ___ suboptimal solid diet, ___ full liquid diet,
   ___ hypocaloric liquids ___ starvation
   Supplement: (circle) none vitamin(s), mineral(s): ________________
   (dose: include amount, frequency/or number per week)
3. Gastrointerestinal symptoms (that persisted for >2 weeks)
   ___ none, ___ nausea, ___ vomiting, ___ diarrhea, ___ anorexia.
4. Functional capacity
   ___ No dysfunction (e.g., full capacity).
   ___ Dysfunction: duration # ___ weeks: Type: ________ working suboptimally,
   ___ ambulatory, ___ bedridden.
5. Disease and its relation to nutritional requirements
   Primary diagnosis (specify): ________________
   Metabolic demand (stress): ___ none ___ low stress ___ moderate stress
   ___ high stress.

B. Physical (for each trait specify: 0 = normal, 1+ = mild, 2+ = moderate, 3+ = severe)

___ Muscle wasting (quadriceps, deltoids, temporalis)
___ Ankle edema
___ Sacral edema
___ Loss of subcutaneous fat (triceps, chest)
___ Mucosal lesions
___ Cutaneous lesions
___ Hair change
___ Ascites

C. SGA rating (select one)

___ Well nourished
___ Moderately (or suspect of being) malnourished
___ Severely malnourished

Source: Adapted with permission from Detsky, McLaughlin and Baker, et al.[80]

has not been reproducible among hospitals.[72] Furthermore, to use the assessment for predicting operative risk, standardization, assessment of reproducibility, and evaluation of underlying medical illnesses is required.[35] Otherwise, SGA appears to be an easy, inexpensive, and useful nutritional assessment tool that can be used by trained clinicians.

Prognostic Nutrition Indices

Several predictive models (also known as multivariable equations) have been developed and are based on anthropometric, biochemical, and immunologic tests.

These models may be used by clinicians to identify patients who may benefit from nutritional support. One of the most common instruments is called the Prognostic Nutrition Index (PNI), designed by Buzby et al.[86] to predict postoperative morbidity and mortality in surgical patients. The PNI is calculated as follows:

$$\text{PNI\%} = 158 - 16.6\ (\text{ALB}) - .78\ (\text{TSF}) - 0.20\ (\text{TFN}) - 5.8\ (\text{DH}),$$

where ALB is albumin(g/dL), TSF is triceps skinfold (mm), TFN is serum transferrin (mg/dL), and DH is delayed cutaneous hypersensitivity skin test (% positive reaction). As PNI increases, the incidence for postoperative complications, major sepsis, and death increase. A score of less than 40 is a low risk, 40 to 49 is an intermediate risk, and more than 50 is a high risk. A similar equation, the Hospital Prognostic Index (HPI) was developed by Harvey et al.[86] The HPI is calculated as follows:

$$\text{HPI\%} = 0.92\ (\text{ALB}) - 1.00\ (\text{DH}) - 1.44\ (\text{SEP}) + 0.98\ (\text{DX}) - 1.09,$$

where SEP is sepsis (present = 1; not present = 2) and DX is diagnosis (cancer = 1; no cancer = 2). This equation is designed to include factors that might predict sepsis and mortality in hospitalized patients. Results of the discriminant function (DF) equation are compared with a DF curve, where DF equals the probability of survival: $-2 = 10\%$, $0 = 50\%$, and $+1 = 75\%$. When investigated in 282 hospitalized patients, the equation had a 75% predictive value for predicting subsequent hospital mortality. Those with an albumin less than 2.2 had a 75% chance of having concurrent sepsis and anergy, and dying. Finally, serial measurements of DH were the most accurate predictor of an improved prognosis (86% predictive value).[86]

SUMMARY

The incidence of malnutrition is high—30% to 50% of medical and surgical patients[87] and up to 53% to 61% of elderly patients (over 70 years old).[88] Because of the prevalence of malnutrition, the nutritional assessment process is vital in identifying patients who may benefit from nutritional intervention. This can best be accomplished through nutritional screening followed by a comprehensive nutritional assessment as appropriate. There are a variety of tools available to aid the clinician in diagnosing malnutrition. However, no single indicator can accurately define nutritional status. Thus, the more indicators that are available, the better the assessment. A registered dietitian is trained in using these indicators and methods along with their strengths and weaknesses. However, there is also much that the whole health care team can do to aid in the nutritional assessment process. This includes completing a nutritional screening form or questionnaire on all patients, obtaining accurate weights, and interviewing and examining the patient for potential indicators of malnutrition. Early identification of malnutrition and appropriate initiation and monitoring of nutrition support may improve the patient's nutrition status and clinical outcome.

REFERENCES

1. Bernard M, Jacobs D, Rombeau J. *Nutritional and metabolic support of hospitalized patients.* Philadelphia: WB Saunders, 1986: 16, 17, 27, 29, 36–39, 233.
2. Roe DA. Medications and nutrition in the elderly. *Primary Care* 1994;21:135–147.
3. Nagel MR. Nutrition screening: identifying patients at risk for malnutrition. *Nutr. Clin. Pract.* 1993;8:171–175.
4. *Report of nutrition screening 1: toward a common view.* Washington DC: Nutrition Screening Initiative, 1991.
5. De Vries JH, Zock PC, Mensink, RP, et al. Underestimation of energy intake by 3-d records compared with energy intake to maintain body weight in 269 nonobese adults. *Am. J Clin. Nutr.* 1994;60:855–860.
6. Mertz W, Tsui JC, Judd JT, et al. What are people really eating? The relation between energy intake derived from estimated diet records and intake determined to maintain body weight. *Am. J. Clin. Nutr.* 1994;54:291–295.
7. Johnson RK, Goran MI, Poehlman ET. Correlates of over- and underreporting of energy intake in healthy older men and women. *Am. J Clin. Nutr.* 1994;59:1286–1290.
8. Breslow, RA, Sorkin JD. Comparison of one-day and three-day calorie counts in hospitalized patients: a pilot study. *J. Am. Geriatr. Soc.* 1993;41:923–927.
9. Heymsfield SB, Tighe A, Wang Z. Nutritional assessment by anthropometric and biochemical methods. In Shils ME, Olson JA, Shike M, eds. *Modern nutrition in health and disease,* Philadelphia: Lea & Febiger, 1994.
10. Shikora SA, Blackburn GL, Forse RA. Nutrition and immunology: clinician's approach. In *Diet, nutrition and immunology.* Boca Raton, FL: CRC Press, 1994; 9–20.
11. Hopkins, B. Assessment of nutritional status. In Gottschlich MM, Matarese, LE, Shronts, EP, eds. *Nutrition support dietetics core curriculum.* Silver Spring, MD: ASPEN, 1993:15–70.
12. McLaren DS. A fresh look at anthropometric classification schemes in protein-energy malnutrition. In Himes JH eds. *Anthropometric Assessment of Nutritional Status.* New York: Wiley-Liss, 1991:273–281.
13. James WPT, Ferro-Luzzi A, Waterlow JC. Definition of chronic energy deficiency in adults. report of a working party of international dietary energy consultative group. *Eur. J. Clin. Nutr.* 1988;42:969–981.

14. Studley, HO. Percentage of weight loss. A basic indicator of surgical risk in patients with chronic peptic ulcer. *JAMA* 1936;106:458–460.
15. Seltzer MH, Slocum BA, Cataldi-Betcher ML, et al. Instant nutritional assessment: absolute weight loss and surgical mortality. *J Parent Ent Nutr* 1982;6:218–221.
16. Roy LB, Edwards PA, Barr LH. The value of nutritional assessment in the surgical patient. *J Parent Ent Nutr* 1985;9:170–172.
17. Halliday AW, Benjamin IS, Blumgart LH. Nutritional risk factors in major hepatobiliary surgery. *J Parent Ent Nutr* 1988;12:43–48.
18. Mullen JL, Gertner MH, Buzby GP, et al. Implications of malnutrition in the surgical patient. *Arch Surg* 1979;114:121–125.
19. Kautz-Osterkamp L. Current perspective on assessment of human body proportions of relevance to amputees. *J Am Diet Assoc* 1995;95:215–218.
20. Tzamaloukas AH, Patron AP, Malhotra D. Body mass index in amputees. *J Paren Ent Nutr* 1994;18:355–358.
21. Lansey S, Waslein C, Mulvihill M, et al. The role of anthropometry in the assessment of malnutrition in the hospitalized frail elderly. *Gerontology* 1993;39:346–353.
22. Fernandez IS, Kurpad AV, Kilpadi AB, et al. 1993. Nutritional assessment of marginally nourished surgical patients. *Nat Med J India* 1993;6:253–256.
23. Smith LC, Mullen JL. Nutritional assessment and indications for nutritional support. *Surg Clin N Am* 1991;71:449–457.
24. Lukaski HC. Methods for the assessment of human body composition: traditional and new. *Am J Clin Nutr* 1987;46:537–556.
25. Van Loan M, Mayclin P. Bioelectrical impedance analysis: is it a reliable estimator of lean body mass and total body water? *Hum Biol* 1987;59:299–309.
26. Kushner RF, Haas A. Estimation of lean body mass by bioelectrical impedance analysis compared to skinfold anthropometry. *Eur J Clin Nutr* 1988;42:101–106.
27. Lukaski HC, Bolonchuk WW, Hall CB, et al. Validation of tetrapolar bioelectrical impedance method to assess human body composition. *J Appl Physiol* 1986;60:1327–1332.
28. Baumgartner RN, Chamlea WC, Roche AF. Estimation of body composition from bioelectrical impedance of body segments. *Am J Clin Nutr* 1989;50:221–226.
29. Vansant G, Van Gaal L, De Leeuw I. Assessment of body composition by skinfold anthropometry and bioelectrical impedance technique: a comparative study. *J Paren Ent Nutr.* 1994;18:427–429.
30. Simons JP, Schols AMWJ, Westerterp KR, et al. The use of bioelectrical impedance analysis to predict total body water in patients with cancer cachexia. *Am J Clin Nutr* 1995;61:741–745.
31. Royall D, Greenberg GR, Allard JP, et al. Critical assessment of body composition measurements in malnourished subjects with crohn's disease: the role of bioelectrical impedance analysis. *Am J Clin Nutr* 1994;59:325–330.
32. Igbal K, Malek MA, Mujibur Rahman M, et al. Changes in body composition of malnourished children after dietary supplementation as measured by bioelectrical impedance. *Am J Clin Nutr* 1994;59:5–9.

33. Sylvie R, Zarowitz BJ, Hyzy R, et al. Bioelectrical impedance assessment of nutritional status in critically ill patients. *Am J Clin Nutr* 1993;57:840–844.
34. Heymsfield, SG, Matthews D. Body composition: research and clinical advances—1993 ASPEN Research Workshop. *J Parent Ent Nutr* 1994;18:94–103.
35. Lipkin EW, Bell S. Assessment of nutritional status: the clinician's perspective. *Clin Lab Med* 1993;13:329–352.
36. Blackburn GL, Bistrian BR, Maini BS, et al. Nutritional and metabolic assessment of the hospitalized patient. *J Parent Ent Nutr.* 1991;1:11–22.
37. Heymsfield SB, Arteaga C, McManus C, et al. Measurement of msucle mass in humans: validity of the 24 hour urinary creatinine method. *Am J Clin Nutr.* 1983;37:478–494.
38. Forbes GB, Bruining GL. Urinary creatinine excretion and lean body mass. *Am J Clin Nutr* 1976;29:1359–1366.
39. Bistrian BR, Blackburn GL, Sherman M, et al. Therapeutic index of nutritional depletion in hospitalized patients. *Surg Obstet Gynecol* 1975;141:512–516.
40. Heding LG. Radioimmunological determination of human c-peptide in serum. *Diabetologia* 1975;11–541–548.
41. Threlfall CJ, Stoner HM, Galasko CSB. Patterns in the secretion of muscle markers after trauma and orthopedic surgery. *J Trauma* 1981;21:140–147.
42. Rosenfalck AM, Snorgaard O, Almdal T, et al. Creatinine height index and lean body mass in adult patients with insulin-dependent diabetes mellitus followed for 7 years from onset. *J. Parent Ent Nutr* 1994;18:50–54.
43. Jeguier E. Measurement of energy expenditure in clinical nutrition assessment. *J Parent Ent Nutr* 1987;11(suppl):86–89.
44. Rennie MJ, Milward DJ. 3-methylhistidine excretion and the urinary 3-methylhistidine/creatinine ratio are poor indicators of skeletal muscle breakdown. *Clin Sci* 1983;65:217–225.
45. Long CL, Birkhahn RH, Geiger JW, et al. Urinary excretion of 3-methylhistidine: an assessment of muscle protein catabolism in adult normal subjects and during malnutrition, sepsis, and skeletal trauma. *Metabolism* 1981;30:765–776.
46. Bozetti F, Migliavacca S, Gallus G, et al. Nutritional markers as prognostic indicators of postoperative sepsis in cancer patients. *J Parent Enter Nutr* 1985;9:464–470.
47. Grant JP, Thurlow J. Current techniques of nutritional assessment. *Surg Clin N Am* 1981;61:437–463.
48. Keller HH. Use of serum albumin for diagnosing nutritional status in the elderly—is it worth it? *Clin Biochem* 1993;26:435–437.
49. Mullen JL, Buzby GP, Waldeman MT, et al. Prediction of operative morbidity and mortality by preoperative nutrition assessment. *Surg Forum* 1979;30:80–82.
50. Rainy-Macdonal CG, Holliday RL, Wells GA, et al. Validity of a two-variable nutritional index for use in selecting candidates for nutritional support. *J Parent Ent Nutr* 1983;7:15–20.

51. Georgieff MK, Amarnath UM, Murphy EL, et al. Serum transferrin levels in the longitudinal assessment of protein-energy status in preterm infants. *J Pediatr Gastroenterol Nutr* 1989;6:234–239.
52. Spiekerman, M.A. Proteins used in nutritional assessment. *Clin Lab Med* 193;13:353–369.
53. Bernstein LH, Leukhardt-Fairfield CJ, Pleban W, et al. Usefulness of data on albumin and prealbumin concentrations in determining effectiveness of nutritional support. *Clin Chem* 1989;35:271–274.
54. Fletcher JP, Little JM, Guest PK, et al. A comparison of serum transferrin and serum albumin as nutritional parameters. *J Parent Ent Nutr* 1987;11:144–148.
55. Winkler MF, Pomp A, Caldwell MD, et al. Transitional feeding: the relationship between nutritional intake and plasma protein concentrations. *J Am Diet Assoc* 1989;89:969–970.
56. Cavarocchi NC, Au FC, Dalal FR, et al. Rapid turnover proteins as nutritional indicators. *World J Surg* 1986;10:468–473.
57. Clemmons DR, Underwood LE, Dickerson RN, et al. Use of plasma somatomedin-c/insulin-like growth factor I measurements to monitor response to nutritional repletion in malnourished patients. *Am J Clin Nutr* 1985;41:191–198.
58. Donahue SP, Phillips LS. Response of IGF-I to nutritional support in malnourished hospital patients: A possible indicator of short term changes in nutritional status. *Am J Clin Nutr* 1989;50:962–969.
59. Buonpane EA, Brown RO, Boucher BA, et al. Use of fibronectin and somatomedin-C as nutritional markers in the enteral nutrition support of traumatized patients. *Crit Care Med* 1989;17:126–132.
60. Chadwick SJD, Sim AJW, Dudley HAF. Changes in plasma fibronectin during acute nutritional deprivation in healthy human subjects. *Br J Nutr* 1986;55:7–12.
61. McKone TK, Davis AT, Dean RE. Fibronectin: a new nutritional parameter. *Am Surgeon* 1985;51:336–339.
62. Shronts EP. Basic concepts of immunology and its application to clinical nutrition. *Nutr Clin Pract* 1993;8:177–182.
63. Gottschlich MM, Jenkins M, Warden GD. Differential effects of three enteral dietary regimens on selected outcome variables in burn patients. *J Parent Ent Nutr* 1990;14:225–236.
64. Marcos A, Varela, P, Santacruz I, et al. Evaluation of immunocompetence and nutritional status in patient with bulemia nervosa. *Am J Clin Nutr* 1993;57:65–69.
65. Chandra RK, Sarchielli P. Nutritional status and immune response. *Clin Lab Med* 1993;13:455–461.
66. Twomey P, Ziegler D, Rombeau J. Utility of skin testing in nutritional assessment: a critical review. *J Parent Ent Nutr* 1982;6:50–58.
67. Windsor JA, Hill GL. Weight loss with physiologic impairment: a basic indicator of surgical risk. *Ann Surg* 1988;207:290–296.
68. Lopes J, Russell D, Whitwell JJ, et al. Skeletal muscle function in malnutrition. *Am J Clin Nutr* 1982;36:602–610.

69. Russell DM, Leiter LA, Whitwell J, et al. Skeletal muscle function during hypocaloric diets and fasting: a comparison with standard nutritional assessment parameters. *Am J Clin Nutr* 1983;37:133–138.
70. Windsor JA, Hill GL. Grip strength: a measure of the extent of protein loss in surgical patients. *Br J Surg* 1988;75:880–882.
71. Klidjan AM, Foster KJ, Kammerlin RM, et al. Relation of anthropometric and dynamometric variables to serious post-operative complications. *Br Med J* 1980;281:899–901.
72. Windsor JA. Underweight patients and the risks of major surgery. *World J Surg* 1993;17:165–172.
73. Jeejeebhoy KN. Bulk or bounce: the object of nutritional support. *J Parent Ent Nutr* 1988;12:539–549.
74. Fraser IM, Russell DMCR, Whittaker S, et al. Skeletal and diagphragmatic muscle function in malnourished chronic obstructive pulmonary disease (Abstract). *Am Rev Respir Disease* 1984;129:A269.
75. Christie PM, Hill GL. Effects of intravenous nutrition on nutrition and function in acute attacks of inflammatory bowel disease. *Gastroenterology* 1990;99:730–736.
76. Driver AG. Iatrogenic malnutrition in patients receiving ventilator support. *J Am Med Assoc* 1980;244:2195–2196.
77. Berkelhammer LH, Leiter LA, Jeejeebhoy KN, et al. Skeletal muscle function in chronic renal failure: an index of nutritional status. *Am J Clin Nutr* 1985;42:845–854.
78. Brough W, Horne G, Blount A, et al. Effects of nutrient intake, surgery, sepsis, and long term administration of steroids on muscle function. *Br Med J* 1986;293:983–988.
79. Jeejeebhoy KN. Clinical and functional assessments. In Shils ME, Olson JA, Shike M. eds. *Modern nutrition in health and disease.* Philadelphia: Lea & Febiger, 1994:805–811.
80. Detsky AS, McLaughlin JR, Baker JP, et al. What is subjective global assessment of nutritional status? *J Parent Ent Nutr* 1987;11:8–13.
81. Baker JP, Detsky AS, Wesson DE, et al. Nutritional assessment: a comparison of clinical judgment and objective measurements. *New Engl J Med* 1982;306:969–972.
82. Hill GL. 1992. Body composition research: implications for the practice of clinical nutrition. *J Parent Ent Nutr* 1992;16:197–218.
83. Lupo L, Pannarale D, Altomare D, et al. Reliability of clinical judgment in evaluation of nutritional status of surgical patients. *Br J Surg* 1993;80:1553–1556.
84. Enia G, Siscuso C, Alati G, et al. Subjective global assessment of nutrition in dialysis patients. *Nephrol Dial Transplant* 1993;8:1094–1098.
85. Buzby G, Mullen, P, Matthew JL, et al. Prognostic nutritional index in gastrointestinal surgery. *Am J Surg* 1980;139:160–167.
86. Harvey KB, Moldawer LL, Bistrian BR, et al. Biological measures for the formulation of a hospital prognostic index. *Am J Clin Nutr* 1981;34:2013–2022.

87. Gamble-Coats K, Morgan SL, Bartolucci AA, et al. Hospital-associated malnutrition: reevaluation 12 years later. *J Am Diet Assoc* 1993;93:27–33.

88. Mowe M, Bohmer T, Kindt E. Reduced nutritional status in an elderly population (>70 y) is probable before disease and possibly contributes to the development of disease. *Am J Clin Nutr* 1994;59:317–324.

CHAPTER

4

Macronutrient Requirements

Angela M. Ogawa, R.D., C.N.S.D.

INTRODUCTION

In 1974 when Charles Butterworth published "The Skeleton in the Hospital Closet," as many as 50% of surgical patients were malnourished, resulting in significant morbidity and mortality.[1,2] Twenty years later, the science of nutrition support is a respected specialty in itself. Many devices exist for administering the plethora of enteral and parenteral formulas and solutions available. As a result, malnutrition in hospitalized patients is much more likely to be recognized and treated if not prevented. The realization that nutrition improved outcome led to the assumption, however, that more must be better. The challenge to nutrition support professionals today is not so often coaxing medical staff to nourish patients but rather to prevent them from overfeeding. This chapter will define the energy, protein, carbohydrate, and fat requirements for adult, hospitalized patients. Potential deleterious effects of excess substrates are discussed as well.

MACRONUTRIENT UTILIZATION

In normal physiological states, a mixed diet containing protein, carbohydrate, and fat maintains weight, muscle mass, and organ protein. Carbohydrate is the primary energy fuel, while protein maintains, repairs, and builds muscles and other tissues. Fat, with a caloric density more than twice that of carbohydrate or protein is the primary storage fuel. Humans typically have sufficient endogenous fat to sustain life as long as 60 days in nonstressed starvation. Once muscle glycogen is depleted (usually after one day of fasting), muscle protein is oxidized to meet 8% to 12% of energy needs during the next 10 days. Then, an adaptive response occurs and protein degradation decreases to the point that as much as 97% of energy requirements are met with adipose tissue.[3] This protein-sparing effect is vital to preserving muscle mass, organ tissue, and enzyme and hormone production. These very structures and functions protected by the protein sparing effect of adapted starvation, however, are threatened when the human is subjected to metabolic stress.

Traumatic injury, burns, infection, and disease initiate a cascade of events known as the metabolic stress response. Described by Cuthbertson in 1932, the response is now known to be mediated by catabolic hormones, cytokines, and the parasympathetic and sympathetic nervous system and is an effort to fight the insult to which the body has been subjected.[4] Systemic effects of the stress response include hyperglycemia, insulin resistance, and protein catabolism, which peak after 5 to 7 days. The peak is characterized by the highest energy expenditure and greatest nitrogen loss, after which, the stress response lessens and the patient begins to recover. If the extent of the insult(s) is too great or the patient has poor reserves due to preexisting malnutrition, total recovery may not occur. Residual organ dysfunction or organ system failure, such as ventilator dependence, is the result. Some patients never enter the recovery phase because of continued metabolic stress and deplete their nitrogen reserves, resulting in multisystem organ failure (MSOF) and eventually death.[5] Early initiation of nutrition support and special attention to the macronutrient composition is needed to maximize effectiveness without exacerbating the metabolic status of the patient.

It is well documented that the rate of metabolism is elevated during critical illness or injury. It is important to note, however, that while basal metabolic rate (BMR) is greater, external energy expenditure may be little to none. Such is the case in the sedated, mechanically ventilated, intensive care unit (ICU) patient. This decrease in energy expenditure from absence of muscular contraction may blunt the overall increase in metabolic rate. The increased basal metabolism of the critically ill should not be confused with total energy expenditure (TEE).

ENERGY

Prior to discussing specific energy requirements a point needs to be clarified. Energy requirements will be expressed in terms of total calories from carbohydrate, protein, and fat rather than nonprotein calories. Some nutrition support references exclude the caloric value of protein in a mixed-fuel system. The rationale is that protein is to be reserved for tissue anabolism rather than as a calorie source. Lab studies show, however, that a significant portion of the protein molecule following deamination of the nitrogenous portion is oxidized, providing carbon dioxide, water, and adenosine triphosphate (ATP).[6] Failure to consider these protein calories will likely result in overfeeding patients. Calculations for parenteral nutrition support regimens should have protein calories included at 4 kcal/g. The energy content of enteral formulas is generally expressed in terms of total calories per liter of formula. Consensus on this point among clinicians helps prevent misinformation.

Many calculations exist to predict energy expenditure. The most widely known is the Harris Benedict equation (HBE), which predicts basal energy expenditure (BEE), and is based on height, weight, age, and sex. Stress factors and activity factors can be added to predict resting energy expenditure (REE).[7] Because indirect calorimetry is widely available in medical centers now, the Harris Benedict equation has come under scrutiny. The Harris Benedict equation is calculated as follows:

$$\text{Males: } 66 + 13.7W\ (\text{kg}) + 5H\ (\text{cm}) - 6.8A\ (\text{years}) = \text{BEE}$$
$$\text{Females: } 655 + 9W\ (\text{kg}) + 1.8H\ (\text{cm}) - 4.7A\ (\text{years}) = \text{BEE}$$

where W is weight, H is height, and A is age.

Studies of patients with congestive heart failure (CHF) showed that the Harris Benedict equation underestimated REE, most likely from the increased effort of breathing not represented in the height, weight, age and sex equation.[8] A study of patients with pancreatitis by Dickerson et al. demonstrated that the use of the stress factors would have overestimated REE for some and underestimated for others.[9] In that study, the measured average REE was 26 kcal/kg, but individuals ranged from 77% to 139% predicted REE from HBE. Although defined as determining BEE, the Harris Benedict equation seems to more closely approximate REE. Therefore, without the additional stress and injury factors, it could be used to estimate REE.[10]

Indirect calorimetry calculates REE by measuring oxygen consumption, carbon dioxide production, and minute ventilation with 95% accuracy.[11] Justly, it has become the gold standard in the critical care setting for determining patient's energy needs. Activity factors can be added to the REE to account for passive motion and variable activity levels of patients. It must be emphasized that hypermetabolism of injury and illness will be reflected in a measured REE and that critically ill patients are not anabolic. Therefore overfeeding is unnecessary. As clinical status improves, anabolism and repletion become the central objective; additional calories then may be beneficial. Evidence of clinical improvement include the resolution of pneumonia, successful ventilator weaning, healing of fistulas and wounds, positive nitrogen balance, normal white blood counts, and absence of infection. The administration of additional calories can be considered for patients with activity-related energy expenditure that will not be measured during an indirect calorimetry study at rest.

Generally, mechanically ventilated patients do not require greater than 125% of their REE, because of their obvious low activity and sedation. Exceptions include patients with burns as well as patients with closed head injuries that exhibit posturing.[11] Burn injuries present patients with the highest degree of catabolism and hypermetabolism of any illness or injury. Adequate nutrition support is vital for preserving, treating, and or preventing secondary infections and rebuilding injured tissues. The Curreri formula is the most widely known for estimating energy requirements of burn patients. It was initially found to overestimate calorie needs, especially in small children, and has since been modified.[12,13]

The modified Curreri formula is as follows:

$$20 \text{ kcal/kg} + (40 \times \%\text{BSA burned}) \text{ (with maximum 50\% BSA)}$$

where BSA is body surface area. Formulas provide guidelines for energy requirements; however, at a calorie level exceeding 35 kcal/kg, attention must be given to the possibility of overfeeding. Even in these obviously hypermetabolic patients, glucose infusion rates exceeding 7 mg/kg $\cdot$ min^{-2} caused increased carbon dioxide production and hepatic steatosis.[14] In clinical practice, it is advisable to limit carbohydrate calories to a maximum of 5 mg/kg $\cdot$ min^{-2}.

Victims of chronic malnutrition may present with marasmus characterized by depletion of fat stores but with somewhat normal visceral protein status. If traumatic injury or illness occurs to a patient with this condition, survival is much less likely than in a patient with normal tissue stores. In the absence of stress, however, the treatment for severe maramus can be lethal if special attention is not given to the nutritional status of the patient. A set of complications called the *refeeding syndrome* can occur and is characterized by electrolyte abnormalities (hypophosphatemia, hypokalemia, hypomagnesemia), volume overload, and CHF.[15] Untreated, it can result in cardiac and respiratory failure. These sequelae can be prevented by initially restricting volume to 1 L of fluid/day, sodium load to a minimum, and carbohydrate to 150 g/day. Nitrogen can be given at normal anabolic levels of 1.5 to 2.0 g/kg, and the need to supplement potassium, phosphorus, and magnesium should be anticipated. As the patient is repleted, strict attention to weight gain can prevent volume overload. Weight gain exceeding 1 kg/week is likely due to fluid gain and should be avoided.[15]

Recognizably, indirect calorimetry is not available at every facility; however, data from studies has provided great insight into the energy requirements of patients based on diagnosis and clinical state. Empirical formulas have been derived that express energy needs in terms of kilo calories per kilogram. Since 80% of the HBE predicted REE is determined by weight, it is not surprising that these simplified empirical formulas correlate well with measured and predicted REE.[16] Studies of the critically ill, both young people with traumatic injuries and elderly postsurgical patients, show REE of 21 to 25 kcal/kg. In fact, 90% of these patients would have been fed adequately with 30 kcal/kg.[16] Using the guidelines in Table 4-1 should provide calories (without significantly exceeding) TEE.

Use of empirical formulas is dependent upon using an appropriate weight in the calculation that best represents the patients metabolically active body cell mass. Most patients are not 100% of their ideal body weight (IBW) but rather slightly more or significantly less.[17] Using IBW may over- or underestimate energy requirements; therefore, the patient's actual dry weight should be used. An exception is the obese patient whose energy requirements should be calculated using an adjusted body weight which more closely approximates the patient's metabolically active mass. This equation should be used to calculate adjusted body weight when a patient weighs more than 130% IBW[18]:

$$(\text{Actual Body Weight} - \text{IBW}) \times 0.25 + \text{IBW} = \text{Adjusted IBW}$$

Once REE has been determined, the activity level of the patient may warrant additional calories. Traditional stress and activity factors intended to be used with the calculated BEE to estimate TEE did not correlate well with measured REE. In some cases, they grossly overestimated TEE.[19] Total energy expenditures may only exceed REE by 5% to 10% in sedated patients or as much as 50% in a patient who ambulates frequently. In severely obese patients (greater than 150%

Table 4.1. Energy Requirements for Hospitalized Patients

kcal/kg	Injury/Illness/Condition
20–22	Morbid obesity
22–25	Marasmus
25–27	Uncomplicated surgery, prolonged illness
27–30	Sepsis, penetrating trauma
30–33	Skeletal trauma, closed head injury
30–35	Burns < 30% BSA, anabolism
35–40	Burns > 30% BSA

Table 4.2. Activity Factors For TEE for Hospitalized Adult Patients

REE%	Activity Factors
REE × 5%	Mechanically ventilated, pharmacologically paralyzed patient
REE × 10%	Critically ill, sedated patient
REE × 20%	Critically ill, some voluntary movement or combativeness
REE × 25%	Conscious patient, resting comfortably
REE × 30%	Conscious patient, assists with ADLs
REE × 50%	Alert patient, ambulates frequently, daily physical therapy

IBW), underfeeding calories promotes mobilization of endogenous fat. Beneficial metabolic effects including diuresis and improved blood glucose control are associated with this, so that activity-related calories should not be added for hospitalized morbidly obese patients.[20] It is imperative, however, to provide the severely obese patient with 1.5 to 1.8 g protein/kg adjusted IBW to achieve the protein sparing effect. Table 4.2 shows activity factors that can be used to meet without significantly exceeding TEE.

Overfeeding patients increases metabolic stress and can be identified by the respiratory quotient (RQ) if indirect calorimetry is available. Respiratory quotient depicts the ratio of carbon dioxide production to oxygen consumption, which varies depending on the substrate utilized. Oxidation of fatty acids produces an RQ of 0.71. Protein oxidation produces an RQ of 0.82, and the oxidation of pure glucose yields an RQ of 1.0. This occurs when glucose infusion is at the maximum oxidation rate for the patient, usually 6 mg glucose/kg · min^{-1}. Infusing glucose at rates higher than this results in lipogenesis and increased REE from the energy cost of storing substrate as fat. The excess carbon dioxide produced in this process may increase the work of breathing for patients and delay ventilator weaning. Ventilated patients who are difficult to wean should have meticulous attention to their nutrition support and calorie level. Indirect calorimetry is extremely valuable in these patients, because it provides the RQ. Exceeding the glucose oxidation rate in critically ill patients also exacerbates stress-related hyperglycemia. Maintenance of lean mass rather than fat deposition is the priority in critically ill patients, another reason to avoid overfeeding.[21]

If injury and activity factors are added to a patient's REE in excess of 30%, it is advisable to reassess the patient within 96 hr to evaluate for evidence of overfeeding. Wasting of lean body mass may actually decrease REE in patients with a critical illness of 2 weeks or longer duration.

PROTEIN

The word protein is derived from the Greek, meaning "of first importance." These nitrogen-containing compounds, which yield amino acids upon hydrolysis, are the key components of every living organism. Unlike calories, protein needs during critical illness are roughly twice the normal requirements. Healthy individuals have approximately 100 g surplus protein nitrogen available. This reserve is quickly depleted with 24 to 48 hr of metabolic stress. Gluconeogenesis occurs then to provide additional glucose and amino acids at the expense of skeletal muscle. If exogenous nitrogen is not administered in seriously and critically ill patients, catabolism of organ protein occurs, increasing the risk of multisystem organ failure.[5] An obvious goal of nutrition support is to provide adequate nitrogen preserving muscle and organ protein and thereby avoiding the associated comorbidity and potential mortality.

Dextrose infusion alone has almost no protein-sparing effect.[22] It is important to remember this when even mildly stressed, postsurgical patients are NPO (Null Per Os") for greater than 4 to 5 days. Nitrogen retention improves dramatically when amino acids are included with peripherally infused dextrose solutions.[23] A plateau effect is observed in nitrogen accretion when amino acid administration reaches 1.4 to 1.75 g protein/kg or approximately 20% of the energy needs.[24] Unless the patient has exogenous protein losses, such as draining wounds or fistulae, exceeding 1.5 to 2.0 g/kg (excepting burn patients) in adults constitutes overfeeding and may unduly tax the renal and hepatic function of critically ill patients. At this point, additional protein becomes an expensive and inefficient calorie source. Table 4.3 shows the protein requirements for adult patients.

The quantity of protein intake has long been recognized as a key factor in growth, healing, and response to stress. More recently, however, the quality and form of protein substrate are being evaluated for maximum nutritional benefit. Quality of protein fractions or protein efficiency ratio (PER) is often referred to as biological value. It is equal to the ratio of weight gained to weight of protein consumed. In humans, the following protein sources have PER of 0.95 or greater, making them very high quality protein: whole egg (lactalbumin), breast milk, and cow's milk (casein and whey). Animal protein sources contain all the essential amino acids (EAA) and are therefore superior in quality to plant protein sources. Soy protein with a PER of 0.75 is the highest quality of all vegetable proteins, yet is a distant second compared to animal protein.[3] Commercial enteral formulas

Table 4.3. Protein Requirements for Adult Patients

Protein (g/kg $\cdot$ d^{-1})	Patient Status
0.8–1.0	Healthy, nonstressed
1.0–1.1	Elderly, nonstressed patients
1.2–1.5	Recovery phase of metabolic stress, discharge
1.5–1.8	Metabolic stress, sepsis, postsurgical, trauma
1.7–2.0	Severe stress, burns up to ×% surface area
2.0–2.5	Burns > ×% surface area, abnormal protein losses (fistula, wound)

Source: Data from Food and Nutrition Board[25] and Campbell, Crim, and Dallal, et al.[26]

are comprised of the high PER proteins. It should be recognized, then, that patients whose nutrition therapy is managed with an oral diet regimen must consume significant amounts of animal protein to reap the benefits that commercial formulas would provide. Anorexia that accompanies illness may preclude consumption of adequate high-quality protein necessitating nutritional support.

The form of protein also appears to affect nutritional status. Studies in humans and animals have shown more rapid changes in serum amino acid profiles, improved nitrogen balance, and enhanced growth when protein was fed as peptides rather than as free amino acids.[27,28] It is not clear yet whether intact protein is as efficacious as peptides, although it is superior to free amino acids. Studies are needed in this area since intact protein is less costly than peptides and may be utilized as well or better. Until a consensus is reached, expensive peptide formulas should be reserved for carefully selected critically ill patients such as those with preexisting malabsorption or malnutrition. As a patient's clinical and nutritional status improves, changing to an intact protein enteral formula is an appropriate, cost-effective treatment measure.

In the past 25 years special amino acid solutions have been evaluated for benefits in treating some disease states, namely, acute renal failure, hepatic failure, and trauma. It was thought that delivering EAA to patients with acute renal failure would decrease ureagenesis, thereby improving the blood urea nitrogen concentration while contributing to the nitrogen balance. Studies have failed to show improved outcome from the use of these formulas.[29] Commercial formulas enriched with EAA remain available, despite lack of data to support their efficacy. Because of their high cost and little known benefit, use of these solutions is not warranted.

In hepatic failure with encephalopathy, the amino acid profile shows decreased branched chain amino acids (BCAA) and higher levels of aromatic amino acids (AAA). Infusion of BCAA-enriched solutions normalize the serum amino acid profile, yet do not always improve nitrogen balance, morbidity, or mortality compared to standard amino acids.[30] A possible benefit, however, may be the faster resolution of hepatic encephalopathy. Because of high costs and inconclusive data, use of BCAA formulas should be reserved for patients with grade III or IV encephalopathy.

CARBOHYDRATES

Carbohydrate is the primary energy fuel in normal nutrition yielding 4 kcal/g. Glucose oxidation continues at a normal rate in metabolic stress. Its uptake, however, is compromised by peripheral insulin resistance resulting in hyperglycemia.[31] Not surprisingly, hyperglycemia is a common complication of nutrition support. Maintaining moderate glucose control in the range of 150 to 180 g/dL is advisable to improve nitrogen balance and decrease the risk of wound infection. Initially, dextrose administration of 150 to 200 g/day helps to attenuate gluconeogenesis without complicating hyperglycemia. As glucose control is attained, dextrose calories can be gradually increased to meet 45% to 70% of energy requirements.

Ill effects may occur when excessive amounts of carbohydrates are administered to critically ill patients. As carbohydrate is metabolized, carbon dioxide is produced. Excess carbon dioxide production in a mechanically ventilated patient increases the work of breathing as the individual attempts to exhale the retained carbon dioxide. This may result in fatigue and difficulty weaning. In patients with compromised respiratory function, this fatigue may contribute to respiratory failure. Adding lipid to meet up to 55% of the energy requirements has been promoted recently to help to decrease carbon dioxide production and the associated respiratory difficulty. This is true, in part, since lipid metabolism yields less carbon dioxide; however, total calorie administration should be reviewed in these cases. Often, if total calories from carbohydrate are reduced, the excess carbon dioxide production resolves.[32]

If carbohydrate overfeeding occurs via total parenteral nutrition, hepatic steatosis is a common complication. Histologic changes range in severity from asymptomatic periportal fat distribution to a panlobular and centrilobular pattern associated with hepatomegally and right-upper-quadrant pain.[33] Alkaline phosphatase levels as well as the serum amino transferases are mildly to moderately elevated. This condition is most often observed when dextrose infusion exceeds the glucose oxidative capacity of the liver (4 to 6 $mg/kg \cdot min^{-1}$) or when dextrose was the sole source of nonprotein calories. Studies show that rats fed 25% dextrose had hepatic fat deposition, whereas those fed 17% dextrose or 25% dextrose with fat did not.[34]

LIPIDS

Endogenous fat is the major energy reserve for nonstressed, starvation-adapted patients. In stressed patients, the protein-sparing effect of fat oxidation is lost. Even infusion of exogenous lipids to stressed patients does not spare protein.[35] Lipids and glucose are equally effective at protein sparing, however, when administered with adequate protein. Essential fatty acid deficiency can be prevented by feeding 4% of the energy requirements as linoleic fatty acid.[3] Lipid administration to meet up to 30% of energy needs is beneficial, because it provides a concentrated calorie source in frequently fluid-restricted critically ill patients and helps avoid the complications of carbohydrate overfeeding. Optimal fat dosing is unknown; however, excessive levels are associated with complications. Known ill effects of lipid overfeeding are related to cardiopulmonary, immune function, and platelet aggregation.

The infusion of long-chain triglycerides (LCT) causes *in vivo* reticuloendothelial system (RES) dysfunction in humans and animals. Specifically, there is decreased clearance of bacteria from phagocytosis of lipid globules resulting in increased risk of bacteremia and sepsis.[36] This effect is amplified when large volumes of intravenous lipid emulsion (IVLE) are infused over 7 to 10 h/day. Intravenous lipid emulsion may cause pulmonary dysfunction due to fat embolism especially in infants.[37] This effect is not observed when medium-chain triglycerides (MCT) replace part of the LCT.[36] Since MCT oil is not approved for parenteral infusion, septic patients may benefit from keeping LCT calories less than 25% to 30% total calories.

Limiting the LCT calories or substituting it with MCT oil may be additionally beneficial considering that commercial IVLEs are comprised of LCT from vegetable oils consisting of omega-6 fatty acids. The metabolism of omega-6 fatty acids lend themselves to prostaglandin synthesis, which serves to fuel the inflammatory response seen in metabolic stress.[38]

When fish oils are infused as part of the lipid emulsion, the omega 3 fatty acids reduce the eicosanoid synthesis of inflammatory agents and form less active prostaglandins.[38] Medium-chain triglycerides and fish oil have been shown to minimize the potential ill effects of LCT oil. A recent novel approach is to structure a lipid comprised of various fatty acid chains on one glycerol backbone in hopes of gaining the benefits of each. Mixtures of MCT, LCT, and fish oil have been hydrolyzed and reesterified to randomly produce triglycerides of mixed omega-3 fatty acids, medium-chain fatty acids, and long-chain fatty acids. Absorption of these structured lipids is efficient, and they may represent the lipid of choice in the future.[37]

CONCLUSION

The optimal nutritional support regimen is one that meets (without significantly exceeding) energy requirements, provides a high-quality protein source, and does not adversely affect respiratory or immune function. Careful attention to macronutrient provision in hospitalized patients can achieve positive results with minimal deleterious effects.

REFERENCES

1. Butterworth CE. The skeleton in the hospital closet. *Nutr Today* 1974;9:4–8.
2. Bistrian BR, Blackburn GL, Hallowell E, et al. Protein status of general surgical patients. *JAMA* 1974;230:858–860.
3. Krause MV, Mahan LK. The metabolic stress response and methods for providing nutrition care for stressed patients. In Krause MV, Mahan LK, eds. *Food, Nutrition and Diet Therapy,* 7th edition. Philadelphia: W. B. Saunders, 1984:707.

4. Cuthbertson DP. Observations on the disturbances of metabolism associated with energy and sepsis. *Q J Med.* 1932;1:233–244.
5. Baue AE. Nutrition and metabolism in sepsis and multiorgan failure. *Surg Clin North Am* 1991;71:549–566.
6. Duke JH, Jorgensen SB, Broell JR, et al. Contribution of protein to calorie expenditure following injury. *Surgery* 1970;68:168–174.
7. Harris JA, Benedict FG. *Biometric standards of basal metabolism in man.* Publication 279. Washington, DC: Carnegie Institute 1919.
8. Poehlman ET, Scheffers J, Gottlieb SS, et al. Increased resting metabolic rate in patients with congestive heart failure. *Ann Intern Med* 1994;121:860–862.
9. Dickerson RN, Vehe KL, Mullen JL, et al. Resting energy expenditure in patients with pancreatitis. *Crit Care Med* 1991;19:484–490.
10. Ireton-Jones C. *Energy requirements of hospitalized patients* Audio teleconference delivered by the University Texas Teleconference Network, 1990.
11. Hester DD. Neurologic impairment. In *Nutrition support dietetics core curriculum,* 2nd edition. Rockville, MD: ASPEN, 1993:230.
12. Van Way CW. Nutritional support in the injured patient. *Surg Clin North Am* 1991;71:537–548.
13. Waymack JP, Herndon DN. Nutritional support of the burned patient. *World J Surg* 1992;16:380–386.
14. Burke JF, Wolfe RR, Mullaney CJ, et al. Glucose requirements following burn injury. *Arch Surg* 1979;190:274–285.
15. Apovian CM, McMahon MM, Bistrian BR. Guidelines for refeeding the marasmic patient. *Crit Care Med* 1990;18:1030–1033.
16. Hunter DC, Jaksic T, Lewis D, et al. Critically ill estimations versus measurement. *Br J Surg* 1988;75:875–878.
17. Lang CE, Shutte CV. Nutrition Assessment: Adult patient. In Lang EE, ed. Nutrition Support in Critical Care, Rockville, MD: ASPEN, 1987.
18. Wilkens K. Adjustment for obesity, *ADA Renal Practice Group Newslett,* Winter, 1984.
19. Cortes V, Nelson L. Errors in estimating energy expenditure in critically ill surgical patients. *Arch Surg* 1989;124:287–290.
20. Dickerson RN, Rosato EF, Mullen JL. Net protein anabolism with hypocaloric parenteral nutrition in obese stressed patients. *Am J Clin Nutr* 1986;44:747–755.
21. Zaloga GP. In Zaloga GP, ed. *Nutrition support in critical care.* St Louis: Mosby Yearbook, 1994.
22. Shaw JHF, Klein S, Wolfe RR. Assessments of alanine, urea, and glucose interrelationships in normal subjects and in patients with sepsis with stable isotopic tracers. *Surgery* 1985;97:557–568.
23. Askanazi J, Carpentia YA, Jeevanardon J, et al. Energy expenditure, nitrogen balance and norepinephrine excretion after injury. *Surgery* 1981;89:478–484.

24. Iapichino G, Gattinini L, Solea M, et al. Protein sparing and protein replacement in acutely injured patients during infusion of TPN with and without amino acid supplementation. *Intern Care Med* 1982;8:25–31.
25. Food and Nutrition Board, National Research Council. *Recommended Daily Allowances,* 9th ed. Washington, DC: National Academy of the Sciences, 1989:39–51.
26. Campbell WW, Crim C, Dallal GE, et al. Increased protein requirements in the elderly: new data and retrospective reassessments. *Am J Clin Nutr* 1994;60:501–509.
27. Beer WH, Fan A, Halstead CH. Clinical and nutritional implications of radiation enteritis. *Am J Clin Nutr* 1985;41:85–90.
28. Poullain MG, Cezard JP, Roger L, et al. Effect of whey proteins, their oligopeptide hydrolysates and free amino acid mixtures on growth and nitrogen retention in fed and starved rats. *J Parent Ent Nutr* 1989;13:382–386.
29. Rapp RP. The use and abuse of specialized amino acid solutions, Presented at the Challenges in Nutrition Support Symposium, Alamo Area Society for Parenteral and Enteral Nutrition, San Antonio, TX, 1990.
30. Bower RH. Nutritional and metabolic support of critically ill patients. *J Parent Ent Nutr* 1990;14(suppl.);257–259.
31. Wolfe RR, Jahoor F, Herndon D. Isotopic evaluation of the metabolism of pyruvate and related substrates in normal adult volunteers and severely burned children. *Surgery* 1991;110:54–67.
32. Talpers SS, Romberger DJ, Bunce SB, et al. Nutritionally associated increased carbon dioxide production. Excess calories vs high proportion of carbohydrate calories. *Chest* 1992;102:551–555.
33. Baker AL, Rosenberg IH. Hepatic complications of TPN. *Am J Med* 1987;82:489–497.
34. Nussbaum MS, Li S, Bowen RH, et al. Addition of lipid to total parenteral nutrition prevents hepatic steatosis in rats by lowering the portal venous insulin/glucagon ratio. *J Parent Ent Nutr* 1992;16:106–109.
35. Steat SJ, Beddoe AJ, Hill GL. Aggressive nutritional support does not prevent protein loss despite fat gain in septic intensive care patients. J Trauma 1987;27:262–266.
36. Hamawy KJ, Moldawer LL, Georgieff M, et al. The Henry M Vars Award: The effect of lipid emulsions on reticuloendothelial system function in the injured animal. *J Parent Ent Nutr* 1985;9:559–565.
37. Mascioli EA, Bistrian BR, Babayan VE, et al. MCT and structured lipids as a unique non-glucose energy source in hyperalimentation. *Lipids* 1987;22:421–423.
38. Wan JM-F, Teo TC, Babayan VK, et al. Invited comments: lipids and the development of immune dysfunction and infection. *J Parent Ent Nutr* (suppl) 1988;12:43–52.

CHAPTER

5 Micronutrients in Clinical Nutrition

Thomas G. Baumgartner, Pharm.D., M.Ed., B.C.N.S.P., George Henderson, Ph.D., and Susan L. Baumgartner, Pharm.D., M.B.A.

INTRODUCTION

Micronutrients are the metabolic glue that is behind virtually all anabolic and catabolic processes in the human. Electrolytes are responsible for seemingly limitless metabolic reactions, play key roles in structure, and maintain acid–base balance. Trace elements also facilitate complex metabolic reactions and, not unlike vitamins, play key roles in antioxidant activities. Free radicals and their metabolic impact are receiving increasing attention in scientific arenas as modulators of disease. Radical oxidants have been shown to damage cellular and intracellular proteins, nucleic acids and lipids. In fact, in older animals and humans, as much as 40% of intracellular protein may be oxidized with corresponding reductions in the activity of many enzymes. It is likely that the damage from these reactive species accumulates with aging and becomes particularly prominant after age 60 in humans, the so-called age of apoptosis or cellular demise. Both nuclear and, more so, mitochondrial deoxyribonucleic acid (DNA) undergo cumulative damage. Mitochondria are at the center of respiration and are constantly exposed to reactive oxygen species. Even the protection of intracellular antioxidants, however, does not prevent constant and cumulative damage from reactive active species. It has been estimated that free-radical interactions with DNA modify approximately 10,000 DNA bases/cell · day.$^{-1}$ [1]

Increased production of reactive species can overwhelm antioxidant pathways. This is thought to be secondary to inadequate antioxidants (i.e., vitamin C, vitamin E, beta-carotene, and selenium), or to a genetic abnormality that increases reactive species or decreases availability of antioxidants (i.e., heritable malabsorption syndromes that decrease or block the absorption of fat-soluble vitamin E).

One need not look far in the literature to illustrate the importance of free radicals. Neoplasia is common in continuously replicating cells (bronchial epithelium, colon and marrow) but rare in intermittent replicators (smooth muscle), which are involved in degenerative phenomena such as atherosclerosis. The nonreplicating adult neuron can give rise to age-related degenerative diseases such as Alzheimer's or Parkinson's disease. Greater than four million cases of Alzheimer's disease have been identified in the United States with 20% to 30% aged 80 years or more. The unique pattern of Alzheimer's disease at autopsy is intriguing and suggests support for the hypothesis that a small beta-amyloid peptide from a precursor of a gene on chromosome 21 is a contributing cause. Although this form of Alzheimer's disease may have a strong genetic component, a cofactor may be inflammatory. In its aggregated form, the beta-peptide can, on its own, initiate the classical complement pathway. Neurons around the plaque show signs of attack by and defense against the complement pathway end product, the membrane attack complex. Microglia, the macrophages of the brain, cluster around the so-called senile plaque and are then activated to produce cytokines and free radicals. Beta-amyloid may also be able to initiate aging or apoptosis.[2]

In addition to reactive oxygen species and cytokines, other highly potent substances are produced by the immune system that include oxidant molecules, such as hydrogen peroxide and hypochlorous acid. The purpose of immunity, of course, is to destroy invading organisms and damaged tissue, bringing about recovery. However, oxidants and cytokines can damage healthy tissue. Excessive or inappropriate production of these substances is associated with mortality and morbidity after inflammation, infection, and trauma. Oxidants enhance interleukin-1, interleukin-8, and tumor necrosis factor production in response to inflammatory stimuli by activating the nuclear transcription factor, NF kappa B. Sophisticated antioxidant defenses directly and indirectly protect the host against the damaging influence of cytokines and oxidants. Indirect protection is afforded by antioxidants, which reduce activation of NF kappa B, thereby preventing up-regulation of cytokine production by oxidants. Cytokines increase both oxidant production and antioxidant defenses, thereby minimizing damage to the host. Although antioxidant defenses interact when a component is compromised, the nature and extent of the defenses are influenced by dietary intake of sulfur amino acids that are used, for example, in glutathione synthesis, and vitamins E and C. In animal studies, *in vivo* and *in vitro* responses to inflammatory stimuli are influenced by dietary intake of copper, zinc, selenium, N-acetylcysteine, cysteine, methionine, taurine, and vitamin E. *N*-Acetylcysteine, vitamin E, and a cocktail of antioxidant nutrients have reduced inflammatory symptoms in inflammatory joint disease, acute and chronic pancreatitis, and adult respiratory distress syndrome. Impaired antioxidant defenses also contribute to disease progression after infection with human immunodeficiency virus.[3]

Therefore, the contribution of electrolytes, trace elements, and vitamins to homeostasis is pivotal in maintaining health, particularly as antoxidants, suggesting that micronutrients may, indeed, be one of the primary messengers of health as we move into the 21st century.

MICRONUTRIENTS

Acid–base considerations must be assessed before any electrolytes can be provided to patients. If a patient is alkalemic, acid salts (chloride, phosphate, and sulfate) should be used, whereas an acidemic patient will require the use of bicarbonate precursors, such as acetate, citrate, gluconate, or lactate. Each salt must be carefully selected with pathophysiologic insight. The pH, followed by an appreciation of base excess must be coupled with metabolic electrolyte appraisal that will include principally chloride and bicarbonate status. Respiratory indices (i.e., carbon dioxide, oxygen), if available, will provide information about mixed acid–base status and compensatory mechanisms.

Electrolytes

Electrolytes, trace elements, and vitamins must be provided daily in clinical nutrition to provide effective catabolism and anabolism. The role of electrolytes in disease is probably best appreciated in the area of cardiovascular disease. Historically, the sodium ion has been given prominence in relation to cardiovascular disease, perhaps to the exclusion of other ions. However, recently, other ions, including chloride, potassium, magnesium, and calcium have received increasing attention with regard to hypertension, cardiac arrhythmias, and metabolic derangements. Endocrine factors controlling these ions include classic hormonal actions as well as neurotransmission and paracrine hormonal actions. Indeed, studies indicate that control of the renin–angiotensin–aldosterone system resides in cytosolic calcium ion levels in the juxtaglomerular cell, as well as chloride ion and prostaglandins at the macula densa. Renin release is stimulated by hyperpolarization of the juxtaglomerular cell induced by beta-1-agonists, parathyroid hormone, glucagon, magnesium, and low cytosol calcium. Renin release is inhibited by high calcium, potassium and angiotension II. Subsequent to renin release, hormonal regulation includes stimulation of converting enzyme activity by cortisol and prostaglandin (PGE_2).

Other hormonal control includes antidiuretic hormone-producing dilution of extracellular electrolytes and augmented peripheral resistance. A natriuretic factor isolated from cardiac atria appears to be a potent diuretic with actions similar to that of furosemide, a loop diuretic. Chloride may play a dominant role in renal sodium reabsorption, responding to prostaglandin levels.

Calcium has been recognized as a basic regulator of the secretion of such hormones as norepinephrine, renin, and aldosterone. Also, calcium ion changes are the means by which smooth muscle contraction is effected. Parathyroid hormone and vitamin D regulate the level of this ion in the body. In addition, a high dietary calcium intake appears to play a protective role against hypertension, and calcium channel blockers appear also to reduce blood pressure.

Endocrine systems play a major role in the protection against acute elevations in serum potassium by means of insulin action and adrenergic modulation of extrarenal potassium disposal. Aldosterone is recognized as the delayed regulator of potassium excretion. Magnesium levels fall in hyperaldosteronism, hyperparathyroidism, and diabetic ketoacidosis, as well as in malnutrition states. A coexisting potassium deficiency may be refractory to therapy until hypomagnesemia is corrected. The integrated action of these hormones and electrolytes are, thus, of major importance in regulation of the cardiovascular system.[4]

Sodium

The statement "where sodium goes, so too goes water" depicts the close relationship of sodium to fluid balance. The normal recommended sodium chloride intake is 1 g/L of water. However, arduous physical work in the hot weather may require a daily intake of up to four teaspoonsfuls or 300 mEq of sodium chloride. Salt, which is used routinely to treat alkalemia, is approximately 40% sodium and 60% chloride. Therefore, a normal 4-g sodium diet will be approximately 2 teaspoonsfuls or 10 g of sodium chloride/day. If one places 10 g of sodium chloride in a liter, normal saline (i.e., 154 mEq/L of sodium and 154 mEq/L chloride) results. It follows that a low-sodium diet is 2 g sodium or 5 g sodium chloride per day. Sodium is the chief extracellular cation and exerts more intravascular oncotic draw than any other electrolyte.

The relationship of the chief extracellular cation to eicosanoid and cytokine shower has been studied recently. Cytokine and sodium channel investigation has highlighted the importance of this electrolyte in the mediation of proarrhythmic effects. Human recombinant interleukin-2 (rIL-2) has been bath-applied to isolated human cardiocytes while sodium currents were triggered and registered using the whole-cell recording technique. In the presence of the cytokine, sodium currents were reversibly blocked, with 50% peak current reduction occurring at a concentration of 500 U/mL. In contrast to rIL-2, recombinant tumor necrosis factor-alpha rTNF$_\alpha$ did not affect the sodium currents. It was concluded that rIL-2 acts like a class I antiarrhythmic drug on human cardiac sodium channels.[5] The interaction of IL-2 with the Na^+ channels is very fast (within msec), and it has been suggested that it occurs when the Na^+ channels are in the state of fast inactivation. The recovery from inactivation was only slightly slowed by IL-2, in agreement with the absence of any use dependence. All effects were readily reversible on washout of the cytokine, suggesting that an inhibitory effect of IL-2 on the sodium currents has been found. It was concluded that the cytokine blocks the voltage-dependent muscular sodium channels by keeping the sodium channels in a state of fast inactivation.[6]

Sodium also plays major roles in other organ systems, such as gut homeostasis. For example, it has been demonstrated that sodium-dependent brush border glutamine transport is diminished in septic patients. To examine the potential regulation of this decreased transport by endotoxin, cytokines, or glucocorticoids, the human intestinal Caco-2 cell line was studied *in vitro*. Sodium-dependent glutamine transport across the apical brush-border membrane was assayed in confluent monolayers of differentiated cells that were 10 days old. Results suggested that cytokines and glucocorticoids may work independently and synergistically in regulating sodium-dependent brush-border glutamine transport in human intestinal cells.[7]

Potassium

Potassium, the chief intracellular cation, is closely related to sodium balance and acid–base status. Potassium serum levels are inversely related to pH (i.e., 0.4 to 0.6 mEq/L for every 0.1 pH). Concurrent hypercalciuria and hyperkalemia are not uncommon in the setting of acidemia. Alkalinization using acetate, lactate, citrate, or gluconate salts will curtail urinary losses of calcium and impact upon the hydronium and potassium renal ion exchanges. Fluid and electrolyte losses from either end of the gastrointestinal tract impact both on potassium and the acid–base status. Intracellular (where 99% of potassium resides) status is quite another matter and, as with sodium, potassium's relationship to cytokine actions are currently being aggressively investigated.

The effects of rTNF$_\alpha$, secreted by activated macrophages, on the electrical membrane properties of cultured oligodendrocytes have also been investigated. Several studies have shown that rTNF$_\alpha$ inhibited the inwardly rectifying K^+ current and membrane potential in cells and appeared morphologically normal. This suggested that abnormal ion channel expression in oligodendrocytes may precede and contribute to eventual myelin swelling and damage.[8,9]

With the initiation of TPN therapy, monovalent (i.e., sodium, potassium, bicarbonate, chloride) and divalent (e.g., calcium, magnesium, and phosphate) cations should be monitored daily. After the baseline, periodic draws will be empirical depending on the stability of the patient. Ventilator adjustments can change the acid–base status abruptly (within minutes) remembering the alkalinization promoted with oxygenation and acidification fostered with increased carbon dioxide retention. However, a slower metabolic acid–base impact is possible with the use of chloride salts for metabolic alkalemia treatment and acetate, citrate, gluconate, or lactate salts for the treatment of acidemia. Patients receiving enteral intake must be closely monitored to ensure that the diet is a balanced one, because acid or basic ash foods can influence acid–base balance. An excess anionic enteral intake will promote acidification. Plasma levels of electrolytes (i.e., serum calcium and potassium) may be influenced by this kind of acid–base impact and others (i.e., a nocturnal increase in urine pH; an increase in pH with an increased intermittent mandatory ventilation rate) and may need to be drawn several times a day in a rapidly changing setting. Ionized levels may provide more precise information about the actual plasma concentration and minimize the influences of the acid–base status and of protein binding. Plasma levels, however, may not always be a reliable monitoring tool for electrolyte status, suggesting that more sensitive monitoring tools should be used when available (e.g., electrocardiograms or leukocyte and erythrocyte content).

Calcium

Calcium is intimately associated with electrophysiologic mechanisms and plays a key role in hormonal regulation. It is principally found in the bone and is resorbed via parathormone (PTH) or sent back to the bone via calcitonin as needed. Daily repletion ranges are wide and monitoring with serum levels may be quite unreliable, because it is bound about 50% to albumin. Hypercalciuria usually occurs in the setting of acidemia and has been reported to occur with long-term cyclic TPN therapy. Ionized calcium levels may be desired when there are clinical interferences (i.e., acidemia that causes hypercalciuria) with the interpretation of bound calcium in the serum.

Once more, much remains to be learned about the implications of calcium and cytokine activities. Activation of polymorphonuclear leukocytes by most soluble stimulants is associated with a marked increase in cytosolic free Ca^{2+} ($[Ca^2+]i$). Interleukin-8 (IL-8), a monocyte-derived neutrophil chemotactic factor and potent neutrophil-activating cytokine, effectively enhanced the resting free $[Ca^{2+}]i$ within human polymorphonuclear leukocytes in a dose-dependent manner. The increase in $[Ca^{2+}]i$ was substantially (55%) inhibited in the absence of extracellular Ca^{2+}. Thus, the increase was due to extra- and intracellular cooperative mobilization of Ca^{2+}, as supported by the reduced effect of IL-8 on $[Ca^{2+}]i$ after quenching with Mn^{2+}. Granulocyte-macrophage colony-stimulating factor and interferon-γ failed to induce a change in $[Ca^{2+}]i$, suggesting that they may operate through different signal pathways. In conclusion, study results have shown that IL-8 induced cooperative mobilization of intra- and extracellular Ca^{2+} leads to a net Ca^{2+} influx into the cytoplasm through a process mediated by a guanosine triphosphate-binding protein.[10]

Calcium ion release from mitochondria can be induced by a variety of chemically different prooxidants. Release induced by these compounds is possibly regulated by protein mono(ADP)ribosylation and leaves mitochondria initially intact. Excessive cycling (continuous release and uptake) of Ca^{2+} by mitochondria leads to their damage, as shown by a decreased membrane potential, fast Ca^{2+} release, and impairment of ATP synthesis. When cycling is prevented by Ca^{2+} chelators or by inhibition of the uptake route with ruthenium red, prooxidants still induce Ca^{2+} release, but mitochondria remain intact. It has recently been suggested that formation of a *pore* in the inner mitochondrial membrane participates in the Ca^{2+} release mechanism. The prooxidant-induced Ca^{2+} release doesn't appear to parallel by sucrose entry into, or K^+ release from, or swelling of mitochondria, provided Ca^{2+} cycling is prevented. Thus, the prooxidant-induced Ca^{2+} release does not require formation of a pore and its' release occurs via a specific pathway.[11]

Finally, although 99% of total body calcium is found in bone, labile and intracellular stores have a dramatic impact on human homeostasis. Approximately a third of oral calcium is absorbed from duodenal and proximal jejunum, about one half is ionized in the serum and the other half is bound to albumin, and up to 400 mg/day is excreted in the urine. There are also sweat, bile, pancreatic juice, saliva, feces, and milk losses.

Magnesium

The effects of magnesium infusions on urinary and fecal magnesium excretion, serum magnesium, and nitrogen balance have been examined in seven well-nourished and three nutritionally depleted adult surgical patients receiving TPN. The participants were maintained on isonitrogenous/equicaloric intake for approximately 2 weeks. Magnesium doses ranged from 0 to 664 mg/day and were given in varying crossover patterns. In both groups, urinary magnesium excretion increased as the amount of infused magnesium increased. At comparable magnesium infusions, depleted patients excreted significantly less magnesium. Renal conservation was most pronounced in well-nourished patients on magnesium-free intake and in depleted patients given 70 mg magnesium daily. Urinary magnesium losses were 40 ± 5 mg/day and 33 ± 8 mg/day, respectively, in these two groups. Endogenous fecal magnesium excretion was minimal and ranged from 2 to 38 mg/day. At each level of magnesium intake, serum levels of well-nourished patients were normal. With infusions of less than 200 mg/day, serum magnesium concentrations in depleted subjects averaged 1.6 mg/day. In both groups, a positive correlation between magnesium and nitrogen balance was noted. An average dosing for parenteral magnesium is approximately 24 mEq/day, but requirements increase in the settings of alcoholism, stress, excess carbohydrate, or high protein.[12] Magnesium supplementation in the setting of hypomagnesemia and hypocalcemia has been shown to raise both magnesium and calcium serum levels without additional calcium supplementation.

Although an electrocardiogram may be the most sensitive way to monitor plasma levels of electrolytes, tissue stores may be more reliably followed with urine excretion (i.e., urine concentrations greater than 50% of magnesium intake may reflect adequate magnesium tissue stores). Bone contains approximately 50% of total body magnesium; the remainder is distributed between muscle and nonmuscular soft tissue. Approximately 30% to 50% is absorbed from jejunoileal areas of the gut and approximately a third of oral magnesium is absorbed and excreted in the urine depending on the magnesium salt.

Phosphate

Phosphate must be followed closely and if plasma levels fall below 2 mg%, a bolus dose of 0.16 mmol phosphate/kg must be given (generally over several hours to minimize abrupt changes in calcium status, unless calcium is simultaneously repleted) and the level verified with a bone battery or phosphate level immediately after infusion. Parenteral electrolyte adjustments should always be done in a gradual fashion, as possible, to avoid iatrogenic complications associated with imbalances. There are also circadian rhythms of electrolytes (i.e., phosphate has as much as a 2 mg/dL daily rhythm) that have been defined. Therefore, monitoring electrolyte parameters at the same time of the day will be important.[13] Glucose loads are a major reason for a phosphate shift from the intravascular compartment (part of the so-called refeeding syndrome), which may deplete erythrocytes of 2,3-diphosphoglycerate, tighten hemoglobin–oxygen affinity and shift the oxygen dissociation curve to the left.

Approximately 70% of phosphate is absorbed, primarily from the jejunum (it may increase up to 90% in dietary phosphate restriction states). Between 80% and 85% resides in soft tissues and approximately 85% to 90% of the 500 mg of phosphate that is filtered is reabsorbed by the kidney. There is an approximate excretion of 100 to 200 mg/day in the stool. Of all of the electrolyte products (see Table 5.1), phosphate is the most confusing because it is a mono- and diabasic buffer and is measured in millimolar concentrations rather than milliequivalents.

Trace Elements

There are approximately 17 trace elements that are considered to be essential (Table 5.2). The assessment of trace element status is fraught with problems.

Table 5.1. Commercially Available Parenteral Electrolytes

Sodium acetate	2.5 mEq/mL
Sodium chloride	2.5 mEq/mL; 4mEq/mL
Potassium acetate	2 mEq/mL
Potassium chloride	2 mEq/mL
Magnesium sulfate	4 mEq/mL
Calcium gluconate	0.5 mEq/mL
Potassium phosphate	4.4 mEq/mL with 3 mM phosphate/mL
Sodium phosphate	4 mEq/mL with 3 mM phosphate/mL

Unreliable histories, contamination concentrations in medications and nutrients, analytical techniques (Table 5.3), bioavailability, and excretion are all problematic. Since pragmatic serum levels leave much to be desired, supplementation from day one of feeding may be in the best interests of the patient. A variety of tissues have been proposed to assess regulatory, functional, transport, complexed, or storage sites (Table 5.4).

The activity of trace elements is intimately related to metabolic function in the human (Table 5.5). As with vitamins, anabolism or effective catabolism is influenced by adequate and appropriate supply and effective use of these micronutrients. All macronutrients are affected by trace element status in some manner. Furthermore, for optimum health, micronutrients also interact with each other (i.e., tryptophan will provide niacin under the drive of pyridoxine).

Finally, the kinetics of trace elements (Table 5.6) must be appreciated. Meticulous review of biological systems must take place. Needs, toxicities, nutritional state, and disease all play a role in determining dosing.

The original parenteral trace element foursome proposed by the American Medical Association was chromium, copper, manganese, and zinc (see Table 5.7). In addition to the combination and single additive packaging of the parenteral products, other trace elements have emerged, such as molybdenum and selenium. Trace elements should be added to all parenteral formulas to ensure supplementation, particularly if patients are receiving long-term parenteral therapy.

Baseline and periodic assessment of both signs and symptoms should continue

Table 5.2. Seventeen Essential Trace Elements in Humans

Arsenic	Iron	Selenium
Cadmium	Lead	Silicon
Chromium	Lithium	Tin
Cobalt	Manganese	Vanadium
Copper	Molybdenum	Zinc
Fluorine	Nickel	

Source: From Burch and Hahn.[14]

Table 5.3. Sources of Error in the Analysis of Trace Elements in Biological Materials

Precision or Accuracy of Analytical Process
Calibration
Instrument errors
Individual Biological Factors
Genetic factors (e.g., sex, race)
Timing of sample withdrawal and circadian rhythm
Nutrient intake
Medication use
Other factors (e.g., age, pregnancy)
Interaction with material or equipment that are used to obtain or process sample
Sampling procedure
Sample site
Sampling technique
Sample handling during collection or storage (e.g., temperature, light, irradiation, material contact/leaching)

Source: From Mertz.[15]

throughout the course of parenteral therapy and during the transitional period to enteral use since bioavailability may not yet be in place. Oral requirements (Table 5.8 and Table 5.9 can only be generally met with at least 1 to 2 L of an enteral formula.

Other Trace Elements

Aluminum is toxic at the cellular level and pathological symptoms follow its entry into organisms (plants, fish, humans) when the normal exclusion mecha-

Table 5.4. Tissues for Trace Element Analysis[16]

Regulatory sites	Sequestrian tissues
Intestinal mucosa and iron	Lung
Thyroid and iodine	Kidney
Liver and manganese or chromium	Reticuloendothelial
Essential function sites	Nails
Hemoglobin and iron	Hair
Liver and zinc	Stratum corneum of the skin
Transport and storage sites	
Serum concentration	
Urine concentration	
Bone marrow	

Source: From Laker.[16]

Table 5.5. Trace Element Activity in Humans

Element	Activity
Chromium	Maintains protein, carbohydrate, and lipid metabolism Glucose Tolerance Factor (GTF)–cofactor with insulin Important in peripheral nerve function
Copper	Maintains iron utilization because it is a cofactor for ceruloplasmin (94% serum binding) Oxidase necessary for the formation of transferrin (carrier of iron) Associated with the maintenance of normal synthetic rates for red and white blood cells
Iodine	Sole function in man, incorporation into thyroid hormones (T3 & T4): regulates cellular metabolism, temperature, and normal growth
Iron	Oxygen transport—component of heme which is the nonprotein portion of hemoglobin (Hgb); combines with oxygen in lungs and distributes to the tissue; also found in myoglobin Cytochrome content—normal cellular respiration; ATP storage through iron-coupling reactions T-cell immunity and cognitive function roles
Manganese	Activator for several enzymes—glycotransferase activation (mucopolysaccharides formation) Found in association with vitamin K deficiencies—glycotransferase activation in prothrombin synthesis Found in association with squalene (precursor of cholesterol and sex hormones) Found in association with protein synthesis, carbohydrate, and lipid (activates lipoproteinlipase) metabolism
Molybdenum	Constituent of three enzymes—aldehyde oxidase, sulfite oxidase, and xanthine oxidase Causes increased mobilization of copper from tissue and increases urinary excretion
Selenium	Glutathione peroxidase (antioxidant) component—protects the cell from lipid oxidation; functions with and can substitute for Vitamin E
Zinc	Functions with hundreds of enzymes (e.g., carbonic anhydrase, alkaline phosphatase, lactic acid dehydrogenase, alcohol dehydrogenase, RNA and DNA polymerases, and superoxide dismutase), wound healing, taste, increases oxygen affinity in normal and sickle cell patients, component of bone, synthesis of nucleic acid, glutathione, connective tissue, and collagen precursors.

Source: From Ulmer.[17]

Table 5.6. Trace Element Kinetics

Element	Kinetics
Chromium	Absorption is minimal (i.e., 0.5–2%). Chromium, not unlike iron uses a transferrin vehicle for transport to target sites. Mean serum concentrations are 0.16 ng/mL. The excretion of chromium is 0.2–0.6 μg/L urine and diets high in simple carbohydrates may increase urinary excretion up to 3-fold.
Copper	Absorption is about 30–60%. Distribution is principally (94%) via binding to ceruloplasmin; however, there also is binding to transcuprein albumin and amino acids. Plasma concentrations are approximately 1 μg/mL with principal excretion in bile (80%). Intestinal and urine excretion account for 16% and 4% excretion, respectively.
Iodine	Absorption is greater than 50% and up to 100% from gut; Iodine is also absorbed from skin. Dietary iodized salt is a major source of this trace element. The distribution of iodine is mainly via triiodothyronine (T3) and thyroxine (T4). Plasma concentrations are about 60 ng/mL and excretion is principally in urine with rapid turnover (i.e., 2/3 of absorbed iodide is excreted within 2–3 days)
Iron	Absorption is approximately 5–15% as the ferrous ion. Once absorbed, it is converted to the ferric ion in the mucosal cells of the intestine and binds to apoferritin that forms ferritin (regulates absorption of iron); transferrin binds iron in the plasma and transports iron to the liver, spleen, and bone marrow (where it is stored again as ferritin). Nascent distribution is bound to transferrin and is influenced by IL-1 (leucocyte endogenous mediator). The plasma concentration is approximately 1 μg/mL and is excreted principally in bile, but also via skin and urine.
Manganese	Absorption is about 3–4%; however absorption may be decreased with iron, cobalt, calcium, and phosphate. The distribution of manganese is bound extensively to transferrin or transmanganin. There is a small amount bound to $\beta_1$globulin. Plasma concentrations are about 0.6–2 ng/mL. Excretion is extensive in bile (greater than 99%) and excretion (not absorption) regulates homeostasis.
Molybdenum	Absorption is about 40–100% in duodenum. Plasma concentration is about 2–6 ng/mL, and, although excretion is principally in the urine, great gastrointestinal losses can occur.
Selenium	Absorption is 35–85% in duodenum; however absorption is dependent on selenium solubility and the ratio of selenium to sulfur. Plasma concentrations are approximately 100–130 ng/mL and excretion is principally in the urine; however, as with molydenum, great intestinal losses can occur. Burn patients (greater than 20%) will need 200μg selenium for at least 2 weeks.

continued on next page

Table 5.6. *Continued*

Element	Kinetics
Zinc	Absorption is about 10–40% with 2.5–4 mg absorbed/day in duodenum and proximal jejunum. Absorption will be diminished if given with copper and vitamin D may increase zinc bioavailability. Zinc-metallothionein regulates zinc metabolism and absorption. Distribution is via binding to albumin, transferrin, ceruloplasmin, gamma-globulin. Plasma concentrations are approximately 1 μg/mL and excretion in biliary and pancreatic losses may account for up to 25% of daily losses (12 mg/L fistula, 17 mg/kg stool). Sweat losses can be as high as 1 μg/mL.

Source: From Baumgartner.[18]

nisms fail or are bypassed, as for example in renal dialysis. It is also a frequent contaminant in medicinals, such as citrated (Shohl's) solutions, albumin, calcium, and phosphate additives. The present debate concerns the availability of environmental aluminum and the possible impact of its slow and insidious absorption and accumulation in vulnerable individuals. Aluminum has been mired in the pathogenesis of parenteral nutrition-related bone disease (so-called metabolic bone disease). The etiology is further complicated by the effect of underlying

Table 5.7. Suggested Daily Intravenous Intake of Essential Trace Elements

		Adult		
	Pediatric (μg/kg)[a]	Stable	Acute Catabolic[b]	GI Losses[b]
Zinc	300[c] 100[d]	2.5–4 mg	Additional 2 mg	Add 12.2 mg/L small bowel fluid lost; 17.1 mg/kg of stool or ileostomy output[e]
Copper	20	0.5–1.5 mg	—	—
Chromium	0.14–0.2	10–15 μg	—	20 μg
Manganese	2	0.15–0.8 mg	—	—

[a]Limited data are available for infants weighing less than 1500 g. Their requirements may be more than the recommendations because of their low body reserves and increased requirements for growth.

[b]Frequent monitoring of blood concentrations in these patients is essential to provide proper dosage.

[c]Premature infants (weight less than 1500 g) up to 3 kg of body weight. Thereafter, the recommendation for full-term infants apply.

[d]Values derived by mathematical fitting of balance data from a 71-patient-week study in 24 patients.

[e]Mean from balance study.

Source: From American Medical Association.[19]

Table 5.8. Recommended Daily Dietary Allowances of Trace Elements

Trace Element	Daily Dose Range[a]
Iodine	40–200 μg
Iron	6–15 mg
Selenium	10–75 μg
Zinc	5–16 mg

[a]Dosing is dependent on age, sex, pregnancy, and the duration of female lactation

Source: From *Recommended Daily Allowances.*[20]

illnesses, therapeutic interventions, and preexisting nutrition deficiencies before the initiation of parenteral nutrition therapy. Aluminum toxicity may be common to both adult and pediatric populations.[21] Silicon is considered an essential element, but the mechanisms underlying its essentiality remain unknown, and binding of the element (through oxygen) with biomolecules has not been demonstrated. There is, however, a unique affinity between aluminum and silicon that mediates the bioavailability and cellular toxicity of aluminum. The observed effects of silicon deficiency can be attributed to consequential aluminum availability. There are important implications for the epidemiology and biochemistry of aluminum-induced disorders and any consideration of one element must include the other.[22] Boron may be of nutritional significance relative to selenium, copper, and optimal calcium and, thus, bone metabolism. Zirconium, again, like selenium enters the human system to the extent of environmental concentrations (i.e., soil, plant content). Retention is initially in soft tissues and then slowly into bone. The metal is able to cross the blood–brain barrier and is deposited in the brain and the placental barrier to enter milk. The daily human uptake has been known to be as high as 125 mg. The level of toxicity has been found to be moderately low, both in histological and cytological studies. The toxic effects induced by

Table 5.9. Estimated Safe and Adequate Daily Dietary Intakes of Trace Elements

Trace Element	Daily Dose Range[a]
Chromium	10–200 μg
Copper	0.4–3 mg
Fluoride	0.1–4 mg
Manganese	0.3–5 mg
Molybdenum	15–250 μg

[a]Dosing is dependent on solely age

Source: From *Recommended Dietary Allowances.*[20]

very high concentrations are nonspecific in nature. Despite the presence and retention in relatively high quantities in biological systems, zirconium has not yet been associated with any specific metabolic function.[23] Another duo, nickel and magnesium are also biologically active elements in higher animals. Recent studies *in vivo* as well as *in vitro* point to interactions in some of the same enzyme and endocrine systems, body and cell structures, and transport systems. Magnesium status has determined some responses to dietary nickel. Again, evidence to date, however, warrants further investigation of the nutritional and metabolic relationships between these elements, over wide ranges of dietary intakes or exposure concentrations.[24] Interestingly, in spite of the improved awareness of the potential for nickel, cobalt, and chromium to cause skin allergy, the incidence of sensitization to them is generally on the increase, especially for nickel. A person sensitized to these metals may have many other more significant sources of daily contact with metal objects (e.g., jewelry). Therefore, it may be necessary to focus on decreasing the high exposure to these transition metals from other sources rather than on possible trace amounts found in consumer products. Current good manufacturing practice ensures that trace nickel, cobalt, and chromium concentrations in consumer products are less than 5 ppm of each metal. It has been recommended that this be accepted as a standard for maximum concentrations and that the target should be to achieve concentrations as low as 1 ppm.[25] Vanadium is still another pervasive element of biological systems, being widely distributed across the food supply. Food refining and processing appear to increase vanadium content. At higher intakes, it accumulates in body tissues such as liver, kidney, and bone. Essentiality of the nutrient has been established in lower life forms, but the significance and extent of vanadium's role in humans has been overshadowed by the absence of deficiency symptoms in man. Even though the pharmacologic properties of vanadium have stimulated much interest, knowledge of basic metabolic processes regulating vanadium remains incomplete. Ultimate determination of essentiality for humans will, of course, depend on greater understanding of the fundamental biochemical roles of vanadium.[26]

Vitamins

One cannot anabolize or effectively catabolize without vitamins. There must be a constant awareness with regard to vitamin deficiency or toxicity. The clinician must have an appreciation of kinetics (Table 5.10) as it relates to organ integrity. The role that vitamins play as antioxidants has received much attention in studies of the physiologic or pathophysiologic imbalances induced by prooxidants or antioxidants. Vitamin E in the membrane compartment and vitamin C in the aqueous compartment exhibit direct reactivity with radicals, and there are repair pathways for the tocopheryl and ascorbyl radicals that are generated. Glutathione can exhibit repair capacity as well as direct reactivity with radicals. Singlet

Table 5.10. Vitamin Kinetics

Vitamin A	Absorption is principally by the duodenal and upper jejunal areas. Vitamin A is stored as enterohepatic b-glucuronide, and bound extensively to retinol-binding protein. It is excreted in the urine and feces as metabolite and parent vitamin. Corticosteroids and renal failure cause an hepatic efflux of vitamin A.
Vitamin D	Absorption is as ergocalciferol and cholecalciferol from the small intestine. Distribution is rapid into chylomicrons in lymph, kidneys, adrenals, bones, and intestines. The vitamin is activated by the kidney and the liver and about 40% is excreted in 10 days in bile. Only a small amount is excreted in the urine.
Vitamin E	Absorption is about 35% from the small intestine with a distribution into lymph and all tissues. Metabolites (urinary tocopheronic acid and γ-lactone glucuronides) are found in the urine; however, the excretion is principally via the liver (70–80%).
Vitamin K	Absorption is via the small intestine and colon and requires bile (it is estimated that up to 50% of total body vitamin K is synthesized in the bowel). Distribution is in lymph fluid, and metabolites are glucuronide and sulfate conjugates that are excreted in the bile and urine.
Vitamin B_1 (thiamine)	Absorption is in the small intestine (maximum absorption, 8–15 mg/day), distribution heart, brain, liver, kidneys and skeletal muscles. Metabolism is in all tissues and excretion is via the kidney.
Vitamin B_2 (riboflavin)	Absorption is in the small intestine, but distribution is in all tissues. There is little storage, and it is excreted up to 9% (increases with larger doses) unchanged in urine.
Vitamin B_3 (niacin)	Absorption is throughout the gastrointestinal tract with distribution to all tissues, particularly the liver. Niacin has metabolites (*n*-methyl-2-pyridone-5-carboxamide; *n*-methyl-4-pyridone-3-carboxamide) that are excreted unchanged in urine.
Vitamin B_6 (pyridoxine)	Absorption occurs throughout the gastrointestinal tract with accumulation in brain, liver, and kidneys. The metabolite, 4-pyridoxic acid, is found in liver and is excreted via feces and urine.
Vitamin B_{12} (cyanocobalamin)	Absorption is in the distal ileum and distribution is to liver, heart, kidney, spleen, and brain. After acidification and with calcium for intrinsic factor action, the cobalt complex is excreted in bile (principally), feces, and urine.
Pantothenic Acid (Vitamin B_5)	Absorption occurs throughout the gastrointestinal tract. Distribution is in all tissues, however 70% is excreted unchanged in the urine.

continued on next page

Table 5.10. *Continued*

Inositol	Absorption occurs throughout the gastrointestinal tract with distribution to all tissues, particularly the brain, heart and skeletal muscle. Inositol can be metabolized to glucose and is excreted in small amounts through the kidney.
Choline	Choline is incompletely absorbed as lecithin. Distribution is to the liver and peripheral tissues. Trimethylamine (metabolized by intestinal bacteria) is excreted via feces.
Biotin	Absorption is throughout the gastrointestinal tract with distribution primarily to the liver and brain. Biotin is metabolized in the kidney and excreted via urine.
Folic Acid	Absorption is in the ileum with storage in the liver. Metabolites are excreted into urine.
Vitamin C (ascorbic acid)	Absorption is via the intestine with distribution into plasma and body cells. Vitamin C metabolites are excreted in the urine when serum levels rise above 2 mg/dL.

Source: From Baumgartner.[28]

molecular oxygen is an electronically excited state of oxygen with considerable chemical reactivity. It is a nonradical compound, generated by photochemical reactions or by the process of lipid peroxidation of biomembranes (that is, by photoexcitation or by chemiexcitation). Singlet oxygen can be inactivated by a number of biological compounds known as quenchers (beta-carotene)[27].

Vitamin A. The typical lesions of vitamin A deficiency are night blindness, Bitot's spots on the conjunctiva, xerosis or keratinization of various membranes (particularly xerophthalmia), and the formation of defective bony tissue and dentine during growth. Measurement of dark adaptation of the eye is the most sensitive test for vitamin A deficiency. Vitamin A serum levels are less reliable since the serum level does not fall until the body's reserves are fully depleted. Xerophthalmia is still one of the most common causes of blindness in children in many countries. Although the requirement for vitamin A is proportional to body weight, parenteral multivitamins contain 3300IU/day supplementation. Good enteral sources of vitamin A are the fish oils (e.g., cod 1000, herring 5000, halibut and tuna 50,000–100,000 IU/g), liver, milkfat, and egg yolk. Green vegetables and carrots are rich in carotenes. Vitamin A injection has a pH of 6.5–7.1.

Inadequate dietary intake, impaired absorption (fat deficiency) or storage, disturbances in the conversion of carotene into vitamin A, or rapid depletion of the body's reserves are associated with vitamin A deficiency. Celiac disease, cystic fibrosis of the pancreas, ulcerative colitis, pancreatectomy, obstruction of the biliary ducts, and cirrhosis of the liver generally decrease vitamin A bioavailability. Conversion of carotenes may also be impaired in diabetes and hyperthyroidism.

Anorexia, alopecia, skin and mucosa changes, swelling of the bones and diaphyses of the limbs, anemia, enlargement of the liver and spleen, as well as, headache can result if high doses of vitamin A are used for long-term. These symptoms are reversible and disappear rapidly on cessation of the treatment. In children, overdosage of vitamin A may interfere with bone development and lead to premature fusion of the epiphyses.

Lycopene, a biologically occurring carotenoid, exhibits the highest physical quenching rate constant with singlet oxygen, and its plasma level is slightly higher than that of beta-carotene. This is of considerable general interest, since nutritional carotenoids, particularly beta-carotene, and other antioxidants such as α-tocopherol have been implicated in the defense against prooxidant states. Epidemiological evidence reveals that such compounds exert a protective action against certain types of cancer. Also, albumin-bound bilirubin is a known singlet oxygen quencher. Vitamin A has long been recognized as a key nutrient in maintaining and restoring epithelial barriers and is, therefore, important in maintaining resistance to respiratory and gastrointestinal infections. Vitamin A deficiency impairs secretory immunoglobulin A (IgA) production, decreases mucus production, and leads to keratinization of secretory epithelia. It has also been shown that vitamin A, bound to its physiologic carrier retinol-binding protein (RBP) and to chylomicron remnants, modulates normal B-cell activation, cytokine production, and cell differentiation. Other studies have cited the importance of retinoids for B-cell production of antibodies. Beta-carotene, one of approximately 50 carotenoids that can be metabolized to vitamin A, has been shown to enhance T-cell and B-cell generation in animals.[29] Prealbumin transports the RBP–retinoid moiety after it leaves the liver and one must remember that patients that have high vitamin A serum levels (i.e., acute or chronic renal failure) will also have markedly higher serum prealbumins.

Vitamin D. In adults, vitamin D is endogenously synthesized or ingested. Deficiencies of vitamin D cause rickets in children or osteomalacia in the adult. Progressive demineralization and weakening of the skeleton results. Primary vitamin-D-resistant rickets, which is characterized by lower plasma phosphate levels and an increase in the phosphate excretion index is congenital and usually hereditary. The clinical symptoms of rickets (generally pain related to epiphyseal joints and tetany) are consequences of calcium losses. However, there is only a slight lowering of the plasma calcium level along with a lower than normal calcium phosphate product (i.e., less than 30 to 40 mg%). There is also a marked lowering of the plasma phosphate level, reduced urinary calcium excretion, increased phosphate clearance, and a rise in the plasma concentration of alkaline phosphatase. Interestingly, an increase in the amino acid content of the urine is one of the first signs of vitamin D deficiency.

The calcium mobilized from the bones by toxic intake of vitamin D is taken up in the soft tissues, particularly the kidneys and vascular bundles. Upon cessation of intake, the symptoms of vitamin D poisoning are reversible.

Vitamin E. Healthy adults require 10 to 30 mg/day of α-tocopherol, depending on the intake of polyene fatty acids. In human beings, the signs of vitamin E deficiency are not marked. The peroxide hemolysis test is used to measure tocopherol status because a lowered serum tocopherol level is associated with an increase in *in vitro* hemolysis. A low serum level is common in newborn infants, particularly premature infants, and vitamin E deficiency is a possible cause of macrocytic anemia and hemolytic anemia in infants. Vitamin E deficiency may also be caused by impaired absorption of fats. In fact, low tocopherol serum levels, with creatinuria and the deposition of ceroid pigments in the gastrointestinal tract smooth muscle, have been observed in sprue, celiac disease, biliary cirrhosis, pancreatitis, and, particularly, cystic fibrosis of the pancreas. Vitamin E, the major lipid-soluble membrane antioxidant, protects unsaturated fatty acids from oxidation. Vitamin E deficiency in animals is associated with T-cell and B-cell dysfunction, phagocytic cell dysfunction, and suppressed NK ("Natural Killer") cell response. T-cell dysfunction induced by vitamin E deficiency was reversed by vitamin E supplementation in a study in human models. The optimal level of supplementation, however, is controversial. Recent research supports a need for vitamin E in amounts much higher than the recommended dietary allowance (RDA) to maintain optimal function of the immune system. Although the research is compelling, the deleterious effects of excessive intakes of micronutrients must be recognized. Vegetable oils, wheat germ oil, cereals, and eggs are good sources of tocopherols; whereas, animal tissues contain little α-tocopherol.

Vitamin K. Spinach, cabbage, tomatoes, liver, fruits, milk, and meat contain vitamin K. Medications (e.g., antibiotics and altered bowel flora), steatorrhea, and bile losses secondary to fistula losses or biliary obstruction, are primary reasons for prolonged prothrombin time. Prothrombin time is an ideal test for hepatic function, because the hypoprothrombinemia associated with severe injury to the liver parenchyma is not due to vitamin K deficiency and is not reversed by administration of the vitamin. It must be remembered that intravenous lipid products contain a therapeutic dose of vitamin K in a small volume (i.e., approximately 60 μg vitamin K in 20% lipid and 30 μg in 10%). Vitamin K is generally added to the intravenous nutrition once per week rather than given subcutaneously or intramuscularly.

Thiamine (vitamin B_1). The Food and Nutrition Board recommended daily dietary allowance for thiamine is based on a daily intake of 0.5 mg thiamine per 1000 kcal. The requirement for thiamine increases with the metabolic rate. Additional thiamine may be needed in intensive care, because there is only 3 mg of thiamine in standard multivitamin infusions with body stores of only 2 weeks in the erythrocyte and 3 weeks in the leukocyte.

Riboflavin (Vitamin B_2). The metabolic body size was used by the Food and Nutrition Board to determine riboflavin needs. Daily allowances based on the 75th power of body weight is 0.07 mg B_2/kg. Increased intake of protein may also increase riboflavin requirements. Liver, kidneys, heart, protein, green vegetables, and milk are sources of riboflavin. Gastrointestinal lesions (e.g., glossitis, inflammation of the buccopharygeal mucosa, and cheilitis), as well as skin and bone marrow injury, are all thought to be associated with riboflavin deficiency. One of the reasons for this might be the fact that vitamin B_2 is stored in the liver, spleen, kidneys, and heart. Vitamin B_2 is converted to the coenzyme flavin mononucleotide, which, in turn, is converted to flavin adenine dinucleotide in the liver. Large doses may cause yellow urine. Sunlight has been shown to degrade riboflavin.

Pyridoxine (Vitamin B_6). As with riboflavin, pyridoxine requirements increase with increased intake of protein. The Food and Nutrition Board recommends a vitamin B_6 intake of 2 mg/day when the daily protein intake is 100 g or more. The conversion of tryptophan to niacin requires pyridoxine as well as the monosaccharide-to-carbohydrate interplay. Pyridoxine is ubiquitously found in all vegetable, meats, yeast, liver, and cereals.

The central nervous system neurologic effects of B_6 are associated with long-term large-dose administration. Deficiency carries with it symptoms of irritability, depression, somnolence, nausea, impairment of vibrational and positional change sense and, rarely, peripheral neuritis.

Vitamin B_6 or pyridoxine is a vitamin of special interest because, like vitamin E, it is suggested that concentrations in excess of the RDA are needed for maintaining immune function. Deficiency is associated with both impaired cell-mediated (T-cell) and humoral immune (B-cell) responses. In humans with poor vitamin B_6 intake, immunological findings include decreased circulating lymphocytes, reduced IgD levels, and a decreased percentage of T-helper cells. It is proposed that the immunodeficiency common to alcoholics and chronic renal failure patients may be attributed to vitamin B_6 deficiency.

Niacin. The Food and Nutrition Board recommends a daily intake of 6.6 mg per 1000 kcal. Sixty milligrams of tryptophan, with pyridoxine, provides about 1 mg of nicotinic acid. Nicotinic acid deficiency causes pellagra, which is promoted with stress (e.g., sunlight and heavy physical work). Approximately 1% of trytophan is converted to serotonin. In carcinoid tumors, up to 60% of tryptophan is diverted away from nicotinic acid formation and driven to serotonin, which is no longer available as a source of nicotinic acid. Several medications (e.g., isoniazid), large amounts of leucine or diseases that affect pyridoxal phosphate will, in turn, affect the conversion of tryptophan to niacin. Again, dementia, gastrointestinal lesions and skin lesions are the hallmarks of nicotinic acid deficiency (e.g., dementia, diarrhea, dermatitis).

Folic acid. As with other vitamins, inadequate dietary intake, malabsorption, and increased requirements may precipitate deficiency states. In addition, disturbances of folic acid metabolism may be associated with deficiency. Megaloblastic marrow, macrocytic anemia, leukopenia, excessive segmentation of the leukocytes, thrombocytopenia, glossitis, and gastrointestinal disturbances are indicative of folic acid deficiency. Vitamin B_{12} is also necessary for the conversion of 5-methyltetrahydrofolic acid to tetrahydrofolic acid and may be a secondary cause of folic acid deficiency. Folic acid may obscure the diagnosis of pernicious anemia; that is, it may reverse the hematologic evidence for anemia and the neurologic disease process may ensue.

Vitamin B_{12}. As with folic acid, cyanocobalamin is absorbed principally in the distal ileum so that malabsorption in this area may usher in a deficiency state. A strict vegetarian diet will not contain sufficient B_{12}. The absence of intrinsic factor secretion will cause B_{12} deficiency. Calcium and a pH from 5.4 to 8 are required for intrinsic factor-B_{12} receptor site attachment to the ileum for vitamin B_{12} absorption. In contrast to folic acid (absorbed in the proximal intestine), vitamin B_{12} is absorbed in the distal ileum, as are bile acids.

Macrocytic anemia, megaloblastosis of the bone marrow, leucopenia, thrombocytopenia, glossitis, morphologic changes in the gastrointestinal tract, and (in contrast to folic acid deficiency) progressive degeneration of the axis cylinders of the spinal-cord neurons (pernicious anemia) are associated with cyanocobalamin deficiency. Folic acid, vitamin B_{12}, and biotin are not as water soluble as the other vitamins that are packaged as multivitamin products. Two (cyanocobalamin and folic acid) of the three vitamins are generally available as single additives or all three are packaged together.

Biotin. The biotin requirement of humans is unknown. The primary action of biotin is as a cofactor in elongating carbon chains through carboxylation reactions. Nervous disturbances, alopecia, seborrheic dermatitis, lethargy, loss of appetite, nausea, muscular pain, and localized paresthesia are associated with biotin deficiency. Increased intake of raw eggs can precipitate biotin deficiency. Organ foods, yeast, egg-yolk, vegetables, nuts, and cereals contain biotin.

Pantothenic acid. Pantothenic acid is of importance in all macronutrient metabolism because it is a constituent of coenzyme A (CoA), a component of acetyl CoA, which is one of the entry sites for the Krebs (tricarboxylic acid) cycle. Neuromuscular degradation, adrenal insufficiency, mild fatigue, headache, sleeplessness, nausea, epigastric pain, paresthesia of the limbs, muscle spasms, coordination disturbances, and death may result from pantothenic acid deficiency.

Pantothenic acid is present in vegetables (not fruits), cereals, meats (particularly organ foods), and yeast.

Ascorbic acid. Scurvy was the first syndrome associated with vitamin C deficiency. Decreased vitamin C is now known to be associated with depressed cell-mediated immunity, poor bactericidal activity, and impaired macrophage mobilization. Supplementation with vitamin C enhances T-cell and B-cell proliferation and bacterial phagocytosis by macrophages. Studies of vitamin C in the elderly have shown decreased infection with supplementation. Antioxidants like vitamin C lower the free radical burden and protect membrane lipid receptors and other components of immune cells from oxidation. Because vitamin C is a major water-soluble antioxidant, it is important in decreasing free radicals in intra- and extracellular fluids. Dosing greater than 0.5 g/day may be associated with a prooxidant action and should be avoided for extended periods of time. Cabbage, spinach, potatoes, milk, citrus fruits, tomatoes, strawberries, and red currants contain vitamin C. In vegetables, the ascorbic acid falls rapidly during withering.

The Food and Nutrition Board recommends about 50 mg/day. The most important symptoms of ascorbic acid deficiency are capillary fragility, impairment of connective tissue formation with changes in bone structure and growth, defective tooth formation and fissuring of the skin. This vitamin is converted to oxalate and must be used with caution in renally compromised patients.

Vitamin deficiencies and their signs and symptoms are replete in the literature. However, their kinetics are not as easily referenced. As is true with other nutrients, an appreciation of any surgical resection or areas of enteral compromise will better define the degree of parenteral supplementation. Intravenous vitamin products should be included (Table 5.11) with the initiation of every nutritional regimen, because body stores will be unclear and effective anabolism (or catabolism) will not take place without vitamins.

In conclusion, patients receiving micronutrition must be fastidiously assessed and meticulously monitored to provide optimal patient care and avoid medicolegal liability. Amino acid, carbohydrate and lipid formulas must be carefully concocted to respect renal solute load, carbon dioxide production, nitrogen balances and substrate imbalances relative to micronutrient dosing. Parenteral micronutrients must be a part of every nutritional formula for the efficient catabolic and anabolic processes from day one because histories may be unclear. Enteral formulas must be given in a volume of 1 to 2 L to meet the recommended oral requirements. Electrolyte status should be monitored using approximate intravascular and interstitial (third compartment) appraisal. Acid–base assessment must preclude electrolyte supplementation to determine the salt required. Trace element and vitamin deficiencies and toxicities must be monitored and pertain directly on the disease state. In addition, serial monitoring of signs and symptoms should be in place.

Table 5.11. Adult and Pediatric Intravenous Vitamins—Manufacturer: Astra (MVI-12®)

Vitamin	Dosage			Units
ADULT AND PEDIATRIC INFUSION[a]				
Ascorbic Acid(C)	100			mg
Vitamin A	3300			IU
Vitamin D	200			IU
Thiamine (B_1)	3			mg
Riboflavin (B_2)	3.6			mg
Pyridoxine HCl (B_6)	4			mg
Niacinamide	40			mg
Pantothenic acid	15			mg
Vitamin E	10			IU
Biotin	60			μg
Vitamin B_{12}	5			μg
Folic acid	400			μg
PEDIATRIC INFUSION[b]				
	2 mL	3.3 mL	5 mL	
Ascorbic Acid	32	52.8	80	mg
Vitamin A	920	1520	2300	USP units
Ergocalciferol (D)	160	264	400	USP units
Thiamine (B_1)	0.48	0.79	1.2	mg
Riboflavin (B_2)	0.56	0.92	1.4	mg
Pyridoxine (B_6)	0.4	0.66	1	mg
Niacinamide	6.8	11.2	17	mg
Pantothenic Acid	2	3.3	5	mg
Vitamin E	2.8	4.6	7	USP units
Biotin	8	13	20	μg
Folic Acid	56	92.4	140	μg
Vitamin B_{12}	0.4	0.66	1	μg
Vitamin K	80	132	200	μg

[a] Available in two 5 mL or 50 mL sets (nine water-soluble vitamins with three semi-water-soluble vitamins) or one 10 mL two-chambered vial. Available in single and multidose vials.

[b] Suggested daily volumes of Multivitamin Infusion for Pediatrics are 2 mL for less than 1 kg, 3.3 mL for 1 to 3 kg; 5 mL for greater than 3 kg but under 11 years old.

If there are extensive bowel fluid losses secondary to upper or lower bowel transit or if bowel integrity is not appropriate (i.e., bowel inflammation, infection, risk for aspiration, peritonitis, or translocation), avoid its use. If used appropriately, parenteral nutrition, encompassing both macronutrition and micronutrition therapy, can buy time needed to permit bowel hyperplasia or hypertrophy that will enable return to gut homeostasis.

REFERENCES

1. Edgington SM. As we live and breathe: free radicals and aging. *Biotechnology* 1994;12:37–40.
2. Schehr RS. Therapeutic approaches to Alzheimer's disease. *Biotechnology* 1994;12:140–144.
3. Grimble RF. Nutritional antioxidants and the modulation of inflammation: theory and practice. *New Horiz* 1994;2(2):175–85.
4. Dawson KG. Endocrine physiology of electrolyte metabolism. *Drugs.* 1984;28 (suppl 1):98–111.
5. Proebstle T, Mitrovics M, Schneider M, et al. Recombinant interleukin-2 acts like a class I antiarrhythmic drug on human cardiac sodium channels. *Pflugers Arch* 1995;429(4):462–469.
6. Kaspar A, Brinkmeier H, Rudel R. Local anaesthetic-like effect of interleukin-2 on muscular Na^+ channels: no evidence for involvement of the IL-2 receptor. *Pflugers Arch* 1994;426(1–2):61–67.
7. Souba WW, Copeland EM. Cytokine modulation of Na(+)-dependent glutamine transport across the brush border membrane of monolayers of human intestinal Caco-2 cells. *Ann Surg* 1992;215(5):536–544; discussion 544–545.
8. Soliven B, Szuchet S, Nelson DJ. Tumor necrosis factor inhibits K^+ current expression in cultured oligodendrocytes. *J Membr Biol* 1991;124(2):127–137.
9. McLarnon JG, Michikawa M, Kim SU. Effects of tumor necrosis factor on inward potassium current and cell morphology in cultured human oligodendrocytes. *Glia* 1993;9(2):120–126.
10. Liu JH, Blanchard DK, Wei S, Dieu TY. Recombinant interleukin-8 induces changes in cytosolic Ca^{2+} in human neutrophils. *J Infect Dis* 1992;166(5):1089–1096.
11. Richter C, Schlegel J, Schweizer M. Prooxidant-induced Ca^{2+} release from liver mitochondria. Specific versus nonspecific pathways. *Ann NY Acad Sci* 1992;663:262–268.
12. Freeman JB, Wittine MF, Stegink LD, et al. Effects of magnesium infusions on magnesium and nitrogen balance during parenteral nutrition. *Can J Surg* 1982;25(5):70–72, 574.

13. Muratani H, Kawasaki T, Ueno M, et al. Circadian rhythms of urinary excretions of water and electrolytes in patients receiving total parenteral nutrition (TPN). *Life Sci* 1985;37(7):645–649.
14. Burch RE, Hahn HK. Trace elements in human nutrition. *Med Clin North Am* 1979;63(5):1057–1068.
15. Mertz W. Trace elements nutrition and disease: contributions and problems of analysis. *Clin Chem* 1975;21(4):468–475.
16. Laker M. On determining trace element levels in man: the uses of blood and hair. *Lancet* 1982;260–262.
17. Ulmer DD. Trace elements; *N Engl J Med.* 1977;297:318–321.
18. Abstracted from Baumgartner TG, ed., Clinical Guide to Parenteral Micronutrition. Deerfield, III: Fujisawa, 1991: appendix.
19. American Medical Association, Department of Foods and Nutrition. Guidelines for essential trace element preparations for parenteral use. *JAMA* 1979;241(19):2051–2054.
20. National Research Council Food and Nutrition Board. *Recommended Dietary Allowances.* 10th ed. Washington, DC: National Academy of Sciences, 1989.
21. Koo WW. Parenteral nutrition-related bone disease. *J Parent Ent Nutr* 1992; 16(4):386–394.
22. Birchall JD. The interrelationship between silicon and aluminum in the biological effects of aluminum. *Ciba Found Symposium* 1992;169:50–61; discussion 61–68.
24. Kenney MA, McCoy H. A review of biointeractions of Ni and Mg. I. Enzyme, endocrine, transport, and skeletal systems. *Magnes Res* 1992;5(3):215–222.
23. Ghosh S, Sharma A, Talukder G. Zirconium. An abnormal trace element in biology. *Biol Trace Elem Res* 1992;35(3):247–271.
25. Basketter DA, Briatico-Vangosa G. Nickel, cobalt and chromium in consumer products: a role in allergic contact dermatitis? *Contact Dermatitis* 1993;28(1):15–25.
26. French RJ, Jones PJ. Role of vanadium in nutrition: metabolism, essentiality and dietary considerations. *Life Sci* 1993;52(4):339–346.
27. Sies H. Relationship between free radicals and vitamins: an overview: an overview. *Int J Vitam Nutr Res Suppl* 1989;30:215–223.
28. Abstracted from Baumgartner TG, ed. *Clinical Guide to Parenteral Micronutrition.* Deerfield, Ill: Fujisawa, 1991: appendix.
29. DiMascio P, Kaiser S, Sies H. Lycopene as the most efficient carotenoid singlet oxygen quencher. *Arch Biochem Biophys* 1989; 274(2):532–538.

Perioperative Nutritional Interventions

Robert C. Gorman, M.D., and
Gordon P. Buzby, M.D.

INTRODUCTION AND HISTORICAL PERSPECTIVE

The nutritional status of surgical patients and their metabolic response to injury are currently well recognized as important factors influencing wound healing, rates of postoperative complications, risk of infection, and overall recovery from surgical stress.[1] This fact was appreciated by surgeons half a century ago[2] and has been further substantiated by recent clinical and basic scientific research.

Prior to the late 1960s, alternatives for complete perioperative nutritional support of patients who could not eat were limited to forced enteral feedings delivered by nasogastric or surgically placed gastric or jejunal feeding tubes. These techniques were often poorly tolerated in ill surgical patients with compromised gastrointestinal function.

Armed with the pioneering work of Rhode et al.[3] and Wretlind[4,5] of the late 1940s and 1950s, Dudrick, et al. demonstrated the feasibility of central venous feedings of all nutrients required for growth or maintenance of nutritional well-being in pure-bred beagles[6] and subsequently in humans.[7] The development of total parenteral nutrition (TPN) provided a potentially powerful new tool for the treatment and prevention of malnutrition in surgical patients.

The initial use of TPN was limited to patients with permanent or prolonged gut dysfunction incompatible with survival as a result of resection, malabsorption, inflammatory bowel disease, or fistula. In this application, TPN has clearly been successful in preventing death from starvation.

During the late 1970s and 1980s, numerous reports appeared in the nutrition and surgical literature documenting a previously unrecognized high incidence of malnutrition in surgical patients. In addition, a close association between malnutrition and postoperative morbidity was also clearly demonstrated.[8–25] The availability of TPN and the increased concern about nutrition-related complications led to the wide application of perioperative TPN with the intent to reduce postoperative complications. This approach was intuitively attractive and was supported by the results of several nonrandomized studies of perioperative TPN in patients with varying degrees of nutritional depletion.[20–24] Many clinicians shared the bias that a well-designed prospectively randomized clinical trial would eventually conclusively demonstrate the widespread efficacy of perioperative nutritional support in reducing operative morbidity.

By the late 1980s, several randomized protocols had been completed; however, the majority of these studies were not able to demonstrate any benefit from perioperative nutritional support in terms of improved outcome. This failure to document efficacy occurred at a time when fiscal pressures made application of expensive clinical technologies difficult unless these technologies were of clear-cut benefit. In part, because of these cost containment forces, perioperative nutritional support (especially parenteral) fell largely out of favor. During the last several years, reassessment of existing data and publication of additional well-designed clinical trials has further clarified the appropriate use of perioperative nutritional support.

The intense scrutiny with which perioperative nutritional support has been studied has resulted in an extensive database. Analysis of this data now allows an accurate assessment of the efficacy, cost, and cost effectiveness of this treatment modality. In this chapter, these issues are discussed, and clinically applicable guidelines are presented for the appropriate use of perioperative nutritional support.

PERIOPERATIVE NUTRITIONAL SUPPORT: REQUIREMENTS FOR EFFICACY

The goal of perioperative nutritional support is to improve operative results by avoiding or reversing significant nutritional deficiencies. Although better nutritional status may be a legitimate goal in and of itself, it does not justify the potential morbidity and cost of aggressive nutritional support unless there is a clearly demonstrable benefit to overall surgical outcome.

The rationale for the use of perioperative nutritional support is, therefore, dependent on the validity of five assumptions:

1. Malnutrition is a problem that affects a significant number of surgical patients.
2. Malnutrition can be adequately defined and diagnosed.
3. Malnutrition adversely affects surgical outcome by increasing postoperative morbidity and/or mortality.
4. Adequate nutritional support can improve and/or maintain the nutritional status of patients as determined by objective measurements.
5. The reversal and/or prevention of malnutrition significantly reduces postoperative morbidity and/or mortality.

The validity of the first four of these assumptions is well supported in the surgical literature, and studies documenting their authenticity are briefly reviewed here. The final assumption has been the subject of debate; its validity or lack thereof is evaluated subsequently in the context of the most recently completed prospectively randomized trials evaluating the efficacy of perioperative nutritional support.

Numerous studies that have demonstrated that malnutrition is common in hospitalized patients are our justification for using perioperative TPN in more than an occasional patient. Protein-calorie malnutrition is most prevalent and affects both affluent and indigent populations equally.[8–11] In a survey of surgical patients in a large urban hospital, the incidence of moderate to severe protein-calorie malnutrition was nearly 50%.[8] Subsequent studies have confirmed the magnitude of this problem.

Several diagnostic tests to identify patients with nutritional deficits have been proposed. A *perfect* test should be accurate, inexpensive, easy to apply, and unaffected by nonnutritional comorbidities such as renal, hepatic, or cardiac insufficiency. In addition, such an assessment should be sensitive to and rapidly reflect changes that occur in nutritional status. Physical examination, degree of weight loss, serum levels of various circulating proteins, lymphocyte count, skin test anergy, various measures of muscular strength and/or function, and anthropometric measurements have all been correlated with nutritional status; however, none of these tests adequately detects and quantifies nutritional deficiencies in all patients, and all fall far short of our perfect test.

In several studies, multivariant analyses of multiple nutritional parameters have been performed in attempts to develop more objective and quantifiable methods for assessment of nutritional status. An example of such a relationship is the prognostic nutrition index (PNI) which was developed as a means of relating the risk of postoperative complications and death to preoperative nutritional status. The PNI is calculated using the following equation.

$$PNI = 158 - 16.6\ (ALB) - 0.78\ (TSF) - 0.2\ (TFN) - 5.8\ (DH)$$

Where Alb is serum albumin (g/dL), TSF is triceps skinfold (mm), TFN is serum transferrin (mg/dL), and DH is delayed cutaneous hypersensitivity graded from 0 to 2. Application of this assessment prospectively to 100 nonemergent surgical patients was highly predictive of postoperative complications.[22]

The success of the PNI and other statistically derived relationships between nutritional assessment parameters and operative outcome is important for two reasons. Firstly, it demonstrates that nutritional status can be evaluated objectively, quantitatively, and reproducibly. Secondly, such statistical methods have helped to delineate the association of malnutrition with poor postoperative results and have permitted objective stratification of severity of malnutrition in subsequent interventional protocols. The widespread clinical utility of these relationships is limited, however, because they are somewhat cumbersome, time-consuming (especially if skin tests ar required), and quite sensitive to laboratory technique. Small variations in normal range between laboratories for components of the model will substantially alter what would be considered the normal range for the statistical model itself. This, in turn, will shift upward or downward the threshold value at which malnutrition is diagnosed. Nevertheless, despite the limitations of both individual parameters and multivariate models as measures of nutritional status, a considerable body of experience now exists supporting the validity of assumptions 2 and 3 (from the list above) regarding rational use of perioperative TPN. Malnutrition can be diagnosed, although imperfectly, and, when present, it has an adverse effect on clinical outcome.

Assumption 4, the ability of intensive nutritional support regimens to maintain or improve nutritional status in perioperative patients who are not severely stressed or septic, has been repeatedly demonstrated dating back to the pioneering studies of TPN.[7] Increases in weight, serum albumin levels, serum transferrin levels, and repletion of intracellular electrolytes have all been documented to result from aggressive nutritional support. The remaining question is whether improvements in these indices of nutritional status translate into improved clinical outcomes. This key issue will be explored in the next section.

PREOPERATIVE NUTRITIONAL SUPPORT FOR THE TREATMENT OF PREEXISTING MALNUTRITION

In patients who could go to surgery immediately if it were not for preexisting nutritional deficits, surgeons have disagreed whether surgery should be delayed to permit preoperative nutritional support. Five retrospective studies conducted during the late 1970s and reported before 1983 all demonstrated a benefit of preoperative TPN in malnourished patients.[26–30] Although these studies were subject to all the pitfalls associated with retrospective analyses, they did generate strong support for elective preoperative nutritional support. Most surgeons believed that these findings would ultimately be confirmed by prospective randomized trials.

Between 1976 and 1990, fourteen prospective randomized trials evaluating the efficacy of perioperative TPN were published.[31–44] The majority of these studies failed to demonstrate any improvement in mortality and major postoperative morbidity in TPN-treated patients relative to untreated controls. In one prospective study by Heatly et al.,[32] mortality, anastomotic leaks, wound infections, and other infectious complications appeared to be reduced in TPN-treated patients; however, only in the case of wound infections did the difference reach statistical significance. In another prospective study by Bellantone et al.[33] TPN-treated patients who were severely malnourished (as assessed by albumin levels and total lymphocyte count) experienced a significant reduction in postoperative complications (21% vs 53%) but no change in postoperative mortality. Although this study did show a statistical reduction in postoperative complications in severely malnourished patients, it suffered from a relatively small sample size (100 patients), lack of randomization for disease (significantly more colon and gastric surgery in the control group), and failure to stratify patients for nutritional status prior to randomization.

The only prospective randomized study published prior to 1990 to demonstrate a reduction in both major complications and mortality for patients receiving preoperative TPN was reported by Muller et al.[31] This study evaluated the effect of 10 days of preoperative TPN on postoperative complication rates and mortality in patients with cancers of the gastro-intestinal tract. The favorable results of this study have been questioned for several reasons. Some patients in both groups were well nourished (40%), and patients were not stratified prospectively for degree of nutritional depletion. In addition, a third group of patients not included in the initial report of this study has subsequently been described in a more recent publication.[40] This third group was administered a regimen of TPN that included lipid and experienced greater morbidity and mortality relative to both the groups described in the initial report of the study. The authors speculate that these results were due to the immunosuppressive effects of the lipid. Regardless of the cause, the poor results obtained in patients receiving lipid-containing TPN call into question the strongly positive results of the initial study.

The results of these randomized trials as a whole quelled much of the optimism for perioperative nutritional support that had been stimulated by the retrospective analyses of the 1970s. Reviews on the subject appearing in the late 1980s were much less enthusiastic.[44,46]

The failure of these prospective trials to demonstrate conclusively the efficacy of perioperative nutritional support fell short, however, of ruling out a possible beneficial effect because of defects in the design of these studies. Many of the early clinical trials of TPN were not designed to determine the impact of TPN on operative morbidity and mortality but rather to assess its feasibility, safety, and effect on nutritional status. It is therefore somewhat unfair to criticize the authors of these studies for permitting design flaws that compromise the validity of these studies with regard to their ability to assess the impact of TPN on operative outcome. However, many of these studies have been quoted in various contexts by both proponents and opponents of the use of perioperative TPN, and it is helpful to critically assess the design of these studies if one hopes to better define appropriate indications for perioperative TPN.

Many of the early studies (those published prior to 1982–1986) were reviewed during protocol development for a large multiinstitutional Veterans Affairs Cooperative Study of perioperative TPN[45] and/or as part of a meta-analysis of perioperative TPN.[46] In these reviews, numerous defects in the design of previous studies were identified that compromised their usefulness in assessing the impact of TPN on operative outcome. These defects included one or more of the following: inadequate sample size, lack of appropriate randomization procedures, failure to exclude well-nourished patients, inadequate treatment regimens, and imprecise definition of complications and other endpoint criteria. With these considerations, it was found that an accurate assessment of the efficacy of perioperative TPN was not possible based on data available through 1986.

An important study was published in 1987[46] in which a meta-analysis was used to evaluate the results of all available trials of perioperative TPN (as listed in the Medline database from 1966 to August 1986). In this meta-analysis, results of 18 studies were reviewed, and results of 11 that met certain minimal criteria for adequate design were pooled (two of these evaluated only postoperative TPN). In this analysis, a possible small benefit to TPN was suggested, but the 95% confidence interval for rates of complications was wide, ranging from 12.8% better for TPN-treated patients to 2.3% worse. The authors concluded that any possible benefit of perioperative TPN in well-nourished patients must be small and is probably not clinically important, whereas the efficacy in mildly or severely malnourished patients may be greater but required further confirmation. This study was useful, but the efficacy of perioperative TPN in the large population of patients with degrees of malnutrition ranging from mild to severe remained uncertain.

Two well-designed protocols designed to evaluate the efficacy of perioperative nutritional support, the Veterans Administration (VA)[47] and the Maastricht[48] trials considered this question. These studies were specifically and carefully designed to avoid the deficiencies of previous studies. The results of these two trials resolve many of the questions raised by prior studies and better define the indications for preoperative nutritional support of nonemergency surgical candidates.

The VA multiinstitutional trial[47] provides data that clarifies the utility of TPN in this large population. Patients for this study were recruited from a population of adult veterans requiring nonemergent laparotomy or thoracotomy. Patients who were expected to die within 90 days (even if they survived the operation), who had received TPN within the preceding 15 days, or who had undergone an operation within 30 days of presentation were ineligible. Exclusion of these patients was necessary to permit valid interpretation of the study results, but may render these results inapplicable to certain defined populations of patients (e.g., patients undergoing a series of staged operations or other sequential multimodal therapies, which may include TPN).

Of the 3259 patients admitted to participating hospitals for eligible operations and not excluded for one of the above reasons, slightly less than a quarter (24%) met all subsequent eligibility criteria. Patients in whom an operative delay of 7 days was contraindicated or in whom TPN was impossible or dangerous were ineligible. Exclusion of these patients (8.5% of otherwise eligible patients) probably did not bias the study results, because a course of preoperative TPN was not an option even if it were proven efficacious. The presence of one or more major concurrent organ system derangements precluded study eligibility in 14.7% of patients. The rationale for this exclusion was to eliminate patients with diseases that may be a major determinant of operative outcome independent of nutritional status and thereby obscure any potential benefit associated with correcting nutritional risk factors. Although this exclusion may render more convincing a negative overall result from the clinical trial (by avoiding one possible explanation for this negative result), it also means that extrapolation of the results to the population of patients with multiorgan-system derangements may be invalid.

The final reason for excluding patients was if TPN was considered essential by local standard of care. Inappropriate application of this exclusion criterion had the potential to severely bias the results if it were used to exclude patients on the basis of malnutrition alone, because this is the population most likely to benefit if preoperative TPN is effective. The protocol limited use of this exclusion criterion to those patients who needed bowel rest and/or gastrointestinal decompression for more than 3 days prior to operation. Although they were severely malnourished as a group, an indication other than malnutrition (e.g., pancreatitis or gastric outlet obstruction) existed for the use of preoperative TPN in these 97 patients (3.0% of otherwise eligible patients). Thus, the results of the clinical trial cannot be extrapolated to clinical situations where a specific indication for preoperative TPN exists or where prompt operation is undesirable in patients who cannot eat.

After application of these exclusion criteria, 2448 patients remained (75% of patients undergoing eligible operations). Nutritional assessment identified 38% of these as malnourished using criteria developed in a previous study based on body weight, degree of weight loss, serum albumin, and prealbumin.[45] In a pilot study prior to the clinical trial, patients identified as malnourished by these criteria demonstrated complication and mortality rates two and seven times greater, respectively, than patients identified as well nourished.[45]

Randomization was offered to 782 malnourished patients and accepted by 459 (59%). These 459 patients were randomly assigned to receive TPN for 7 to 15 days prior to operation and 3 days after operation (TPN group) or no perioperative TPN (control group). After operation, patients were monitored for complications for 90 days. Rates of major complications (TPN group: 25.5%; control group: 24.6%) and mortality (TPN group: 13.4%; control group: 10%) were similar. More infectious complications occurred in the TPN group (14.1% vs. 6.4%; $p = 0.01$), but more noninfectious complications occurred in the control group (22.2% vs 16.7%; $p = 0.20$). The increased infection rate in TPN-treated patients was not explained simply by the presence of a catheter (catheter sepsis or unexplained bacteremia) but reflected a higher frequency of common postoperative infections (especially pneumonia and wound infections). Likewise, differences in the type of operations performed did not readily explain this observation, with procedures usually associated with increased risk of infection (colon procedures) being performed with somewhat greater frequency in control patients.

It is unclear whether the TPN was causally related to this increased rate of infection. TPN-treated patients were hospitalized an average of 5 days longer prior to operation than controls, perhaps providing greater opportunity for colonization with resistant pathogens. Alternatively, TPN per se may have in some way predisposed to clinically apparent infection as a result of the route of feeding, the energy source, the quality of calories provided, or the decreased oral intake associated with parenteral feeding. A possible role of lipid in this process was suggested by Muller et al.,[40] as described previously in this chapter.

In contrast, noninfectious complications were less common in TPN-treated patients. These noninfectious complications were primarily those that reflect the ability to heal wounds and maintain normal organ function, with the largest numerical differences between groups occurring in complications associated with failure to heal wounds (anastomotic leaks and bronchopleural fistulae). The response to TPN (beneficial or deleterious) was dependent on the baseline nutritional status of the patient. The increased rate of infections was confined to patients categorized as well nourished or mildly malnourished by Subjective Global Assessment (SGA)[49] or objective nutritional assessment (based on serum albumin and weight loss), and these patients experienced no demonstrable benefit from preoperative TPN. Severely malnourished patients experienced significantly fewer noninfectious complications with TPN (5% vs 43%; $p = 0.03$) with a reduction in overall major complications (21% vs 47%) approaching statistical significance ($p = 0.11$).

The divergence of the findings of this trial depending on the baseline nutritional status of the patient was consistent with the pooled results of earlier trials as presented in the meta-analysis previously described.[46] Furthermore, it provided new information clarifying the utility of TPN in patients with moderate to severe degrees of malnutrition. This trial confirmed the lack of benefit of TPN in minimally malnourished patients, provided strong evidence against clinically important efficacy in mildly or moderately malnourished patients, and suggested but did not confirm efficacy in severely malnourished patients. In the VA study, the severely malnourished population was small, representing approximately 5% of patients undergoing major laparotomy or noncardiac thoracotomy after excluding patients with major concurrent illnesses and patients with clear-cut indications for preoperative TPN (3% of surgical population). In the absence of severe malnutrition or other specific indications for preoperative TPN, this study would indicate that most patients are best served by prompt operation.

In the Maastricht trial reported by von Meyenfeldt et al.,[48] 150 nutritionally depleted patients with gastric or colorectal cancer were enrolled in the study. Fifty nondepleted patients with the same malignancies served as a reference group. Nutritional assessment was performed using a statistically derived nutritional index (NI) which had been developed and validated in a previous publication.[50] The equation for NI is given below:

$$\text{NI} = 0.14\ (\text{ALB}) + 0.3\ (\text{PIW}) + 73\ (\text{TLC}) - 8.9,$$

where ALB is serum albumin (g/dL), PIW is the percent of ideal body weight; and TLC is total lymphocyte count (cells/cc). An NI value of less than 1.31 was considered evidence of nutritional depletion. After stratification for percent weight loss, age, and tumor site, patients were randomly assigned to receive preoperative TPN for at least 10 days, to receive preoperative total enteral nutrition (TEN) for at least 10 days, or to undergo surgery without delay. The reference group of 50 nondepleted patients also underwent surgery without delay. When the number and type of complications and the mortality rates in each group were compared, no significant differences could be detected. However, stratification for weight loss greater than 10% allowed for subset analysis of the patient group displaying the most severe depletion. This analysis demonstrated a significant decrease in the number of patients developing intraabdominal sepsis in both the TPN and total enteral nutrition (TEN) groups. The difference between complication rates in the two groups receiving nutritional support compared to the untreated control group became more pronounced in the subset of patients suffering major blood loss (> 500 cc) during the operation.

The observation that nutritional support reduced complications in severely depleted patients was consistent with the results of the VA trial that had approached but did not reach statistical significance, perhaps due to the small number of severely depleted patients in the VA study. The statistical significance reached by the Maastricht study in contrast to the VA trial might also be explained by the fact that a more homogenous patient population was operated on by a small number of surgeons in a single hospital with a lower mortality rate (5% vs 10%) in the Maastricht trial.[51]

Both trials support prompt operation in all but the most severely depleted patients (> 10% weight loss). In contrast, severely malnourished patients appear to benefit from 7 to 10 days of preoperative TPN or TEN.

PERIOPERATIVE NUTRITIONAL SUPPORT TO PREVENT NUTRITIONAL DEPLETION

Many surgical patients cannot eat or are not permitted to eat during the days immediately before and after surgery. Given the high morbidity and mortality associated with malnutrition, it is unacceptable to permit malnutrition to develop or progress during preoperative starvation in a patient who is not a candidate for immediate surgery. Prolonged postoperative starvation is equally damaging. Definitive studies assessing how long a surgical patient can starve before significant deficiencies develop are lacking. However, some insight can be derived from starvation studies in otherwise healthy young adults.

Thirty nine individuals undergoing hunger strikes lost a mean of 38% of their body weight in 60 days with a resulting 33% mortality rate. (Figure 6.1) During the first 10 days of starvation, the average weight loss was 7%, and by 15 days it was 10%.[52] This rate of weight loss is consistent with that reported by Keys in 1950 in his now classic publication.[53] Keys described profound physical and mental changes associated with more than 10% to 15% weight loss. In a more recent study, Kinney demonstrated the relationship between weight loss and reduction in organ function, immune status, wound healing, and muscle strength. He concluded that clinically significant changes begin to appear with rapid weight loss in association with disease when the degree of weight loss is between 5% and 10%.[54] These deficits develop more rapidly in stressed and/or catabolic patients.

It is difficult to set an upper limit on the duration of perioperative starvation that is acceptable. Guidelines developed by the American Society for Parenteral and Enteral Nutrition in 1986[55] and updated in 1993[56] suggest a maximum of 7 days of severely limited nutrient intake for hospitalized patients in general. For patients undergoing a major operation that is likely to be followed by a period of mandatory postoperative starvation, a reasonable upper limit for an acceptable duration of preoperative starvation is 3 to 4 days in a stressed patient or 4 to 5 days in an unstressed patient. It is reasonable to provide preoperative nutritional support in those circumstances when the duration of preoperative starvation is likely to exceed these levels. The relative benefits of enteral versus parenteral feeding are discussed in greater detail later in this chapter. Suffice it to say here

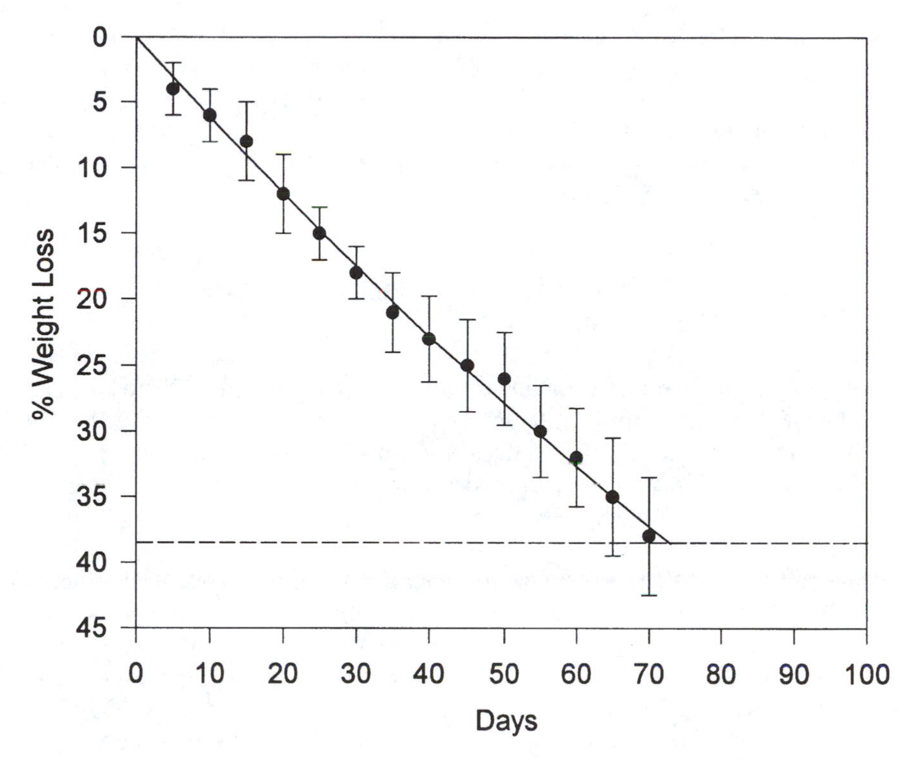

Figure 6.1. Mean weight loss with complete starvation in 30 hunger strikers. Mortality was 33% with a 38% weight loss at 70 days. Data from Allison.[52]

that the enteral route is in general preferable during the preoperative period if the patient can tolerate enteral feedings.

During the postoperative period, feeding should be initiated if the total duration of starvation (preoperative plus postoperative) is likely to exceed 7 days. In patients who are catabolic or severely stressed or who exhibit preexisting malnutrition, nutritional support should be considered earlier, probably if postoperative starvation will exceed 4 or 5 days. In those situations where postoperative nutritional support is anticipated, enteral access should be established at the time of the initial operation, and feeding should be instituted within 3 days if possible. If a feeding tube has been placed at the time of surgery but the gastrointestinal tract is temporarily not functioning adequately in the early postoperative period, it is perfectly acceptable to initiate TPN with a gradual transition to enteral feeding as the gastrointestinal function returns.

This approach is consistent with the ASPEN practice guidelines[56] and also with guidelines published from the TPN Technology Assessment and Practice Guidelines Forum held at Georgetown University and supported by the U.S.

Department of Health and Human Services.[57] Although this forum did not provide specific recommendations for the use of postoperative TPN, it did consider TPN use in critically ill patients. The forum recognized the lack of definitive data in this area but did conclude that TPN may be of benefit in critically ill hypermetabolic patients who were previously well nourished who will not attain adequate enteral intake within 5–7 days.

COST AND COST-EFFECTIVENESS OF PERIOPERATIVE TOTAL PARENTERAL NUTRITION

Until recently, a clear understanding of what it costs to provide TPN to hospitalized patients has been difficult to achieve. Early estimates of the daily costs of TPN ranged between $75 and $800 (based on the value of the dollar in the 1980s). Without a reasonably precise determination of the cost to provide TPN, an assessment of cost-effectiveness was not possible.

The issue of TPN cost and cost-effectiveness was addressed in detail in an economic analysis done in conjunction with the VA Cooperative TPN efficacy trial.[58] In this analysis, costs were determined by detailed cost-finding techniques including time and motion studies, chart reviews, and analysis of hospital accounting records. This type of analysis was a substantial improvement over earlier studies, which were usually based on hospital charges. In the VA study, the increased cost associated with providing perioperative TPN to malnourished patients was $2405 when compared to the cost to treat control patients with similar illnesses and undergoing similar operations but not receiving TPN. Because the TPN patients were hospitalized approximately 4 days longer prior to operation than control patients, the overall increased cost was $3169 when the cost of the longer hospital stay was included (or an additional $764; all dollar amounts based on early-1980s dollar valuations).

The VA study, as previously noted, only demonstrated efficacy for perioperative TPN in the most severely malnourished patients. In these patients, TPN reduced the number of complications. If one considers the magnitude of reduction in complication rates as well as the incremental cost associated with total parenteral nutrition, a reasonable estimate of the cost effectiveness of perioperative TPN is approximately $14,000 per major complication avoided. In more meaningful terms, it costs $14,000 to avoid a major complication by providing TPN to severely malnourished patients who require major operations (one complication was avoided for every 4 to 5 severely malnourished patients receiving TPN). This does not consider, however, the potential cost savings achieved by avoiding treatment of these complications. For many serious complications requiring operative and/or intensive medical interventions, this may be quite substantial and may exceed the cost associated with avoiding the complication initially. In these circumstances TPN use would actually reduce costs. The $14,000 estimate per complication avoided should therefore be viewed as an upper limit and compares favorably with the economic impact of other medical interventions where these have been determined with reasonable accuracy.

Although our knowledge of the cost and cost effectiveness of perioperative TPN remains imperfect, it has clearly been subjected to greater scrutiny and is better understood than are the costs and cost-effectiveness of many other complex treatment modalities that are commonly employed in clinical care.

COMPARISON OF ENTERAL AND PARENTERAL FEEDINGS

There is a general consensus that if the gastrointestinal tract is functional, enteral feedings are preferred to parenteral feedings. The relative risks and benefits of the two routes nearly always favor enteral feeds when a functional gut is accessible and nutrients are delivered skillfully and carefully. Patients fed enterally benefit from stimulation of gut trophic hormones, improved immune function, reduced infection rates, and better tolerance of septic insults when compared to parenterally fed patients.[59]

In the preoperative period, patients who are able and willing to take food by mouth should be encouraged to do so with a nutrient goal of at least 1.3 to 1.5 times the resting energy expenditure. When voluntary oral intake cannot achieve this goal, it is unclear whether supplemental nutrition via the enteral or parenteral route is superior. It is not known whether the clear advantages of enteral feeding over strict parenteral feeding (with nothing by mouth) including maintenance of gut integrity), less translocation, better immune function, and so on, are also a factor when forced enteral feeding is used to supplement suboptimal oral intake. Most studies comparing preoperative enteral and parenteral nutrition have been limited by small sample size. Studies by Lim et al.[37] in patients with carcinoma of the esophagus and Sako et al.[39] in patients with head and neck cancer suggested a minimally superior efficacy for TPN in terms of improving nutritional status but failed to show any differences in clinical outcome between the two feeding modalities.

Complication rates are generally considered to be lower for enterally fed patients. However, patients receiving forced enteral feeding preoperatively usually require nasoenteric tubes, which can result in aspiration. This catastrophic complication can be minimized by careful attention to postpyloric placement of the feeding tube and diligent patient monitoring for signs of intolerance. Based on currently available data, physician preference and patient comfort largely determine which route is clinically utilized to supplement oral intake during the preoperative period.

In the postoperative period, enteral feeding is clearly superior to the parenteral route. Patient discomfort and the risk of aspiration, which are major concerns during preoperative forced enteral feeding, are largely avoided by the placement of a feeding jejunostomy at the time of surgery.

A recent meta-analysis[60] attempts to deal with the issue of the limited sample size and inadequate statistical power of previously reported studies comparing postoperative enteral and parenteral feeding by pooling data from eight prospective randomized clinical trials comparing postoperative enteral and parenteral nutritional support in high-risk surgical patients. This study is subject to the same criticisms as any meta-analysis and is of further concern because six of the eight trials included in the analysis had not been previously published. Nevertheless, patients fed enterally experienced fewer complications than those fed parenterally (41% vs 52%) although this difference did not achieve statistical significance ($p = 0.09$). When only infectious complications were considered, twice as many parenterally fed patients experienced complications when compared to patients fed enterally (16% vs 35%, $p < 0.05$). This increased incidence of infectious complications in TPN-fed patients may be similar to the higher rate of infections in TPN-treated patients compared to controls seen in the VA TPN trial.[47]

Postoperative enteral feedings, however, are associated with specific problems. Most importantly, as many as a third of patients are intolerant to feedings at some point in the immediate postoperative period. This problem is usually manifest by diarrhea or abdominal distention.[61] Intolerance to forced enteral feeding is seen most commonly in the severely traumatized patient and in patients with uncontrolled sepsis.

There is some suggestive evidence that early postoperative jejunal feedings may be linked to small bowel nonocclusive ischemia. Gaddy[62] has reported five patients with extensive small bowel gangrene, occurring 4 to 15 days after initiating enteral feedings. All of these patients expired. Smith-Choban[63] also described five patients with small bowel necrosis developing during jejunal feedings with no evidence for any mechanical or primary vascular compromise. At our institution, we have observed an additional four patients developing a similar clinical picture.[64] Although no causal relationship has been established between nonocclusive ischemia and early jejunal feedings, one must be aware of this potential relationship. This syndrome usually presents with relatively painless abdominal distention and obstipation. Plane films of the abdomen may reveal evidence of inspissated tube feeds within the small bowel. Pneumatosis is a late finding. If there is no evidence clinically or radiographically of perforation or ischemia, these patients should be treated initially with bowel rest and TPN. However, clinicians must maintain a high index of suspicion that transmural necrosis may be present and respond expeditiously to clinical deterioration with operative intervention.

Although it is well documented that early postoperative enteral feedings can be done with safety, it is clear that patients receiving such therapy should be monitored closely. The excellent results reported in published clinical series may reflect the intense monitoring of these patients dictated by study protocols. In clinical practice, similar intense monitoring is required to insure comparable safety.

SUMMARY AND GUIDELINES

One can argue that the efficacy, cost, and cost-effectiveness of perioperative nutritional support has been subjected to as intensive an analysis as any currently available major treatment modality for surgical patients. Although some experts in the field of nutritional support argue for further studies to evaluate the efficacy of currently available parenteral and enteral products in this setting,[65] it is unclear whether such studies will be or should be undertaken pending more basic scientific research of nutritional pharmacology and modulators of nutrient utilization in specific disease states. The data available regarding perioperative nutritional support with clinically available enteral and parenteral nutrient prescriptions is extensive, although admittedly, imperfect. Its analysis allows for the development of reasonable clinical guidelines.

The following recommendations are proposed:

1. Generally, nutritional deficits should not be allowed to develop in previously well-nourished patients who are being prepared for surgery or who have recently undergone surgery. While this point seems obvious, maintenance of nutritional status is often neglected in perioperative patients.
2. Preoperative nutritional counseling and/or support should be considered for patients whose operation must be delayed for more than 3 to 5 days. The goal is to prevent depletion of well-nourished patients or to prevent further deterioration of the malnourished patient. The oral route including supplements is preferred in this group. Patients should be encouraged to eat as much as possible even if adequate amounts cannot be achieved (1.3–1.5 times the resting energy expenditure). Oral intake can be supplemented by forced enteral or parenteral feeding depending on patient tolerance and clinical preference.
3. Preoperative enteral or parenteral nutritional support should be considered in the most severely malnourished patients (> 10% weight loss) if a delay in operation is not contraindicated. Data suggest that a period of at least 7 to 10 days of feeding is required to produce any beneficial effect. By limiting preoperative TPN to the most severely depleted patients, the cost-effectiveness of this treatment modality is maximized. Appropriate use of preoperative nutritional support to replete severely malnourished patients may actually reduce overall costs in this population by reducing the need to treat major complications.

4. Postoperative nutritional support should be considered when oral feedings are not anticipated within 7 to 10 days in a previously well-nourished patient or within 5 to 7 days in a previously malnourished or critically ill patient. The enteral route is preferred to maintain gut integrity and reduce septic complications. Patients who receive early postoperative enteral feedings must be monitored closely for any evidence of feeding intolerance.

REFERENCES

1. Mequid MM, Compose AC, Hammond WA. Nutritional support in clinical practice. *Am J Surg* 1990;159:345–358.
2. Studley HO. Percentage of weight loss: a basic indicator of surgical risk in patients with chronic peptic ulcer disease. JAMA 1936;106:458–560.
3. Rhode CM, Parkins W, Tourtellote D, Vars HM. Method for continuous intravenous administration of Nutritive Solution suitable for prolonged metabolic studies in dogs. *Am J Physiol* 1949;159:409–414.
4. Wretlind A. Free amino acides in dialyzed casein digest. *Acta Physiol Scand* 1947;13:45–54.
5. Schubert O, Wretlind A. Intravenous infusion of fat emulsion, phosphatides and emulsifying agents. *Acta Clin Scand* 1961;278 (suppl):1–21.
6. Dudrick SJ, Wilmore DW, Vars HM, Rhoads JE. Longterm total parenteral nutrition with growth, development and positive nitrogen balance. *Surgery* 1965;64:134–141.
7. Wilmore DW, Dudrick JJ. Growth and development of an infant receiving all nutrients exclusively by vein. *JAMA* 1968;203:860–864.
8. Hill GL, Blackett RL, Pickford I, et al. Malnutrition in surgical patients. An unrecognized problem. *Lancet* 1977;1:689–692.
9. Seltzer MH, Bastidas JA, Cooper DM, et al. Instant nutritional assessment. *J Parent Ent Nutr* 1979;3:157–159.
10. Reinhardt GF, Myscofski JW, Wilkens DB, et al. Incidence and mortality of hypoalbuminemic patients in hospitalized veterans. *J Parent Ent Nutr* 1980;4:357–359.
11. Bistrian BR, Blackburn GI, Hallowell E, Heddle R. Protein status of general surgical patients. *JAMA* 1974;230:858–860.
12. Young GA, Hill GL. Assessment of protein-calorie malnutrition in surgical patients from plasma proteins and anthropometric measures. *Am J Clin Nutr* 1978;31:429–435.
13. Kaminski MV, Fitzgerald MJ, Murphy RJ. Correlation of mortality with serum transferrin and anergy. *J Parent Ent Nutr* 1977,1:27.
14. Blackburn GL, Bistrian BR, Harvey K. Indices of protein-calorie malnutrition as predictors of survival. In SM Levenson, ed. *Nutritional assessment—present status. Future directions and prospects.* Columbus, OH: Ross Laboratories, 1991:131–137.

15. Buzby GP, Foster J, Rosato EF, Mullen JL. Transferrin dynamics in total parenteral nutrition. *J Parent Ent Nutr* 1979,3:34.
16. Mullen JL, Gertner MH, Buzby GP, et al. Implications of malnutrition in the surgical patient. *Arch Surg* 1979;114:121–125.
17. Hickman DM, Miller RA, Rombeau JL, et al. Serum albumin and body weight as predictors of postoperative course in colorectal cancer. *J Parent Ent Nutr* 1980;4:314–316.
18. MacLean LD, Meakins JL, Taguchi K et al. Host resistance in sepsis and trauma. *Ann Surg* 1975;182:207–217.
19. Meakins JL, Pietsch JB, Bubenick O. Delayed hypersensitivity: indicator of acquired failure of host defenses in sepsis and trauma. *Ann Surg* 1977;186:241–250.
20. Harvey KB, Moldawer BS, Bistrian BR. Biologic measures for the formation of a hospital prognostic index. *Am J Clin Nutr* 1981;34:2013–2022.
21. Mullen JL, Buzby GP, Waldman TF, et al. Prediction of operative morbidity and mortality by preoperative nutritional assessment. *Surg Forum* 1979;30:80–82.
22. Buzby GP, Mullen JL, Matthews DC, et al. Prognostic nutritional index in gastrointestinal surgery. *Am J Surg* 1980;139:160–167.
23. Freund H, Atamian S, Holroyde J. Plasma amino acids as predictors of the severity and outcome of sepsis. Ann Surg 1979;190:571–576.
24. Shizgal H. Body composition and nutritional support. Surg Clin North Am 1981;6:729–741.
25. Dempsey DT, Mullen JL, Buzby GP. The link between nutritional status and clinical outcome: can nutritional intervention modify it? *Am J Clin Nutr* 1988;47:352–356.
26. Mullen JL, Buzby GP, Matthews DC, et al. Reduction of operative morbidity and mortality by combined preoperative and postoperative nutritional support. *Ann Surg* 1980;192:604–613.
27. Smale BF, Mullen JL, Buzby GP, et al. The efficacy of nutritional assessment and support in cancer surgery. *Cancer* 1981;47:2375–2381.
28. Copeland EM, Daly JM, Dudrick SJ. Nutrition as an adjunct to cancer treatment in the adult. *Cancer Res* 1977:37:2451–2456.
29. Daly JM, Massar E, Giacco G. Parenteral nutrition in esophageal cancer patients. *Ann Surg* 1982;196:203–208.
30. Grimes CJ, Younathan MT, Lee WC. The effect of preoperative total parenteral nutrition on surgery outcome. *Research* 1987;87:1202–1206.
31. Muller JM, Brenner U, Dienst C, Pichlmaier H. Preoperative parenteral feeding in patients with gastrointestinal carcinoma. *Lancet* 1982;1:68–71.
32. Heatley RV, Williams RHP, Lewis MH. Preoperative intravenous feeding—a controlled trial. *Postgrad Med J* 1979;55:541–545.
33. Bellantone R, Doglietto GB, Bossola M, et al. Preoperative parenteral nutrition in high risk surgical patients. *J Parent Ent Nutr* 1988;12:195–197.
34. Moghissi K, Hornsaw J, Teasdale PR, Dawes EA. Parenteral nutrition in carcinoma of the oesophagus treated by surgery: nitrogen balance and clinical studies. *Br J Surg* 1977;64:125–128.

35. Holter AR, Fischer JE. The effects of perioperative hyperalimentation on complications in patients with carcinoma and weight loss. *J Surg Res* 1977;23:31–34.
36. Preshaw RM, Attisha RP, Hollingworth WF. Randomized sequential trial of parenteral nutrition in healing of colonic anastomoses in man. *Can J Surg* 1979;22:437–439.
37. Lim STK, Chou RG, Lam KH, et al. Total parenteral nutrition versus gastrostomy in the preoperative preparation of patients with carcinoma of the oesophagus. *Br J Surg* 1981;68:69–72.
38. Thompson BR, Julian TB, Stremple JF. Perioperative total parenteral nutrition in patients with gastrointestinal cancer. *J Surg Res* 1981;30:497–500.
39. Sako K, Lore JM, Kaufman S, et al. Parenteral hyperalimentation in surgical patients with head and neck cancer: a randomized study. *J Surg Oncol* 1981;16:391–402.
40. Muller JM, Keller HW, Brenner U, et al. Indications and effects of preoperative parenteral nutrition. *World J Surg* 1986;10:53–63.
41. Simms JM, Oliver E, Smith JAR. A study of total parenteral nutrition (TPN) in major gastric and esophageal resection for neoplasia. *J Parent Ent Nutr* 1980;422.
42. Jensen S. Parenteral nutrition and cancer surgery. *J Parent Ent Nutr* 1982;6:335.
43. Moghissi M, Teasdale P, Dench M. Comparison between preoperative nasogastric feeding and parenteral feeding in patients with cancer of the esophagus undergoing surgery. *J Parent Ent Nutr* 1982;6:335.
44. Schildt B., Groth O., Larsson J, et al. Failure of preoperative TPN to improve nutritional status in gastric carcinoma. *J. Parent. Ent. Nutr* 1981, 360–365.
45. Buzby GP, Williford WO, Peterson OL, et al. A randomized clinical trial of total parenteral nutrition in malnourished surgical patients: the rationale and impact of previous clinical trials and pilot study on protocol design. *Am J Clin Nutr* 1988;47:357–365.
46. Detsky AS, Baker JP, O'Rourke K, Goel V. Perioperative parenteral nutrition: a meta-analysis. *Ann Intern Med* 1987;107:195–203.
47. Buzby GP, Blouin G, Colling CL, et al. Perioperative total parenteral nutrition in surgical patients. *N Eng J Med* 1991;325:525–532.
48. Von Meyenfeldt MF, Meijeriuk WJHJ, Rouflart MMJ, et al. Perioperative nutritional support: a randomized trial. *Clin Nutr* 1992;11:180–186.
49. Detsky AS, McLaughlin JR, Baker JP. What is subjective global assessment of nutritional status? *J Parent Ent Nutr* 1987;11:8–13.
50. deJong PCM, Wesdrop RIC, Volvies, et al. The value of objective measurements to select patients who are malnourished. *Clin Nutr* 1985;4:61–66.
51. Meijerink WJ, Von Meyenfeldt MF, Rouflart MMJ, Soeters PB. The efficacy of perioperative nutritional support. *Lancet* 1992;340:187–188.
52. Allison SP. The uses and limitations of nutritional support. *Clin Nutr* 1992;11:319–330.
53. Keys A, Brozek J, Henschel A. *The biology of human starvation.* Minneapolis: Minneapolis University Press, 1950.

54. Kinney JM. The influence of calorie and nitrogen balance on weight loss. *Brit J Clin Pract Symp* 1988;63:114–120.
55. ASPEN Board of Directors. Guidelines for the use of parenteral nutrition in hospitalized patients. *J Parent Ent Nutr* 1986;10:441–445.
56. ASPEN Board of Directors. Guidelines for the use of parenteral and enteral nutrition in adult and pediatric patients. *J Parent Ent Nutr* 1993;17 (suppl):1SA–52SA.
57. Pillar B, Perry S. Evaluating total parenteral nutrition: final report and statement of the technology assessment and practice guidelines forum. *Nutrition* 1990;6:314–318.
58. Eisenberg JM, Glick HA, Buzby GP, et al. Does perioperative total parenteral nutrition reduce medical costs? *J Parent Ent Nutr* 1993;17:201–209.
59. Kudsk, KA, Croce MA, Fabian MC, et al. Enteral vs parenteral feeding. Effects on septic morbidity after blunt and penetrating abdominal trauma. *Ann Surg* 1992;215:503–510.
60. Moore FA, Feliciano DV, Andrassy RJ, et al. Early enteral feeding, compared with parenteral, reduces postoperative septic complications. The results of a meta-analysis. *Ann Surg* 1992;216:172–183.
61. Jones TN, Moore FA, Moore EE, et al. Gastrointestinal symptoms attributed to jejunostomy feeding after major abdominal trauma—a critical analysis. *Crit Care Med* 1989;17:1146–1150.
62. Gaddy MC, Max MH, Schwab CW, et al. Small bowel ischemia: a consequence of feeding jejunostomy. *South Med J* 1986;79:180–182.
63. Smith-Choban P, Max MH. Feeding jejunostomy: a small bowel stress test? *Am J Surg* 1988;155:112–116.
64. Schunn CD, Daly JM. Small bowel necrosis—associated with postoperative jejunal tube feeding. *J Am Coll Surg* 1995;80:410–416.
65. Campos AC, Meguid MM. A critical appraisal of the usefulness of perioperative nutritional support. *Am J Clin Nutr* 1992;55:117–130.

PARENTERAL NUTRITION

CHAPTER 7

Indications for Parenteral Nutrition

Carl T. Bergren, M.D.

INTRODUCTION

Since the late 1960s when Dudrick first demonstrated the efficacy of long-term parenteral nutrition,[1] the indications for this therapy have expanded as our knowledge has increased. Better technology has led to improved administration and reduction of complications, which has enabled this service to be applied beyond the hospital setting. Originally developed for the pediatric population with short gut and failure to thrive, total parenteral nutrition (TPN) is now in widespread use with renal dialysis, cancer therapy, and inflammatory bowel disease. This chapter reviews the general and specific indications for the provision of parenteral nutritional support.

GENERAL CONSIDERATIONS

The first indication for TPN is the interruption of oral intake (see Table 7.1). The human body is designed to withstand the loss of oral alimentation for a

Table 7.1. General Considerations for the Use of TPN

1. Interruption of oral alimentation for greater than 7–10 days (less than 7 days for patients with compromised nutritional stores and/or severe catabolic illness.)
2. The inability to utilize the gastrointestinal tract to provide nutritional support.
3. The ability to establish stable intravenous access.
4. The ability to formulate an appropriate TPN solution.

period of time without adverse consequences. It is generally believed that a healthy person with normal nutritional reserves can easily undergo about 7 to 10 days of fasting.[2] This interval of time is obviously shorter for patients with preexisting malnutrition. Ongoing inadequate intake can lead to a multitude of complications. It is therefore prudent to initiate nutritional support prior to the development of malnutrition-related complications.

One of the major considerations in determining the appropriateness of parenteral nutrition is the status of the function of the gut. The functioning gastrointestinal tract should be utilized, if at all accessible, prior to institution of parenteral therapy. The enteral route is safer, more physiologic, cheaper, and appears to improve the immune function of the patient.[3] However, if the gut is not capable of adequate absorption, or an appropriate administration route cannot be established, then parenteral nutrition can supplement or replace enteral feedings.

Other considerations before parenteral nutrition can be established are the site of access. Most parenteral nutrition is administered using a percutaneous central venous access via the subclavian or jugular veins. Short-term administration can often suffice with a single or multiple lumen catheter placed at the bedside. Other catheters are available for long-term administration but usually require a more specialized procedure room and fluoroscopy for insertion. Occasionally, peripheral venous administration can be given. This method, however, is limited in the concentration of nutrients that can be safely administered and is often reserved as a supplement for oral or enteral intake.

Formulation of an appropriate solution requires careful consideration especially in the initiation stage of TPN. Changes in electrolyte, mineral, and glucose levels occur on a daily basis, particularly in critically ill patients and careful monitoring by a knowledgeable staff is essential. Individuals with various disease processes also require special consideration since renal, hepatic, pulmonary, and endocrine dysfunction all create unique problems with parenteral nutrition.

Finally, cost and length of therapy need to be considered. Patients who can be expected to have an adequate oral or enteral intake within 7 days may not need parenteral nutrition. Total parenteral nutrition is an expensive therapy, and, in patients with situations where other therapies have been withheld or withdrawn, parenteral nutrition should not be instituted or continued.

SPECIFIC INDICATIONS

Short-Gut Syndrome

Short-gut syndrome in pediatric patients was one of the first indications for use of this newly developed technology by Dudrick.[4] Extensive intestinal loss in the pediatric population may result from intestinal atresias, malrotation with a midgut volvulus, necrotizing enterocolitis, or complicated meconium ileus. Other newborn abnormalities can also lead to intestinal loss and include gastroschisis, omphalocele, and cloacal exstrophy.

In the adult, short gut is usually a result of midgut volvulus, multiple surgical resections from Crohn's disease, or a mesenteric vascular catastrophe. Fortunately, many patients will be able to resume near-normal gut function over time if an intact pylorus and ileocecal valve are present. Sections of bowel as short as 30.5 cm of jejunum have been reported to eventually support adequate enteral functions.[3] Without an ileocecal valve, however, an estimated 100 cm of bowel is necessary for adequate support. This absorptive function may take 2 to 3 years to become established, and parenteral nutrition can support the patient in combination with enteral feedings during this time.

Patients with more than 80% of bowel removed often require lifelong support. In the pediatric population, this long-term support can lead to hepatic dysfunction from cholestasis. This is a particularly serious consequence, which may lead to liver failure and death.[5] Other problems that can be encountered in all patients with short-gut syndrome include deficiencies in fat-soluble vitamins, diarrhea, and cholelithiasis.

Perioperative Nutrition Support

It is well known that no disease state benefits from starvation. Morbidity and mortality have been shown to increase in the malnourished surgical patient.[6] Other studies have shown that up to 40% of hospitalized patients are malnourished, often because of preexisting lifestyles or disease process.[7] The VA TPN Cooperative Study Group evaluated the hypothesis that perioperative parenteral nutrition decreases the incidence of complications after major surgery.[8] They evaluated 395 patients and assigned them either to 7 to 15 days of TPN preoperatively and 3 days afterward or no TPN. Results showed no difference in complication rates between the groups with even more infectious complications in the TPN group. However, when only severely malnourished patients were evaluated, the results favored the use of preoperative TPN. In this study, severely malnourished patients were defined as those patients who had lost at least 15% of premorbid weight or had an albumin level of 2.8 g% or less. In this group, noninfectious complications were noted to be significantly decreased from controls. The conclusions of this study were that preoperative TPN should be limited to severely malnourished patients unless there were specific other indications.

Postoperatively, most healthy patients can tolerate 7 to 10 days of no feeding without undue consequences.[2] Patients with borderline nutritional status and a possible prolonged interval prior to adequate enteral intake may need parenteral supplements for a short period of time. However, parenteral nutrition for less than 7 days is unlikely to be beneficial.[9]

Enterocutaneous Fistulae

Enterocutaneous fistulae are abnormal communications between the lumen of the small intestine and the skin. In most cases, oral alimentation must be discontinued. Significant losses of fluid and electrolytes may occur, and, if persistent, may decrease the likelihood of the fistula closing spontaneously. In addition, if the fistula is located in the proximal small intestine, oral intake may be ineffective for maintaining a healthy nutritional status.

Although the overall mortality associated with fistulae has not decreased with the use of TPN, the rate of spontaneous closure has improved with this modality.[10] Most fistulae will close with the help of TPN if malignancy, infection, distal obstruction, foreign body, end fistulae, or diseased bowel are not involved. In those patients who ultimately require surgery, the addition of parenteral nutrition has helped to improve the overall protein and caloric status of the individual.

Inflammatory Bowel Disease

Inflammatory bowel disease is one of the most common indications for use of parenteral nutrition.[11] In Crohn's disease, acute exacerbations may respond to conservative treatment of bowel rest, TPN, and antibiotics. After surgical resections for Crohn's disease with resulting short gut, home therapy with parenteral nutrition may be indicated until the enteral absorption has improved sufficiently to allow for oral intake. In pediatric and adolescent patients, parenteral nutrition may be indicated for failure to thrive. Dramatic responses in growth and development have occurred after the institution of TPN. Fistulae may close with the use of TPN, yet they tend to reopen, so parenteral nutrition is often an adjuvant to surgical treatment.[3] Ulcerative colitis usually fails to respond to parenteral nutrition but TPN is useful in the perioperative period and for supporting a patient's nutritional status during the acute exacerbation.

Miscellaneous Bowel Disorders

Patients who are undergoing radiation or chemotherapy may suffer debilitating diarrhea, nausea, or vomiting from generalized enteritis. TPN may help alleviate some of these symptoms and support the patient during the acute phase of their therapy. One study has even suggested that home TPN can improve tolerance to chemotherapy.[12] Occasionally, TPN may be used in patients with severe stomatitis, but the enteral tract should be used whenever feasible. In the setting of chronic intestinal malabsorption secondary to end-stage radiation enteritis, TPN may represent the only means of nutritional support.

Patients with acquired immunodeficiency syndrome (AIDS) also may suffer from severe enteritis, which may respond to TPN. Home therapy may provide hydration as well as nutrition but should only be initiated if life expectancy is anticipated to exceed 6 months.

Other systemic and intestinal disease processes may result in malabsorption syndromes secondary to inflammation and ischemia. Parenteral nutrition may help in both short- and long-term situations to support the patient until gut function is adequate. Several of these processes are listed in Table 7.2.

Renal Failure

Patients with renal failure on dialysis present unique problems with nutrition. Electrolytes and fluid shifts occur on a daily basis. Fluid restriction may be required or hypovolemia may result from poor intake from anorexia or vomiting secondary to uremia. These problems may require increased dextrose concentrations. Protein losses through the dialysate leads to continued catabolic state. The malnutrition resulting from the poor nutritional intake and the chronic catabolic state may require TPN in the setting of intestinal dysfunction. The addition of essential amino acids may be necessary, especially in postoperative renal failure.[13] In patients with chronic renal failure, standard amino acid solutions are adequate; however, reduction in total protein concentration may be needed and adjusted to the blood urea nitrogen level.

Hepatic Disease

Like patients with renal disease, those with hepatic insufficiency tend to be malnourished from the disease itself and from the accompanying poor oral intake.

Table 7.2. Causes of Intestinal Malabsorption

Scleroderma
Systemic Lupus Erythematosis
Collagen Vascular Diseases
Sprue
Pancreatitis
Intestinal Ischemia
Intestinal Pseudoobstruction
Radiation Enteritis

When enteral nutrition is contraindicated in this population, TPN may be necessary. Patients in hepatic failure have alterations that result in decreased clearance of the aromatic and sulfur-containing amino acids (phenylalanine and methionine). The build up of trytophan and aromatic amino acids that cross the blood–brain barrier may be the etiology of hepatic encephalopathy. Formulations of TPN with branched-chain amino acids (BCAA), leucine, isoleucine, and valine require little hepatic metabolism and have been used in the treatment of hepatic insufficiency. The use of BCAA formulations in patients with encephalopathy has been shown to cause some improvement in wake up time.[14] However, the use of these expensive formulas in prophylactic treatment of hepatic encephalopathy has not been shown to be effective.

MISCELLANEOUS CONDITIONS

Occasionally during pregnancy, hyperemesis gravidarum is so severe as to be deleterious to the fetus and the mother. Short-term parenteral therapy may be helpful in this situation and result in improved fetal growth. In mothers who were malnourished prior to pregnancy, the use of parenteral nutrition is indicated to restore the mother to nutritional health.

Liver and bone marrow transplantation patients may have short periods of time when parenteral nutrition is required. These episodes may be in response to graft-versus-host reaction or secondary to immunosuppressive therapy that results in severe nausea and/or diarrhea.

CONCLUSION

As improved administration and increased knowledge is gained of parenteral nutrition, the indications for its use will undoubtedly grow. Consideration should always be given first to the use of the enteral tract. Meticulous care and attention to details by skilled professionals helps to decrease complications. Careful patient selection helps to make the most of this expensive therapy.

REFERENCES

1. Dudrick SJ, Wilmore DW, Vars HM, et al. Long term parenteral nutrition with growth and development and positive nitrogen balance. *Surgery* 1968;64:134–142.
2. Buzby GP. Overview of randomized clinical trials of total parenteral nutrition for malnourished surgical patients. *World J Surg* 1993;17:173–177.

3. Sax HC, Souba WW. Enteral and parenteral feedings: guidelines and recommendations. *Med Clin North Am* 1993;77:863–880.
4. Wilmore D, Dudrick S. Growth and development of an infant receiving all nutrients exclusively by vein. *JAMA* 1968;203:860–864.
5. Taylor L, O'Neill, Jr., JA. Total parenteral nutrition in the pediatric patient. *Surg Clin North Am* 1991;71:477–492.
6. Dempsey DT, Mullen JL, Buzby GP. The link between nutritional status and clinical outcome: can nutritional intervention modify it? *Am J Clin Nutr* 1988;47:352–356.
7. Pinchcofsky GD, Kaminski MV. Increasing malnutrition during hospitalization: documentation by a nutritional screening program. *J Am Coll Nutr* 1985;4:471–479.
8. VA TPN Cooperative Study Group. Perioperative total parenteral nutrition in surgical patients. *N Engl J Med* 1991;325:525–543.
9. Katz SJ, Oye RK. Parenteral nutrition use at a university hospital. Factors associated with inappropriate use. *West J Med* 1990;152:683–686.
10. Soeters PB, Ebeid Am, Fischer JE. Review of 404 patients with gastrointestinal fistula. Impact of parenteral nutrition. *Ann Surg* 1979;190:189–202.
11. Daly JM, Steiger E. Introduction to the report prepared for ASPEN by the Johns Hopkins Medical Institutions. *J Parent Ent Nutr* 1986;10:1.
12. Maliakkal RJ, Blackburn GL, Willcutts HD, et al. Optimal design of clinical outcome studies in nutrition and cancer: future directions. *J Parent Ent Nutr* 1992;16:112s.
13. Abel RM, Beck CM, Abbott AM, et al. Improved survival from acute renal failure after treatment with intravenous essential L-amino acids and glucose: Results of a prospective, double blind study *N Engl J Med* 1973;288:695–699.
14. Cerra FB, Cheung NK, Fischer JE, et al. Disease-specific amino acid infusion (F080) in hepatic encephalopathy: a prospective, randomized, double-blind, controlled trial. *J Parent Ent Nutr* 1985;9:288–295.

CHAPTER 8

Intravenous Access

E. Kerry Gallivan, M.D., M.P.H. and
Peter N. Benotti, M.D.

Central Venous Access

The ability to provide safe and durable central venous access using bedside techniques is an essential technical skill for surgeons, anesthesiologists, and all others who manage critically ill patients. Central venous access is often indicated as part of the perioperative management of complicated surgery in healthy individuals, for diagnostic right-heart catheterization, and, for urgent placement of transvenous pacing devices. In critical care units, this technology facilitates the provision of fluid resuscitation to optimize perfusion and hypertonic intravenous nutrition for feeding.

HISTORY

The first available report of intravenous access in humans is in 1665 by Escholtz.[1] The technique of percutaneous infraclavicular subclavian venipuncture was first described by Aubaniac in 1952.[2–4] Lepp from Germany and Villafane from Argentina reported similar techniques one year later.[5–7] These early reports outlined the technical aspects of percutaneous central vein catheterization. The first published experience in the English literature was in 1956 by Keeri-Szanto.[8] He strongly recommended the procedure as a therapeutic option, because he found that successful cannulation of the subclavian vein could be accomplished within 60 sec in 90% of attempts.

In the early 1960s, worldwide interest in central venous access was stimulated. In 1962, Wilson[9] reported on practical uses of catheterization in clinical medicine using the technique first introduced by Aubaniac. Subsequently, the risks and benefits of this procedure became delineated when complications were reviewed by Smith and Matz.[10,11] Dudrick and Wilmore revolutionized the use of central venous catheters when they described their utility for long-term parenteral nutrition in a female infant in 1968 and then in 25 adults with gastrointestinal disease in 1969.[12–14] Since then, the technique of central venous access has become widespread and percutaneous access to the central veins of the thorax has become routine.

INDICATIONS

The ability to provide safe bedside central venous access has contributed to major advances in medicine. The ability to measure central venous pressure, to perform diagnostic right heart catheterization, to administer rapid fluid resuscitation, to place cardiac pacemakers, and to administer hyperalimentation all constitute major therapeutic advances facilitated by central venous access. These advances have contributed to overall improved survival in many medical and surgical illnesses.

The fundamental principles of critical care include rapid and accurate resuscitation, eradication of infection and dead tissue, optimal support of oxygen transport, and, provision of appropriate nutritional therapy. These fundamentals are facilitated by the ability to achieve rapid bedside central venous access. Central venous catheters can direct treatment choices, aid resuscitation, and provide direct access for medications. Often, when resuscitation is completed, access systems can be converted to parenteral feeding systems with additional outcome benefit. The development of multilumen catheters allows simultaneous uses of central venous access systems including monitoring, feeding, and treatment.

In other situations, safe central access may be a major source of patient comfort. Patients with difficult peripheral venous access who need long-term intravenous medication often require central venous access systems. Examples include the morbidly obese and oncology patients who have limited peripheral venous access. These patients are helped significantly by placement of central venous access systems in the perioperative period. Another indication for central venous access is intravenous infusion of toxic, hypertonic, or irritant medications.[15] In this situation, the high flow in the central veins minimizes the risk of thrombosis and allows prompt mixing of infusates.

ANATOMY

The two great veins most often used for central venous access are the subclavian and internal jugular veins. These two sites provide relatively low-risk, easy access to the central venous system. Access to the femoral area is also relatively low risk and simple. The disadvantages of femoral venous access include an increased infection rate, because of the proximity to contaminated areas and the perceived necessity for bedrest[16]

A thorough knowledge of the anatomy of the vein to be cannulated is critical for effective access with minimal complications. Important structures in relationship to the subclavian vein are the first rib, clavicle, subclavian artery, brachial plexus, and the apex of the lung. The subclavian vein is an extension of the axillary vein. It passes under the middle third of the clavicle, over the first rib, anterior to the scaleneus anticus muscle, and then joins the internal jugular vein in the anterior mediastinum. Its widest point is behind the clavicle, making this the entrance point of choice.

The internal jugular vein lies lateral to the carotid artery within the carotid sheath and is also in close proximity to the lung. It runs under the sternocleidomastoid muscle and can easily be located between the sternal and clavicular heads of the muscle just above the clavicle. Denys and Uretsky reported that, in 92% of 183 patients, the internal jugular vein was lateral and anterior to the carotid artery.[17] Examination with ultrasound revealed that, in 5.5% of the patients, the location of their internal jugular vein could not be predicted by external landmarks. Variations included thrombosis, lateral or medial displacement, and small diameter.[17]

TECHNIQUE

Many of the technical aspects of percutaneous central venous catheter insertion are similar for both the subclavian and internal jugular venipunctures (Figures 8.1 and 8.2). This is rarely an emergent procedure, and thus, several preparatory measures can be undertaken to reduce complications and increase the likelihood of success. Patients undergoing central venous catheter placement should be well hydrated. Dehydrated patients often have collapsed veins which are more difficult to cannulate. Preparatory hydration can be very helpful in terms of facilitating safe access when indicated. Examination of the external jugular vein and its

Figure 8.1. The subclavian approach.

meniscus with the patient in the Trendelenberg position gives a bedside index of distension of the great veins.

A cooperative patient will significantly reduce the risk of technical complication; therefore, light sedation often makes the procedure easier, both for the patient and the operator. Adequate assistance by nursing personnel who are familiar with the techniques increases patient comfort and operator success. Patients must be positioned in the Trendelenberg position such that the external jugular vein is distended to the level of the mandible.

For infraclavicular percutaneous subclavian access a rolled towel placed longitudinally beneath the thoracic spine will elevate the spine and allow the shoulders

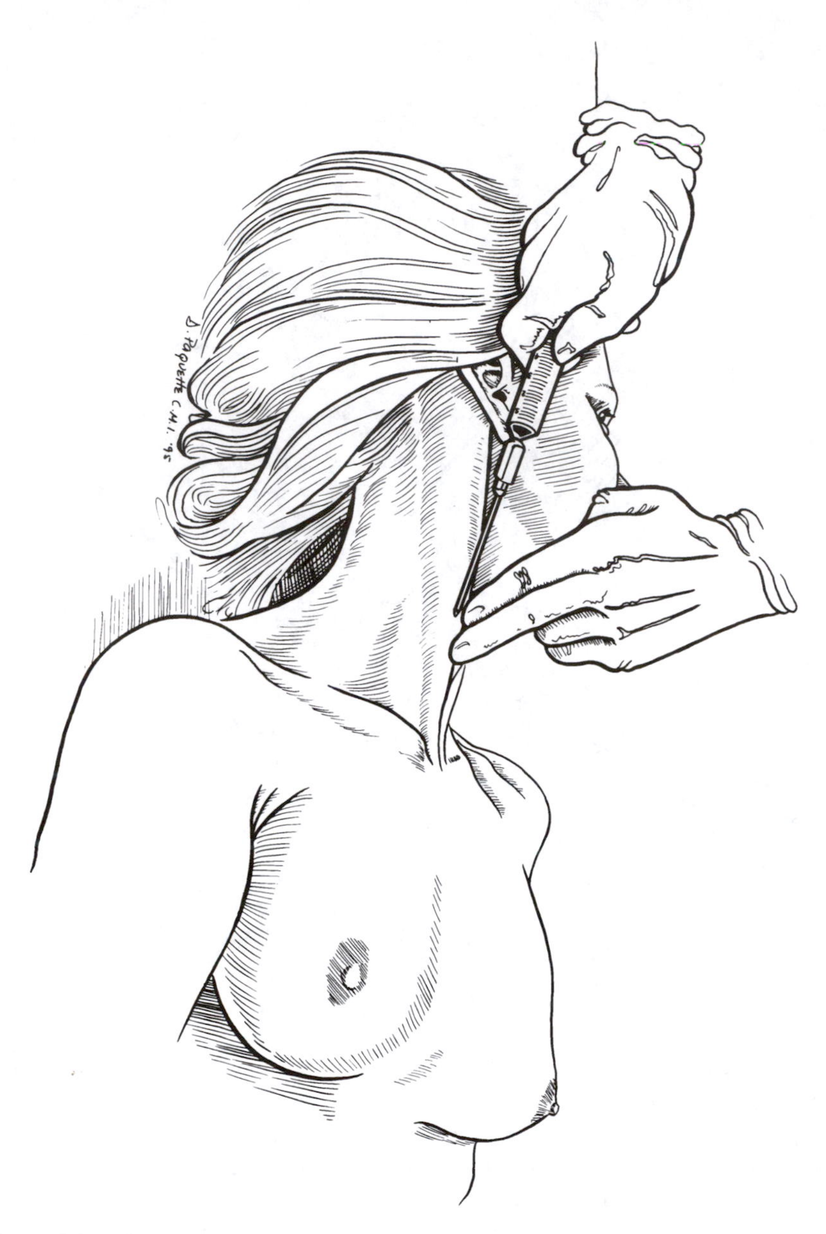

Figure 8.2. The internal jugular approach.

to fall back so as to facilitate a horizontal trajectory. The patient's arm should be kept down at his or her side. The neck and chest should be draped in sterile fashion to avoid the necessity of re-prepping if first attempts at access are unsuccessful and the need for another approach arises. The instrumentor should also be sterile in cap, gown, and gloves to help decrease the risk of infection.[18]

The skillful use of local anesthesia is very important in terms of maximizing safety and patient comfort. Generous and liberal dermal wheals should be produced with slow, steady infiltration followed by extension into the subcutaneous tissue. Application of anesthesia to the periosteum of the clavicle tends to be the most uncomfortable part of the entire procedure. The patient ought to be forewarned prior to this infiltration. Advising the patient at each stage of the procedure is recommended, so as to improve patient comfort and cooperation. The patient is often helpless with head and eyes covered creating a possibly terrifying situation. The periosteum and ligamentous structures involving the inferior aspect of the clavicle are infiltrated generously. The infiltrating needle is advanced along the horizontal trajectory with a short advance followed by liberal infiltration. Intermittent aspiration will help localize the vein. Once the vein is localized, liberal anesthetic should be introduced as the infiltrating needle is withdrawn. The needle is then introduced along the trajectory toward the vein. The needle path is under the clavicle at the junction between the middle and inner third. The course is then in a horizontal plane directed toward a fingernail of an index finger firmly placed in the sternal notch (Figure 8.1) The needle is cautiously advanced with continuous negative pressure until free flow of blood is obtained. Once assured of free flow, the operator removes the syringe and advances a flexible tip guidewire through the needle into the venous system.

During the period of guidewire positioning, cardiac monitoring is extremely helpful in detecting atrial and ventricular arrhythmias.[19] The detection of arrhythmias confirms a normal course of the guidewire. The onset of arrhythmia is an indicator to immediately withdraw the guidewire a short distance from the ventricle.

The needle is then withdrawn and the guidewire held in place. Some advocate securing the wire with several hemostat clamps to the adjacent drape. The dermal puncture site is enlarged with a knife and the tract is dilated. A dilator is passed over the wire, removed, and the catheter advanced over the wire. The wire is then extricated and the catheter aspirated and flushed. Any resistance encountered during the passage of the guidewire through the needle requires removal of the wire and reaspiration to confirm intravascular placement.

An alternate procedure after the wire is positioned in the venous system involves the construction of a subcutaneous tunnel.[20,21] A tunnel is made from the percutaneous site under the skin, usually in an inferior and medial direction. The advantage of a subcutaneous tunnel is that it allows for greater distance between the insertion site and the actual catheter entry into the vein. Some feel that this may reduce the likelihood of subsequent catheter infection.[22]

The intravenous catheter should be secured and positioned using a monofilament suture with a loose knot tied through the full thickness of the skin and subsequent knot securely tied around the shaft of the catheter. Care must be taken to secure the catheter to prevent dislodgement while making sure that the knot is not too tight, because this may cause ischemia and necrosis of the skin. A chest X-ray must be obtained and reviewed by the person who performed the procedure. Very often policies in intensive care units and recovery rooms require a physician review of the X-ray with confirmation of position and determination of complications, such as pneumothorax, before routine use of the central venous catheter is allowed.

Access to the internal jugular vein is similar to that of the subclavian (Figure 8.2). Attempts to cannulate the internal jugular vein are often made as a first choice by anesthesiologists, because they have easy access from the head of the table. The right side is generally preferred, because it is easier to access for someone who is right-handed and has a straight trajectory to the superior vena cava (SVC). Jugular venous access is less attractive than the subclavian access for long-term use for two reasons. Firstly, patient movement often disrupts catheter dressing, thereby possibly increasing local trauma and the rate of infection.[23] Secondly, patients are generally more comfortable with catheter exit sites dressed on their chest rather than their neck.

Preparatory steps for access to the internal jugular vein are the same as those for the subclavian vein, except the patient lies flat without a thoracic spinal roll. The patient's face is turned contralateral to the side of entry and the neck is slightly extended. The main landmark used for this approach is the sternocleidomastoid muscle (SCM). Three approaches to the internal jugular vein have been described: the anterior, posterior, and middle approaches. In the posterior approach, which was described by Brinkman in 1973,[24] the puncture site through the skin is located at the lateral border of the SCM just cephalad to the point where the external jugular vein crosses the SCM. In the middle approach, the needle enters at the apex of the triangle created by the separation of the two heads of the SCM: the sternal and the clavicular heads.[25] The anterior approach punctures through at the anterior border of the SCM. In all approaches, the operator stands at the head of the bed with the patients head at the edge of the mattress. Care must be taken to identify the carotid artery. Bothe advocates a using a needle trajectory approximately 30 degrees downward from the vertical and 30 degrees in a lateral direction.[23] In the posterior approach, this often corresponds to the sternal notch. In the anterior and middle approaches aiming in the direction of the ipsilateral nipple often helps find the correct trajectory. To prevent artery cannulation, the operator should palpate the artery and separate it from the SCM and the interior jugular.

The particular site used for central venous access is largely a function of operator preference and experience. However, certain facts are worthy of mention.

The subclavian vein is often preferred for long-term total parenteral nutrition (TPN), because infection risk is lower and patient comfort is higher.[20] Relative contraindications to the subclavian route include previous clavicular fracture, radical mastectomy, and radical neck dissection.[26] In some published series, the use of the internal jugular vein has proven safer with fewer risks of technical complications.[27] In addition, a dressing on the neck is more likely to be disrupted by movement and perspiration, which may influence the sterility of the exit site and increase the risk of infection.[26] As noted previously, the right internal jugular vein access is preferred for overall ease of access for most right-handed operators. In addition the right jugular vein is wider and further away from the common carotid artery.[27] Use of the left internal jugular vein involves a longer distance and a less direct route to access the SVC, thus increasing the risks of catheter misdirection. Finally, percutaneous access of the left internal jugular vein involves the additional potential risk of injury to the thoracic duct.

FACTORS THAT AFFECT LINE PLACEMENT

In some patients percutaneous central venous access can be technically challenging. These patients include those who have had central access previously and those who are cachectic or obese. Patients with previous central venous access systems often have distorted anatomy because of thrombosis and scarring. In the morbidly obese, it is often difficult to identify usual anatomic landmarks secondary to body habitus. Furthermore, cancer patients and others who suffer from advanced cachexia loose subcutaneous fat and these patients frequently are hypovolemic. Both of these factors contribute to an increased risk of injury to surrounding structures. Radiologic imaging techniques may be used to help increase success of line placement and decrease risk of complications. Postprocedure chest X-rays are taken to look for complications and to check for accurate placement.[28] Catheter tips should be near the superior vena cava/right atrial junction, usually approximately 13 to 17 cm from the exit site.[29]

Fluoroscopy has traditionally been used to help locate and position catheter tips in the large vessels. Recently, ultrasound had been advocated to aid in vein localization and to detect venous thrombosis. Armstrong reported in 1993 that ultrasound helped cannulate central veins with greater speed and less attempts than blind cannulation.[30] Ultrasound did not decrease the rate of carotid artery cannulation. Color Doppler ultrasound can be used in difficult cases as a bedside aid in cannulation. High-resolution sonographic equipment is recommended. Uses include identification of anatomy of the vessels and the presence of thrombosis or stenosis. Once the anatomy is delineated, the ultrasonographic visualization also can be used to aid placement of a central venous catheter.[31]

Another radiologic technique that offers assistance is venography. This technique should be used in patients who have had multiple previous central venous catheters to ensure patency of the great vessels. Venous mapping can help select an appropriate site for catheter placement.

TYPES OF CATHETERS

Various types of central vein catheters are available and the type of catheter selected relates to anticipated length of use, patient comfort, and indication for placement. Single, double, and triple lumen catheters are equally efficacious for administration of parenteral nutrition, medications, and blood products.[32] Quantity of ports needed determines which of these lines are placed.

Catheters are made of stiff plastic material such as Teflon or polyurethane. The stiffness helps prevent kinking and compression of the lumen. Teflon and polyurethane catheters cause an inflammatory reaction.[27] This may contribute to scarring and thrombosis of vessels leading to clotted lines and vessel distortion. Polyurethane catheters are most commonly used in hospitals for routine use.Silastic, another type of plastic, has low thrombogenicity and is used for long-term access catheters. However, because silastic is so soft, these catheters may be more difficult to insert and are more prone to displacement.[33] Polyethylene is a stiffer substance and therefore has been found to be more likely than other materials to cause vessel shearing and perforation. At the other end of the spectrum is silicone, which is more flexible but tends to kink and move more easily.[34] Polyethylene and silicone catheters are used infrequently.

Tunneled silastic catheters as described by Hickman[35] and Broviac[36] are often utilized for long-term access, particularly when used outside of the hospital. This is supported by data confirming the durability of the lines and the advantage of minimal daily care.[37] These catheters tend to be longer (approximately 90 cm) and more flexible, with a Dacron cuff attached in the midportion. This cuff allows for ingrowth of fibrous tissue in the tunneled subcutaneous extension of the line and is thought to prevent bacteria from migrating up the catheter.[38] These more permanent devices are usually placed in the operating room. The central veins are accessed using previously described percutaneous techniques or via open exposure of the cephalic, external jugular, or internal jugular veins. A subcutaneous tunnel is fashioned from the site of vein entry usually to the anterior chest wall, the standard catheter exit site. The catheter is filled with heparinized saline and advanced through the subcutaneous tunnel so as to position the Dacron cuff 2 cm from the exit site for future easy removal. The catheter is then cut to allow for final position in the SVC. If open venous cutdown is used, the catheter is advanced to the SVC under fluoroscopic control. If percutaneous access is used, a peel-away sheath introducer and dilator are advanced over a guidewire into the venous system. The wire and dilator are removed, and the catheter is advanced

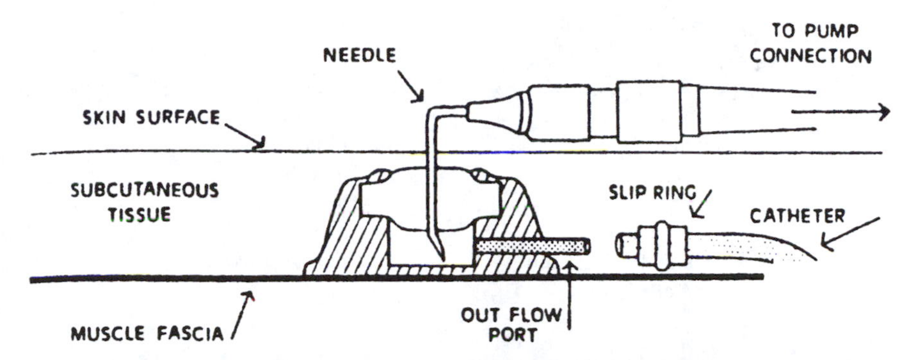

Figure 8.3. Port-A-Cath system with Huber needle in place. Slip ring fits over the outflow port to secure the catheter to the port. Reprinted from Lokich JJ, Bothe A, Jr, Benotti PN et al. Complications and Management of Implanted Venous Access Catheters pages 710–717. *J Clin Oncol* 1985;3(5), with permission from W.B. Saunders, Co.

into the venous system through the sheath. The sheath is then removed leaving the catheter in place. Appropriate line position is confirmed via fluoroscopy.

A totally implanted subcutaneous venous access port can also be used for more permanent central venous access. This consists of a stainless steel port with a septum that can be accessed by a special needle (Huber) (Figure 8.3). The port is positioned deep in the subcutaneous tissue of the chest and anchored with several sutures to the pectoral fascia to prevent migration or rotation. The port must be placed close enough to the skin to facilitate access, but placement that is too superficial, especially in malnourished patients, will predispose to erosion through the skin.[27] Venous access ports possess the advantage of being totally implanted. This allows the patient freedom from dressing changes and the ability to bathe normally. Potential problems include scar formation over the port, which may make access with the needle difficult and the need for a pump for infusion because of increased resistance to flow.[38] Furthermore, many patients need assistance accessing their ports because of the location on the upper, anterior chest wall.

COMPLICATIONS

Complications related to the use of central venous access systems can be divided into technical problems associated with placement of central venous catheters and problems associated with long-term catheter use. (Table 8.1). Generally, the incidence of complications after percutaneous vein catheterization range from 0.4 to 9.9%.[39] Complications of subclavian vein catheterization were noted as early as 1965 and included descriptions of hemothorax, pneumothorax, brachial plexus injuries, hematoma formation, subcutaneous emphysema, and, septicemia.[11]

Table 8.1. Complications

Technical	Long-Term Use
Pneumothorax	Infection
Hemothorax	Colonization
Hydrothorax	Thrombosis
Chylothorax	Thrombophlebitis
Arterial puncture	Erosion
Myocardial perforation	Catheter kink/knot
Pleural effusion	
Improper position	
Arrhythmia	
Air embolism	
Catheter embolism	

Technical Complications

In 1978 Herbst conducted a retrospective review to study percutaneous subclavian vein access complications.[40] He described an 11% complication rate, with pneumothorax making up half of the complications, whereas the balance included arterial puncture, pleural effusion, and improper positioning of the catheter. Most importantly, he showed that inexperience of the operator had the greatest influence on the incidence of technical complications. The overall reported technical complication rate associated with placement of central venous catheters varies from 0.4 to 12% as confirmed by prospective trials.[40,41] However, operator inexperience can increase the incidence of complications from 6% to 20% in difficult procedures.[42]

Technical complications related to catheter insertion are often influenced by patient cooperation, adequacy of anesthesia, variations from standard techniques, and, unpredictable anatomic variations. The most common technical complication is pneumothorax, which occurs from placement of a needle through the apical pleura. This risk can be minimized by maintaining a horizontal projectory during percutaneous infraclavicular subclavian vein access and by the use of a small needle to locate the vein. Passage through the posterior wall of the subclavian vein increases the risk of pneumothorax, because the lung apex is directly posterior to the subclavian vein. In addition, movement of the needle once inserted, especially changing the plane of access, can result in tissue injury with increased risk of technical complications. For patients on mechanical ventilators some advocate manual ventilation with temporary discontinuation of ventilation at the time of needle insertion to decrease lung expansion and therefore decrease the risk of pneumothorax. To diagnose and treat pneumothorax, it is important to obtain an upright, expiratory chest X-ray after the catheter is in place or before attempts are made on the other side. Occasionally, a delayed pneumothorax occurs. These are usually small pneumothoraces, which resolve without intervention.[43]

Catheter malposition is another technical complication that reportedly occurs 1.7% to 13.2% of the time. For example, placement of a catheter into the subclavian vein can traverse either inferiorly into the SVC as desired or superiorly retrograde into the internal jugular vein. This occurs more commonly on the right side because the junction of the SVC and subclavian vein is more acutely angled on the right than the left. Malposition most commonly occurs with the left internal jugular approach. The catheter can pass into the SVC, the left or right subclavian, or, the left internal mammary vein. This can be managed by replacement under fluoroscopy or just mere repositioning at the bedside. Extravascular malposition may be caused by stiff catheters, multiple catheterizations, or unsafe use of guidewires.[44]

Arrhythmia can occur from introduction of a central venous catheter. This usually happens during the insertion of the guidewire. Incidence of atrial arrhythmia is reported to be as high as 41% and ventricular arrhythmia, 11%.[19] Therefore, cardiac monitoring, especially in patients with a cardiac history, is advisable when placing venous catheters centrally.

Vascular complications include artery puncture with significant bleeding complications such as hemothorax. The subclavian artery is inadvertently punctured up to 20% of the time.[45,47–50] Arterial puncture is recognized by a range of clinical findings extending from hematoma formation to hypotension. Rarely, decreased distal flow from a severely injured artery can be identified.[42] Treatment consists of local pressure, observation, and surgical repair if needed. Hemothorax occurs infrequently, 0 to 2% of the time.[45] This occurs from shearing or puncture of the artery. The most important factor relating to this complication is prevention. Serious mechanical complications such as cardiac tamponade occur rarely but lead to a 65% to 78% mortality.[29,46] Proper technique as outlined above is the most crucial factor. Patients with bleeding risks, such as those in renal failure or patients on anticoagulation therapy, should receive platelets and/or other clotting factors just before or during line placement. This minimizes bleeding in high-risk patients. If hemothorax or hemomediastinum occurs, close observation of airway status and vital signs in an intensive care setting is indicated. If bleeding is not controlled with transfusions and correction of coagulopathy, closed tube thoracostomy or thoracotomy with vascular repair may be indicated.

Long-Term Complications

Infection. Infectious complications may range from skin site infections to sepsis. Catheter colonization and infection may be caused by bacterial or fungal organisms. Colonization is defined as the process of establishing colonies and is usually used in reference to the formation of colonies at the catheter exit site in the skin. *Infection* is defined as "invasion by and multiplication of pathogenic microorganisms in a bodily part or tissue, which may produce subsequent tissue injury and progress to overt disease through a variety of cellular or toxic mechanisms."[51] Thus, catheter colonization does not cause systemic problems until the organism invades.

Various definitions of catheter-related infection exist. Controversy surrounds this area and makes it difficult for evaluation of published data and for planning treatment strategies.[52,53] Gosbell states that "the most accepted definition is clinical sepsis in the presence of a central venous catheter, with the same organism isolated from cultures of peripheral blood and catheter tip, in the absence of another likely source of infection."[54] Reported incidence rates for subclavian and jugular catheters range from 3% to 13%, but reporting is as high as 80% depending upon the definition used.[52–55] Many patient-related factors are associated with an increased risk of catheter-related infection including age greater than 60 and less than 12, neutropenia, and loss of skin integrity.[52] Total parenteral nutrition has been found to increase the risk of central line sepsis.[56] This is perhaps mediated by the immunosuppressive effects of hyperglycemia.[57]

The organisms most commonly isolated in catheter-related sepsis are *Staphylococcus aureus,* coagulase-negative staphylococci, *Streptococcus viridans,* gram-negative rods, and fungi.[54,58] Peripheral blood cultures are said to be the most important test to obtain, but, unfortunately they are often negative.[53] Andremont et al. retrospectively examined hub blood cultures and catheter tip cultures from patients presumed to have catheter-related infections.[59] They found that only 18% of blood cultures and 29% of catheter tips were positive. When cultures are positive, they accurately and specifically diagnose the causative agent. Unfortunately, cultures are not helpful in the majority of acute cases, because their yield is so low. Collignon et al. studied central vein catheter tip cultures and their usefulness in the diagnosis of catheter-associated bacteremia.[60] They found a much better correlation of catheter tip cultures and bacteremic episodes and, therefore, recommend semiquantitative culturing of catheter tips to help diagnose and treat central venous catheter bacteremia.

FACTORS THAT CONTRIBUTE TO INFECTION. Optimal central venous catheter care requires specialized care by a team of professionals familiar with central catheter insertion and maintenance. Maki et al. studied factors that may relate to the risk of bacteremia from central venous catheters.[61] His studies demonstrate that aseptic technique is important while placing these catheters.[18] Hence, the recommendations of cap, gown, gloves with prepping, and draping. He has found that cutaneous colonization has a greater incidence than bacteremia and that the risks of infection are related to colonization of the catheter insertion site because the organisms that lead to infection live on the patients' skin. Unfortunately it is often difficult to determine if an isolate from an exit site culture is a contaminant or a pathogen.[59,60]

Maintenance of central venous catheters is, therefore, important because colonization of the skin site can lead to sepsis. Dressing changes are the hallmark of catheter maintenance. Sterile technique must be used to apply all dressings. In a randomized prospective trial in 1994, Maki et al. determined that no statistical difference in infection rates can be found with different types of dressings conventionally used.[62] Polyurethane to gauze may be used to protect the exit site, the best dressing material has not been identified. No time line for optimal dressing changes is known.[63] Furthermore, further research needs to be done to elucidate the most cost-efficient dressing change schedule.

Location of the central venous catheter affects infection risk as well. Insertion into the internal jugular vein has a higher risk of infection than placement into the subclavian vein.[61] This may be because maintenance of a dressing on the neck is more difficult than on the chest because of increased movement and stress on the dressing.[26] Suturing the line close to the entry point is thought to help this by decreasing movement and thereby lessening mechanical introduction of organisms through the skin site.[64]

In 1952, the technique of changing catheters over a wire was described by Seldinger, and its application to central venous catheterization was described in 1974.[65,66] This was initially felt to decrease the incidence of catheter-related sepsis but recent studies have disputed this claim. Neither frequent guidewire changes nor frequent resiting have been found to decrease the incidence of catheter-related sepsis but may increase the number of mechanically related complications such as pneumothorax and hemothorax.[67,68] If a catheter needs to be changed, some advocate it be done via guidewire because guidewire replacement is associated with less complications than initial venipuncture.[69]

Eyer et al. examined whether changing lines once weekly would affect the occurrence of infection.[70] These investigators prospectively studied surgical intensive care patients who required a triple lumen catheter and randomized them into three groups. One group had their catheters changed once a week via guidewire changes, another underwent weekly percutaneous changes to a new site, and, the last had no weekly changes at all. Interestingly the incidence of catheter-related sepsis was no different in any of the groups. This implies that catheters do not need to be changed unless signs or symptoms of infection occur. Eyer found no relationship between the length of time catheters were in place and increased risk of infection. Savage et al. confirmed these findings.[71] Therefore, generally accepted current indications for resiting a central venous catheter are line sepsis, bacteremia, or persistent colonization. Interestingly, Eyer did find a correlation between APACHE II score and risk of infection, implying that severity of disease may be a significant indicator of incidence of line sepsis not duration of line placement.

The number of ports has not been found to correlate with infection incidence. Earlier reports in the evolution of central venous catheters suggested that infection and colonization rates are higher with multilumen catheters.[72,73] However, recent literature shows that infection rates in patients with multilumen catheters are about equal to the frequency observed with single lumen catheters if standard protocols regarding dressing changes and catheter care site are followed. Furthermore, it may be that the sicker patients are receiving catheters with multiple ports, thereby skewing results for sepsis to those already at higher risk.[73] Overall, the important maintenance issues are catheter care, nursing care, and catheter changes as discussed above.[74] If a catheter does have more than one port, care should be taken to use only one lumen exclusively for parenteral nutrition. This can decrease the incidence of infection and contamination.[73,74]

TREATMENT OF INFECTION. The gold standard for treating infection is to remove the catheter as quickly as possible.[64] Ideally, central venous access should be eliminated for 48 hours to allow treatment with intravenous antibiotics to clear the infection. Removing the catheter usually removes the source of infection.[75] Appropriate antibiotics should cover the commonly infecting organisms; therefore, an anti-staphylococcal beta-lactam with a broad-spectrum gram-negative drug such as gentamicin should be used as first line.[54] When a specific organism is identified, antibiotic therapy should be modified appropriately.

There has been recent interest in leaving the infected catheter in place and treating through the line with antibiotics.[58,76,77] This is especially important in patients with very difficult access and may help prevent technical complications associated with line insertion. Flynn et al.[77] reported a prospective study done first in rabbits and then in patients with confirmed catheter-related sepsis. They found that antibiotic therapy was successful in eliminating bacteremia in 65% of their patients. This enabled these patients to keep their central lines thereby avoiding the need for another line placement. Others have reported even higher rates of successful *in situ* antibiotic therapy, as high as 80% to 90%.[76,77] Therefore, at this time, many patients with catheter sepsis are allowed a trial of antibiotics infused through the catheter.[77] This technique is most often used in patients with poor intravenous access, especially children and patients who have had multiple catheterizations in the past. Routine treatment of an antibiotic trial rather than immediate catheter removal remains controversial.

PREVENTION OF INFECTION. The concept of coating lines with antibiotics or heparin has been the subject of much interest in regard to preventing infection and colonization of central venous access catheters. Mermel and Maki found that heparin-bonded catheters may contribute to a low incidence of catheter-related bacteremia in prospective testing of the bactericidal activity of Swan–Ganz catheters.[78] The benzalkonium chloride to which the heparin is bound on most heparinized catheters has antimicrobial activity. The amount of activity against individual organisms directly relates to their susceptibility to this agent. Exposure to serum causes these catheters to lose most of their antimicrobial activity within 24 hr. Antibiotic-impregnated catheters have also been tested for bactericidal activity. These catheters have greater activity against susceptible bacteria and do not loose their activity when exposed to serum. Although benzalkonium-heparin coated lines help decrease the risk of catheter-related bacteremia, further study regarding heparin-induced thrombocytopenia and cost-benefit analyses are needed prior to their widespread use. Antibiotic-coated lines are not advocated because the risks of the development of resistance is likely higher than the benefit achieved.

Needleless systems to access central venous catheters are now commonly used to reduce the risk of needlestick injury. Danzig et al. designed a case-control and retrospective cohort analysis to determine if needleless systems affected the risk of bloodstream infection in patients receiving home infusion therapy.[79] The study suggests that patients with central venous catheters using a needleless system are at higher risk for line infection. Total parenteral nutrition may also increase the risk of infection.

Thrombosis. McDonough and Altemeier first described catheter-related thrombosis in 1971.[80] Most thrombosis is not clinically evident unless venous obstruction occurs or the catheter clots and looses function. Clotting is clinically apparent 0 to 5% of the time.[81–83] Subclavian vein thrombosis most commonly occurs in children secondary to the small diameter vasculature.[84] When venous obstruction does occur, manifestations of venous insufficiency are seen in the neck, face, and chest: ipsilateral edema, superficial vein distension, pain, tightness, and red-blue discoloration.[85] One of the most important complicating factors from thrombosis is loss of the vessel and, therefore, loss of future access sites.[86] Other complications of upper extremity deep venous thrombosis include septic thrombophlebitis, venous gangrene, extravasation of infusate, death, and pulmonary embolism.[87] The incidence of pulmonary embolism may be as high as 12%.[87]

The true incidence of central vein thrombosis from catheterization is an area of controversy. The incidence of subclavian vein thrombosis from catheterization may be as high as 38% or as low as 4%.[83,88] Many feel that the incidence is not really known. Although it is generally reported as low, it has been found that thrombosis occurs much more frequently than realized but is often occult.[81,82] The difficulty determining the true incidence of catheter-related thrombosis relates to difficulty with clearly defining catheter thrombosis. Definitions range from the fibrin sheath around the catheter to catheter lumen thrombosis to great vein occlusion. A fibrin sheath occurs around all catheters within one week.[89] However, the clinical significance of fibrin coating of the catheters remains uncertain.

The diagnosis of thrombosis is made either with ultrasonography or venography. Venography is the gold-standard. Several prospective studies have shown that venography detects thrombosis centrally in 28% to 54% of patients with central venous catheters.[87,90,91] However, very few of these patients have clinically recognizable thrombosis.[89]

Virchow first postulated in 1856 that the causes of thrombosis include low flow, hypercoagulable state, and inherent catheter thrombogenecity or trauma.[80] Patients who receive central venous catheters are prone to all three of these findings: trauma to the endothelium with introduction of the catheter, low flow from low cardiac output and hypovolemia, and, hypercoagulability from underlying disease, such as cancer.[92] Infection has also been found to be associated with thrombosis, but a cause–effect relationship has not been elucidated.[93] The development of thrombosis seems to correlate with the duration of catheterization. Brismar et al. found that up to 73% of thoracic venous catheters cause mural thrombus after 14 days.[94]

Treatment is aimed at clearing thrombus and restoring line patency. Modalities used include thrombolytic therapy, anticoagulation, catheter removal, and clot retrieval.[95] Thrombolytic therapies such as urokinase, streptokinase, and tissue plasminogen activator have been used with approximately 95% of catheters cleared of clot.[89,96–99]

In the pediatric population, much attention has focused on attempts to clear thrombosis and infection using thrombolytics. Although a few report success with the use of antibiotics and urokinase, others have found no benefit with either clearing bacteremia or improving catheter salvage.[100,101,102] All of this data is from small groups of patients with variation in treatment from bolus therapy to constant 24 to 48 infusions. Urokinase therapy has a high likelihood of recannulating thrombosis, but its usefulness as an adjunct to clear catheter-related infection is not clear. In general, thrombolytic therapy achieves complete thrombolysis in approximately 50% of pediatric patients and partial thrombolysis in approximately 12%.[103] However, treatment with tissue plasminogen activator (tPA) or infusion therapy with urokinase increases patency of catheters.[104] Currently, aggressive intravenous thrombolytic therapy is advocated for great vein occlusion. Thrombosed catheters that loose function should be treated with 5000 U/mL of urokinase to restore patency.[96] The urokinase injected should stand for at least 5 min in the catheter and then should be aspirated until 3 cc of blood is readily obtained.[96]

Prevention should be geared toward decreasing coagulability through decreased trauma, treating hypercoagulability, and minimizing infection. Low-dose anticoagulants may be effective, especially in high-risk populations. Bern et al.[90] showed in an open, randomized, prospective study that low-dose warfarin (1 mg/day) reduced thrombosis associated with indwelling catheters in cancer patients. Silastic catheters in conjunction with this therapy reduces the risk of thrombosis without prolonging measurable bleeding factors. Catheter flushes have traditionally been carried out with heparinized saline to help decrease thrombosis incidence; however, recent evidence suggests that heparin and saline flushes have equal efficacy in maintaining catheter patency.[105,106] Because of the added cost of heparin many hospitals are changing to saline flushes alone.

Other long-term complications. Erosion of the catheter through a vessel wall or through the skin may also occur.[27,38,44] Most often, if erosion occurs through a vessel, the patient will present with signs and symptoms of hemorrhagic thoracic effusions. Attempts to confirm placement by aspirating blood are not accurate in these cases because the lumen is often extravascular. This confounding variable makes diagnosis difficult until respiratory distress or hypovolemia ensue.[107]

Air embolism can occur and can be fatal.[108] Aubaniac, one of the first to write about subclavian venipuncture, felt that "the risk of embolization does not exist."[3,48] Symptoms and signs include hypotension, shortness of breath, tachypnea, tachycardia, and the classic "mill wheel" murmur over the precordium.[48] Classically, a sucking sound may be heard. If the diagnosis is suspected, treatment is supportive. The patient should be placed in the Trendelenburg position with the left side down. Many advocate then aspirating the air embolus with a syringe through the catheter. Prevention includes checking materials for holes and cracks. Most importantly, when an operator places a central venous catheter, a finger must be held over the inserting needle while passing the guidewire and catheter.[105]

CONCLUSION

Intravenous access provides many benefits to clinicians and patients. The ability to rapidly cannulate the great veins of the thorax constitutes a major advance in medicine and has greatly enhanced many therapies. Since Escholtz' treatise in 1665, intravenous access has come from a novel, rarely used intervention to a mainstay in clinical medicine. Continued investigations will lead to improved access devices, insertion techniques, and maintenance methods, further enhancing treatment potential and decreasing complications.

REFERENCES

1. Clysmata Nova: sive ratio qua in venam sectam medicamenta immitti possint, etc. Berolini, 1665.
2. Aubaniac R. L'injection intraveineuse sous-claviculaire: avantages et technique. *Presse Med* 1952;60:1456.

3. Aubaniac R. Une nouvelle voie d'injection ou de ponction veineuse: la voie sous-claviculaire: veine sous-claviere, tronc brachio-cephalique. *Semin Hop Paris* 1952;28:3445–3447.
4. Aubaniac R. L'intraveineuse sous-claviculaire: avantages et technique. *Afr Fr Chir (Alger)* 1952;3–4:131–135.
5. Lepp H. Uber eine neue intravenose injektions und punktions-methode: die infra-klavikulare punktion der vena subclavia. *Dtsch Zahnaerztl Z* 1953;8:511–512.
6. Lepp M. Die infraklavikulare punktion der vena subclavia nach Aubaniac. *Munch Med Wschr* 1954;47:1392–1393.
7. Villafane EP. Technica de la transfusion por via subclavicular. *Prens Med Argentina* 1953;40:2379–2381.
8. Keeri-Szanto M. The subclavian vein, a constant and convenient intravenous injection site. *AMA Arch Surg* 1956;72:179–181.
9. Wilson JN, Grow JB, Demong CV, et al. Central venous pressure in optimal blood volume maintenance. *Arch Surg* 1962;85:563–578.
10. Matz R. Complications of determining central venous pressure. *N Engl J Med* 1965;273:703.
11. Smith SE, Modell JH, Gano ML, et al. Complications of subclavian vein catheterization. *Arch Surg* 1965;90:228–229.
12. Wilmore DW, Dudrick SJ. Growth and development of an infant receiving all nutrients exclusively by vein. *JAMA* 1968;203(10):860–864.
13. Dudrick SJ, Wilmore DW, Vars HM, et al. Can intravenous feeding as the sole means of nutrition support growth in the child and restore weight loss in an adult? An affirmative answer. *Ann Surg* 1969;169(6):974–984.
14. Wilmore DW, Dudrick SJ. Safe long-term venous catheterization. *Arch Surg* 1969;98:256–258.
15. Yoffa D. Supraclavicular subclavian venepuncture and catheterisation. *Lancet* 1965;2:614–617.
16. Kemp L, Burge J, Choban P, et al. The effect of catheter type and site on infection rates in total parenteral nutrition patients. *J Parent Ent Nutr* 1993; 18(1):71–74.
17. Denys BG, Uretsky BF. Anatomical variations of internal jugular vein location: Impact on central venous access. *Crit Care Med* 1991;19(12):1516–1519.
18. Maki DG. Yes, Virginia, aseptic technique is very important: maximal barrier precautions during insertion reduce the risk of central venous catheter-related bacteremia. *Infect Cont Hosp Epidemiol* 1994;15(4,pt1):227–230.
19. Stuart RK, Shikora SA, Akerman P, et al. Incidence of arrhythmia with central venous catheter insertion and exchange. *J Parent Ent Nutr* 1990;14(2):152–155.
20. Benotti PN, Bothe A, Miller JD, Blackburn GL. Safe cannulation of the internal jugular vein for long term hyperalimentation. *Surg Gynecol Obstet* 1977;144:574–576.
21. Sriram K, Kaminski MV, Berger R. A safe technique of central venous catheterization. *J Parent Ent Nutr* 1982;6(3):245–248.

22. Severien C, Nelson JD. Frequency of infections associated with implanted systems vs. cuffed, tunneled silastic catheters in patients with acute leukemia. *Am J Dis Child* 1991;145(12):1433–1438.
23. Bothe, Jr., A. Intravenous access. In Daly JM, Cady B, eds. *Atlas of surgical oncology.* St. Louis: Mosby, 1993:3–18.
24. Brinkman AJ, Costley DO. Internal jugular venipuncture. *JAMA* 1973;223(2):182–183.
25. Mostert JW, Kenny GM, Murphy GP. Safe placement of central venous catheter into internal jugular veins. *Arch Surg* 1970;101:431–432.
26. Benotti PN, Bistrian BR. Practical aspects and complications of total parenteral nutrition. *Crit Care Clin* 1987;3(1):115–131.
27. van Way CW, Allen JA. Intravenous nutrition. In van Way, ed. *Handbook of surgical nutrition.* New York: Lippincott, 1992:73–92.
28. Henschke CI, Pasternack GS, Hart KK, et al. Bedside chest radiography: diagnostic efficacy. *Radiology* 1983;149:23–26.
29. McGee WT, Ackerman BL, Rouben LR, et al. Accurate placement of central venous catheters: a prospective, randomized, multicenter trial. *Crit Care Med* 1993;21(8):1118–1123.
30. Armstrong PJ, Cullen M, Scott DHT. The "Site Rite" ultrasound machine—an aide to internal jugular vein cannulation. *Anaesthesia* 1993;48(4):319–323.
31. Longley DG, Finlay DE, Letourneau JG. Sonography of the upper extremity and jugular veins. *Am J Roentgenol* 1993;160(5):957–962.
32. Johnson BV, Rypins EB. Single-lumen versus double-lumen catheters for total parenteral nutrition. *Arch Surg* 1990;125:990–992.
33. Sitzmann JV, Townsend TR, Siler MC, et al. Septic and technical complications of central venous catheterization: a prospective study of 200 consecutive patients. *Ann Surg* 1985;202:766–770.
34. Wilson JN, Stabile BE, Williams RA, et al. Current status of vascular access techniques. *Surg Clin North Am* 1982;62(3):531–551.
35. Broviac JW, Cole JJ, Scribner BH. A silicone rubber atrial catheter for prolonged parenteral alimentation. *Surg Gynecol Obstet* 1973;136:602–606.
36. Hickman RO, Buckner CD, Clift RA, et al. A modified right atrial catheter for access to the venous system in marrow transplant recipients. *Surg Gynecol Obstet* 1979;148:871–875.
37. Stanislav GV, Fitzgibbons RJ, Bailey RT, et al. Reliability of implantable central venous access devices in patients with cancer. *Arch Surg* 1987;122:1280–1283.
38. Steiger E, Grundfest-Broniatowski S, Misny TJ. Intravenous hyperalimentation: temporary and permanent vascular access and administration. In Deitel M, ed. Nutrition in Clinical Surgery, 2nd ed. Baltimore: Williams and Wilkins, 1985:88–104.
39. Borja AR. Current status of infraclavicular subclavian vein catheterization: review of the English literature. *Ann Thoracic Surg* 1972;13(6):615–624.

40. Herbst CA. Indications, management, and complications of percutaneous subclavian catheters: an audit. *Arch Surg* 1978;113:1421–1425.
41. Eisenhauer ED, Derveloy RJ, Hastings PR. Prospective evaluation of central venous pressure catheters (CVP) in a large city-county hospital. *Ann Surg* 1982;196(5):560–564.
42. Puri VK, Carlson RW, Bander JJ, et al. Complications of vascular catheterization in the critically ill: a prospective study. *Crit Care Med* 1980;8(9):495–499.
43. Mitchell A, Steer HW. Late appearance of pneumothorax after subclavian vein catheterisation: an anaesthetic hazard. *Br Med J* 1980;281(6251):1339.
44. Lumb PD. Complications of central venous catheters. *Crit Care Med* 1993;21:1105–1106.
45. Bernard RW, Stahl WM. Subclavian vein catheterizations: a prospective study. I. noninfectious complications. *Ann Surg* 1971;173(2):184–190.
46. Krauss D, Schmidt GA. Cardiac tamponade and contralateral hemothorax after subclavian vein catheterization. *Chest* 1991;99(2):517–518.
47. Christensen KH, Nerstrom B, Baden H. Complications of percutaneous catheterization of the subclavian vein in 129 cases. *Acta Chir Scand* 1967;133:615–620.
48. Feliciano DV, Mattox KL, Graham JM, et al. Major complications of percutaneous subclavian vein catheters. *Am J Surg* 1979;138:869–874.
49. Merk EA, Rush RF. Emergency subclavian vein catheterization and intravenous hyperalimentation. *Am J Surg* 1975;129:266–268.
50. Mogil RA, DeLaurentis DA, Rosemond GP. The infraclavicular venipuncture. *Arch Surg* 1967;95:320–324.
51. *American Heritge Dictionary of the English Language,* 3rd ed. New York: Houghton Mifflin Company, 1992 (on-line).
52. Norwood S, Ruby A, Civetta J, et al. Catheter-related infections and associated septicemia. *Chest* 1991;99:968–975.
53. Hiemenz J, Skelton J, Pizzo PA. Perspective on the management of catheter-related infections in cancer patients. *Ped Infect Dis* 1986;5(1):6–11.
54. Gosbell IB. Central venous catheter-related sepsis: epidemiology, pathogenesis, diagnosis, treatment and prevention. *Intensive Care World Monitor* 1994;11(2):52–58.
55. Maki DG, Ringer M, Alvarado CJ. Prospective randomized trial of povidone-iodine, alcohol, and chlorhexidine for prevention of infection associated with central venous and arterial catheters. *Lancet* 1991;338(8763):339–343.
56. Maki DG. Nosocomial bacteraemia: an epidemiologic overview. *Am J Med* 1981;70:719–732.
57. Pomposelli JJ, Bistrian BR. Is total parenteral nutrition immunosuppressive? *N Horizons* 1994;2(2):224–229.
58. Buchman AL, Moukarzel A, Goodson B, et al. Catheter-related infections associated with home parenteral nutrition and predictive factors for the need for catheter removal in their treatment. *J Parent Ent Nutr* 1994;18(4):297–302.

59. Andremont A, Paulet R, Nitenberg G, et al. Value of semiquantitative cultures of blood drawn through catheter hubs for estimating the risk of catheter tip colonization in cancer patients. *J Clin Microbiol* 1988;26(11):2297–2299.
60. Collignon PJ, Soni N, Pearson IY, et al. Is semiquantitative culture of central vein catheter tips useful in the diagnosis of catheter-associated bacteremia? *J Clin Microbiol* 1986;24(4):532–535.
61. Mermel LA, McCormick RD, Springman SR, et al. The pathogenesis and epidemiology of catheter-related infection with pulmonary artery Swan-Ganz catheters: a prospective study utilizing molecular subtyping. *Am J Med* 1991;91(3B):197S–205S.
62. Maki DG, Stolz SS, Wheeler S, et al. A prospective, randomized trial of gauze and two polyurethane dressings for site care of pulmonary artery catheters: implications for catheter management. *Crit Care Med* 1994;22(11):1729–1737.
63. Murphy LM, Lipman TO. Central venous catheter care in parenteral nutrition: a review. *J Parent Ent Nutr* 1987;11(2):190–201.
64. Henderson DK. Bacteremia due to percutaneous intravascular devices. In Mandell GL, Douglas RG, Bennett JE, eds. *Principles and practice of infectious diseases,* 3rd, ed. New York: Churchill Livingstone, 1990:2189–2199.
65. Seldinger SI. Catheter replacement of the needle in percutaneous arteriography: a new technique. *Acta Radiol* 1952;39:367–376.
66. Blewett JH, Kyger ER, Patterson LT. Subclavian vein catheter replacement without venipuncture. *Arch Surg* 1974;108:241.
67. Powell C, Kudsk KA, Kulich PA, et al. Effect of frequent guidewire changes on triple-lumen catheter sepsis. *J Parent Ent Nutr* 1988;12(5):462–464.
68. Mantese VA, German DS, Kaminski DL, et al. Colonization and sepsis from triple-lumen catheters in critically ill patients. *Am J Surg* 1987;154:597–601.
69. Hagley MT, Martin B, Gast P, et al. Infectious and mechanical complications of central venous catheters placed by percutaneous venipuncture and over guidewires. *Crit Care Med* 1992;20(10):1426–1430.
70. Eyer S, Brummitt C, Crossley K, et al. Catheter related sepsis: prospective, randomized study of three methods of long-term catheter maintenance. *Crit Care Med* 1990;18:1073–1079.
71. Savage AP, Picard M, Hopkins CC, et al. Complications and survival of multilumen central venous catheters used for total parenteral nutrition. *Br J Surg* 1993;80(10):1287–1290.
72. Kelly CS, Ligas JR, Smith CA, et al. Sepsis due to triple lumen central venous catheters. *Surg Gynecol Obstet* 1986;163:14–16.
73. Pemberton LB, Lyman B, Lander V, et al. Sepsis from triple versus single lumen catheters during total parenteral nutrition in surgically or critically ill patients. *Arch Surg* 1986;121:591–594.
74. Williams WW. Infection control during parenteral nutrition therapy. *J Parent Ent Nutr* 1985;9:735–741.

75. Collignon PJ, Munro R, Sorrell TC. Systemic sepsis and intravenous devices: a prospective surgery. *Med J Austr* 1984;141:345–348.
76. Wang EEL, Prober CG, Ford-Jones L, et al. The management of central intravenous catheter infections. *Ped Infect Dis J* 1984;3:110–113.
77. Flynn PM, Shenep JL, Stokes DC, et al. In situ management of confirmed central venous catheter-related bacteremia. *Ped Infect Dis J* 1987;6:(8):729–734.
78. Mermal LA, Stolz SM, Maki DG. Surface antimicrobial activity of heparin-bonded and antiseptic-impregnated vascular catheters. *J Infect Dis* 1993;167(4):920–924.
79. Danzig LE, Short LJ, Collins K, et al. Bloodstream infections associated with a needless intravenous infusion system in patients receiving home infusion therapy. *JAMA* 1995;273(23):1862–1864.
80. McDonough JJ, Altemeier WA. Subclavian venous thrombosis secondary to indwelling catheters. *Surg Gynecol Obstet* 1971;133:397–400.
81. Padberg FT, Ruggiero J, Blackburn GL, et al. Central venous catheterization for parenteral nutrition. *Ann Surg* 1981;193(3):264–270.
82. Ryan JA, Abel RM, Abbott WM, et al. Catheter complications in total parenteral nutrition: a prospective study of 200 consecutive patients. *N Engl J Med* 1974;290(14):757–761.
83. Valerio D, Hussey JK, Smith FW. Central vein thrombosis associated with intravenous feeding: a prospective study. *J Parent Ent Nutr* 1981;5:240–242.
84. Andrew M, Marzinotto V, Pencharz P, et al. A cross-sectional study of catheter-related thrombosis in children receiving total parenteral nutrition at home. *J Ped* 1995;126(3):358–363.
85. Greenfield LJ. Venous and lymphatic disease. In Schwartz SS, ed. Principles of Surgery, 6th ed. McGraw-Hill, New York; 1966:997.
86. Hill SL, Berry RE. Subclavian vein thrombosis: a continuing challenge. *Surgery* 1990;108(1):1–9.
87. Horattas MC, Wright DJ, Fenton AH, et al. Changing concepts of deep venous thrombosis of the upper extremity: report of a series and review of the literature. *Surgery* 1988;104(3):561–567.
88. Haire WD, Lynch TG, Lieberman RP, et al. Duplex scans before subclavian vein catheterization predict unsuccessful catheter placement. *Arch Surg* 1992;127:229–230.
89. Lowell JA, Bothe A. Venous access: preoperative, operative and postoperative dilemmas. *Surg Clin North Am* 1991;71(6):1231–1246.
90. Bern NM, Lokich JJ, Wallach SR, et al. Very low dose of warfarin can prevent thrombosis in central venous catheters: a randomized prospective trial. *Ann Intern Med* 1990;112:423–428.
91. Fabri PJ, Mirtallo JM, Rutberg RL, et al. Incidence and prevention of thrombosis of the subclavian vein during total parenteral nutrition. *Surg Gynecol Obstet* 1982;155:238–240.

92. Wechsler RJ, Spirn PW, Conant EF, et al. Thrombosis and infection caused by thoracic venous catheters: pathogenesis and imaging findings. *Am J Roentgenol* 1993;160(3):467–471.
93. Radd II, Luna M, Khalil SAM, et al. The relationship between the thrombotic and infectious complications of central venous catheters. *JAMA* 1994;271(13):1014–1016.
94. Brismar B, Hardstedt C, Jacobson S. Diagnosis of thrombosis by catheter phlebography after prolonged central venous catheterization. *Ann Surg* 1981;194(6):779–783.
95. Torosian MH, Meranze S, McLean G, et al. Central venous access with occlusive superior central venous thrombosis. *Ann Surg* 1986;203(1):30–33.
96. Gale GB, O'Connor DM, Chu JY, et al. Restoring patency of thrombosed catheters with cryopreserved urokinase. *J Parent Ent Nutr* 1984;8(3):298–299.
97. Glynn MF, Langer B, Jeejeebhoy KN. Therapy for thrombotic occlusion of long-term intravenous alimentation catheters. *J Parent Ent Nutr* 1980;4(4):387–390.
98. Rubin RN. Local installation of small doses of streptokinase for treatment of thrombotic occlusions of long-term access catheters. *J Clin Oncol* 1983;1(9):572–573.
99. Smith NL, Ravo B, Soroff HS, et al. Successful fibrinolytic therapy for superior vena cava thrombosis secondary to long-term total parenteral nutrition. *J Parent Ent Nutr* 1985;9(1):55–57.
100. Ascher DP, Shoupe BA, Maybee D, et al. Persistent catheter-related bacteremia: clearance with antibiotics and urokinase. *J Ped Surg* 1993;28(4):627–629.
101. Jones GR, Konsler GK, Dunaway RP, et al. Prospective analysis of urokinase in the treatment of catheter sepsis in pediatric hematology-oncology patients. *J Ped Surg* 1993;28(3):360–367.
102. LaQuaglia MP, Caldwell C, Lucas A, et al. A prospective, randomized double-blind trial of bolus urokinase in the treatment of established hickman catheter sepsis in children. *J Ped Surg* 1994;29(6):742–745.
103. Wever MLG, Liem KD, Geven WB, et al. Urokinase therapy in neonates with catheter-related central venous thrombosis. *Thromb Haemostas* 1995;73(2):180–185.
104. Haire WD, Atkinson JB, Stephens LC, et al. Urokinase versus recombinant tissue plasminogen activator in thrombosed central venous catheters: a double-blinded, randomized trial. *Thromb Haemostas* 1994;72(4):543–547.
105. Krafte-Jacobs B, Sivit CJ, Mejia R, et al. Catheter-related thrombosis in critically ill children: comparison of catheters with and without heparin bonding. *J Ped* 1995;126(1):50–54.
106. Smith S, Dawson S, Hennessey R, et al. Maintenance of the patency of indwelling central venous catheters: is heparin necessary? *Am J Ped Hematol Oncol* 1991;13(2):141–143.
107. Kollef MH. Fallibility of persistent blood return for confirmation of intravascular catheter placement in patients with hemorrhagic thoracic effusions. *Chest* 1994;106(6):1906–1908.
108. Flanagan JP, Gradisar IV, Gross RJ, et al. Air embolus—a lethal complication of subclavian venipuncture. *N Engl J Med* 1969;281:488–489.

CHAPTER

9

Admixture Considerations

Yoshimi L. Clark, Pharm.D., B.C.N.S.P., and Vernon T. Lew, R.Ph., B.C.N.S.P.

INTRODUCTION

Although gut feeding is fundamental to nutrition, the importance of parenteral nutrition therapy (including home care parenteral nutrition) for patients in whom gut feeding is not feasible, has been well documented, and the specialty of therapeutic nutrition is advancing rapidly. To provide successful parenteral nutrition therapy the pharmacist holds the responsibility to compound a sterile and safe admixture that will provide maximal benefits to the patient. Compounding a safe parenteral nutrition admixture begins with an understanding of the physical and chemical characteristics of available stock solutions and their components. This chapter does not attempt to present an exhaustive review of the tremendous volume of published literature. Rather, it will focus on practical issues and recent developments that have direct relevance to practitioners who are involved with nutrition support services. In this chapter, general information including a history of parenteral nutrition, currently available products and equipment, and quality control issues in compounding an admixture are briefly discussed. We hope to provide practical information concerning available stock solutions of base nutrients, osmolarity issues, automated compounders, and helpful references for the pharmacist. Furthermore, parenteral nutrition additives, factors that influence the compatibility and stability of drugs in parenteral nutrition admixtures, and compounding procedures (both automated and manual) are discussed, emphasizing safe mixing order. Specific examples of mixing procedures are provided.

GENERAL INFORMATION

Background and Definitions

Parenteral nutrition was first administered as cow's milk in the late 1800s. The cow was brought to the hospital, and its freshly collected milk was infused into the patient.[1] Since then, parenteral nutrition admixtures and their availability have dramatically evolved to the currently prepared complete nutrient admixtures, often individualized to meet each patient's need based on his or her specific medical condition. Parenteral nutrition was first compounded as a total parenteral nutrition (TPN) 2-in-1 which is a mixture of two base nutrients, dextrose (D) and amino acids (AA) in the same bag. When safe infusion of intravenous fat emulsion (IVFE) was established,[2–5] a total nutrient admixture (TNA) 3-in-1 (also called triple mix) in which IVFE is added to the D and AA mixture became widely available.

Compounding and Base Solutions

A parenteral nutrition compounding process can be divided into four steps: (a) receiving a parenteral nutrition order and reviewing its appropriateness; (b) mixing substrates or base solutions that include AA, D, and IVFE; (c) adding electrolytes and other additives (compatible drugs) to the base nutrients mixture, and (d) checking the final product. The PN admixtures can be compounded manually using a gravity transfer method or automatically using a computerized compounder. In either case, when compounding a parenteral nutrition solution, various issues need to be considered: concentration of available stock solutions and their chemical characteristics; mixing orders; chemical compatibilities of additives and their stabilities in the TNA solutions; and storage of the admixture. Knowledge of available concentrations of base stock solutions (AA, D, and IVFE) and other additives is a basic requirement for pharmacists and pharmacy technicians in compounding these admixtures, and it becomes extremely important when designing a parenteral nutrition formula for a patient with fluid restriction or high electrolyte requirement. The selection of these products with specific concentrations determines the fluid-restricted patient's minimal parenteral restriction volume that can be compounded. Therefore, the pharmacy personnel must have knowledge of the available concentrations of each stock solution and carefully select them to meet the needs of their patient population. For compounding parenteral nutrition admixtures, glucose in the form of concentrated dextrose (50%, 60% or 70%) in water is the most commonly used carbohydrate caloric source. Amino acid preparations are available in two types, standard and modified formulations, and their concentrations range from 3.5% to 15%. Intravenous fat emulsions are available in three concentrations: 10%, 20%, and 30%.

Base Solutions: Characteristics and Products Information

Amino acid stock solution. Amino acids (AA) in the form of crystalline AA solutions serve as the protein source in parenteral nutrition admixtures. Although AA have caloric value, some practitioners do not include it in calculating the daily energy requirement.[1,2] Amino acid solutions can be broadly divided into two types; standard AA and modified (specialized) AA solutions that include age-specific and disease-specific AA formulations. They differ in composition of essential, semiessential, and nonessential AAs and nitrogen content per gram weight of the total AA. In addition, each AA solution differs in pH, electrolytes content (sodium, acetate, chloride, etc.), available concentrations, and the antioxidant used to stabilize the solution (see Tables 9.1, 9.2, 9.3, and 9.4). The most commonly used is the standard AA solution (10%–15%) that contains both essential and nonessential amino acids in balanced proportion. The age-specific or pediatric AA solutions (6%–10%) are designed to be used for both neonatal and pediatric patients. The disease-specific AA solutions are available for renal failure, hepatic failure, or hypermetabolic patients. The renal formulations were developed based on the theory of urea recycling[6] and are available in 5.2% to 6.5%, which contain primarily essential amino acids (EAA). Hepatic formulations are available in 8%, and they contain increased ratios of branched chain amino acids (BCAA) to the total amino acids (TAA) and reduced aromatic amino acids (AAA) and methionine. Hypermetabolic formulations are available in 6.9% to 15%, and they contain increased ratios of both BCAA/TAA and EAA/TAA. These AA solutions are manufactured by various pharmaceutical companies (Table 9.2). The clinical value of these disease- (hepatic and renal) specific AA formulations is controversial; various studies and reviews that have been published show inconsistent results and improved clinical outcomes have not been proven.[7–12] In addition, the majority of the patients with these disease states requires fluid restriction as a part of their medical therapy; however, these disease-specific AA solutions are available in more dilute form (5.2%–8%) than the standard AA solutions (10%–15%) resulting in a larger volume per gram protein. Some AA products are available as premixed solutions ready for infusion; these often contain various amounts of dextrose (5%, 10%, 25%) and electrolytes. In this chapter, single AA stock solutions are discussed. The AA solutions are available in 250 mL, 500 mL, and 1000 mL volumes depending on the type of

Table 9.1. Selected Standard Amino Acid Stock Solutions, their Concentrations, Components, and Chemistry

	Aminosyn (Abbott)	Aminosyn II (Abbott)		FreAmine III (McGaw)	Novamine (Clintec)		Travasol (Clintec)
Available concentrations (%)	10	10	15	10	11.4	15	10
Nitrogen concentration (g/100 ml)	1.57	1.53	2.3	1.53	1.8	2.37	1.65
Essential AA (mg/100 ml)							
Histidine	300	300	450	280	680	894	480
Isoleucine	720	660	990	690	570	749	600
Leucine	940	1000	1500	910	790	1040	730
Lysine	720	1050	1575	730	900	1180	580
Methionine	400	172	258	530	570	749	400
Phenylalanine	440	298	447	560	790	1070	560
Threonine	520	400	600	400	570	749	420
Tryptophan	160	200	300	150	190	250	180
Valine	800	500	750	660	730	960	580
Semiessential AA (mg/100 ml)							
Arginine	980	1018	1527	950	1120	1470	1150
Cysteine	—	—	—	<24	—	—	—
Taurine	—	—	—	—	—	—	—
Tyrosine	44	270	405	—	30	39	40
Nonessential AA (mg/100 ml)							
Alanine	1280	993	1490	710	1650	2170	2070
Proline	860	722	1083	1120	680	894	680
Serine	420	530	795	590	450	592	500
Glycine (Aminoacetic Acid)	1280	500	750	1400	790	1040	1030
Glutamic Acid	—	738	1107	—	570	749	—
Aspartic Acid	—	700	1050	—	330	434	—

continued on next page

Table 9.1. *Continued*

	Aminosyn (Abbott)	Aminosyn II (Abbott)		FreAmine III (McGaw)	Novamine (Clintec)		Travasol (Clintec)
Electrolytes (mEq/L)							
Sodium	—	45.3	62.7	10	—	—	—
Potassium	5.4	—	—	—	—	—	—
Acetate	148	71.8	107.6	~89	114	151	87
Chloride	—	—	—	<3	—	—	40
Phosphate (mM/L)	—	—	—	10	—	—	—
Antioxidant (mg/100 mL)	230	20	60	<100	30		
	Na hydrosulfite	Na hydrosulfite		Na bisulfite	Na metabisulfite		Na bisulfite
Ratio of composed AA (%)							
Essential AA/Total AA	50	46		49	50		45
Branched-chain AA/Total AA	25	22		23	18		19
Osmolarity (mOsm/L)	1000	873	1300	~950	1057	1388	1000
pH	5.3	5–6.5		6.5	5.2–6		6
Approved Indications							
Peripheral PN	yes	yes		yes	yes		yes
Central PN	yes	yes		yes	yes		yes
Protein Sparing	yes	yes	no	yes	yes	no	yes
How supplied (mL)	500	500		500	500		250/500
	1000	1000		1000	1000		1000/2000

Source: Data from Teasley-Strausburg et al.[2] American Society of Hospital Pharmacy,[14] and *Drug Facts and Comparison.*[15]

Table 9.2. Available Modified (Specialty) Amino Acid Stock Solutions

Type of Amino Acid Solution	Product (manufacturer)[a]	Concentration (%)	How Supplied (mL)
Pediatric formulation	a. Aminosyn-PF (A)	7, 10	250/500
	b. TrophAmine (M)	6, 10	500
Renal formulation	a. Aminosyn-RF (A)	5.2	300
	b. Amiess (C)	5.2	400
	c. RenAmin (C)	6.5	250/500
	d. NephrAmine (M)	5.4	250
Hepatic formulation	a. HepatAmine (M)	8	500
Hypermetabolic formulation	a. Aminosyn-HBC (A)	7	500/1000
	b. BranchAmin (C)	4	500
	c. FreAmine-HBC (M)	6.9	750

[a]A=Abbott; C=Clintec; M=McGaw.

Source: From American Society of Hospital Pharmacy.[14]

the AA solution. Another issue worth mentioning is that some manufacturers' studies reported that 3-in-1 parenteral nutrition admixtures made with AA solutions containing high concentrations of antioxidant (e.g., sodium hydrosulfite, 60 mg/L) have relatively decreased stability (stable up to 48 h).[13] Therefore, when compounding a 3-in-1 parenteral nutrition admixture and long-term stability is a concern, the use of AA solutions containing less antioxidant is recommended. The characteristics of selected standard and modified AA solutions are summarized in Tables 9.1, 9.2, 9.3, and 9.4).

Dextrose stock solution. Dextrose solution (D) in the form of dextrose monohydrate serves as the carbohydrate source in parenteral nutrition admixtures. Currently concentrated D solutions for use in parenteral nutrition are available as 50%, 60%, and 70% in water ($D_{50}W$, $D_{60}W$, and $D_{70}W$, respectively). When compounding parenteral nutrition admixtures chemical characteristics of dextrose solutions play an important role in the stability of the admixture. The dextrose solutions are acidic; pH ranges from 3.5 to 6.5, and osmolarities of the solutions for $D_{50}W$, $D_{60}W$, and $D_{70}W$ are 2525, 3030, and 3535 mOsm/L, respectively. The solution specific gravities may be required when compounding parenteral nutrition admixtures using a gravimetric computerized compounder. The specific gravities of $D_{50}W$, $D_{60}W$, and $D_{70}W$ are 1.17, 1.20, 1.24, respectively. The dextrose solutions are available in 500, 1000, and 2000 mL, packaged in viaflex containers. The chemical characteristics in relation to the issues on its compatibility and stability in the admixtures are further discussed later in this chapter.

Intravenous fat emulsion. Intravenous fat emulsion (IVFE) in the form of sterile fat emulsion serves as a source of calories and essential fatty acids in

Table 9.3. Pediatric Amino Acid Stock Solutions, their Concentrations, Components, and Chemistry[2,14–15]

	Amino Acid Product (manufacturer)			
	Aminosyn-PF (Abbott)		TrophAmine (McGaw)	
Available concentrations (%)	7	10	6	10
Nitrogen concentration (g/100 ml)	1.07	1.52	0.93	1.55
Essential AA (mg/100 ml)				
Histidine	220	312	290	480
Isoleucine	534	760	490	820
Leucine	831	1200	840	1400
Lysine	475	677	490	820
Methionine	125	180	200	340
Phenylalanine	300	427	290	480
Threonine	360	512	250	420
Tryptophan	125	180	120	200
Valine	452	673	470	780
Semiessential AA (mg/100 ml)				
Arginine	861	1227	730	1200
Cysteine	0	0	<140	<160
Taurine	50	70	150	250
Tyrosine	44	44	140	240
Nonessential AA (mg/100 ml)				
Alanine	490	698	320	540
Proline	570	812	410	680
Serine	347	495	230	380
Glycine (Aminoacetic Acid)	270	385	220	360
Glutamic Acid	576	820	300	500
Aspartic Acid	370	527	190	320
Electrolytes (mEq/L)				
Sodium	3.4	3.4	5	5
Acetate	33	46	56	97
Chloride	0	0	<3	<3
Antioxidant (mg/100 mL)	30 Na hydrosulfite		<50 Na metabisulfite	
Osmolarity (mOsm/L)	586	834	525	875
pH	5.4 (5–6.5)		5–6	
Approved Indications				
Peripheral and Central parenteral nutrition	yes	yes	yes	yes
How supplied (mL)	250/500	1000	500	500

Source: Data from Teasley-Strausberg et al.,[2] American Society of Hospital Pharmacy,[14] and *Drug Facts and Comparison.*[15]

Table 9.4. Disease-specific Amino Acid Stock Solutions, their Concentrations, Components, and Chemistry

	Amino Acid Products				
	Aminosyn-RF	Amiess	RenAmin	NephrAmine	HepatAmine
Available concentration (%)	5.2%	5.2%	6.5%	5.4%	8%
Manufacturer	Abbott	Clintec	Clintec	McGaw	McGaw
Nitrogen concentration (g/100 mL)	0.77	0.66	1	0.65	1.2
Essential AA (mg/100 mL)					
Histidine	429	412	420	250	240
Isoleucine	462	525	500	560	900
Leucine	726	825	600	880	110
Lysine	535	600	450	640	610
Methionine	726	825	500	880	100
Phenylalanine	726	825	490	880	100
Threonine	330	375	380	400	450
Tryptophan	165	188	160	200	66
Valine	528	600	820	640	840
Semiessential AA (mg/100 mL)					
Arginine	600	0	630	0	600
Cysteine	0	0	0	20	<20
Tyrosine	0	0	40	0	0
Nonessential AA (mg/100 mL)					
Alanine	0	0	560	0	770
Proline	0	0	350	0	800
Serine	0	0	300	0	500
Glycine (aminoacetic acid)	0	0	300	0	900

continued on next page

Table 9.4. *Continued*

	Amino Acid Products				
	Aminosyn-RF	Amiess	RenAmin	NephrAmine	HepatAmine
Electrolytes (mEq/L)					
Sodium	0	0	0	5	10
Potassium	5.4	0	0	0	0
Acetate	105	50	60	44	62
Chloride	0	0	31	<3	<3
Phosphate (mM/L)	0	0	0	0	10
Antioxidant (mg/100 mL)	60 K metabisulfite	not available	(3 mEq) Na bisulfite	<50 Na bisulfite	<100 Na bisulfite
Ratio of composed AA (%)					
Essential AA/Total AA	89	100	67	99	55
Branched-chain AA/Total AA	33	38	30	39	36
Aromatic AA/Total AA	22	19	11	20	3
Osmolarity (mOsm/L)	475	416	600	435	785
pH	5.2	6.4	5–7	6.5	6.5
How supplied (ml)	300	400	250/500	250	500

Source: Data from Teasley–Strausberg et al.,[2] American Society of Hospital Pharmacy,[14] and *Drug Facts and Comparison.*[15]

the parenteral nutrition admixture. As an emulsion derived from castor oil and cottonseed oil, IVFE was first developed in Europe and became available in the United States around the late 1950s. However, the Food and Drug Administration (FDA) banned its use because of the harmful effects frequently reported.[1] The newly formulated IVFE (currently available IVFE) that are derived from soybean oil or a combination of soybean oil and safflower oil gained FDA approval in 1981 and its safe administration has been well established.[2–5] Currently available IVFE are oil-in-water emulsions and are composed of long-chain triglycerides (derived from soybean or a combination of soybean and safflower oils), phospholipids (obtained from egg yolks as an emulsifying agent), water, glycerin (to adjust the emulsion osmolarity), and sodium hydroxide (to adjust the pH). The lipid particle size of currently available IVFE ranges from 0.3 to 0.5 μm. This becomes an important consideration for the selection of an in-line filter when administering the 3-in-1 admixture.[16–17] The IVFE is available in 10%, 20%, and 30%. The IVFE (10% and 20%) can be administered undiluted via a peripheral line; however, 30% IVFE is hypotonic (200 mOsm/L) and must be mixed as a TNA or a 3-in-1 parenteral nutrition admixture (or its osmolarity must be raised) for administration. The IVFE products differ in their fatty acid composition, osmolarity, and pH. The IVFE osmolarity ranges from 260 to 315 mOsm/L (an exception to this is Intralipid 30% which has an osmolarity of 200 mOsm/L)[18] and pH ranges from 6 to 9. Available in 50, 100, 250, and 500 mL, IVFE are packaged in glass containers. Further details are summarized in Tables 9.5 and 9.6.

Table 9.5. Currently Available Intravenous Fat Emulsions and their Components[a]

Products (Manufacturer[b])	Linoleic Acid	Linolenic Acid	Oleic Acid	Palmitic Acid	Stearic Acid	Egg Phospholipid	Glycerin
Intralipid (C) 10, 20, 30%	50	9	26	10	3.5	1.2	2.25 1.7[c]
LiposynII (A) 10, 20%	65.8	4.2	17.7	8.8	3.4	1.2	
LiposynIII (A) 10, 20%	54.5	8.3	22.4	10.5	4.2	1.2	2.5
NutriLipid (M) 10, 20%	49–60	6–9	21–26	9–13	3–5	1.2	2.21
Soyacal (AT) 10, 20%	49–60	6–9	21–26	9–13	3–5	1.2	2.21

[a]Sources of fatty acids: soybean and safflower oil for Liposyn II, soybean oil for all other products; fatty acid contents = % of total fatty acid present; egg phospholipid and glycerin contents = % by weight.

[b] A=Abbott, AT=Alpha Therapeutic, C=Clintec, M=McGaw

[c] This value is for Intralipid 30%.

Source: Data from Teasley-Strausberg et al.,[2] American Society of Hospital Pharmacy,[14] *Drug Facts and Comparison,*[15] and Clintec Nutrition Company.[18]

ADMIXTURE OSMOLARITY

Osmolarity in Peripheral Parenteral Nutrition

A parenteral nutrition admixture can be administered by a central or peripheral line. When administering via a peripheral line, the osmolarity of the parenteral nutrition admixture must be carefully evaluated for safe administration to minimize possible phlebitis. In current standard practice, maximum osmolarity for peripheral infusion is 900 mOsm/L for 2-in-1 admixture and 1200 mOsm/L for 3-in-1 admixtures (IVFE has a buffering effect).[19–21] However, if the total osmolarity exceeds 600 mOsm/L, 5 to 10 mg of hydrocortisone can be added to each liter of PN admixture to minimize venous irritation or phlebitis.[2,22]

How to Estimate the Final Osmolarity for Peripheral PN

Final osmolarity of the admixture can be easily estimated by using the individual product osmolarities (Table 9.7). Each pharmacist should be familiar with the estimation method and the final admixture osmolarity should be checked prior to compounding peripheral parenteral nutrition admixtures:

Total admixture osmolarity (mOsm/L)
= sum of the osmolarities of each product present in the admixture

REFERENCE SOURCES FOR PHARMACISTS

Parenteral nutrition admixture compatibility questions are one of the most frequently asked questions by other health care professionals such as physicians, nurses, dietitians, and other trained caregivers. It is important for the pharmacist to be familiar with factors that determine compatibility of drugs in parenteral nutrition admixtures. Various studies have been conducted to evaluate drug compatibilities and their stabilities for commonly used drugs.[2,23–27] Compatibility and stability issues are discussed in detail later in this chapter. There are many excellent references; selected references for pharmacists are listed (Table 9.8).

QUALITY ASSURANCE CONSIDERATIONS

Infection Sources and Prevention

Although parenteral nutrition therapy has been widely used since the 1980s and infections related to microbial contamination of the admixtures are uncommon, the

Table 9.6. Chemical Characteristics and Caloric Contents of Intravenous Fat Emulsion

Products and Manufacturer		pH	Osmolarity (mOsm/L)	Particle Size (micron)	Total Calorie (kcal/ml)	EFA Calorie (kcal/ml)	How Supplied (ml)
Intralipid	10%	6–8.9	280	0.5	1.1	0.45	50, 100, 500
(C)	20%	6–8.9	330	0.5	2	0.9	50, 100, 250, 500
	30%	8	200	0.5	3	1.8	500
Liposyn II	10%	6–9	258	0.4	1.1	0.59	100, 200, 500
(A)	20%	6–9	276	0.4	2	1.18	200, 500
LiposynIII	10%	6–9	284	0.4	1.1	0.49	100, 200, 500
(A)	20%	6–9	292	0.4	2	0.98	200, 500
NutriLipid	10%	6–7.9	280	0.33	1.1	0.44–0.54	50, 100, 250, 500
(M)	20%	6–7.9	315	0.33	2	0.88–1.08	250, 500
Soyacal	10%	6–7.9	280	0.33	1.1	0.44–0.54	50, 100, 250, 500
(AT)	20%	6–7.9	315	0.33	2	0.88–1.08	50, 100, 250, 500

[a]A=Abbott; AT=Alpha Therapeutic; C=Clintec; M=McGaw. [b]EFA=essential fatty acids.

Source: Data from Teasley–Strausberg et al.,[2] "American Society of Hospital Pharmacy,[14] *Drug Facts and Comparison,*[15] and Clintec Nutrition Company.[18]

Table 9.7. Approximate Osmolarity of Base Solutions and Additives of Parenteral Nutrition Admixture

Component	Osmolarity (mOsm)
Amino acid	100 per %*
Dextrose	50 per %*
Lipid 10%	2.8 per g
Lipid 20%	1.5 per g
Lipid 30%	1 per g
Calcium gluconate	1.4 per mEq
Magnesium (sulfate, gluconate)	1.7 per mEq
Potassium (acetate, chloride, phosphate)	2 per mEq
Sodium (acetate, chloride, phosphate)	2 per mEq

*final concentrations of D or AA in the PN admixture.

Source: Data from Teasley–Strausberg et al.[2] and *Remington's Pharmaceutical Sciences.*[19]

most frequent serious complication during parenteral nutrition therapy is infection. The resulting consequences not only increase health care costs but also lead to increased morbidity and mortality.[28–32] Infections result from contamination of the catheter and other attached devices during their manipulations or from the infusion of contaminated parenteral nutrition admixture. Infections related to the infusion of contaminated admixtures occur much less frequently than those related to manipulation of catheter or other devices; however, outbreaks of polymicrobial infections resulting from contaminated parenteral nutrition admixtures or IVFE infusions have been reported.[33–34] In either case the contamination hazard can be minimized by training the involved personnel, employing aseptic compounding techniques, handling and administering correctly, using filters, and storing the admixtures properly. Basically all parenteral products, especially as complex as parenteral nutrition admixtures, must be prepared in the pharmacy aseptically using laminar flow hoods.[35–37] When unstable additives, vitamins, and drugs need to be added outside of the pharmacy where a laminar flow hood is not available, the personnel must be trained in aseptic techniques to do so.

Other Problems Associated with Parenteral Nutrition Admixture Compounding

The basic goal for the pharmacist is to compound and provide an admixture that contains appropriate ingredients in appropriate amounts without contaminants or particulate matter. With such a complex process, there is increased potential for human error during the parenteral nutrition compounding. Various types of monitoring systems can be used to minimize these possible human errors. At

Table 9.8. Selected References for Pharmacists

Type	Reference
Comprehensive	a) *Nutrition support handbook: a compendium of products with guidelines for usage* Teasley-Strausberg KM, Cerra FB, Lehmann S, Shronts EP. Cincinnati: Harvey Whitney 1992.
Products and pharmacology	a) *Drug facts and comparison.* Facts and Comparisons, Inc., Updated annually.
	b) *American hospital formulary service (AHFS) drug information.* American Society of Hospital Pharmacy. Updated annually.
Physicochemical issues of compatibility	a) Niemiec PW, et al. Compatibility considerations in PN solutions. *Am J Hose Pharmacy* 1984;1:893–911
	b) Brown R, et al. *Total nutrient admixture: review J Parent Ent Nutr* 1986;10(6):650–658.
Compatibility and stability	a) Trissel LA. *Handbook on injectable drugs.* 6th Ed. Bethesda, MD: American Society of Hospital Pharmacy Special Project Division. 1994.
	b) King JC, Catania PN. *King guide to parenteral admixtures.* St. Louis: Pacemarq, 1994.
Clinical nutrition therapy	a) *Clinical nutrition: parenteral nutrition.* 2nd ed. Rombeau JL, Caldwell MD. Philadelphia: W.B. Saunders, 1993.
Guidelines and recommendations	a) ASPEN Board of Directors. Guidelines for the use of parenteral and enteral nutrition in adult and pediatric patients. *J Parent Ent Nutr* 1993;17(4, suppl):1SA–51SA.
	b) *The United States pharmacopeia: drug information for the health care professional.* Rockville, MD: USP Convention, 1996.

our facility in collaboration with intravenous room pharmacists, one of us (YLC) conducted a 1-year study of parenteral nutrition compounding errors to identify the areas that are frequently involved with errors (*unpublished*). Some important areas where errors were made are as follows:

(1) Errors in transcription (when inputting the formula into the computer).

(2) Stock solutions (containers) were inadvertently switched.

(3) Parenteral nutrition orders were written wrongly:
  i. Additive dose per liter vs per day
  ii. Additive ordered in wrong units (mEq, mM, mg)

iii. Prescribed concentration for peripheral parenteral nutrition exceeds osmolarity limitations
iv. Ca^{2+}/PO_4^{3-} exceeded solubility limit

Based on the results of this study, a specific compounding process monitoring and documentation system was developed in an attempt to minimize the compounding errors. Since the implementation of the monitoring system, each pharmacist became more alert with the areas that were identified as critical steps and errors were reduced significantly. The parenteral nutrition compounding process monitoring form developed at our hospital is shown in Figure 9.1.

Quality Assurance Plans

Developing personnel training and periodic credentialing programs may be helpful.[38] Quality assurance guidelines for prevention of infections should be developed to comply with the individual institution's infection-control policy. General guidelines are listed in the Table 9.9. Furthermore, parenteral nutrition products and equipment are being continually improved, and the nutrition support pharmacist should review them periodically and update when needed to provide maximal cost-effective parenteral nutrition therapy.

COMPOUNDING METHOD OPTIONS

Manual Compounding

Parenteral nutrition admixtures may be prepared manually using a volumetric method. In the past, pharmacy personnel have been compounding parenteral nutrition admixtures manually under a laminar flow hood. This method simply employs gravity transfers of the measured base solutions and additives into the parenteral nutrition container. The advantages of this method include no additional devices or extensive training needed, procedure easily learned, and less costly than the automated method. The disadvantages of manual compounding include the fact that it is a time-consuming process (very slow process), there is an increased contamination risk from the increased manipulation of sterile products, accuracy is less than computerized delivery systems, and no safeguard system (prone to more human errors) are available. This method can be used for a pharmacy that does not routinely have many parenteral nutrition orders.

Automated Compounding

Parenteral nutrition admixtures may be prepared automatically using a computerized compounder. A computerized automatic parenteral nutrition compounder

PHARMACIST CHECK LIST

for Parenteral Nutrition (PN) Preparation
WHMC, San Antonio, TX
May 1995

DATE:__________ **PATIENT NAME:**________________________**WARD:**______

CHECKED & CORRECT: **Initials: RPh-1/RPh-2**

Patient Name & Location ______/______

( ) **Central** Infusion or ( ) **Peripheral** Infusion:
if Peripheral --Dextrose concentration **≤ 12 %** ______/______
--Osmolarity ≤ 1200 mOsm/L (hydrocortisone dose= mg/day)

Rate/Total Volume: Continuous/Cyclic infusion (circle one) ______/______
Total vol. = (ml/day); Continuous rate = (ml/hr X 24 hrs) or
Cyclic rate = ( )

Calcium & Phosphate (K-phos & Na-phos) within limit: ______/______
( )Adult & Pediatric patient (with Aminosyn);
AA conc. **≥ 3%** ------- PO_4 (mM/L) + Ca (mEq/L) -->≤ **35** ( )
AA conc. **> 1.8%** ---- check w/authorized RPh
( )Neonatal & Pediatric Patient (with Aminosyn-PF):
See Calcium & Phosphate solubility curves posted

Insulin dose appropriate: ______/______
Previous day = (units/day) Today = (units/day)
if **>40 units/day** and **increased by >25% in 24 hrs** check w/MD or RPh

Electrolytes/Additives ordered by appropriate **Units**: ______/______
(**mEq, mM, mg, units**, per **day** or per **L**)

Compounded PN admixture check:
Variance between ordered & delivered Vol. (%) (must be ≤ ±3 %) ______

MVI & Folic Acid volume delivered: ______
MVI = (ml/**bag**) **Folic Acid** = (ml/**bag**)

Final PN admixture free from particulate matter: ______

Filter attached: ( ) 0.22 micron for 2-in-1, ( ) 1.2 micron for 3-in-1 ______

Time completed: ( )=first check ( )=final check

COMMENTS or FOLLOWUP:

____

____

Figure 9.1. Parenteral nutrition monitoring form.

Table 9.9. Recommended Quality-Assurance Plans

Preparation method	Aseptic technique: hand washing, laminar flow hood use, staff training, minimal manipulations, develop policies and procedures—see ASHP/FDA/JCAHO/NCCLVP standards[35,39–40]
End-product testing	Recommendation by CDC—no routine testing, except when infection suspected[39] Recommendations by the National Coordinating Committee on Large Volume Parenterals (NCCLVP)[40]
Addition of unstable additives for home parenteral nutrition[41]	Patient and caregivers training for proper procedures
Administration	Filter use: 0.22 μm for 2-in-1, 1.2 μm for 3-in-1, hand washing and minimal manipulations[15–16]
Storage	Store in the refrigerator at 2–8°C (36–46°F)[42–44]
Stability and sterility	Recommendation by FDA–use within 24 hr Studies conducted by home care institutions and manufacturers have shown 7–14 days when compounded aseptically and kept at 2–8°C[45–47]

first became available around the 1980s and use of an automated compounder is becoming more popular as cost-effective pharmacy practice and pharmacists' clinical roles have been emphasized. Since its initial development, the computerized compounders have become more sophisticated and reliable in terms of time saving, sterility, and most importantly its accuracy in compounding.[48–52] Furthermore, they have the capability to generate patients' clinical data, compounding reports, and other relevant information as programmed for the institution. Its disadvantages include not being cost-effective for small institutions and requiring specific training to operate the compounders and computers. Mechanical problems can also occur with a computerized compounder. The decision to employ an automated compounding system must be based on the numbers of parenteral nutrition admixtures prepared daily, frequency, patient populations (neonatal patient, critically ill patients, or routine parenteral-nutrition-requiring patients etc.), and the pharmacy's manning condition. If the patient population includes primarily neonatal or critically ill patients, parenteral nutrition admixtures will most likely be individualized and will require complex formula compounding. In this case, using a computerized automated compounder may be more beneficial because of the reduction in compounding time and contamination potential. In addition, the currently available computerized compounders are able to compound any ingredient as little as 0.2 to 1 mL within a 2% to 5% error. Currently several types of automated compounders are available; Table 9.10 compares their functions and capabilities.

Table 9.10. Comparison of Available Computerized Compounders

	Compounder (Manufacturer)		
	Nutrimix-Micro/Macro (Abbott)	MicroMacro-12/23 (Baxa)	Automix 3+3/Micromix (Clintec)
Number of ingredients	14	12/23	16
Automated microaddition	yes	yes	yes
Minimum pumping volume	1 mL	0.2 mL	1 mL/0.3 mL
Ingredient barcode check	yes	yes	yes
QA report per bag	yes	yes	no
Flush requirement-3-in-1	no	yes	yes
Pre-assembled tube sets	no	yes	no
High speed microaddition	no	yes	no
Solution delivery system	volumetric	both volumetric and gravimetric	gravimetric
Macro/Micro compounding	2 separate compounders	1 unit does both	2 separate compounders
Comments	available on a variety of placement options Tel: (708) 937-6100	purchasing options flexible to fit budget Tel: (800) 525-9567	Nutrition/health education support provided Tel: (800) 422-2751

Source: Ref. 53.

MIXING PROCEDURES

Background and Basic Principles

When mixing additives to an admixture we must consider the complex chemical milieu that exists. Parenteral nutrition admixtures are complex solutions containing various ingredients: amino acids (up to 19 different amino acids); dextrose; IVFE (5 different fatty acids and an emulsifying agent); electrolytes (up to 8); multiple vitamins (up to 12); and trace elements (up to 7). The possibilities for physicochemical incompatibilities are innumerable. Relying on visual criteria such as precipitation, color change, turbidity, and so on, as was done in many earlier compatibility studies, may lead to false negative conclusions,[36] because (a) many chemical reactions (e.g., hydrolysis, oxidation, reduction, and complexation) that occur may not necessarily be visually apparent[26]; (b) it has been reported that even the highly trained unaided human vision can only detect particles down to 50 μm in size, whereas 6-μm-sized particles can occlude the human pulmonary capillaries; and (c) some reactions that lead to incompatibility may take hours to occur, long after the parenteral nutrition admixtures have left the pharmacy. To avoid these problems, the pharmacist must recognize the potential for incompatibility using professional judgment based on experience, the knowledge of basic chemical reactions with emphasis on common reactive and labile groups, and interpreting and utilizing information from well-established references.

Factors Influencing Compatibility and Stability

There are many factors that influence parenteral nutrition admixture compatibility and stability: pH of each ingredient; chemical characteristics (reactivities and affinities), temperature, amino acid product, concentration and salt form of each additive, sequence of additives, and concentration of each base solution (AA, D, and IVFE).[54–56] Because AAs have the unique quality of titratable acidity, parenteral nutrition admixtures possess a relatively high buffering capacity and are generally able to maintain a consistent pH in spite of additive insult. In contrast, being manufactured at an acidic pH (3.5–6.5), $D_{50}W$, $D_{60}W$, and $D_{70}W$ solutions possess very little buffering capacity. Therefore, the additives' pH can have a significant influence on the final pH of the admixtures containing high concentration of dextrose.

Commonly Added Medications

A short list of medications generally acknowledged to be both compatible and therapeutically available for both TPN (2-in-1) and TNA (3-in-1) are

1. H_2 blockers (cimetidine, famotidine, ranitidine)
2. Heparin
3. Hydrocortisone
4. Regular insulin
5. Metoclopramide

It should be remembered that many medication compatibility studies rely upon visual assessment only as their study end point. The stability and pharmacologic effectiveness of the medications themselves in the parenteral nutrition admixtures is usually not addressed. This may unknowingly result in a therapeutic failure in some patients. Parenteral nutrient admixtures are primarily meant to be nutrition delivery vehicles, not drug delivery vehicles. The patient's nutritional needs should not be sacrificed because the medication in their parenteral nutrition admixture requires titration. Therefore, in general, medications requiring titration should not be added to parenteral nutrition admixtures. When considering the addition of drugs to admixtures, there is a question of incompatibility; one should scrutinize the primary literature and consult with the manufacturers of the medications and the admixture components.

Mixing Order

The mixing sequence may be a critical factor affecting the compatibility of an admixture, as detailed in the Food and Drug Administration's 1994 Safety Alert[17]. The alert included some mixing sequence recommendations to avoid potentially life threatening events. There are some well-documented incompatibilities that may occur during parenteral nutrition admixture compounding, and precautions should be taken to avoid them.

Mixing order base components. The mixing sequence of admixture base components, although not normally an issue with TPN (2-in-1), may be a crucial factor when compounding TNA (3-in-1). Totally nutrition admixtures are thermodynamically unstable because they are emulsions. Dextrose has been shown to rapidly destabilize IVFE when directly added without dilution.[57] Another important fact to keep in mind with regard to TNA is the final concentration of the IVFE. An IVFE final concentration of not less than 2% is thought to contain the minimum amount of emulsifier necessary to maintain emulsion stability. Also considered to be important are the AA final concentration of not less than 2% and a dextrose final concentration of not less than 10%.[58] Some acceptable base component mixing sequences are as follows.

1. AA → Dextrose → IVFE
2. IVFE → AA → Dextrose
3. AA/Dextrose/IVFE (all mixed simultaneously)

If possible, the admixture container should be gently agitated or swirled during and after the mixing sequence to produce a homogeneous mixture. It has also been suggested that the IVFE be added late in the mixing sequence.[52] This supposedly would preserve the visual component of compatibility analysis. There are other mixing sequences that may be acceptable and one should consult either the literature on the admixture component or the manufacturer for additional information.

Mixing order of additives. As a general rule, for TNA, electrolytes, especially divalent cations (e.g., calcium gluconate, magnesium sulfate, and zinc chloride), trace elements, or undiluted dextrose should not be directly added to a minimally diluted or undiluted IVFE.[52] Compounding lines should be flushed between the addition of any potentially incompatible components.

At present, the only electrolytes for which the mixing sequence seems critical are calcium and phosphate. It is recommended that phosphate be added early and that calcium be added late in the mixing sequence, in attempts to add calcium to a maximally diluted admixture.[17] For compatibility calculations, utilize the admixture volume at the time these electrolytes are actually added.[52]

It has been suggested that colored substances, such as parenteral multivitamins, be added late in the mixing sequence.[59] This would supposedly preserve the visual component of compatibility analysis as with IVFE. As previously mentioned, the admixtures should be periodically agitated and visually checked for precipitates during the mixing sequence.

Example. Utilizing the study parenteral nutrition admixture formulation from the institution involved in the 1994 safety alert,[60] we can compare the study mixing sequence against a recommended mixing sequence which was subsequently published[52] (Table 9-11). The study mixing sequence utilized was

7 — 8 — 6 — 2 — 9 — 10 — 4 — 5 — 3 — 1 — 11,

where no flush water was used. The recommended mixing sequence is

6 — 7 — 8 — 11 — 2 — 3 — 2 — 1 — 5 — 9 — 4 — 12 — 10 — 12,

where number 2 additives were split to flush the line; and number 12 used flush water twice. The recommended mixing sequence follows several of the previously detailed recommendations by the FDA and resulted in a compatible parenteral nutrition admixture, whereas the study mixing sequence resulted in the formation of a precipitate in all test TNA.

Table 9.11. Study Parenteral Nutrition Admixture Formulation

1. Dextrose 70%	100.00 mL/L
2. FreAmine 10%	330.00 mL/L
3. Lipids 20%	195.00 mL/L
4. Calcium gluconate 10%	22.22 mL/L
5. $MgSO_4$ 50%	0.96 mL/L
6. Potassium phosphate	3.90 mL/L
7. KCl	3.07 mL/L
8. NaCl 23.4%	5.97 mL/L
9. MVI-12	10.52 mL/bag
10. MTE-4	2.63 mL/bag
11. Sterile water	325.73 mL/L
12. Flush water	20.00 mL per flush

Source: From Hill et al.[60]

CONCLUSION

Parenteral nutrition is an important advance for patients who are unable to be fed adequately via the gut. Compounding parenteral nutrition admixtures that are complex infusates, containing up to 50 (or more) different chemical entities, is an intensive process and challenging to the pharmacist. Providing successful nutritional therapy by compounding appropriate, safe, and optimally effective admixtures is the ultimate goal of the pharmacist as a member of the nutrition support team. To accomplish this goal, the pharmacist must have a working knowledge of admixture components and their chemistries. The pharmacy personnel must also be cognizant of possible interactions among the components in the parenteral nutrition admixture, as well as of proper mixing order to ensure their compatibility and stability. Finally, the admixture must be compounded aseptically and the pharmacist should be familiar with both manual and automated compounding techniques. By advancing our knowledge and skills, we hope that we can accomplish our tasks and contribute to successful nutritional therapy.

REFERENCES

1. Grant JP. *Handbook of total parenteral nutrition.* 2nd ed. Baltimore: W.B. Saunders, 1992.
2. Teasley-Strausburg KM, Cerra FB, Lehmann S, Shronts EP. *Nutrition support handbook: a compendium of products with guidelines for usage.* Whitney Harvey, Cincinnati: 1992.

3. Black CD, Popovich NG. Stability of intravenous fat emulsions. *Arch Surg.* 1980;115:891.
4. Macfie J, Smith RC, Hill GL. Glucose or fat as a nonprotein energy source. *Gastroenterology* 1981;80:103–107.
5. Brown R, Quercia RA, Sigman RT. Total nutrient admixture: a review. *J Parent Ent Nutr* 1986;10(6):650–658.
6. Giordano C. Use of exogenous and endogenous urea for protein synthesis in normal and uremic patients. *J Lab Clin Med* 1963;62:231–246.
7. Calvey H, Davis M, Williams R. Controlled trial of nutritional supplementations, with and without branched-chain amino acid enrichment, in treatment of acute alcoholic hepatitis. *J Hepatol* 1985;1:141–51.
8. Freund H, Atamian S, Fischer J. Comparative study of parenteral nutrition in renal failure using essential and nonessential amino acid containing solutions. *Surg Gynecol Obstet* 1980;151:652–656.
9. Hiyama DT, Fischer JE. Nutritional support in hepatic failure. Current thought in practice. *Nutr Clin Pract* 1988;3:96–105.
10. Jimenez-Jimenez FJ, Leyba CO, Mendez SM, et al. Prospective study on the efficacy of branched-chain amino acids in septic patients. *J Parent Ent Nutr* 1991;15:252–261.
11. Mirtallo JM, Schneider PJ, Mavko K, et al. A comparison of essential and general amino acid infusions in the nutritional support of patients with compromised renal fusion. *J Parent Ent Nutr* 1982;6:109–113.
12. Melnik G. Value of specialty intravenous amino acid solutions. *Am J Hosp Pharm* 1996;53(6):671–674.
13. Shaw HL. Important drug information-appropriate instructions for the storage and use of 3-in-1 TPN admixtures (information letter). Abbott Park, IL: Medical/Regulatory Affairs & Advanced Research, Abbott Hospital Product Division, Oct. 2, 1992.
14. Caloric agents. In *American hospital formulary service drug information.* Bethesda, MD: American Society of Hospital Pharmacy, 1993:1594–1603.
15. Intravenous nutritional therapy. In *Drug Facts and Comparison.* St. Louis: *Facts and Comparisons,* 1995:100–171.
16. Lewis JS. Justification for use of 1.2 micron end-line filters on total nutrient admixtures. *Hosp Pharm* 1993;28(7):656–658.
17. Food and Drug Administration. Safety alert: hazards of precipitation associated with parenteral nutrition. *Am J Hosp Pharm* 1994;51:1427–1428.
18. Products information. Clintec Nutrition Company, Mar. 1994.
19. *Remington's Pharmaceutical Sciences.* 17th ed. Easton, PA: Mack Publishing, 1985:1456–1465.
20. Nordenstrom J, Jeppsson B, Loven L, et al. Peripheral nutrition effect of a standardized compounded mixture on infusion phlebitis. *Br J Surg* 1991;78:1391–1394.
21. Kehoe JE, Mihranaian MH, Masser EL, et al. Use of 20% fat emulsion in peripheral parenteral nutrition. *J Parent Ent Nutr* 1984;8:647–651.

22. Messing B, Leverve X, Regand D, et al. Peripheral venous complications of a hyperosmolar (960 mOsm) nutritive mixture: the effect of heparin and hydrocortisone. A multicenter double-blinded random study in 98 patients. *Clin Nutr* 1986;5:57–61.
23. Athanikar N, Boyer B, Deamer R, et al. Visual compatibility of 30 additives with a parenteral nutrition solution. *Am J Hosp Pharm* 1979;36:511–513.
24. Baptista RJ, Dumas GJ, Bistrian BR, et al. Compatibility of total nutrient admixtures and secondary cardiovascular medications. *Am J Hosp Pharm* 1985;42:777–778.
25. Baptista RJ, Lawrence RW. Compatibility of total nutrient admixtures and secondary antibiotic infusions. *Am J Hosp Pharm* 1985;42:362–363.
26. Bullock L, Fitzgerald JE, Glick MR. Stability of famotidine 20 and 50 mg/L in total nutrient admixtures. *Am J Hosp Pharm* 1989;46:2326–2328.
27. Newton DW. Physicochemical determinants of incompatibility and instability in injectable drug solutions and admixtures. *Am J Hosp Pharm* 1978;35:1213–1222.
28. Gilbert M, Gallagher SC, Eads M, et al. Microbial growth patterns in a total parenteral nutrition formulation containing lipid emulsion. *J Parent Ent Nutr* 1986;10(5):494–497.
29. Danzing LE, Short LJ, Collins K, et al. Bloodstream infections associated with a needleless intravenous infusion system in patients receiving home infusion therapy. *JAMA*. 1995;273(23):1862–1864.
30. Goldmann DA, Maki DG. Infection control in total parenteral nutrition. *JAMA* 1973;223:1360–1364.
31. Thompson B, Robinson LA. Infection control of parenteral nutrition solutions. *NCP* 1991;6(2):49–54.
32. Williams WW. Infection control during parenteral nutrition therapy. *J Parent Ent Nutr* 1985;9:735–746.
33. Maki DG. Infections due to infusion therapy. In Bennett JV, Brachman PS, eds. *Hospital Infections,* 3rd ed. Boston, MA: Little Brown, 1992:849–898.
34. Mershon J, Nogami W, Williams J, et al. Bacterial/fungi growth in a combined parenteral nutrition solution. *J Parent Ent Nutr* 1986;10(5):498–450.
35. National Coordination Committee on Large Volume Parenterals. Recommended methods for compounding intravenous admixtures in hospitals. *Am J Hosp Pharm* 1975;32:261–270.
36. ASHP Invitational conference on quality assurance for pharmacy-prepared sterile products. *Am J Hosp Pharm* 1991;48:2391–2397.
37. Driscoll DF. Total nutrient admixtures: theory and practice. *NCP* 1995;10(3):114–119.
38. Morris BG, Avis KE, Bowles GC. Quality-control plan for intravenous admixture programs II: validation of operator technique. *Am J Hosp Pharm* 1980;37:668–672.
39. National Coordination Committee on Large Volume Parenterals (NCCLVP). Recommended standards of practice, policies, and procedures for intravenous therapy. *Am J Hosp Pharm* 1980;37:660–663.

40. National Coordination Committee on Large Volume Parenterals (NCCLVP). Recommended guidelines for quality assurance in hospital centralized intravenous admixture services. *Am J Hosp Pharm* 1980;37:645–655.
41. Simmons BP. *Guidelines for prevention of intravascular infections.* Atlanta: Centers for Disease Control, 1981.
42. Smith JL, Canham JE, Wells PA. Effect of phototherapy light, sodium bisulfite, and pH on vitamin stability in total parenteral nutrition admixtures. *J Parent Ent Nutr* 1988;12(4):394–402.
43. Bettner FS, Stennett DJ. Effects of pH, temperature, concentration, and time on particle counts in lipid-containing total parenteral nutrition admixtures. *J Parent Ent Nutr* 1986;10(4):375–380.
44. Sayfed FA, Tripp MG, Sukumaran KB, et al. Stability of total nutrient admixtures using various intravenous fat emulsions. *Am J Hosp Pharm* 1987;44:2271–2280.
45. Driscoll DF. Practical considerations regarding the use of total nutrient admixtures. *Am J Hosp Pharm* 1986;43:416–419.
46. Driscoll DF. Physicochemical stability of total nutrient admixtures. *Am J Hosp Pharm* 1995;52:623–634.
47. Jurgens RW, Henry RS, Welco A. Amino acid stability in a mixed parenteral nutrition solution. *Am J Hosp Pharm* 1981;38:1358–1359.
48. Seidel AM, Woller TW, Somani S, et al. Effect of computer software on time required to prepare parenteral nutrient solutions. *Am J Hosp Pharm* 1991;48:270–275.
49. McClendon RR. Comparative evaluation of methods used to compound parenteral nutrition solutions. *Nutr Sup Serv* 1983;3(12):46–54.
50. Mackenzie N, Poole RL. The use of computers in parenteral nutrition. In Kerner, Jr., JA, ed. *Manual of Pediatric Parenteral Nutrition* New York: Wiley, 1983:329–333.
51. Karnack RA, Karnack CC. Computerized operations manual for sterile products. *Am J Hosp Pharm* 1984;41:2586–2588.
52. Driscoll DF. Automated compounders for parenteral nutrition admixtures. *J Parent Ent Nutr* 1994;18(4):385–386.
53. 9th annual IV devices buyer's guide. *Pharm Pract News* Apr. 1995;9(Apr):18–19.
54. Niemiec PW, Vanderveen TW. Compatibility considerations in parenteral nutrient solutions. *Am J Hosp Pharm* 1984;41:893–911.
55. Sturgeon RJ, Athanikar NK, Henry RS, et al. Titratable acidities of crystalline amino acid admixtures. *Am J Hosp Pharm* 1980;37:388–390.
56. Chan JCM, Malekzadeh M, Hurley J. pH and titratable acidity of amino acid mixtures used in hyperalimentation. *JAMA* 1972;220:1119–1120.
57. Wells PA. *Guide to total nutrient admixtures.* Chicago: Precept Press, 1992.
58. Driscoll DF. Nutrient admixtures. pp. 1–2. Sep., 1995. Hyperalimentation course. Boston: New England Deaconess Hospital, Harvard Medical School. 1995:1–2.
59. Keck-Jones L. Clinical alert. *ASPEN Commun.* May 27, 1994.
60. Hill SE, Heldman LS, Goo EDH, et al. Fatal microvascular pulmonary emboli from precipitation of a total nutrient admixture solution. *J Parent Ent Nutr* 1996;20:81–87.

Peripheral Parenteral Nutrition

CHAPTER 10

Linda M. Codina, R.D., C.N.S.D.

INTRODUCTION

Peripheral parenteral nutrition (PPN) is the administration of amino acids, calories, maintenance electrolytes, minerals, vitamins, and trace elements via the peripheral venous system. This form of nutrition support differs from central total parenteral nutrition (TPN) in that, although it can also safely provide nutrition support, its uses are more limited. It is generally accepted that the peripheral vein can only tolerate solutions of up to 900 mOsm/L for a limited amount of time. Although increased levels of osmolarity may be tolerated with the administration of lipids admixed with the formula, the nutrient solutions cannot be as concentrated as with central parenteral nutrition. Therefore, relatively larger volumes of fluid are required for PPN to meet the nutritional needs of most patients. In many cases, PPN can provide 100% of the protein requirements of nutritionally compromised, hospitalized patients, but only a portion of the total energy needs. In addition, the limited functional durability of the peripheral veins leads to an access shortage within a short period of time. Because of these inherent limitations, PPN is not the appropriate choice of intravenous feeding for patients with severe metabolic stress or significant preexisting malnutrition.

In addition, it is a poor choice for patients anticipated to be unable to take oral nutrition for a prolonged period of time. In such situations intensive central parenteral nutritional therapy would be necessary. Most appropriately, PPN should be reserved for less stressed patients who will soon resume eating or who require intravenous feeding for less than 10 to 14 days.[1]

Historically, patients unable to eat have been supported with 5% dextrose and water administered by the peripheral vein to provide glucose and reduce exogenous protein catabolism for gluconeogenesis. However, secondary to the limited calories and lack of protein provided by such dextrose solutions, the need became evident for an intravenous solution that could be delivered peripherally and would provide more efficient protein sparing than the standard intravenous solutions.

In the early 1970s Blackburn et al. first advocated the use of isotonic amino acids for peripheral intravenous feeding.[2] It was believed at that time that such regimens would enable patients to produce an insulin-suppressed state, similar to that of fasting. There would be a mobilization of fat stores and subsequent development of a starvation-like ketosis with a reduction in net protein catabolism.[3,4]

Other investigators including Greenberg et al.[5] and Gazzaniga et al.[6] have also studied the effects of exogenous energy substrates and have concluded that a mixed-fuel substrate causes better protein sparing than a pure isotonic amino acid solution. Today, it is well-recognized that an ideal PPN regimen should provide all nutrient requirements in physiologic amounts.[1]

This chapter reviews the indications, contraindications, and benefits of PPN and discusses the general considerations for its proper prescription.

INDICATIONS

Peripheral parenteral nutrition is used to partially or totally meet nutrient needs in patients unable to do so by the oral or enteral route or when central access is not feasible or in the best interest of the patient. Candidates for PPN may include patients with one or more of the following clinical conditions:

- Gastrointestinal disorders, such as bowel obstruction or ileus or inflammatory bowel disease where oral or enteral feeding is contraindicated
- To adjunct enteral or oral nutrition in the patient with limited nutrient intake on a short-term basis
- Where parenteral nutrition support is anticipated for less than 10–14 days until oral feeding can be resumed or central TPN is needed[7,8]
- When there is a high risk of catheter sepsis

- If the patient is able to tolerate lipid emulsions
- Patients with mild metabolic stress or mild preexisting malnutrition
- For patients previously on central TPN who had central access removed secondary to sepsis or complications
- As a temporary measure until more aggressive nutrition support can be initiated
- Where patients have good peripheral venous access

CONTRAINDICATIONS

As mentioned previously, total nutritional repletion, especially in the long term, is difficult to achieve with PPN because of the limited concentration of nutrients that the peripheral vein can tolerate, the quick exhaustion of peripheral venous access, and the large fluid volume usually required to deliver adequate nutrition. Therefore, PPN is not the appropriate choice for patients with hypermetabolism, severe metabolic stress, significant preexisting malnutrition, hemodynamic instability, or increased fluid and electrolyte requirements. Also PPN is contraindicated for patients with poor venous access, renal, cardiac, or pulmonary insufficiency[8] and those with egg yolk allergy unable to tolerate lipid emulsions[1].

BENEFITS OF PERIPHERAL PARENTERAL NUTRITION

There are several advantages associated with the use of PPN:

1. Peripheral access is easily obtained by nonsurgical personnel
2. The risk of serious complications is minimal while avoiding risks of TPN including hyperglycemia, hyperinsulinemia, catheter sepsis, and insertion complications
3. PPN does not require tapering when discontinuing, because low concentrations of dextrose are used[9]
4. PPN is less expensive and less complex than TPN
5. PPN can provide preoperative nutrition support
6. PPN provides more efficient protein sparing than the more traditional administration of 5% dextrose in water[4,8]

Despite the many advantages associated with the use of PPN, there are also some disadvantages. Although PPN may meet nutrition needs in lower ranges, the necessary volume needed to meet the entire requirements of most patients may not be tolerated particularly for patients that must be volume restricted. The larger fluid volumes necessary to meet nutritional needs may cause congestive heart failure, pulmonary edema, or worsen ascites in susceptible patients. Additionally, the use of PPN is associated with the risk of development of thrombophlebitis.[10] Rypins et al. found that the use of standard amino acid–glucose PPN solutions led to a 65% incidence of phlebitis.[11] Causes include the catheter material, duration of therapy, composition of the solution (pH, osmolarity, etc), location of the catheter site, decreased blood flow, and the infusion rate.[11] In general, a peripheral vein can tolerate peripheral parenteral solutions for only 1 to 2 weeks before vein irritation and phlebitis develop. As confirmed by Rypins and colleagues, the major factor in the development of phlebitis is the osmolarity of the solution, which will be discussed later.

TYPICAL PPN REGIMENS

The most common PPN solutions contain 5% to 10% dextrose, 3% to 5% amino acids (AA), and sufficient lipid to provide a significant source of calories and essential fatty acids. The lipid can be admixed in the PPN solution or given as a separate 10% to 20% intravenous lipid emulsion.[8] The daily amount of lipid should not exceed 2.5 g/kg of body weight.[4] When calculating the nutrients to be provided in the PPN, it is important to account for calorie intake from other sources such as tube feeding, intravenous fluids, and oral diet.

Although the most common carbohydrate source is dextrose, there is a commercially available product, Procalamine (Kendall-McGaw) that utilizes glycerol instead of dextrose. This premixed peripheral parenteral nutrition solution contains 3% AA and 3% glycerol, is heat stable, contains a mixture of maintenance electrolytes, and has an osmolarity of 735 mOsm. This solution can greatly reduce pharmacy mixing time and may be reasonable or even preferable when compounding of parenteral solutions is limited.[1,12] However in most cases, it would require exceedingly large volumes of this product to meet nutrition needs, especially protein.[1,8,13] Procalamine, however, can be supplemented with 10% to 20% intravenous lipid emulsion to attempt to more closely meet calorie needs.

In general, the current practice of PPN administration to adult patients includes the provision of 100 to 150 g dextrose combined with 1.0 to 1.5 g protein/kg of body weight, and usually 1 to 2 g lipid/kg or 500 mL of 10 or 20% lipid emulsion per day. Such regimens can be expected to minimize the depletion of lean body mass in mild to moderately stressed patients for approximately 1 week.[1] Maintenance levels of electrolytes and standard doses of multivitamins, minerals, and trace elements should also be added.

GENERAL CONSIDERATIONS

A major limitation in the provision of PPN is the osmolarity of the solution. Osmolarity is contributed primarily by amino acids (100 mOsm/1% final AA concentration), secondarily by dextrose (50 mOsm/1% final dextrose concentration), and minimally by the lipid, which is isotonic and neutral in pH.[1] The addition of electrolyte and mineral additives further contribute to the osmolarity of the base solution, approximately 1 mEq per ion or usually an additional 100 to 150 mOsm.[1,8] Solutions with an osmolarity of greater than 600 mOsm/L have been associated with a higher incidence of phlebitis.[8,14] Generally, solutions of greater than 10% dextrose and 5% amino acids exceed the recommended osmolarity limits. To adequately nourish most patients with solutions that do not exceed the osmolarity limits, large volumes of fluid are required. Unfortunately, 3L/day or 125 mL/hr is the maximum volume tolerated by most patients.

To circumvent the problems of fluid overload, attempts have been made to decrease the risk of phlebitis caused by PPN solutions of higher osmolarity. Additives such as hydrocortisone, heparin, and sodium hydroxide have been shown to be effective. Hydrocortisone was selected as an additive to PPN for its ability to produce an antiinflammatory effect on the vascular epithelium. Sodium heparin is added to prevent the formation of a fibrin clot that can form around the tip of the cannula in the vein, leading to thrombus formation, catheter occlusion, and infusion thrombophlebitis.[7,8,10] Makarewicz and Freeman demonstrated a marked reduction in the incidence of phlebitis when a mixture of hydrocortisone, heparin, and sodium hydroxide was added to the PPN solutions.[15] In a second prospective, randomized study by Payne-James and Khawaja, the addition of heparin and hydrocortisone reduced the incidence of peripheral thrombophlebitis by approximately 50%.[7] When lipids are infused separately, it is now a common practice to accept an upper limit of 900 to 1100 mOsm/L provided that hydrocortisone 5 mg/L and heparin 500 to 1000 U/L are included in the solution.[16–18]

Recent experience with total nutrient admixtures (TNA), also known as 3-in-1 solutions, indicates that the incidence of phlebitis may be reduced even with solution osmolarities in the 1100 to 1300 range.[1] In a TNA, lipid is added directly to the bag containing the dextrose and amino acid mixture. These solutions can simplify nursing care and promote better tolerance to the lipid in most patients. The success in the administration of such high-osmolarity admixtures via the peripheral route has been attributed to several factors. The concommitant infusion of lipid may limit phlebitis by protecting the lining of the vein.[8] Recent studies suggest that the lipid may also promote improved tolerance from the buffering and dilutional effects of the isoosmolar lipid with the higher pH of the AA solution.[8,16,19]

Our institution utilizes a standard 5% dextrose–4% amino acid–3% lipid PPN admixture to promote maximum delivery of nutrients while minimizing the risk of development of thrombophlebitis. Additionally, 5 mg/L hydrocortisone and 1000 U/L heparin are routinely added. Using this PPN formula as an example, the osmolarity of a solution can be calculated from the two major substrates:

$$5\%\ \text{dextrose} + 4\%\ \text{AA} = 250 + 400\ \text{mOsm/L}$$
$$= 650\ \text{mOsm/L}$$

Even with the addition of approximately 150 mOsm/L contributed from other additives, the osmolarity of this formulation still is within the recommended range for solution osmolarity.

Like patients receiving central parenteral nutrition, those on PPN should also be monitored for daily weights, fluid balance lipid tolerance, and electrolyte abnormalities. It is also advisable to rotate the PPN insertion site every 2 to 3 days and monitor for signs of peripheral vein intolerance such as phlebitis or infiltration at the catheter site.[8,16]

FUTURE CONSIDERATIONS

In the near future, the indications and uses of PPN may be expanded to benefit a greater number of patients as a result of the recent developments in metabolic modulators and in venous access technology.

Growth Hormone

Recent studies have supported the use of recombinant growth hormone as an adjunct to PPN. This anabolic hormone has been shown to improve nitrogen balance even with hypocaloric feeding.[20,21] Manson et al.[20] randomized volunteers to receive 10 mg/day of growth hormone or placebo. All volunteers were then fed daily with PPN which provided 1 g of protein/kg and only 30% to 100% of required calories. The patients given the growth hormone demonstrated improved nitrogen balance and increased protein synthesis. Similar results were also reported by Lehmann and coworkers.[21] The use of metabolic modulators such as growth hormone may enable patients with increased nutritional requirements or fluid restrictions to improve the nutritional potency of otherwise insufficient PPN.

Peripherally Inserted Central Catheters

The recent popularity of the peripherally inserted central catheter (PICC) offers the unique opportunity to access the central circulation from a peripheral insertion site. These catheters are introduced into the basilic or cephalic veins of the arm and threaded so that the distal tip is positioned into the superior vena cava. The PICC functions similarly to the centrally inserted catheter but has fewer complications. They can also be placed by specially trained nurses. Using these access devices, more concentrated hypertonic solutions can be infused. Therefore, fluid restriction or the need for a greater delivery of nutrients can be accommodated.

CONCLUSION

The use of PPN may be indicated to improve nitrogen balance and maintain nutritional status in a carefully selected group of hospitalized patients requiring parenteral nutrition support for only 1 to 2 weeks. Daily PPN regimens typically contain 100 to 150 g dextrose, 1.5 g protein/kg of body weight, and 1 to 2 g lipid/kg of body weight per day in a maximum of approximately 3 L of fluid daily. Although this form of nutrition support can be used as a temporary holding regimen to transition to more aggressive nutrition support, it is currently not appropriate for patients with long-term needs, hypermetabolism, fluid restrictions, or poor venous access. Thus, potential PPN candidates should be carefully assessed initially and monitored closely regarding nutrition goals, clinical limitations, and possible alternative feeding modalities to ensure that the use of PPN is appropriate.

REFERENCES

1. Miller SJ. Peripheral parenteral nutrition: theory and practice. *Hosp Pharm* 1991;26:796–801.
2. Blackburn GL, Flatt JP, Clowes GHA Jr, et al. Protein sparing therapy during periods of starvation with sepsis or trauma. *Ann Surg* 1973;177:588–594.
3. Blackburn GL, Flatt JP, Clowes GHA Jr, et al. Peripheral intravenous feeding with isotonic amino acid solutions. *Am J Surg* 1973;125:447–454.
4. Landengham SV, Newmark SR. Peripheral vein nutrition. *Clin Trends Hosp Pharm* 1982;4:26–31.
5. Greenberg GR, Marliss EB, Anderson GH, et al. Protein-sparing therapy in postoperative patients. *N Engl J Med* 1976;294:1411–1416.
6. Gazzaniga AB, Day AT, Bartlett RH, et al. Endogenous caloric sources and nitrogen balance. *Arch Surg* 1976;111:1357–1361.
7. Payne-James JJ, Khawaja HT. First choice for total parenteral nutrition: the peripheral route. *J Parent Ent Nutr* 1993;17:468–418.
8. Barber JR, Hack SL. Clinical considerations in peripheral parenteral nutrition. *Clin Trends Hosp Pharm* 1990;4:57–59.
9. Mahan LK, Krause M. *Krause's food, nutrition, and diet therapy.* Philadelphia: W.B. Saunders, 1992:515.
10. Kohlhardt SR, Smith RC. Peripheral versus central intravenous nutrition: comparison of two delivery systems. *Br J Surg* 1994;81:919–923.
11. Rypins EB, Johnson BH, Reder B, et al. Three-phase study of phlebits in patients receiving peripheral intravenous hyperalimentation. *Am J Surg* 1990;159:222–225.

12. Freeman JB, Fairfull-Smith R, Rodman GH, et al. Safety and efficacy of a new peripheral intravenously administered amino acid solution containing glycerol and electrolytes. *Surg Gynecol Obstet* 1983;156:625–631.
13. Waxman K, Day AT, Stellin GP, et al. Safety and efficacy of glycerol and amino acids combination with lipid emulsion for peripheral parenteral nutrition support. *J Parent Ent Nutr* 1992;16:374–378.
14. Gazitua R, Wilson K, Bistrian BR, et al. Factors determining peripheral tolerance to amino acid infusions. *Arch Surg* 1979;114:897–900.
15. Makarewicz PA, Freeman JB. Prevention of superficial phlebitis during peripheral parenteral nutrition. *Am J Surg* 1986;151:126–129.
16. Alpers DH, Clouse RE, Stenson WF. *Manual of nutrition therapeutics.* Boston: Little, Brown, 1988:257–260.
17. Isaacs JW, Millikan WJ, Stackhouse J, et al. Parenteral nutrition of adults with a 900 milliosmolar solution via peripheral veins. *Am J Clin Nutr* 1977;30:552–559.
18. Ang SD, Daly JM. Potential complications and monitoring of patients receiving total parenteral nutrition. In Rombeau JL, Caldwell MD, eds. *Clinical nutrition, Vol. 2. Parenteral nutrition.* Philadelphia: W.B. Saunders, 1986.
19. Hohein DF, O'Callaghan TA, Joswiak BJ, et al. 1990. Clinical experience with three-in-one admixtures administered peripherally. *Nutr Clin Pract* 1990;5:118–122.
20. Manson JM, Smith RJ, Wilmore DW. Growth hormone stimulates protein synthesis during hypocaloric parenteral nutrition. *Ann Surg* 1988;208:136–142.
21. Lehmann SL, Teasley KM, Konstantinides NN, et al. Growth hormone enables effective nutrition by peripheral vein in postoperative patients: a pilot study. *J Am Coll Nutr* 1990;9:610–615.

Disease-Specific and Patient-Specific Formulations

Wendy Swails Bollinger, R.D., and
Timothy J. Babineau, M.D.

INTRODUCTION

The development and clinical application of total parenteral nutrition (TPN) in the late 1960s by Dudrick and colleagues[1] represented a major advance in the care of hospitalized patients. Over the ensuing 30 years, the practice of administering TPN has been refined. Traditionally, TPN solutions were available only as standardized solutions, such as 4.25% amino acids and 25% dextrose. Hence, clinicians were forced to choose the formula that most closely approximated the patient's nutritional requirements. Unfortunately, such therapy often resulted in the development of metabolic complications related to overfeeding and glucose intolerance. Although some institutions still use standardized TPN solutions, clinicians have come to appreciate the need for parenteral solutions that can be tailored to meet the nutrient needs of the individual (i.e., patient-specific TPN). Since the advent of total nutrient admixtures, patient-specific TPN solutions have become a reality, thus making the use of standardized TPN solutions suboptimal. In addition, evidence suggests that the manipulation of certain nutrient substrates may significantly alter the response to particular illnesses (i.e., disease-specific TPN) and facilitate the healing process. These findings suggest that patient-specific feeding with disease-specific formulas may hold promise for improved patient outcome. The purpose of this chapter is twofold. First, it will provide practical guidelines on how to write a patient-specific TPN order. Secondly, it will describe how the alteration of certain TPN components may be beneficial in various disease and stress states.

PATIENT-SPECIFIC FORMULATIONS

At our institution, parenteral solutions are specifically formulated for each patient. The pharmacy stocks only maximally concentrated base nutrient solutions: 70% dextrose (D), 10% amino acids (AA), and 20% lipids (L). Such a system allows parenteral solutions to be maximally concentrated in the least amount of volume possible, which is essential for fluid-restricted patients. Such stock solutions are then mixed in one container to make up the so-called total nutrient admixture (TNA). Total nutrient admixtures allow clinicians to create a variety of customized formulas that either contain fat (3-in-1 admixtures) such as $A_6\ D_{15}\ L_2$ (6% amino acids, 15% dextrose, and 2% lipids in 1000 mL), $A_7\ D_{14}\ L_2$, and $A_7\ D_{10}\ L_3$, or do not contain fat (2-in-1 admixtures) such as $A_7\ D_{21}$ and $A_8\ D_{14}$.

Energy Requirements

Prior to the development of indirect calorimetry, the Harris-Benedict equation (HBE)[2] (multiplied by a "stress factor" based on the patient's underlying illness) was the most widely accepted method for determining energy requirements. In more recent years, it has been shown through the use of indirect calorimetry that caloric needs in moderately stressed patients can accurately be estimated by using either the baseline HBE (no additional stress factor) or 22 to 25 kcal/kg.[3–5] These results suggest that the practice of multiplying the HBE by a "stress factor" may result in the overfeeding of moderately stressed patients. Overfeeding, in turn, may predispose patients to complications such as poor glucose control, macrophage dysfunction, respiratory decompensation, and liver steatosis. In addition, it is important to note that the average age of the patients in most of these studies was 61 years. Thus, a slightly lower energy expenditure might be expected in the elderly because lean tissue mass decreases with age. In contrast, younger patients and those that are severely stressed (e.g., major burn, multiple trauma, closed head injury) will likely have higher caloric requirements (40 to 45 kcal/kg).[6]

Protein Requirements

One of the primary goals of nutrition support is to reduce protein catabolism while maintaining protein synthesis. Thus, in the absence of significant renal or hepatic disease, the stressed patient should receive at least 1.5 g protein/kg of body weight. Protein provided daily at levels in excess of 2 g/kg exceeds the body's maximal utilization rate and subsequently results in increased ureagenesis.[7]

Protein requirements should be based on actual body weight for patients without significant weight loss, on usual body weight for patients experiencing a recent weight loss, and on ideal body weight for the obese patient.[8] Nitrogen balance can be a useful tool for adjusting protein intake. However, it is important to remember that, due to the accelerated protein catabolism that accompanies stress and injury, nitrogen balance can be difficult to achieve, regardless of protein intake, in a stressed patient.

Carbohydrate Requirements

The rate at which glucose is produced by the liver is approximately 2 $mg/kg \cdot min^{-1}$,[9] which is equivalent roughly to 200 g/day for a 70 kg individual. In stressed patients, the optimal rate of glucose utilization by the liver has been shown to be approximately 4 $mg/kg \cdot min^{-1}$ or 400 g glucose/day in a 70 kg individual.[10] Thus, carbohydrate needs are typically estimated to be between 2 and 4 $mg/kg \cdot min^{-1}$. When glucose infusion exceeds this rate, complications such as hepatic lipogenesis, hepatic dysfunction, and hyperglycemia (with an increased risk of infection) may arise.[11,12] In addition, when the carbohydrate load exceeds the actual energy expenditure, the net lipogenesis significantly increases oxygen consumption and carbon dioxide production, resulting in an increased respiratory quotient (i.e., $RQ > 1$).[13] One way to minimize the development of hepatic and respiratory problems associated with dextrose overfeeding is to provide a portion of the calories in the TPN as lipid.

Lipid Requirements

In the past, lipids were not added directly to the main TPN solution but rather were infused separately (i.e., piggy-backed) over a shorter time period. However, due to refinements in the preparation and delivery of TPN, lipids can now be added to the same container as the carbohydrate and protein (i.e. TNA or 3-in-1 admixture). Although both methods of lipid administration are still used today, it appears to be preferable to administer lipids in a continuous fashion as is done with the TNA rather than in the traditional intermittent or bolus fashion. Seidner and colleagues[14] noted that the ability of the reticuloendothial system (RES) to clear bacteria was impaired when 20% lipid emulsions were infused over 10 hr for three consecutive days at a rate providing 0.13 g $lipid/kg \cdot hr^{-1}$ or the equivalent of approximately 85 g lipid/day. Although the optimal percentage of calories that should be infused as fat is unknown, this study provides clinicians with a maximum limit. At our institution, we recommend that lipids should ideally provide no more than 30% of total energy requirements.

Table 11.1. Normal Fluid and Electrolyte Concentrations in Gastrointestinal Secretions

Fluid	Volume (mL/d)	Electrolyes (mEq/L)			
		Na	K	HCO_3	Cl
Biliary	300–700	120–160	3–12	30–50	80–120
Gastric	1000–2500	10–100	5–40		80–150
Pancreatic	500–1000	100–160	3–10	60–110	50–90
Small bowel	1000–3000	80–150	2–10	20–40	60–130

Source: From Driscoll DF.[66]

Micronutrient and Vitamin Requirements

Micronutrient requirements are discussed in depth elsewhere in this book (see Chapter 5). It is important to emphasize, however, that electrolytes, calcium, magnesium, and phosphate are typically adjusted in the TPN on a daily basis in accordance with the patient's clinical condition and available laboratory values. When possible, all fluid and electrolyte losses (i.e. nasogastric and other drainage outputs, urine, stool, fistulae, etc.) should be quantified to aid in replacement via the TPN (Table 11.1). All patients receiving TPN should receive at least 3 mL of trace minerals (M.T.E.-5 TM) and 10 mL of multivitamins (MVI-12 TM). Vitamin K is not a component of the adult MVI-12 TM formulation and therefore must be given once a week (10 mg) by a subcutaneous injection, unless the patient has a coagulation disorder. Certain clinical circumstances may warrant additional selenium (i.e., long-term home TPN) and/or zinc (i.e., patients with high ileostomy/diarrheal outputs).

Writing a Patient-Specific TPN Order

Tables 11.2 and 11.3 summarize the above guidelines and demonstrate how they can be applied when writing a patient-specific TPN order. Once the patient's nutrient requirements have been determined, the clinician must verify that the quantities of nutrients needed will be able to "fit" into the TPN container; in other words, the TPN order must recognize certain pharmacologic restrictions and be stoichiochemically feasible. To aid in this process our institution developed the equation shown in Table 11.3 (i.e., macronutrient calculator) to define the upper concentration limit when using stock solutions of 70% dextrose, 10% amino acids, and 20% lipids.[15] As shown in Table 11.3, the patient-specific TPN order written for the 60-kg patient adheres to the restrictions of the macronutrient calculator equation. In fact, because the result is less than 70 (i.e., 61.2), it indicates that there is still "room" within the TPN bag to increase the final

Table 11.2. Guidelines to Maximize Benefits and Minimize Complications of Parenteral Nutrition

1. Avoid calorie and glucose overload
   a. 25 to 30 $kcal/kg \cdot day^{-1}$
   b. 2 to 4 mg $dextrose/kg \cdot min^{-1} \cdot day^{-1}$
2. Avoid fat overload
   a. ≤30 % of total energy requirements
   b. Provide as a continuous infusion
3. Avoid protein catabolism
   a. 1.5 to 2.0 g $protein/kg \cdot day^{-1}$
   b. BCAA-enriched formulations may offer certain neurologic and metabolic advantages to patients with hepatic encephalopathy and significant renal failure (i.e., BUN > 100 mg/dL)
4. Avoid micronutrient deficiencies
   a. 10 mL of multivitamins/day (MVI-12 TM)
   b. 3 mL of trace minerals (M.T.E.-5 TM)
   c. 10 mg Vitamin K/week (subcutaneous injection), unless coagulation disorder present

Source: Adapted from Cerra.[67]

concentration of one or more of the macronutrients. However, in reviewing the current TPN order, we see that it provides the desired energy and protein requirements without exceeding the desired dextrose and lipid requirements; subsequently, the TPN order does not need to be altered. If, however, this particular patient required a fluid restriction, then another option would be to maximally concentrate the TPN order as shown in Table 11.4 so that the desired amounts of nutrients are provided in the least amount of fluid possible. Note that it is generally preferable for the pharmacy to have the final volume of TPN ordered in increments of 250 mL (i.e., 1000 mL, 1250 mL, 1500 mL, 1750 mL, 2000 mL, etc.).

DISEASE-SPECIFIC FORMULATIONS

Traditionally, the term disease-specific was reserved for the category of commercially available, specialized products designed to meet the altered needs of patients with specific illnesses. In the case of parenteral nutrition, this category referred to the modified amino acid solutions designed primarily for patients with hepatic and renal failure. Today, the definition of disease-specific has been expanded to also include the alteration of standard TPN components, as well as the addition of nonstandard TPN substrates (i.e., glutamine) in response to various disease and stress states.

Table 11.3. Application of the Patient-Specific TPN Feeding Strategy

Patient characteristics	60-kg male, status: postgastrointestinal surgery; delayed return of bowel function.
Nutritional goals	Calories = (60 kg) (25 kcal/kg) = 1500 kcal/day Protein = (60 kg) (1.5 g/kg) = 90 g protein/day Carbohydrate = (2–4 mg/kg/min) (60 kg) (1.44 min/day) = 173–346 g carbohydrate/day Fat = ≤ 30% total kcal = 450 kcal ÷ 9 g/kcal = ≤ 50 g fat/day Volume = (60 kg) (30 mL/kg) = 1800 mL TPN/day (90 g protein) (4 kcal/g) = 360 kcal from protein (260 g carbohydrate) (3.4 kcal/g) = 884 kcal from carbohydrate 1500 kcal – (360 kcal + 884 kcal) = 256 kcal remaining 256 kcal ÷ 9 g fat/kcal = ~30 g lipid/day
CONVERT ABOVE NUTRIENT NEEDS TO PERCENTAGES WHEN WRITING THE TPN ORDER:	
Protein	$\frac{90\ \text{g}}{1800\ \text{mL}} = \frac{X\ \text{g}}{100\ \text{mL}}$ $X = 5$ g/100 mL or 5%
Carbohydrate	$\frac{345\ \text{g}}{1800\ \text{mL}} = \frac{X\ \text{g}}{100\ \text{mL}}$ $X = 19.2$ g/100 mL or 19.2%
Fat	$\frac{30\ \text{g}}{1800\ \text{mL}} = \frac{X\ \text{g}}{100\ \text{mL}}$ $X = 1.7$ g/100 mL or 1.7%[a]
Pharmacy stock solutions	10% amino acids, 70% dextrose, 20% lipids
Macronutrient calculator	(A dfc × 7) + D dfc + (L dfc × 3.5) ≤ 70
Definitions	A = amino acids; D = dextrose; L = lipid dfc = desired final concentration of each macronutrient as a percentage (g/100 mL) in the final TPN volume. The multiplying factors of the macronutrient calculator vary with the pharmacy inventory and are derived from: $\frac{\text{Most concentrated macronutrient carried by pharmacy (\%)}}{\text{Stock concentration of the individual macronutrient (\%)}}$ Example: A dfc = 70/10 = 7; D dfc = 70/70 = 1; L dfc = 70/20 = 3.5
Macronutrient calculator check for patient-specific formula	(5 × 7) + (19.2) + (2 × 3.5) = 61.2 ≤ 70
Final patient-specific TPN order	1.8 L A_5 $D_{19.2}$ L_2

[a]Three-in-one admixtures must contain at least 2% lipids

Table 11.4. How to Maximally Concentrate a Patient-specific TPN Order for a Fluid-restricted Patient

Current TPN order: 1.8 L A_3 $D_{19.2}$ L_2

1. Determine concentration of the current TPN order:

$$\frac{61.2}{70} \times 100 = 87\% \text{ concentrated}$$

2. Determine percentage of free water in current TPN order:
   100% − 87% = 13% free water
3. Determine minimum volume needed to maximally concentrate current TPN order:
   1800 mL × 0.87 = 1575 mL
4. Convert macronutrients to percentages to determine new TPN order:

Protein: $$\frac{90\text{ g}}{1575\text{ mL}} = \frac{X\text{g}}{100\text{ mL}} \quad X = 5.7\text{ g/100 mL or } 5.7\%$$

Carbohydrate: $$\frac{345\text{ g}}{1575\text{ mL}} = \frac{X\text{g}}{100\text{ mL}} \quad X = 21.9\text{ g/100 mL or } 21.9\%$$

Fat: $$\frac{30\text{ g}}{1575\text{ mL}} = \frac{X\text{g}}{100\text{ mL}} \quad X = 1.9\text{ g/100 mL or } 1.9\%^{a}$$

5. Macronutrient calculator check for new patient-specific TPN formula:

$$(5.7 \times 7) + (21.9) + (2 \times 3.5) = 68.8 \leq 70$$

6. New maximally concentrated patient-specific TPN order:
   1.575 L $A_{5.7}$ $D_{21.9}$ L_2

[a] Three-in-one admixtures must contain at least 2% lipids

Diabetes

Total parenteral nutrition can be a safe and effective means of providing nutritional support in the diabetic patient or the patient with so-called stress-induced diabetes provided a few practical guidelines are followed. First, it is prudent to limit the total dextrose infusion to between 100 and 150 g/day in the initial TPN order for the type I diabetic, and between 150 and 200 g/day for the type II diabetic or patient with stress-induced diabetes.[16] When exogenous dextrose administration exceeds the postabsorptive production rate by the liver (i.e., 2 $mg/kg{\cdot}min^{-1}$ or approximately 200 g/day for a 70 kg individual) blood glucose management can become difficult. At our institution, we strive to maintain blood glucose levels above 100 mg/dL to minimize the risks associated with hypoglycemia and below 220 mg/dL to reduce the risk of infection. This upper limit is based on evidence suggesting that severe hyperglycemia (i.e., blood glucose levels > 220 mg/dL) interferes with phagocytic function[17] and may be associated with the development of nosocomial infections.[18–20]

Maintaining good blood glucose control in this population often requires the administration of insulin in the TPN. If the patient was an insulin-dependent diabetic prior to hospitalization, then the amount of insulin provided in the initial TPN order should be based on the patient's nonstressed prehospitalization requirements. At our institution, we generally add one half of the patient's home daily insulin requirement to the initial TPN order.[21] As the caloric content of the TPN approximates 90 to 120% of the diabetic patient's estimated basal energy expenditure, insulin needs, on average, will be approximately 200% of the patient's prehospitalization insulin requirement.[22] If the patient's prehospitalization insulin requirement is not known, then an estimate of approximately 0.1 U insulin/g dextrose can be used.[23]

For type II diabetics and type I diabetics in whom the optimal amount of insulin has yet to be determined, a sliding scale coverage of subcutaneous regular insulin is frequently required to treat blood glucose levels that are above 200 to 250 mg/dL. On each subsequent day, two-thirds of the amount of insulin administered subcutaneously during the previous 24 hr is added to the TPN along with the amount of insulin originally prescribed.[21] Once the amount of insulin needed to consistently maintain blood glucose levels below 220 mg/dL is determined, the dextrose level in the TPN can be gradually advanced (i.e., 50 g/day) while maintaining the established dextrose-to-insulin ratio. For example, if a patient is receiving 150 g of dextrose with 25 U of insulin, then the dextrose-to-insulin ratio would be 6:1. Thus, increasing the dextrose to 200 g/day would require approximately 33 U of insulin to maintain the dextrose-to-insulin ratio. This dextrose-to-insulin ratio will generally remain stable unless an infection or some other metabolic stress arises. As noted by Hongsermeier and Bistrian[22], infected patients exhibited a lower dextrose-to-insulin ratio (i.e., more insulin per amount of dextrose) than noninfected patients (3.2 ± 1.5 vs. 4.5 ± 1.0 g dextrose/unit insulin; $p<0.05$) due to the increase in insulin resistance that accompanies infection.

Cancer

Many of the clinical trials that have evaluated the efficacy of providing nutrition support to patients with cancer have yielded conflicting results. Thus, the indications for TPN in the cancer patient are often based more on clinical bias rather than on information derived from randomized prospective trials. To be of substantial benefit in this patient population, parenteral nutrition must yield a measurable improvement in clinical parameters. There is now evidence to suggest that certain nutrients (e.g., glutamine, arginine, nucleotides, and omega-3 fatty acids from fish oil) may alter various metabolic and outcome parameters when administered to patients with cancer.[24–27] The majority of these studies, however, examined the effects of enterally administering these so-called immunostimulatory nutrients. Thus, with the exception of glutamine, it is unknown whether the parenteral administration of these nutrients will yield the same promising results in this patient population. To date, omega-3 fatty acids and nucleotides are not commercially available in a parenteral form, and arginine is only present in small amounts (<2% of total amino acid content) in current amino acid solutions.

Perhaps one of the most intriguing findings in this area of research was demonstrated in two recent clinical trials where the administration of TPN supplemented with glutamine significantly altered several outcome parameters in patients undergoing bone marrow transplantation.[22,23] Zielger et al.[24] randomized 45 patients undergoing allogeneic bone marrow transplantation for hematologic malignancies to receive either standard TPN or TPN supplemented daily with 0.57 g of free glutamine/kg body weight. Fewer patients receiving the glutamine-supplemented TPN developed infections (13% versus 43% in the standard TPN group; $p = 0.041$). In addition, the average length of hospital stay following transplantation was 29 ± 1 days in the glutamine-supplemented group compared to 36 ± 2 days in the standard TPN group ($p = 0.017$). Schloerb and Amare[25] also observed a significant reduction in length of hospital stay in patients undergoing allogeneic or autologous bone marrow transplantations who received glutamine-supplemented TPN. Patients[29] were randomized to receive either TPN supplemented with approximately 40 g of free glutamine/day or a standard TPN solution. The group receiving the glutamine-supplemented TPN had an average length of hospital stay of 27 ± 1 days compared to 33 ± 2 days for the group receiving the standard TPN group ($p<0.05$). Unlike Ziegler et al.,[24] Schloerb and Amare[25] did not note a significant difference between the two feeding groups with respect to infectious complications. However, Schloerb and Amare's patient population consisted of patients with hematologic malignancies as well as solid tumors, whereas, in contrast, Ziegler et al.[24] only studied patients with hematologic malignancies. In addition, Schloerb and Amare[25] suggest that their patients had a higher severity of illness than those in Ziegler et al.'s study as evidenced by a higher mortality rate (14% versus 0%, respectively). Taken together, however, these two trials suggest that the supplementation of TPN with approximately 40 g of free glutamine/day may be beneficial in patients undergoing bone marrow transplantations. Additional prospective, randomized trials will help to identify other types of cancer populations that may also favorably respond to parenteral glutamine supplementation.

Currently, commercially available parenteral amino acid solutions do not contain the free amino acid glutamine for several reasons. First, glutamine has traditionally been considered to be a nonessential amino acid, because sufficient quantities are synthesized by the body in healthy subjects.[28] However, during periods of stress, the body's requirement for glutamine exceeds its production.[29–31] Secondly, it was originally thought that glutamine had a shorter shelf-life than the other amino acids commonly used in commercially available parenteral amino acid solutions. However, Schloerb and Amare[25] recently reported that the free-glutamine-supplemented parenteral solution used in their clinical trial was able to be stored at 5° Celsius for up to 6 weeks without loss of more than 5% of the glutamine. In addition, other authors have reported no appreciable changes in glutamine-enriched parenteral formulas stored at room temperature for 24 hr.[32–34]

Liver Dysfunction

Modified amino acid formulations are available for patients with hepatic disease. These formulas differ from conventional amino acids solutions in that they contain a greater concentration of branched-chain amino acids (BCAA) and a lower amount of the aromatic amino acids (AAA). The use of BCAA-enriched, low AAA formulas in patients with encephalopathy is based predominantly upon the AAA/false neurotransmitter theory.[35] Fischer postulated that the decrease in BCAA and increase in AAA plasma concentrations seen in patients with hepatic dysfunction allows a disproportionate amount of AAAs to cross the blood–brain barrier. As a result, there is an increase in serum levels of "false neurotransmitters" (octopamine and phenyethylanine) and a concommitant decrease in serum levels of normal neurotransmitters (dopamine and norepinephrine). In addition, there is an increase in serum serotonin concentration (physiologic neuroinhibitor) due to excess tryptophan. This alteration in the ratio of BCAAs to AAAs is believed to contribute to the development of encephalopathy. The modified amino acid profile of BCAA-enriched solutions is thought to counterbalance the altered plasma amino acid concentrations seen in patients with encephalopathy from liver dysfunction.

There have been a number of clinical trials that have evaluated the efficacy of using parenteral branched-chain amino acid (BCAA) solutions in patients with encephalopathy due to liver failure (36–38). The results of these trials, however, have been inconsistent, most likely because there are differences in experimental design, patient population, type and amount of BCAA solution administered, and diversity of the control group (Table 11.5). Nevertheless, these trials clearly demonstrate that BCAA solutions can be safely administered to patients with liver disease in amounts that provide 1.0 to 1.5 g protein/kg·day^{-1} without exacerbating preexisting encephalopathy. Thus, patients with liver dysfunction who demonstrate a worsening of encephalopathy when fed standard amino acid solutions may benefit from receiving BCAA-enriched solutions. Furthermore, there is evidence to suggest that BCAA-enriched solutions are better utilized by such patients and may improve hepatic protein synthesis when compared to conventional AA solutions.[39] Branched-chain amino acid-enriched solutions may, therefore, potentially offer certain neurologic and metabolic advantages over conventional amino acid solutions when administered to liver disease patients with encephalopathy.

Although clinicians often restrict protein in patients with hepatic encephalopathy, there is evidence that a low protein intake (0.5 g protein/kg dry weight) is associated with worsening hepatic encephalopathy, whereas a higher protein intake (1.1 g protein/kg dry weight) correlates with improvement in hepatic encephalopathy.[40] In view of this recent finding and the fact that the primary goal of nutrition support during hepatic failure is to provide sufficient protein to

Table 11.5. Parenteral BCAA Clinical Trials

Author	Patients	Experimental Group	Control Group	Results/Conclusions
Rossi-Fanelli et al.[36]	34 patients with liver cirrhosis/grade III to IV encephalopathy	100% BCAA (57 g protein/day) + 20% dextrose	Lactulose + 20% dextrose	No significant change in encephalopathy. BCAAs are as effective as lactulose in reversing hepatic encephalopathy.
Wahran et al.[37]	50 patients with liver cirrhosis/grade II to IV encephalopathy	100% BCAA (40 g protein/day) + dextrose + lipid	Dextrose + lipid	No significant change in encephalopathy.
Cerra et al.[38]	75 patients with liver cirrhosis/grade II to IV encephalopathy	35% BCAA-enriched (maximum of 85 g protein/day) + placebo tablets	25% dextrose + oral neomycin	Significant improvement in encephalopathy and survival in BCAA-enriched group.

From Swails, et al.[68]

support protein synthesis and to promote regeneration of liver cells, it seems reasonable that this patient population should receive at least 1 g protein/kg body weight.

Renal Failure

The first specialized parenteral amino acid solution to be developed was a formula containing only essential amino acids (NephrAmine) that was intended to minimize urea accumulation in patients with renal failure.[16] The development of this product was spurred by early studies showing that the administration of essential amino acids (EAA) as the sole dietary source of nitrogen resulted in a lowering of blood urea nitrogen levels and a reduction of uremic symptoms in patients with chronic renal failure who were not being dialyzed.[41,42] Despite the theoretical advantages of essential amino acid therapy, the clinical efficacy of using such products in patients with renal failure remains controversial. Several investigators compared the use of EAAs alone to a combination of EAAs plus nonessential amino acids (NEAAs) in patients with acute renal failure and found no difference between the two feeding groups with respect to survival or recovery of renal function.[43–45]

It is our institution's practice to use a BCAA-enriched solution rather than an EAA-based solution in patients with significant renal dysfunction (i.e., blood urea nitrogen concentration > 100 mg/dL).[16,46] There are several reasons why we prefer using BCAA-enriched solutions over EAA and standard amino acid solutions in this patient population. First, as renal failure progresses, both EAAs and NEAAs are required. To date, only one commercially available renal failure formula contains any appreciable amount of NEAAs (i.e., RenAmin), the rest are virtually devoid of NEAAs (Table 11.6). In contrast, the BCAA-enriched formula used at our institution (made by combining our standard amino acid solution with a commercially available BCAA-enriched formula) provides a more complete balance of amino acids (EAAs plus NEAAs). Secondly, because BCAAs are preferentially used by the body for protein synthesis, there is a greater decrease in the rate of urea production with BCAA-enriched solutions compared to conventional amino acid solutions. The final amino acid composition of the BCAA-enriched solution used at our institution is 50% BCAA and 50% non-BCAA. In comparison, the standard amino acid solution stocked by our pharmacy contains only 19% BCAA.

Another nutrient substrate that requires special attention when prescribing TPN for renal failure patients is glucose. The glucose calories absorbed from the dialysate during peritoneal dialysis are often overlooked and may inadvertently result in overfeeding.[47] Although the amount of glucose absorbed during peritoneal dialysis will depend upon the number of exchanges, as well as the volume and the glucose concentration per exchange, the TPN solution should be altered

Table 11.6. Commercially Available Parenteral Renal Failure Formulations

	Product Name (Manufacturer)			
	Aminess (Clintec)	Aminosyn-RF (Abbott)	NephrAmine (McGaw)	RenAmin (Clintec)
Amino acid concentration	5.2%	5.2%	5.4%	6.5%
EAA (mg/100 mL)				
Isoleucine	525	462	560	500
Leucine	825	726	880	600
Lysine	600	535	640	450
Methionine	825	726	880	500
Phenylalanine	825	726	880	490
Threonine	375	330	400	380
Tryptophan	188	165	200	160
Valine	600	528	640	820
Total EAA	5175	4627	5330	4320
NEAA (mg/100 mL)				
Alanine	—	—	—	560
Arginine	—	600	—	630
Histidine	412[a]	429[a]	250[a]	420[a]
Proline	—	—	—	350
Serine	—	—	—	300
Taurine	—	—	—	—
Tyrosine	—	—	—	40[a]
Glycine	—	—	—	300
Glutamic Acid	—	—	—	—
Aspartic Acid	—	—	—	—
Cysteine	—	—	<20	—
Total NEAA	0[a]	600[a]	<20[a]	2140[a]
EAA%	100	89	100	66

[a]Histidine and tyrosine are considered essential amino acids in infants and patients with renal failure; this is reflected in the total EAA, total NEAA, and EAA%.

Source: Adapted from Parenteral and Enteral Nutrition. In *American Medical Association Drug Evaluations Annual* 1995.

accordingly. At our institution, we have taken a number of aliquot samples from exchanged dialysate and have noted that a fixed glucose concentration of approximately 800 mg% occurs, regardless of the diaysate's dextrose concentration.[16] Table 11.7 illustrates how the dextrose provided in the TPN solution should be adjusted based on the number of dextrose calories being absorbed from the dialysate.

Table 11.7. Adjusting TPN to Account for Calories from Dialysate

Caloric requirements	1800 kcal/day
Diaysis regimen	4.25% dextrose; 2 liter exchanges 4 times/day
Dextrose content of dialysate	85 g dextrose/exchange × 4 exchanges/day = 340 g dextrose
Dextrose content of exchanged dialysate	16 g dextrose/exchange (800 mg%) × 4 exchanges/day = 64 g dextrose
Calories absorbed during dialysis	340 g dextrose in – 64 g dextrose out = 276 g dextrose or 802 kcal
TPN calorie adjustment	1800 kcal required – 802 kcal from dialysate = 998 kcal to be provided in TPN

Source: Adapted from Driscoll and Bistrian.[16]

Pulmonary Dysfunction

Accurate caloric delivery is critical in the ventilator-dependent patient. Inadequate feeding of calories and protein can lead to increased catabolism of respiratory muscles (including the diaphragm) contributing to further respiratory dysfunction. Overfeeding results in increased carbon dioxide production and minute ventilation, which may prolong ventilator dependence. Thus, in the patient requiring prolonged ventilatory support, it is often helpful to use the metabolic cart to determine the patient's resting energy expenditure so that accurate energy delivery can be ensured.

There has been much debate regarding the optimal mix of nutrient substrates (carbohydrate, protein, and fat) when feeding patients with pulmonary dysfunction. Several investigators have shown that respiratory failure may be exacerbated by the administration of high glucose loads to patients with impaired respiratory function.[48–49] Feeding dextrose above the rate at which it can be oxidized leads to glycogen and/or lipid formation. Since glycogen stores are rapidly repleted, hepatic lipogenesis is often the predominate metabolic process. As lipogenesis occurs, carbon dioxide production is increased and the respiratory quotient subsequently increases. As the lung increases its ventilatory capacity in an effort to eliminate the excess carbon dioxide, a greater minute ventilation occurs. Although this increase appears to be insignificant in healthy subjects with normal lung function, it may further exacerbate respiratory failure and potentially complicate weaning from ventilatory support. Thus, when hypercapnea is present, manipulation of the carbohydrate content of the TPN may be useful. Burke and colleagues[10] noted that the optimal rate of glucose utilization in stressed patients is approximately 4 $mg/kg \cdot min^{-1}$. Some authors have suggested that the weaning process may be further facilitated by administering a high-fat, low-carbohydrate diet in this patient population.[49,50]

Although the administration of lipids can aid in decreasing the amount of dextrose provided in the TPN (and subsequent carbon dioxide production), excessive administration of currently available intravenous lipid emulsions can be detrimental. The intravenous lipid emulsions currently available in the United States are derived solely from long-chain triglycerides (LCTs) of the omega-6 polyunsaturated fatty acid family (e.g., safflower and soybean oil). Although omega-6 fatty acids are a rich source of the essential fatty acid linoleic acid, excessive provision of linoleic acid may lead to an increased production of arachidonic acid and its metabolites prostaglandin E_2, thromboxane A_2, and leukotriene B_4. These eicosanoids are known to enhance vasoconstriction, platelet aggregation, neutrophil migration, immunosuppression, cytokine depression, and free-radical formation, all of which may predispose patients to a worsening inflammatory state and sepsis.[51] Most complications associated with intravenous lipid administration have occurred when lipid calories were provided in excess of 1.0 kcal/kg·hr^{-1} (or the equivalent of approximately 0.11 g lipid/kg·hr^{-1}[52] and administered in a bolus fashion (i.e., infused over 10 h) for three consecutive days.[14]

Although not yet commercially available, intravenous lipid emulsions containing omega-3 polyunsaturated fatty acids from fish oil may prove therapeutically beneficial for patients with respiratory dysfunction. Omega-3 fatty acids differ from omega-6 fatty acids in that they produce prostaglandins and thromboxanes of the 3 series, rather than the 2 series, and leukotrienes of the 5 series, rather than the 4 series. These eicosanoids have been shown to exhibit less potent vasoconstrictive, platelet aggregatory, and inflammatory properties than the corresponding omega-6 fatty acid metabolites. In a small pilot study,[53] patients with cystic fibrosis were randomized to receive either an intravenous lipid emulsion containing 10% fish oil or a standard 10% soybean oil intravenous lipid emulsion. Both lipid emulsions provided 150 mg lipid/kg body weight and were intravenously infused daily over a 6 hr period for 4 weeks. Pulmonary function studies revealed that patients receiving the fish-oil-containing lipid emulsion exhibited improved expiratory flow-parameters compared to those receiving the standard intravenous lipid emulsion.

Severe Protein-Calorie Malnutrition

A myriad of hormonal and metabolic derangements occur as a result of prolonged semistarvation. Thus, haphazard refeeding of the severely malnourished patient (i.e., >20% weight loss, serum albumin <2.5 g/dL, or >30% lean tissue loss by anthropometry, creatinine excretion, or body composition measurement) can result in potentially life-threatening complications.[54,55] The primary consideration in refeeding the severely malnourished patient, therefore, is to ensure that the refeeding process is not too aggressive.[56,57]

At our institution, we initially estimate daily energy requirements in this patient population to be approximately 22 kcal/kg (using the patient's actual weight). After 1 week of feeding at or slightly below this level, calories may gradually be advanced to repletion levels (i.e., 25 to 30 kcal/kg). Weight gain should be closely monitored, recognizing that a gain in excess of 1 kg per week should be avoided, as it likely represents water gains rather than lean tissue repletion. The infusion of hypertonic dextrose induces an increase in insulin secretion from the pancreas. In turn, this hyperinsulinemia results in an antinatriuresis and a subsequent expansion of extracellular fluid volume. In some patients, this rapid increase in fluid volume may lead to cardiac problems. Dextrose calories should therefore initially be limited to the basal glucose production rate by the liver: 2 $mg/kg \cdot min^{-1}$ or approximately 150 to 200 g dextrose/day. Fluid retention can be further minimized by restricting fluid intake to 800 mL/day and sodium intake to no more than 20 mEq/day during the first few days to weeks of treatment.[56]

Refeeding the severely malnourished patient can also cause sudden decreases in serum phosphorus, potassium, and magnesium concentrations when supplementation in the TPN is inadequate. During starvation, serum phosphorus levels are maintained by a reduced renal excretion[54] and an increased mobilization of phosphorus from the bone.[58] However, once refeeding is initiated, serum phosphorus levels may precipitously fall as a result of the sudden increase in insulin and carbohydrate serum levels stimulating the intracellular uptake of phosphorus. In addition, phosphorus is needed for lean tissue anabolism at a rate of 0.08 g phosphate/g nitrogen.[59] Untreated hypophosphatemia can result in respiratory failure, neuromuscular and myocardial dysfunction, and ultimately death.[55,60] In the patient with normal renal function, approximately 30 mmol of phosphorus provided in the TPN daily should maintain phosphorus balance.[16]

Hyperinsulinemia and lean tissue anabolism also stimulate the intracellular uptake of potassium.[59,61] In addition, potassium is required for glycogen synthesis at a rate of approximately 1 mEq potassium/3 g of glycogen produced.[62] Thus, the infusion of glucose into the glycogen-depleted patient causes an intracellular potassium shift. Like phosphorus and potassium, magnesium is also required for lean tissue anabolism; approximately 0.5 mEq of magnesium is retained per gram of nitrogen.[59] Hypomagnesemia can also lead to hypocalcemia, hypokalemia, and hypophosphatemia.[63]

Providing TPN to the severely malnourished patient can be safely accomplished if the clinician is familiar with and able to prevent the severe fluid and electrolyte shifts that are associated with refeeding such patients. In turn, the morbidity and mortality associated with refeeding will be reduced.

Critical Illness

The ideal nutritional approach in the critically ill patient is to minimize the metabolic alterations and the catabolism associated with critical illness. In this setting, TPN should be viewed as a vehicle for providing metabolic as well as nutritional support.

Regulation of fluid, electrolyte, and acid-base homeostasis are given first priority in the critically ill patient. Fluid retention and excess extracellular fluid are common in critically ill patients due to a hypersecretion of antidiuretic hormone, aldosterone, and insulin. In addition, an overzealous administration of intravenous fluids, particularly those containing dextrose and sodium, often exacerbates the problem. Thus, TPN is often initially limited to 1000 mL/day, thereby precluding the ability to meet the full protein and calorie requirements of the patient. When the volume of TPN must be restricted, one should use maximally concentrated nutrient solutions to provide as much protein as possible with sufficient energy for the body to efficiently utilize the protein. At our institution, the most concentrated nutrient solutions stocked by the pharmacy are 10% amino acids and 70% dextrose. Thus, the pharmacy is able to deliver 70 g of protein and 210 g of dextrose in 1000 mL, providing approximately 1000 calories. There are, however, 15% amino acid solutions available that allow 105 g of protein to be delivered in 1000 mL while maintaining the same amount of dextrose (i.e., 210 g). Once the patient begins to diurese, the nutrition support volume can be increased, thereby allowing the full nutrient needs of the individual to be met.

Electrolyte abnormalities are common in the critically ill and require daily monitoring and correction. Hyponatremia is commonly seen in patients who have an excess in total body water from liver or renal disease. In addition, H-2 blocking agents (used in combination with continuous nasogastric suction) may lead to excess sodium losses from the stomach, whereas aggressive diuretic therapy results in a decreased reabsorption of sodium and an increased free-water clearance by the kidney. Conversely, hypernatremia often arises because of the provision of parenteral antibiotics as the sodium salt and the mineralocorticoid effect that exogenous steroid administration has on the kidney.[21] It is important to note that correction of serum sodium concentration should not exceed 10 mEq/L within 24-hr period to avoid potentially lethal osmotic changes than can occur within the brain.[64]

As mentioned previously in this chapter, the aggressive administration of dextrose in the TPN can precipitate the development of hypophosphatemia. In addition, certain medications such as aluminum-containing antacids can bind phosphorus and subsequently increase the parenteral requirement for phosphorus from approximately 30 mmol/day to between 45 and 60 mmol/day.[16] Another frequent cause of hypophosphatemia in the critically ill patient is the use of corticosteroids. Because these medications enhance the urinary excretion of phosphorus, parenteral requirements may be as high as 60 to 80 mmol of phosphorus/day.[16]

The use of nephrotoxic antibiotics is commonplace in the critical care setting.

Such medications (amphotericin B and aminoglycosides, in particular) cause damage to the renal tubule, thereby reducing its ability to secrete hydrogen ions. As a result, acid accumulates and a nonanion gap metabolic acidosis develops. Other causes of metabolic acidosis include small bowel diarrhea and excessive bicarbonate losses from gastrointestinal fistulae and ureteral diversions. Mild metabolic acidosis can be managed by adding potassium or sodium acetate to the TPN. Sodium bicarbonate should not be added to the TPN in the presence of calcium, because it leads to the formation of the insoluble calcium carbonate salt.[16]

Like metabolic acidosis, metabolic alkalosis is often associated with renal and gastrointestinal gains or losses. The most common etiologies of metabolic alkalosis in the critical care setting are nasogastric tube losses, aggressive diuresis, large bowel diarrhea, steroid administration, and citrate toxicity from excessive blood product administration. Initially, metabolic alkalosis can be treated by providing chloride in the form of sodium or potassium chloride in the TPN. In the case of severe metabolic alkalosis or when sodium and/or potassium intake is restricted, hydrochloric acid can be added to the TPN. Although hydrochloric acid replacement rarely exceeds 150 mEq/day, it is worth noting that hydrochloric acid should be limited to 100 mEq/L TPN as higher amounts have resulted in central venous line damage.[65] In addition, hydrochloric acid should not be added to TPN containing lipids, because it can disrupt the lipid's emulsifying agent and cause a potentially lethal capillary fat embolism.[16]

SUMMARY

Clinicians now recognize that TPN support offers more to the patient than simply providing protein and calories. Over the past two decades, we have come to realize that various disease and stress states induce a host of nutritional and metabolic disturbances. Investigators have subsequently discovered how the alteration and/or addition of specific nutrients may favorably affect these metabolic disturbances, as well as alter cytokine response, hormone secretion, and immune function. These advances in nutritional and metabolic medicine have led to major changes in recommendations for nutrient intakes and have prompted the development of disease-specific nutritional formulations. In view of these developments, standardized TPN solutions do not appear to efficiently support the vast majority of hospitalized patients. Scientific research and clinical practice suggest that parenteral formulas should be tailored to meet the individual patient's nutritional and clinical requirements. The key to providing patient- and disease-specific therapy is to realize that one diet, one formula, and one feeding route does not fit all patients. By ensuring that patients receive the right nutritional therapy at the right time, we hope to decrease morbidity and improve patient survival.

REFERENCES

1. Dudrick SJ, Wilmore DW, Vars HM. Long-term total parenteral nutrition with growth in puppies and positive nitrogen balance in patients. *Surg For* 1967;18:356–357.
2. Harris JA, Benedict FG. A biometric study of basal metabolism in man. Publication 279. Washington DC, Carnegie Institute of Washington, 1919.
3. Paauw JD, McCamish MA, Dean RE, et al. Assessment of caloric needs in stressed patients. *J Am Coll Nutr* 1984;3:51–59.
4. Mann S, Westenskow DR, Houtchens BA. Measured and predicted caloric expenditure in the acutely ill. *Crit Care Med* 1985;13:173–177.
5. Hunter DC, Jaksic T, Lewis D, et al. Resting energy expenditure in the critically ill: estimations versus measurement. *Br J Surg* 1988;74:875–878.
6. Shils ME. Parenteral nutrition. In Shils ME, Olson JA, Shike M, eds. Modern nutrition in health and disease, 8th ed. Lea & Febiger 1994:1430–1458.
7. Wolfe RR, Goodenough RD, Burke JF, et al. Response of protein on urea kinetics in burn patients to different levels of protein intake. *Ann Surg* 1983;197:163–172.
8. McMahon MM, Bistrian BR. The physiology of nutritional assessment and therapy in protein-calorie malnutrition. *Dis Mon* 1990;36:384–417.
9. Cahill GF. Starvation in man. *N Engl J Med* 1970;282:668–675.
10. Burke JF, Wolfe RR, Mullany CJ, et al. Glucose requirements following burn injury: parameters of optimal glucose infusion and possible hepatic and respiratory abnormalities following excessive glucose intake. *Ann Surg* 1979;190:274–285.
11. Sax HC, Talamini MA, Brackett K, et al. Hepatic steatosis in total parenteral nutrition: failure of fatty infiltration to correlate with abnormal serum hepatic enzyme levels. *Surgery* 1986;100:697–709.
12. Meguid M, Akohoshi M, Jeffers S, et al. Amelioration of metabolic complications of conventional total parenteral nutrition. *Arch Surg* 1984;119:1294–1298.
13. Askanazi J, Rosenbaum S, Hyman A, et al. Respiratory changes induced by the large glucose loads of total parenteral nutrition. *JAMA* 1980;243:1444–1447.
14. Seidner DL, Mascioli EA, Istfan NW, et al. Effects of long-chain triglyceride emulsions on reticuloendothelial system function in humans. *J Parent Ent Nutr* 1989;13:614–619.
15. Driscoll DF, Bristrian BR, Baptista RJ, et al. Base solution limitations and patient-specific TPN admixtures. *Nutr Clin Pract* 1987;(2)160:163.
16. Driscoll DF, Bistrian BR. Clinical issues in the therapeutic monitoring of total parenteral nutrition. *Clin Lab Med* 1987;7(2):699–714.
17. Bagdade JD, Koot RK, Bulger RJ. Impaired leukocyte function in patients with poorly controlled diabetes. *Diabetes* 1974;23:9–15.
18. Baxter JK, Babineau TJ, Apovian CM, et al. Perioperative glucose control predicts increased nosocomial infections in diabetics. *Crit Care Med* 1990;18:s207 (abstract 65).

19. Moore FA, Feliciano DV, Andrassy RJ, et al. Early enteral feeding, compared with parenteral, reduces postoperative septic complications. The results of a meta-analysis. *Annals of Surgery* 1992;216:172–183.
20. Buzby GP, Blouin G, Colling CL, et al. Perioperative total parenteral nutrition in surgical patients. The Veterans Affairs total parenteral nutrition cooperative study group. *N Engl J Med* 1991;325:525–532.
21. McMahon M, Manji N, Driscoll DF, et al. Parenteral nutrition in patients with diabetes mellitus: theoretical and practical considerations. *J Parent Ent Nutr* 1989;13:545–553.
22. Hongsermeier T, Bistrian BR. Evaluation of a practical technique for determining insulin requirements in diabetic patients receiving total parenteral nutrition. *J Parent Ent Nutr* 1993;17:16–19.
23. Kenler AS, Blackburn GL, Babineau TJ. Total parenteral nutrition: Priorities and practice. In Shoemaker WC, Ayres S, Grenvik A., eds. *Textbook of Critical Care.* Baltimore: W.B. Saunders, 1995:1116–1126.
24. Zeigler TR, Young LS, Benfell K, et al. Clinical and metabolic efficacy of glutamine-supplemented parenteral nutrition after bone marrow transplantation: a randomized, double-blind, controlled study. *Ann Intern Med* 1992;116:821–828.
25. Schloerb PR, Amare M. Total parenteral nutrition with glutamine in bone marrow transplantation and other clinical applications (a randomized, double-blind study). *J Parent Ent Nutr,* 1993;17:407–413.
26. Daly JM, Lieberman MD, Goldfine J, et al. Enteral nutrition with supplemental arginine, RNA, and omega-3 fatty acids in patients after operation: immunologic, metabolic, and clinical outcome. *Surgery* 1992;112:56–67.
27. Kenler AS, Swails WS, Driscoll DF, et al. Early enteral feeding in postsurgical cancer patients. Fish oil structured lipid-based polymeric formula versus a standard polymeric formula. *Ann Surg* 1996;223:316–333.
28. Rose WC. Amino acid requirements of man. *Fed Proc* 1949;8:546–552.
29. Roth E, Funovics J, Muhlbacher F, et al. Metabolic disorders in severe abdominal sepsis: glutamine deficiency in skeletal muscle. *Clin Nutr* 1982;1:25–41.
30. Kapadia CR, Muhlbacher F, Smith RJ, et al. Alterations in glutamine metabolism in response to operative stress and food deprivation. *Surg For* 1982;33:19–21.
31. Askanazi J, Furst P, Michelson CB, et al. Muscle and plasma amino acids following injury: influence of intercurrent infection. *Ann Surg* 1980;192:78–85.
32. Hardy G, Bevan SJ. Stability of glutamine in parenteral feeding solutions. *Lancet* 1993;342:186 (letter).
33. Hornsby-Lewis L, Shike M, Brown P, et al. L-Glutamine supplementation in home total parenteral nutrition patients: stability, safety and effects on intestinal absorption. *J Paren Ent Nutr* 1994;18:268–273.
34. Lowe DK, Benfell K, Smith RJ, et al. Safety of glutamine-enriched parenteral nutrition solutions in humans. *Am J Clin Nutr* 1990;52:1101–1116.
35. Fischer JE, Baldessarini RJ. False neurotransmitters and hepatic failure. *Lancet* 1971;2:75–79.

36. Rossi-Fanelli F, Riggio O, Cargiano C, et al. Branched-chain amino acids vs lactulose in the treatment of hepatic coma: a controlled study. *Digest Dis Sci* 1982;27:929–935.
37. Wahren J, Denis J, Desurmont P, et al. Is intravenous administration of branched chain amino acids effective in the treatment of hepatic encephalopathy? A multicenter study. *Hepatology* 1983;3:475–480.
38. Cerra FB, Cheung NK, Fischer JE, et al. Disease-specific amino acid infusion (F080) in hepatic encephalopathy: a prospective, randomized, double-blind, controlled trial. *J Parent Ent Nutr* 1985;9:288–295.
39. Marchesini G, Zoli M, Dondi C, et al. Anticatabolic effect of branched-chain amino acid-enriched solutions in patients with liver cirrhosis. *Hepatology* 1982;2:420–425.
40. Morgan TR, Moritz TE, Mendenhall CL, et al. Protein consumption and heptic encephalopathy in alcoholic heppatitis. *Am Coll Nutr* 1995;14:152–158.
41. Giordano C. Use of exogenous and endogenous urea for protein synthesis in normal and uremic subjects. *J Lab Clin Med* 1963;62:231–246.
42. Giovannetti S, Maggiore Q. A low-nitrogen diet with protein of high biologic value for severe chronic uremia. *Lancet* 1964;1:1000–1003.
43. Feinstein EI, Blumenkrantz MH, Healy M, et al. Clinical and metabolic responses to parenteral nutrition in acute renal failure. A controlled double-blind study. *Medicine* 1981;60:124–137.
44. Feinstein EI, Kopple JD, Silberman H, et al. Total parenteral nutrition with high or low nitrogen intakes in patients with acute renal failure. *Kidn Int* 1983;16:S319–323.
45. Mirtallo JM, Schneider PJ, Marko K, et al. A comparison of essential and general amino acid infusion in the nutritional support of patients with compromised renal function. *J Parent Ent Nutr* 1982;6:109–113.
46. Babineau TJ, Borlase BC, Blackburn GL. Applied total parenteral nutrition in the critically ill. In Rippe, et al., eds. Intensive Care Medicine, 2nd ed. Boston: Little, Brown, 1991:1675–1691.
47. Manji N, Shikora S, McMahon M, et al. Peritoneal dialysis for acute renal failure: overfeeding resulting from dextrose absorbed during dialysis. *Crit Care Med* 1990;18:29–31.
48. Askanazi J, Elwyn DH, Silverbery PA, et al. Respiratory distress secondary to a high carbohydrate load: a case report. *Surgery* 1980;97:596–599.
49. Al-Saady NM, Blackmore CM, Bennett ED. High fat, low carbohydrate, enteral feeding lowers PaCo2 and reduces the period of ventilation in artificially ventilated patients. *Intens Care Med* 1989;15:290–295.
50. Covelli HD, Black JW, Olsen MS, et al. Respiratory failure precipitated by high carbohydrate loads. *Ann Intern Med* 1981;95:579–581.
51. Kinsella JE, Lokesh B, Broughton S, et al. Dietary polyunsaturated fatty acids and eicosanoids: potential effects on the modulation of inflammatory and immune cells: an overview. *Nutrition* 1990;6:24–45.
52. Miles JM. Intravenous fat emulsions in nutritional support. *Curr Opin Gastroenterol* 1991;7:306–311.

53. Manner T, Katz DP, Askanazi J, et al. Parenteral fish-oil administration in patients with cystic fibrosis. *J Parent Ent Nutr* 1993;17:24s (abstract 8).
54. Silvis SE, Paragas Jr, PD. Parasthesias, weakness, seizures and hypophosphatemia in patients receiving hyperalimentation. *Gastroenterology* 1972;62:513–520.
55. Weinsier RL, Krundieack CL. Death resulting from overzealous total parenteral nutrition: the refeeding syndrome revisited. *Am J Clin Nutr* 1981;34:393–399.
56. Apovian CM, McMahon MM, Bistrian BR. Guidelines for refeeding the marasmic patient. *Crit Care Med* 1990;18:1030–1033.
57. Solomon SM, Kirby DF. The refeeding syndrome: a review. *J Parent Ent Nutr* 1990;14:90–97.
58. Knochel JP. The pathophysiology and clinical characteristics of severe hypophosphatemia. *Arch Intern Med* 1977;137:203–220.
59. Rudman D, Millikan WJ, Richardson TJ, et al. Elemental balances during intravenous hyperalimentation of underweight adult subjects. *J Clin Invest* 1975;55:94–104.
60. O'Connor LR, Wheeler WS, Bethune JE. Effect of hypophosphatemia on myocardia performance in man. *N Engl J Med* 1977;297:901–903.
61. Moore RD. Stimulation of Na:H exchange by insulin. *Biophysiol J* 1981;33:203–210.
62. Elwyn DH. Nutritional requirements of adult surgical patients. *Crit Care Med* 1980;8:9–20.
63. Knochel JP. Complications of total parenteral nutrition. *Kidn Int* 1985;27:489–496.
64. Illowsky BP, Laureno R. Encephalopathy and myelinolysis after rapid correction of hyponatremia. *Brain* 1987; 110:855–867.
65. Lopel PF, Durbin CG. Pulmonary artery catheter deterioration during hydrochloric acid infusion for the treatment of severe metabolic alkalosis. *Crit Care Med* 1989;17:688–689.
66. Driscoll DF. Drug-induced metabolic disorders and parenteral nutrition in the intensive care unit: a pharmaceutical and metabolic perspective. *DICP Ann Pharmacother* 1989;23:363–371.
67. Cerra FB. Tissue injury, nutrition, and immune function. In Forse RA, Bell SJ, Blackburn GL, Kabbash LG, eds. *Diet, nutrition and immunity.* Orlando: CRC Press, 1994:39–50.
68. Swails WS, Babineau TJ, Blackburn GL. Nutrient substrates. In Latifi R, Dudrick SJ, eds. *Current Surgical Nutrition.* Boca Raton: RG Landes, 1996:89–117.

Total Parenteral Nutrition and Drug Delivery

CHAPTER 12

George Melnik, Pharm.D., B.C.N.S.P.

INTRODUCTION

This chapter discusses the following aspects of parenteral nutrition and drug delivery: drug delivery via total parenteral nutrition (TPN) and total nutrient admixture (TNA); drug additives to TPN solutions; TNA and drug delivery; compounding solutions with additives; and drug incompatibilities with TPN/TPA. Total parenteral nutrition and TNA are complex intravenous solutions and emulsions. They are composed of numerous intravenous injections and large-volume solutions. As a result of their inherent composition, numerous potential chemical reactions (e.g., precipitation, inactivation, and modification of bioavailability) can occur. Physical and chemical incompatibility of drugs with TPN/TNA have been studied and are limited in scope to initial visual inspection of the nutrient solution. Other studies have examined solutions or emulsions for precipitation, gas formation, turbidity, color change, or pH change that would not necessarily manifest an apparent visual change. Investigations in this area do not adequately address chemical stability or bioavailability (i.e., whether the drugs and nutrients are biologically active and available even though no physical change was evident). With multiple additions of medications to complex parenteral nutrition solutions, multiple compatibility problems are created.

Clinical situations that encourage the use of TPN/TNA as a drug delivery system are fluid restriction, limited intravenous access, and parenteral nutrition support. Other effects of drug addition to TPN/TNA are savings in nursing and pharmacy time, decreased cost, and pharmacokinetic advantages for some drugs (e.g., cimetidine, insulin).

DRUG DELIVERY

Drugs administered by the TPN/TNA system should have the following characteristics:

1. Physical compatibility with TPN/TNA
2. Chemical stability and bioavailable
3. Pharmacologic effectiveness with constant infusion
4. Therapeutic window that allows frequent dosage adjustments
5. Compatible and stable with other drugs and nutrients.

Antiarrhythmics, diuretics, and vasopressors should not be given with TPN/TNA because they do not meet all of the above criteria.

DRUG ADDITIVES TO TOTAL PARENTERAL NUTRITION

Total parenteral nutrition solutions include dextrose and amino acids, which are compatible with a number of additives. Monovalent anions, vitamins, and trace elements are compatible with TPN solutions. In addition, acetate and chloride salts, the monovalent cations, usually do not cause problems when added to TPN.

Sodium bicarbonate when added to TPN has been shown to increase the pH, evolve carbon dioxide gas, and precipitate divalent cations; therefore, should not be added to TPN. When calcium and phosphate are added to TPN they can (and have) precipitated to form insoluble calcium phosphate. This precipitation is dependent upon a number of factors including the concentration of calcium, phosphate, dextrose, and amino acids, the salt form of calcium utilized, the amino acid brand, the volume at time of admixture, the temperature of the solution, the pH of the solution, the time since admixture, the order of admixing, and the presence of other additives.

TOTAL NUTRIENT ADMIXTURE AND DRUG DELIVERY

Total nutrient admixtures, also referred to as 3-in-1 solutions or all-in-one admixtures, contain dextrose, amino acids, and fat emulsion. The addition of excess di- and trivalent cations (Ca^{2+}, Mg^{2+}, Fe^{3+}) will crack the emulsion (i.e., cause the emulsion to separate). Dextrose, drugs, electrolytes, trace elements, and vitamins should not be added directly to the fat emulsion, but may be added to the TNA. Another important feature of TNA is the opaque nature of the emulsion. With the addition of the fat emulsion, the transparent nature of the TPN solution is obliterated. Therefore, the first line of detecting an incompatibility is gone. The inability to visually inspect solutions can be extremely dangerous, as hidden calcium phosphate precipitates have been reported to have contributed to patient deaths.

COMPOUNDING SOLUTIONS WITH ADDITIVES

All additives should be included as part of the compounding process in the aseptic laminar flow environment of the pharmacy. Drugs or any other additive should *never* be added on the nursing units. This ensures that compounding is performed under optimal circumstances by trained personnel who are aware of the complexities of TPN/TNA compatibility and incompatibility.

INCOMPATIBLE DRUGS

Table 12.1 lists a variety of medications that are incompatible with TPN and TNA solutions. Most of these agents are not physically compatible with parenteral nutrition solutions, and others have shown a decrease of drug concentration or pharmacological activity. A number of volumes provide lists of drugs that are compatible with TPN/TNA, but the important point is to know the ones that are incompatible.

CONCLUSION

Knowledge of actual and potential problems with parenteral nutrition and drug delivery affords the practitioner the ability to provide optimal clinical and nutritional therapy. Without such information, provision of one treatment may undermine the other.

Table 12.1 Drug Incompatibilities with Parenteral Nutrition Solutions

Drug	TPN	TNA	Ref.
Amikacin		incompatible	1
Amphotericin B	incompatible		2
Ampicillin	incompatible		2–4
Ascorbic acid	degrades in sunlight		19
Carbenicillin	lowers antibiotic activity		3,5
Cefamandole	incompatible		6
Cephalothin	incompatible		7
Cephradine	incompatible		4
Famotidine	4–6% reduction in concentration	5–7% reduction in concentration	8–11
Folic acid	degrades in sunlight		19
Imipenem	lowers antibiotic stability		16
Kanamycin	lowers antibiotic activity		5
Methyldopate		incompatible	27
Pyridoxine	sunlight lowers concentration 100%		21
Ranitidine	> 10% reduction in concentration	> 10% reduction in concentration	12–14
Riboflavin	sunlight lowers concentration 100%		21
Mezlocillin	lowers antibiotic activity		15
Tetracycline		incompatible	17
Thiamin	lowers activity in glass bottles		20
Thiamin	26% reduction in concentration in sunlight		21
Ticarcillin	lowers antibiotic stability		15
Vancomycin	< 7% reduction in concentration		18
Vitamin A	7–68% reduction in concentration		22–26
Vitamin D	32% reduction in concentration		23
Vitamin E	36% reduction in concentration		23

REFERENCES

1. Bullock L, Clark JH, Fitzdgerald JF, et al. The stability of amikacin, gentamicin, and tobramycin in total nutrient admixtures. *J Parent Ent Nutr* 1989;13:505–509.

2. Anthanikar N, Deamer R, Harbison R, et al. Visual compatibility of 30 additives with a parenteral nutrition solution. *Am J Hosp Pharm* 1979; 36:511–513.
3. Colding H, Anderson GE. Satiability of antibiotics and amino acids in two synthetic l-amino acid solutions commonly used for total parenteral nutrition in children. *Antimicrob Agents Chemother* 1978; 13:555–58.
4. Watson D. Piggyback compatibility of antibiotics with pediatric parenteral nutrition solutions. *J Parent Ent Nutr* 1985; 9:220–224.
5. Feigin RD, Moss KS, Shackelford PG. Antibiotic stability in solutions used for intravenous nutrition and fluid therapy. *Pediatrics* 1973; 51:1016–1026.
6. Bowtle WH, Heasman MJ, Prince AP, et al. Compatibility of the cephalosporin, cefamandole nafate with injections. 1980; *Int J Pharm* 1980; 4:263–265.
7. Schuetz DH, King JC, Compatibility and stability of electrolytes, vitamins and antibiotics in combination with 8% amino acid solution. *Am J Hosp Pharm* 1978; 35:33–34.
8. Bullock L, Fitzgerald JG, Glick MR. Stability of famotidine 20 and 50 mg/L in total nutrient admixtures. *Am J Hosp Pharm* 1989; 46:2326–2329.
9. Bullock L, Fitzgerald JG, Glick MR. Stability of famotidine 20 and 40 mg/L in total parenteral nutrient solutions. *Am J Hosp Pharm* 1989; 46:2321–2325.
10. DiStefano JE, Mitrano FP, Baptista RJ, et al. Long-term stability of famotidine 20 mg/L in a total parenteral nutrient solution. *Am J Hosp Pharm* 1989; 46:2333–2335.
11. Montoro JB, Pou L, Salvador P, et al. Stability of famotidine 20 and 40 mg/L in total nutrient admixtures. *Am J Hosp Pharm* 1989; 46:2329–2332.
12. Bullock L, Parks RB, Lampasona V, et al. Stability of ranitidine hydrochloride and amino acids in parenteral nutrient solutions. *Am J Hosp Pharm* 1985; 42:2683–2687.
13. Walker SE, Bayliff CD. Stability of ranitidine hydrochloride in total parenteral nutrient solution. *Am J Hosp Pharm* 1985; 42:590–592.
14. Cano SM, Montoro JB, Pastor C, et al. Stability of ranitidine hydrochloride in total nutrient admixtures. *Am J Hosp Pharm* 1988; 45:1100–1102.
15. Perry M, Khalidi N, Sanders CA. Stability of penicillins in total parenteral nutrient solution. *Am J Hosp Pharm* 1987; 44:1625–1628.
16. Zaccardelli DS, Krcmarik CS, Wolk R, Khalidi N. Stability of imipenem and cilastatin sodium in total parenteral nutrient solution. *J Parent Ent Nutr* 1990; 14:306–309.
17. Baptista RJ, Lawrence RW. Compatibility of total nutrient admixtures and secondary antibiotic infusions. *Am J Hosp Pharm* 1985; 42:362–363.
18. Nahata MC. Stability of vancomycin in total parenteral nutrient solutions. *Am J Hosp Pharm* 1989; 46:2055–2057.
19. Nordfjeld K, Langpedersen F, Rasmussen M, et al. Storage of mixtures for total parenteral nutrition II: Stability of vitamins in TPN mixtures. *J Clin Hosp Pharm* 1984; 9:293–301.
20. Bowman BB, Nguyen P. Stability of thiamin in parenteral nutrition solutions. *J Parent Ent Nutr* 1983; 7:567–568.

21. Chen MF, Boyce W, Triplett L. Stability of the B vitamins in mixed parenteral nutrition solution. *J Parent Ent Nutr* 1983; 7:462–464.
22. Allwood MC. Influence of light on vitamin A degradation during administration. *Clin Nutr* 1982; 1:63–70.
23. Gillis J, Jones G, Pencharz P. Delivery of vitamins A, D, and E in total parenteral nutrition solutions, *J Parent Ent Nutr* 1983; 7:11–14.
24. Hartline JV, Zachman RD. Vitamin A delivery in total parenteral nutrition solution. *Pediatrics* 1976; 58:448–451.
25. Riggle MA, Frandt RB. Decrease of available vitamin A in parenteral nutrition solutions. *J Parent Ent Nutr* 1986; 10:388–392.
26. Gutcher GR, Lax AA, Farrell PM. Vitamin A losses to plastic intravenous infusion device and an improved method of delivery. *Am J Clin Nutr* 1984; 40:8–13.
27. Baptista RJ, Dumas GJ, Bistrian BR, et al. Compatibility of total nutrient admixtures and secondary cardiovascular medications. *Am J Hosp Pharm* 1985; 42:777–778.

Complications of Total and Peripheral Parenteral Nutrition and Their Prevention

CHAPTER 13

Mitchell V. Kaminski, Jr., M.D.

INTRODUCTION

The first description of a practical technique to parenterally administer nutrients to sustain human life was published in 1968.[1] This technique was termed total parenteral nutrition (TPN). At that time, a hydrolysate of fibrin or casein was used as the nitrogen source. Most clinicians were unaware that it was contaminated with a small amount of minerals and trace metals. The only form of concentrated nonprotein calories for intravenous use was hypertonic dextrose. Therefore, these solutions had to be administered via a high flow vein to avoid phlebothrombosis. New equipment facilitating placement of a subclavian catheter with its tip directed into the superior vena cava gained in popularity. Not long after, crystalline amino acids were developed, parenteral fat became available, and peripheral parenteral nutrition (PPN) offered another option for short-term intravenous nutritional support.[2] Hyperalimentation teams were nonexistent, and a learning curve began regarding complications, which included heretofore rare clinical manifestations of electrolyte, vitamin, trace metal, and substrate imbalances, as well as problems related to chronic indwelling venous catheters and profound hypoproteinemia.

Today, several decades later, we are aware of these complications. They can be classed into four major categories: mechanical, septic, metabolic, and physiologic. The important clinically useful concept is that all of these misadventures are secondary to specific causes. Thus, at least theoretically, although one must be aware of a myriad of details, there should be no complications with either TPN or PPN. To achieve this, protocols are essential, as is division of labor. Despite protocols related to tubes and catheters, an overriding principle remains for nutrient solutions: *There is no such thing as a standard patient; therefore, there can be no such thing as a standard formula.*

This chapter will discuss the various complications related to TPN and PPN and will detail the techniques necessary to prevent them.

CENTRAL AND PERIPHERAL VENOUS CATHETERIZATION

The popularity of subclavian vein catheterization for central line placement rests not only with the relatively simple access to the central venous system to direct the catheter tip into the superior vena cava but also with the fact that the exit site is on the upper chest, a fixed flat surface. The procedure was first described by Aubaniac during the French and Indonesian War.[3] Its use for rapid volume resuscitation was short term with little need for meticulous attention to aseptic technique. By adding meticulous preparation and maintenance of a sterile skin field, it made the ideal catheter technique for prolonged periods of TPN. However, when the subclavian veins are compromised, there are a number of other veins accessible by percutaneous technique for central venous catheter (CVC) placement. (Figure 13.1)

Peripheral parenteral nutrition has became a popular alternative to TPN for short-term nutritional support. Driving this popularity was the well-deserved fear of CVC complications. A PPN solution in which there is approximately 50% fat and 50% carbohydrate as nonprotein calories has an osmolality that will rarely exceed 600 mM. A peripheral vein can tolerate 600 mM; therefore, PPN can be infused safely with an acceptably low incidence of thrombophlebitis.

Placing a peripheral vein catheter has several advantages. Technical hazards such as pneumothorax, hemothorax, line shear embolus, and others are minimized. Developing life-threatening sepsis is obviated in that peripheral vein tenderness develops well before symptoms of sepsis. Finally, the requirement to use a higher fat concentration decreases the risk of metabolic complications associated with concentrated dextrose solutions. PPN should however never be infused into the hand or the foot. Likewise, placing the catheter directly into the antecubital fossa is not advised because of the need to restrict motion.

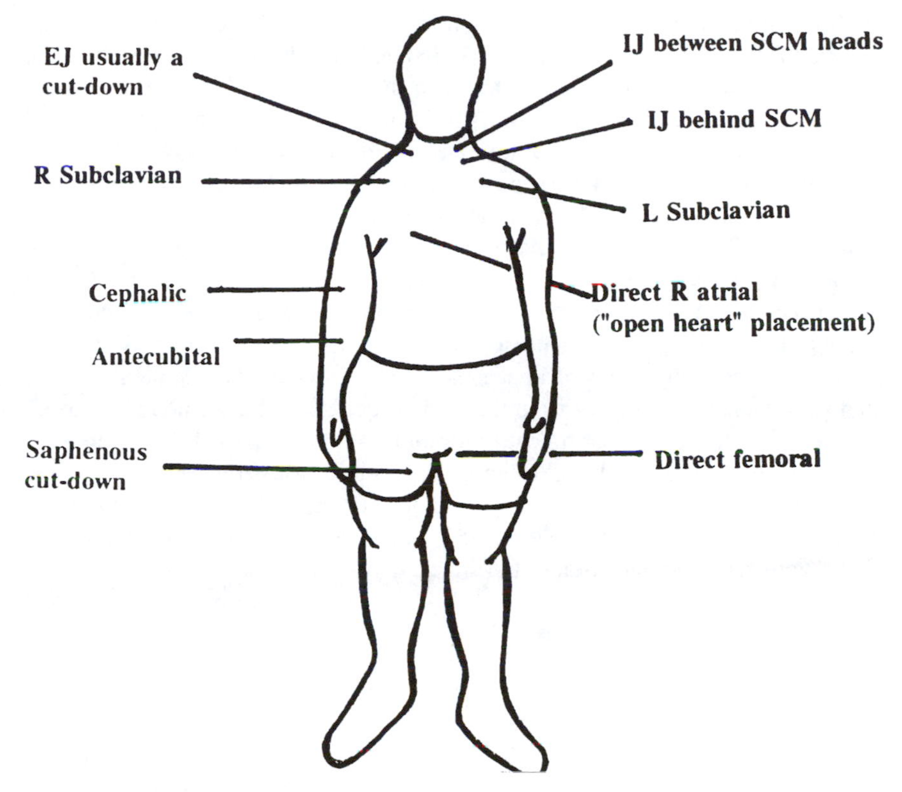

Figure 13.1. The central venous system for TPN includes the superior and inferior vena cava and the right atrium. Long lines can be used from arm or leg sites if the subclavian and jugular veins are compromised. Though infrequently used, the right atrium is always available via a minithoracotomy.

Subclavian CVC Placement

Complications of subclavian vein cannulation can be minimized by following a catheter insertion protocol. Prior to placement, the patient should be put at ease by a description of what is about to take place. Pharmacy or central supply should package supplies and equipment into a single bag. The materials contained in the disposable central line tray are inadequate for the rigorous aseptic technique necessary to achieve placement for prolonged use. This auxiliary bag might contain mask, cap, gown, sterile surgical gloves, and additional skin prep equipment. Supplementary sterile drapes are also helpful, as well as extra lidocaine, needles, syringes, and 2-in. tape to restrain the patient in position during insertion and for the dressing. A small amount of intravenous Valium or Versed can be slowly given to relax a nervous patient and gain full cooperation. Be sure that the bed can be placed in deep Trendelenburg position to distend the subclavian vein and to make it an easier target. The patient's head should be rotated to the opposite side of the procedure and taped to the bedframe. Hands are restrained toward the feet and in women with pendulous breasts, the breast is retracted inferiorly and taped. The dehydrated and malnourished patient will have flat external jugular veins. It can be presumed that the subclavian vein is also flat. Cannulating this relatively tough-walled vein in this nondistended configuration can be difficult and risky. On the other hand, if the external jugular veins are visible and relatively turgid with the patient supine, a deep Trendelenburg position is not necessary. A 22-gauge 1.5-in. needle can be used to simultaneously anesthetize the deeper tissues and then aspirated to act as a vein locator. Before the needle is withdrawn, a mental image of the trajectory of the needle should be made. As the needle is drawn from the skin, a squirt of a small amount of blood-stained Lidocaine along that trajectory onto the sterile drape will help guide the larger cannulating needle back to the same position. Most physicians place the middle finger in the jugular notch, the index finger in the space between the heads of the sternocleidomastoid and the thumb just inferior to the clavicle to give a stereotactic appreciation of the subclavian vein's location relative to other anatomy. It is wise to not remove that hand following use of the 22-gauge locator needle until the cannulating needle is in place.

Internal Jugular CVC Placement

If the central vein is approached via the percutaneous catheterization of the internal jugular (IJ) vein, malposition as well as pneumothorax and arterial injury are rare. The IJ vein can be found at the apex of the triangle formed between the clavicle and the sternal/clavicular head of the sternocleidomastoid muscle. In this location, the local anesthetic/locator needle should be advanced straight back and *laterally 20° to 25°* (Figure 13.2a). The vein lies behind the clavicular portion of the sternocleidomastoid muscle. A lateral sternocleidomastoid muscle approach can also be used.

Unlike the placement of a Swan Ganz catheter, which uses a lateral percutaneous approach starting above the course of the external jugular vein and aiming toward the sternal notch/opposite nipples, placement of a lateral IJ central line for TPN starts inferior to the external jugular vein. The needle is not directed downward toward the opposite nipple, but in a more tangential direction, that is, behind the muscle approximately 15° from the clavicle (Figure 13.2b). After IJ insertion the catheter should be tunneled onto the anterior chest in the subclavian region for simplicity of aftercare. This is facilitated by widening the incision in the neck adjacent to the Seldinger wire and making another 1 cm incision in the subclavian area on the flat surface of the upper chest. A small hemostat is then used to dissect the tissues between these two incisions. This creates a subcutaneous

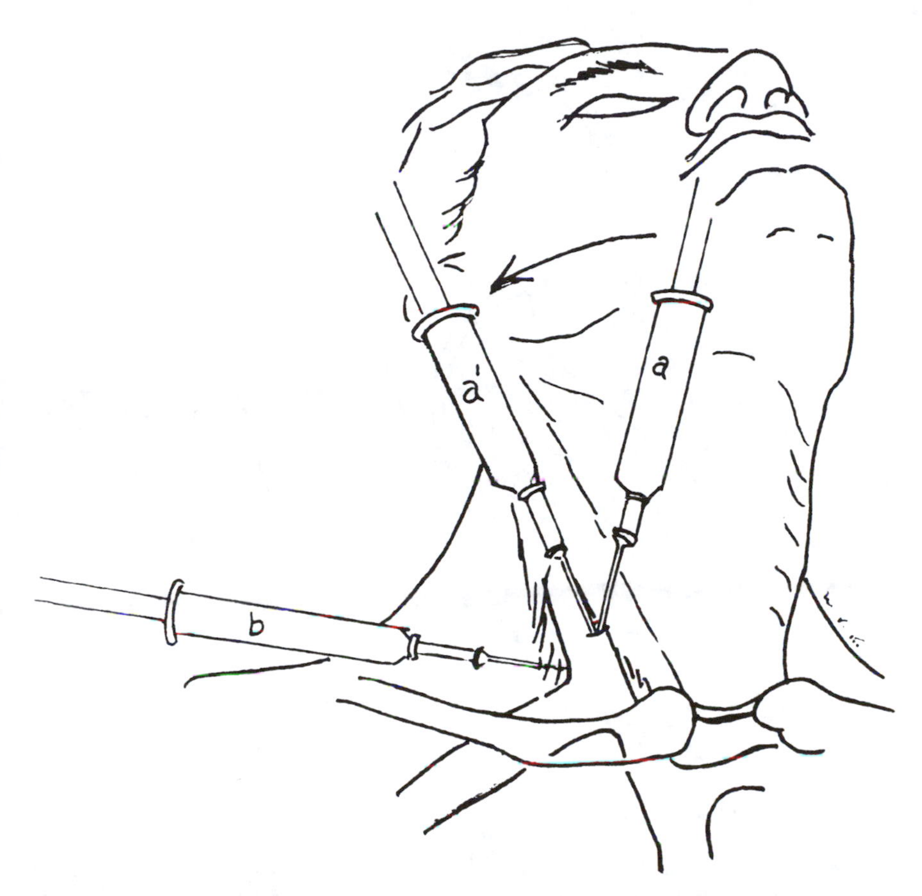

Figure 13.2. The internal jugular approach to the superior vena cava begins with advancing the locator needle straight down and lateral (a). The plastic needle sheath is advanced after moving the syringe to position a.[1] In some patients the carotid artery is found lateral to its usual position and occupies the "between the head location." Here the internal jugular can be cannulated using the lateral approach (b).

tract between the incisions anterior to the clavicle. This subcutaneous tunnel is held open by drawing the clear plastic tube which shields the long needles found in the catheter placement tray through the space (Figure 13.3). The catheter is then advanced through the tube from the incision on the chest and exited at the incision on the neck. The clear plastic tube is removed and the wire passed through the catheter. Remember to lubricate the internal lumen of the catheter by filling it with sterile heparinized saline. When the wire presents itself at the female adapter of the catheter, the cap is removed and the wire clamped with the hemostat. The entire assembly is then advanced into the internal jugular vein

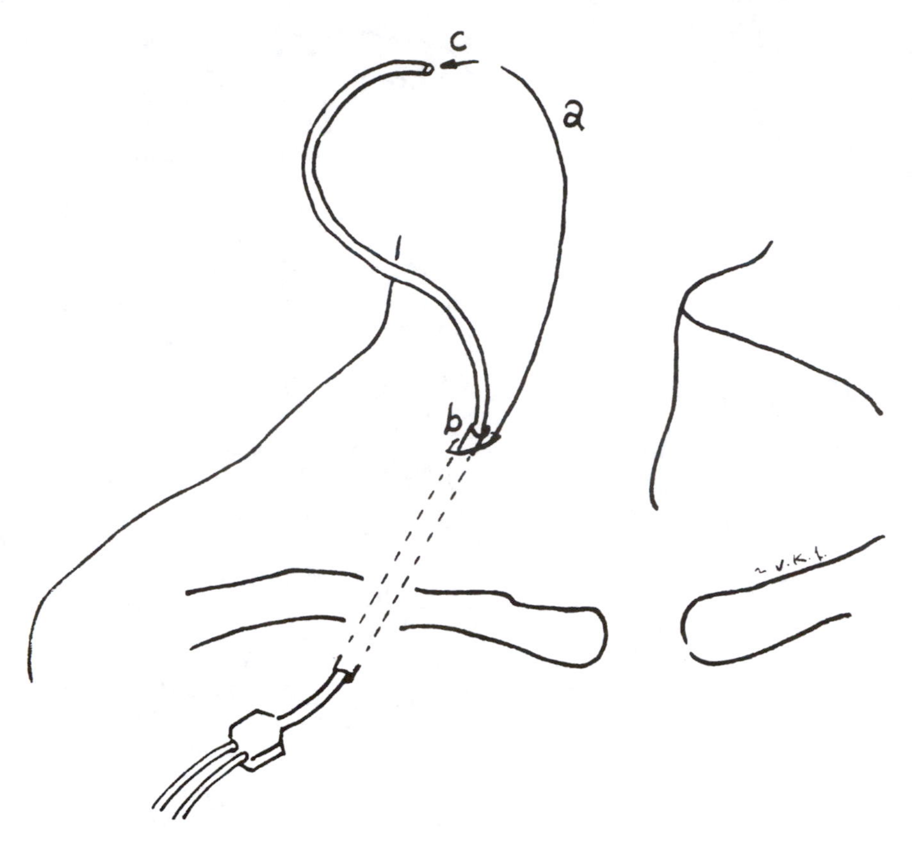

Figure 13.3. Tunneling the percutaneous internal jugular approach from the subclavian area to the neck insertion site combines the advantages of each. The Seldinger wire is placed first and a small incision made next to it (a). The clear plastic tube shielding the angiocath is drawn above the clavicle from the incision in the neck to the incision on the chest (b). The catheter is run through the tube and the tube removed before the wire is advanced and grasped at the female adapter end (c). The catheter wire is gradually withdrawn.

while simultaneously removing the wire at appropriate intervals. Nylon suture should be used to close the wound of the neck as well as to secure the device at the exit site. Silk is an organic fiber and promotes erythema. Although there are usually two eyes on the CVC device to be anchored to the patient's chest, both should not be used. Anchoring only one allows the device to be easily elevated so that the area beneath it can be adequately cleaned during protocol dressing changes.

MECHANICAL COMPLICATIONS

Pneumothorax

Due to the proximity of the apex of the lung to the subclavian vessels, pneumothorax is the most frequent complication associated with subclavian catheter placement. In the best hands, it still occurs 4% of the time.[4] *The first symptom of a pneumothorax is pleuritic pain. The first sign is tachycardia.* The chest x-ray for catheter tip location is to be carefully inspected for a pneumothorax. If a pneumothorax is present and not greater than 10% and the patient is stable, a chest tube is not necessary. Serial chest x-rays and exams are mandatory. The greatest danger is that a pneumothorax will progress into a life-threatening tension pneumothorax. Increasing symptoms require emergency decompression.

Arrhythmias

Most commercial devices contain a guidewire to allow CVC placement using Seldinger's technique. If the wire is advanced into the right atrium, an arrhythmia can result. Although rarely life-threatening arrhythmias need not occur if care is taken not to overadvance the wire and monitor the pulse rhythm. Prefilling the catheter lumen with a heparinized solution serves to eliminate air and also lubricate the catheter lumen, so that it slips over the wire easily. If the CVC kinks, advancing the wire through the catheter will become difficult or impossible.

Subclavian Artery Injury

This complication is more frequent when the patient is dehydrated and the external jugular veins cannot be distended by Trendelenburg position or Valsalva maneuver. The subclavian artery courses directly behind and inferior to the vein. The vein and artery are separated by the scalenus anticus muscle. However, if the vein is soft the needle will push the anterior wall against the posterior wall of the vein. Then, the needle will penetrate through the muscle into the artery before a flashback of venous blood is seen. Furthermore, if the patient is hypooxygenated, blood drawn from the artery can look venous. Once the syringe is removed from the needle, however, the blood return will be pulsatile. Another technique to confirm whether the blood aspirated has been drawn from a vein or the artery is to express 1 or 2 mL onto a white drape or gauze pad. Venous blood is decidedly dark maroon in color, whereas even poorly oxygenated arterial blood is still quite red by comparison as it absorbs through the fibers of the white material. Once recognized, the needle should be withdrawn and direct pressure applied above and below the clavicle by pinching the space between the thumb and forefinger for 3 to 5 mins.

Air Embolism

Air embolism cannot occur if the patient's central venous pressure is higher than atmospheric pressure. This can be verified by observing for distended external jugular veins prior to starting the procedure. If this procedure is not followed, an air embolism can occur when the cannulating device is opened to atmospheric pressure. The first symptom of an air embolism is shortness of breath. The first sign is coughing. If massive, the patient will become severely hypotensive and have a churning cardiac murmur on auscultation.[5] In addition to immediately placing the patient in steep Trendelenburg position, the patient should be rolled into a left lateral decubitus position. Eventually, the air now risen to the apex of the right ventricle will be absorbed. An invaluable lesson will be learned regarding the initial positioning of the patient and observing for distention of the external jugular veins before proceeding with cannulation.

Catheter Malposition

Catheter malpositioning can be avoided by paying attention to the manner in which the Seldinger wire passes through the cannula into the central venous system. The wire should drop in with relative ease. If it does not, something may be wrong. If the wire tracks up the internal jugular vein, it will offer resistance to passage 6 to 10 cm before it does when it is positioned in the superior vena cava/right atrium. To facilitate repositioning from the jugular vein into the vena cava, it is helpful to rotate the head toward the side of cannulation and raise the shoulder toward the ear before withdrawing and readvancing the wire.

Venous Thrombosis

The relatively rigid polyethylene catheters used in the earlier days of TPN and PPN have been replaced by a number of other softer materials. This is an advantage in that the stiffness of the catheter itself can cause intimal injury, thrombosis, and perhaps influence the incidence of catheter-related infection.[6] Bozzetti et al. used venograms to show that venous thrombosis could be demonstrated in 46% of the patients with polyvinylchloride catheters and only 11% of patients with silicone catheters after an average CVC duration of 12 days.[7] None of these patients, however, manifested symptoms. Newer catheter materials are very soft. Elastomeric hydrogel is rigid at the time of insertion but absorbs plasma fluid to become 50 times softer within minutes of placement.[8] Other newer configurations of urethane elastomer are similarly soft. In general, it is desirable to use the softest material possible.[9] Nevertheless, the clinician should check for early signs of phlebitis by *palpating over the site of every vein that contains a catheter every day and observe for tenderness or erythema*. Tenderness precedes redness; redness precedes sepsis. If tenderness develops in a previously nontender cannulation site, it would be wise to change the catheter before erythema and cellulitis further compromise that site precluding future use or threatens the well-being of the patient.

Care is taken not to touch the catheter even with gloved hands. Instruments are used when handling the catheter to avoid depositing talc or other drying agents from the sterile gloves onto the surface of the catheter. These agents are thrombogenic. Subclavian vein thrombosis has been observed in up to 50% of patients with prolonged subclavian catheterization.[4,10] Subclavian vein thrombosis can be suspected when collateral veins over the chest and shoulder began to appear. The thrombose itself has been incriminated as contributing to sepsis. In fact, minidose heparin in the range of 1 to 2 Units/kg·hr^{-1} included in the TPN solution has been advocated.[10] If the thrombose is not contaminated by a break in aseptic technique during catheter placement or aftercare/dressing changes, venous thrombosis is usually of little clinical significance except that recannulating the same vein at a later date may be difficult if not impossible. On rare occasion, the CVC thrombose produces a full blown superior vena cava syndrome. This is a dramatic event and results in facial, head, neck and bilateral arm and upper chest edema. It is uncomfortable but may be treated conservatively. Collateral circulation will soon develop and spontaneously resolve the condition in 3 days to 2 weeks. An option is to treat the patient with streptokinase therapy using 250,000 IU as a loading dose followed by a continuous infusion of 1,000 IU/minute. The course of therapy is monitored by obtaining an hourly thrombin time and fibrinogen level. As therapeutic efficacy is achieved, the infusion should be decreased to 500 IU/min and discontinued as symptoms resolve. Following resolution, the central line should be removed and the patient given 24 to 48 hr before a new CVC is placed in another location.

Central Venous Catheter Clotting and Clogging

If fluid pressure within the catheter lumen is not higher than central venous pressure, blood will back flow into the catheter. There is a 1 or 2 hr window of time during which the catheter can be easily irrigated. Aspirating the clot has rarely worked. Irrigation frees the small clot into the general circulation. This, however, has not been associated with symptoms. The procedure should be done with an assistant to facilitate perfect aseptic technique. Great care is taken for prepping the connection sites. Sterile gloves must be used. To generate sufficient pressure to dislodge the clot a 1-cm insulin or tuberculin syringe is used. The thumb and forefinger of the left hand hold the male–female connection together tightly while the thumb of the right hand is used to force the heparin solution into the catheter.[11] If the catheter is made of silicone, a rupture can occur. Polyethylene or polyurethane catheters rarely rupture. If the luminal clot has been in place for 12 to 24 hr, this technique may not work. In fact, an overvigorous attempt can rupture even polyurethane catheters. Thrombolytic therapy has been successful in these circumstances.[12] Streptokinase or Urokinase 5,000 U/mL is gradually worked into the catheter using a 1-mL syringe. The follow-up flush with heparinized saline should be delayed 60 min or more. Success rates range from 32% to 100%.[13]

Clogging from Precipitate in the Solution

Two common forms of precipitate have been identified. Those associated with calcium phosphate and those associated with lipid emulsion build up. Calcium phosphate precipitates should not occur. They are generally caused by an inappropriate mixing order in the pharmacy, where calcium is not the first ion added and phosphate the last, or where sodium bicarbonate instead of sodium acetate was ordered, producing calcium carbonate. Never use sodium bicarbonate in a TPN solution. In addition to forming a precipitate of chalk, carbon dioxide gas bubbles will form. These errors can be life-threatening.[14] If a calcium precipitate clogs a critical line, a 0.1N hydrochloric acid solution can be instilled into the catheter in an amount that precisely fills the lumen. It is allowed to dwell for 4 to 6 hr and usually produces good results.[15]

Three-in-one solutions have been associated with luminal occlusion that does not respond to thrombolytic or hydrochloric acid therapy. This occlusion might be secondary to lipid build up. Reportedly, it has been cleared with the instillation of 70% ethanol.[16]

The risks of declogging a catheter or of placing a new line should be carefully weighed. As with mechanical complications, septic complications associated with TPN and PPN are a matter of technique.

SEPTIC COMPLICATIONS ASSOCIATED WITH TOTAL PARENTERAL NUTRITION

This section, as well as the next section on metabolic complications, highlights the importance of having a qualified nutritional support team consisting of a nurse, pharmacist, and dietitian to advise the physician who is writing orders. Complication-free nutritional support is labor-intense. The work is divided logically according to professional disciplines among the team personnel. There are economic pressures to shrink or totally abandon the team approach, because the cost of team salaries and overhead cannot be offset by demonstrated revenue. Complications, however, also cost money as well as lives. Within 6 months of disbanding a team that for some years boasted of a zero pneumothorax and catheter infection rate, the cost of complications totaled over one million dollars.[17,18] This far exceeded the cost of the team.

Contamination of Solutions

The role of the pharmacist in TPN and PPN is significant.[19] In addition to compounding solutions so that incompatibilities and precipitates will not occur, a major effort is directed at preventing contamination of the solutions. Use of filters for parenteral nutrition solutions should not be a crutch to support sloppy compounding. Double-blind trials have failed to demonstrate the benefit of the use of filters.[20] Current guidelines recommend in-line filters for PPN for particulate as well as bacterial contamination.[21] The Center for Disease Control (CDC) does not strongly recommend the use of filters but classifies filters as being highly suggestive of having benefit.[22] Filtering for particulates should be done in the pharmacy during compounding under the laminar flow hood. Given that 3-in-1 solutions are popular today and will not pass through a sterilizing 0.22-μm filter, use of filtration devices on infusion tubing is now strictly for particulate. It has been observed that *S. epidermitidis*, *E. coli*, and *C. albicans* grow rapidly in lipid emulsion.[23] This observation gives emphasis to the importance of pharmacy compounding of a perfectly sterile solution.

CVC Infections

Aseptic technique at the time of insertion and an aftercare protocol is mandatory to avoid the complications of infection associated with long-term intravenous foreign bodies. These materials are sterile when they are shipped by the manufacturer. Thus, unless contaminated, peripheral as well as central catheters can be left for an extended period of time. If daily inspection reveals neither tenderness nor early erythema, there is no need to change the catheter. The same skin prep, insertion, and dressing change protocol should be followed for PPN as for TPN catheters.

Catheter infections are either related to cellulitis deriving from contamination at the catheter exit site (type I) or microorganism contamination within the lumen of the catheter (type II). Both are due to a break in aseptic technique.

Type I Line Infection

Contrary to the popular notion that skin is skin and is always heavily contaminated, the skin of the forearm or neck/subclavian area is relatively clean and can be used for prolonged venous access if the surface is appropriately prepared and maintained. Statements such as "no catheter is safe after 48 hr" have been used by some resulting in a defeatist attitude and poor attention to aseptic technique. This increases type I infections. One of the responsibilities of the hyperalimentation team nurse is skin prep and exit site aftercare. This responsibility cannot be overemphasized.[24] The skin prep cannot be routinely expected to be adequate if it only consists of three organic solvent swabs, followed by 1 to 3 swabs of an iodophor solution.[25] A CVC is a foreign body that is going to be placed in or near a chamber of the heart. The same respect for aseptic technique that is held for an intracardiac prosthesis should be extended to the CVC. Skin prep should be equivalent to that used before any major vascular procedure. That is, by no less than a five minute detergent scrub of a wide area followed by an application of an iodophor solution which is allowed to dry.

This process can only be abridged for PPN catheter placement and occasionally central placement. Here defatting and cleaning is accomplished with repeated wipe down with sterile alcohol sponges. The sponges must then be observed for discoloration or debris. If discoloration is noted, the area is cleaned again. This is repeated until no discoloration is seen on the alcohol gauze and then repeated one more time. In teaching this technique, the phrase "clean until the gauze is clean and clean one more time" is imprinted in the student's memory. Iodophor solutions are water soluble and are not cleaning agents. Originally acetone was used for this purpose but was shown to be irritating to the skin and of no adjunctive benefit as a first step in skin preparation. Alcohol is effective against gram-positive and gram-negative organisms, as well as tubercle bacillus, fungi, and viruses.[26] Chlorhexidine gluconate was demonstrated to be superior to other skin prep solutions, because its maximum duration of activity is 6 hr.[27] Rather than focusing attention on cleaning and prepping agents that have a high rate of bacterial kill or prolonged duration of action, attention might better be focused on the individual mechanical technique used for skin prep and aftercare. A professional attitude is the deciding factor: There should never be a Type I infection. If there is, it is due to a break in technique and can be traced to an individual.

Dressings and aftercare. Once the CVC is placed, a variety of materials and kits have been developed to maintain the catheter exit site. It seems to make little difference if the gauze or tape traditional dressing is used or the transparent adherent dressings.[28] What seems to be important is a "pride of ownership" by the individual responsible for preventing Type I infections. Quality assurance and control studies of type I infection are especially helpful, because they increase the institutional level of awareness of the importance of aseptic technique. Checking for venous cannula-related tenderness is simple and can be taught to several people on the care team. The patient should be distracted by conversation while palpating or tapping over the catheter exit site and observing for a painful expression on the patient's brow. Whoever makes this observation should bring it to the attention of the nutritional support team nurse for dressing removal and site inspection. If tenderness is associated with redness, it is wise to change the cannula at that time. This system of intense monitoring has resulted in extremely low catheter infection rates.[17]

Routine line changes. Some have advocated the routine use of line changes to minimize problems associated with type I infections.[29] If attention is paid to details in aseptic technique as described above, this may not be necessary. In this era of medical economic contraction, this approach when routinely followed, might be considered an overutilization of resources.

Single or multiple lumen catheters. As multilumen catheters became popular, the question was asked whether or not multilumen catheters increased the incidence of catheter-related sepsis. This question has little to do with type I infections. Adding extra lumens to the same catheter, however, potentiates the risk for type II infections.

Type II Infection

Anytime the female adapter or cap is manipulated, there is a chance for inadvertent contamination. As TPN teams implemented stringent protocols to control type I complications, type II infections emerged as the dominant catheter problem. This is perhaps because the hookup–disconnect portion of catheter care can no longer be relegated to a single individual.[30] Every person on the healthcare team that starts a parenteral medication can potentially contaminate the system. The configuration of the hub and the fact that colonization with coagulase-negative staphylococci can be seen in 21 out of 23 hub cultures highlights the fact that this can be a difficult problem.[31] Instituting a protocol for intravenous line junction cleaning before manipulation has been shown to improve the incidence of type II infections.[32] Regardless of the protocol used, whether it consists of only using two alcohol swabs followed by an iodophor swab or any other combination, this type of activity heightens the awareness of the nursing staff that bacterial touch contamination can occur. Although not studied, it is perhaps more the focus of attention to careful detail than the type of agents used that prevents touch contamination while the female adapter is being refitted and exposed to the environment. Organisms will never generate spontaneously from the intraluminal or external surface of a CVC device. Therefore, whether it is a type I or type II infection, it always results from a break in technique.

Catheter Flushes

Whether one uses a solution of saline or heparin in concentrations of 1000 U/mL or 10 U/mL does not seem to make as much a difference in the incidence of luminal thrombosis or contamination as does the frequency of flushing.[33,34] Increased flushing is associated with increased rates of contamination. If the catheter lumen has a solution running continuously through it, it need not be interrupted and flushed. If the catheter is heparin- or saline-locked, flushing once a day is sufficient.

METABOLIC COMPLICATIONS

Hyperglycemia

There are two dangers to significant hyperglycemia. One is immediate and the other is delayed. The immediate complication is hyperosmolar nonketotic dehydration, coma, and possibly death secondary to an osmotic diuresis.[35] This will be considered in detail later in the chapter.

Although rare, chromium deficiency has been reported. The chromium ion can assume a quadravalent cationic configuration. Theoretically, this allows chromium to bind insulin to the insulin receptor. If there is chromium deficiency, the patient will respond to increasing insulin doses in an erratic manner, or as though no increase had been made at all.[36] Adding 150 μg/L of chromium to the TPN solution rapidly corrects the response to insulin. In fact, as soon as glucohomeostasis is achieved, the amount of insulin in the solution should be reduced by 30% to 50% to avoid hypoglycemia.

If the patient is prediabetic or diabetic, the TPN or PPN solution can be modified by adding the amount of regular humulin necessary to bring the blood sugar consistently below 170 mg/dL. A blood sugar greater than 170 mg/dL begins to compromise immune function.[37] If the blood sugar is consistently greater than 200, the immune system is flat regarding defense against candida albicans.[38]

Sepsis, stress, and steroids affect cytokine-mediated endocrine responses that result in insulin resistance.[39] As these extraneous variables come under control, the amount of insulin required to maintain blood glucose below 170mg/dL will decrease. This should be anticipated and the solution altered on a daily basis depending on the clinical change. This may require using single liter admixtures as opposed to a 24-hr 3-L bag for fear of the need to discard solution because of hypoglycemia.

There is a helpful rule of thumb regarding TPN solutions that contain insulin. *If you raise the rate of the infusion, you will raise the blood glucose. If you lower the rate of infusion, you will lower the blood glucose*. This follows Michaelis-Menton mechanics, which describes the action of a hormone at the cell receptor.[40]

Acid-Base Balance

Compounding a TPN solution to match the patient's acid–base balance gives the clinician a powerful metabolic tool to correct acidosis or alkalosis. Likewise, serious acid–base balance problems can occur if the TPN solution is misformulated. For example, if all of the cations are added as the chloride salt, the bottle will contain amino acids in a hyperchloremic environment. Expect the patient to develop hyperchloremic metabolic acidosis. However, if all of the cations were added as the acetate salt, the patient will develop hypochloremic alkalosis.

Acetate is rapidly metabolized to bicarbonate. As stated earlier, never use sodium bicarbonate in a TPN solution for fear of a calcium bicarbonate precipitate and the release of microbubbles of carbon dioxide. When making a decision to add or delete sodium or potassium in a TPN formula to correct an aberrant serum electrolyte value, always check serum chloride and bicarbonate first. If the patient is hyperchloremic, the acetate salt should be used. The milliequivalent degree of manipulation of each to correct an electrolyte imbalance is described in the following section. Presentation of deficiencies are listed in Table 13.1.

Respiratory acidosis can be secondary to overfeeding carbohydrate. All calories fed in excess of expense must be stored as fat. This metabolic process produces excess carbon dioxide when the abundant substrate is carbohydrate. The respiratory quotient (RQ) for using a carbohydrate calorie for energy is 1; however, the RQ to process carbohydrate into fat is greater than 1 and increases parallel to the overfeeding. The patient will develop a respiratory acidosis. This is a particularly important consideration when trying to wean a patient from the ventilator. The maximum amount of dextrose that can be metabolized has been determined to be 5 mg/kg·min^{-1} or approximately 25 Kcal/kg·day^{-1}.[41] It might be helpful to underfeed carbohydrate while maintaining adequate nitrogen support while trying to wean a patient from a ventilator.

Extracellular Electrolytes

The principle extracellular electrolytes to be considered are sodium, chloride, and acetate. The average TPN solution in a patient with normal electrolyte values contains approximately 50 mEq/L of each. If a change is necessary, additions or deletions are made in 25 mEq/L increments. If the solution is a 3-in-1 solution mixed to meet the entire day's requirements, the number of liters per day is multiplied by the per liter requirement. The sodium-to-chloride-to-acetate ratio should be approximately 1:1:1. Remember, there is acetate in the amino acid solution component to the TPN formula. When hydrolysates of protein were used, trace metal deficiencies were not seen. Even hypophosphatemia was not described until pure crystalloid amino acid solutions came into use. Then, every ion that played a role in intracellular metabolism had to be added in appropriate amounts.

Intracellular Electrolyte Balance

The intracellular electrolytes are potassium, phosphorous, calcium, magnesium, zinc, copper, chromium, and other trace minerals. Where the extracellular electrolytes are added per volume, the intracellular electrolytes are added to match calories. This is because the intracellular electrolyte profile must remain normal as the cell "grows" and repletes its intracellular substance. The rate of uptake depends on the rate of repletion. The rate of repletion depends on the

Table 13.1. Underfeeding TPN Constituents[a]

Component	Consequence
Nonprotein calories	Failure to recover somatic assessment indices, recovery of somatic mass requires exercise
Amino acids	Relatively essential amino acids like glutamine, arginine and others may cause gut and hepatic loss of function and immune response slowing
Hyponatremia	Less than 120 mg/dL may be associated with lethargy, convulsions, and coma
Hypokalemia	Less than 3.5 mg/dL may see arrhythmia, tachycardia, and cardiac standstill in systole
Hypochloremia	Alkalosis
Hypocarbia	Alkalosis
Hypocalcemia	Tetany, convulsions, hyperreflexia, carpopedal spasms
Hypomagnesemia	Neuromuscular irritability beginning with nightmares and proceeding to hallucinations, first visual, then adding auditory and finally tactile
Hypozincism	Achrodermatosis enteropathica or a seborrheic rash in lateral eye brow, nasolabial fold, and trunk
Chromium deficiency	Unusual insulin resistance
Copper deficiency	Microcytic hypochromium deficiency with normal iron indices
Selenium deficiency	Reversible "cardiomyopathy"
	VITAMINS
Vitamin A	Waxy follecular hyperkeratosis
Vitamin B_1	Wernicke's encephalopathy, Korsakoff's syndrome if prolonged, dry beri-beri with a severely symptomatic intracellular and mild lactic acidosis
Vitamin B_2	Chelosis, angular stomatitis, glossitis, ophthalamitis
Vitamin B_6	Unknown
Vitamin C	Cellophane skin and capillary fragility causing purplish brusing on arms and hands
Vitamin D	Osteoporosis
Vitamin K	Prolonged PT. Naturally in IV lipid preparation
Vitamin E	Unknown
Biotin	Unknown
Folic Acid	Unknown

[a] This table includes TPN constituents that more commonly manifest as observable undesired clinical consequences. The TPN formula should be tailored to the patient's requirements to limit occurrence.

Table 13.2. Overfeeding TPN Constituents[a]

Component	Consequence
Carbohydrate	Possible hyperglycemia, glucosuria, and osmotic diuresis causing dehydration, coma, and death
Fat	Lipid particle clogging of macrophages and the RES
Amino acids	Acidosis and angioneumotic edema
Hypernatremia	Hyperosmolarity, thirst, agitation
Hyperkalemia	Arrhythmia, bradycardia with peaked T waves, cardiac standstill in diastole
Hypercarbia	Acidosis
Hypercalcemia	Lethargy, confusion, unusual complaint of muscle, back, and abdominal pain
Hypermagnesemia	Lethargy, anorexia
Vitamin A	Hepatomegaly, polar bear liver syndrome

[a] This table includes TPN constituents more commonly associated with adverse clinical consequences. The TPN formula can be adjusted to limit occurrence.

protein and nonprotein calories delivered. Therefore, for each 750 to 1000 kcal delivered there should be 40 mEq of potassium, 5 mM of phosphate, 8mEq of magnesium, 10 mEq of calcium, and 1 U of combined trace metals (MTE) and 1 ampule of MV1. Overfeeding any of these constituents can result in a complication (Table 13.2).

Adjustments to the TPN formula to correct a serum abnormality of the intracellular electrolytes are made as follows: Add or delete potassium at increments of 5–10 mEq/750–1000 Kcal. Magnesium is changed in increments of 2–4 mEq/750–1000 kcal, calcium at 5 mEq/750–1000 kcal, and phosphate at 2.5 mm. Daily electrolytes and magnesium profiles may be necessary for a few days, especially if the patient started with or develops an imbalance, but usually a Monday, Wednesday, Friday laboratory analysis is sufficient for most in-patients.

Abnormal Liver Function Test

Peculiar abnormal liver function tests have been associated with TPN. Virtually 100% of patients will have mild elevations in liver enzymes after 6 weeks of therapy.[42] Continuously overfeeding carbohydrate sustains hyperinsulinemia. The combination of excess calories and hyperinsulinemia on a 24-hr/day basis leads to hepatic steatosis.[43] Lipid emulsions given in excess can engorge Kupffer cells.[44] The risk of hepatic steatosis is minimized by placing the patient on an infusion interrupt system. This fed-fast routine more closely resembles normal cycles, allowing insulin to decrease and glucagon to increase during the fast phase, mobilizing hepatic stores of fat and glycogen.

Cholestasis is another significant problem.[45] In addition to bacteria-derived circulating factors that may be absorbed from a leaky gut due to dysbiosis, the lack of stimulation to bile flow by cholecystokinin in patients who are not nourishing themselves by mouth, leads to cholestasis, sludge, and stone formation. Gut trophic hormones are not elaborated unless the gut is stimulated with food. In experiments, utilizing the isolated ileal loop model in rats, structural hypoplasia can be demonstrated during starvation, as well as when the animals are maintained on parenteral nutrition.[46] This is overcome by oral feeding. In addition to cholecystokinin, gastrin,[47] secretin,[48] insulin,[49] enteroglucagon,[50] motilin,[51] neurotensyn,[51] and epidermal growth factor[52] have been shown to improve or maintain gastric mass during TPN. It is therefore best to provide some oral nutrition as soon as possible.

Leaky Gut

Solutions of TPN are glutamine free. Glutamine is a conditionally essential amino acid. Glutamine is the primary fuel for the enterocyte.[53] Short-chain fatty acids are the primary fuel for the colonocyte.[54] Glutamine-free enteral or parenteral diets have an atrophic effect upon small bowel mucosa and removal of soluble fiber from the diet produces mucosal atrophy in the large bowel. Soluble fiber is metabolized to short-chain fatty acids by gut bacteria.[55] In addition to creating a leaky gut, which will result in absorption of luminal antigens and toxins that must be cleared by the liver, the lack of glutamine also leads to a decreased production of glutathione.[56] Glutathione plays a principle role in hepatic phase I and phase II detoxification, as well as serves as the primary hepatic free radical scavenger. Hepatic dysfunction seen during TPN might be worsened by a decrease in the gut's specific immunity. Specific immunity depends on the production of secretory immunoglobulin A (SIgA). Secretory IgA is the most abundant immunoglobulin produced by the body, constituting more than 60% of the daily immunoglobulin production.[57] Several significant physiologic challenges common to patients on TPN have been shown to result in a decrease in SIgA production.[58,59,60] Without specific immunity, the adhesion of enterotoxigenic luminal organisms to the mucosa perpetuates the leaky gut cycle. As organisms adhese to and proliferate on the mucosa, inflammation results in an increased leak. This leak is thought to be responsible for the translocation and increased oxidative stress in the liver.

It has been known for some time that the more a patient can eat orally, the better the liver functions as determined by standard laboratory liver function tests. Usually this correlates with the amount of bowel function and the amount of oral nutrition tolerated.[61] To assist in gut adaptation and avoid these complications, oral feeding and perhaps glutamine supplementation should begin as soon as possible. Furthermore, phase II hepatic detoxification involves conjugation/soluablization requiring some nonessential amino acids not found in amino acids used in TPN. A relative deficiency also can increase oxidative stress in the liver.

PHYSIOLOGIC COMPLICATIONS

Hyperosmolarity Secondary to Hyperglycemia and Glucosuria

The average renal threshold for glucose is 250mg/dL. When the tubular limit for reabsorption of glucose is exceeded, glucosuria will create an osmotic diuresis and pull water from the patient. The intake and output may be deceptive in that the totals might be 3 L in and 3 L out as urine. This means that in addition to the 600 to 800 cc of insensible loss, the urine contained obligate water from the presence of glucose. The input–output balance should reflect the insensible loss. In addition, the patient is not starving, so body mass water will not be excreted. Further 1 L of TPN is not the same volume of water as 1 L of an electrolyte solution. It is, therefore, normal to see an input–output gradient of just over 1000 mL. If not, hyperglycemia and glucose urea are the first things to be ruled out.

Profound Hypoproteinemia and Edema

Fluid balance and compartment distribution of water is also the responsibility of the physician writing TPN orders. Thus, every patient is evaluated daily for intake and output levels and for the presence of jugular venous distention, edema, or dehydration. The patient should not be overloaded, dehydrated, nor edematous. The exam begins by reviewing the intake and output. The patient must be voiding at least 900 mL of urine/day to excrete metabolites. If 900 mL of urine was not produced and intake was adequate, or if there is greater than a 1500 mL difference between intake and output, the patient is losing circulating volume somewhere. Examination of fluid status includes asking the patient if he or she is thirsty. Observation is then made of the tongue for saliva. Following this, the head is turned and the external jugular vein occluded at the clavicle just lateral to the sternocleidomastoid muscle and the jugular vein allowed to fill. The finger is removed and the rapidity of jugular venous emptying or maintenance of jugular venous distention is observed. The patient is usually positioned at a 35 to 45° angle for this observation. The lungs are then auscultated and the calves, thigh, and dependent back checked for pitting edema.

If the patient demonstrates decreased urine output, states he or she is thirsty, has no jugular venous distention, and tinting of the cervical skin plus no edema, this patient is truly volume deficient and needs an increase in crystalloid until these findings are corrected. On the other hand, if the patient has no extracorporeal losses, such as fistula drainage or massive diarrhea, and a low urine output is observed along with *flat neck* veins and *peripheral edema* one must check the laboratory values for the latest total protein. Total protein is the sole factor creating colloid osmotic pressure, not albumin alone. The colloid osmotic pressure holds fluid in circulation. If the total protein is 5.0 g/dL or less, the patient will be unable to hold fluid in circulation even against a normal hydrostatic pressure gradient. That is, a CVP of 4 to 8 cm of water. This dependent edema is called hypooncotic edema. It is not fluid overload. Note, there is no jugular venous distention. This condition is not uncommonly observed in profoundly malnourished patients. It would be an error in judgement to give this patient a fluid challenge in an attempt to increase urine output by increasing circulating volume. That would have a negative physiologic consequence. Crystalloid will further dilute total protein while elevating hydrostatic pressure, thus driving more fluid into the interstitial space. What is needed under these circumstances is colloid. Reviews of Starling's equation, which governs the distribution of fluid in spaces, have been published.[62,63]

Albumin can be added to the TPN solution. Albumin is a highly negatively recharged molecule and can crack a 3-in-1 fat emulsion. Therefore, while albumin is in the TPN solution, fat will have to be delivered via a separate line. As the total protein rises, creating a colloid pressure higher than the hydrostatic pressure, the third-space hypooncotic edema will be recaptured. At this point, circulating volume will expand increasing urine output. In fact, this expansion can be rapid necessitating the use of diuretics to avoid pulmonary edema now secondary to a rebound rise in hydrostatic pressure.

CONCLUSION

The above review of complications and measures to be taken to avoid or correct them is not exhaustive. Rather, it is an attempt to share with the reader the more common elements encountered along the learning curve. Nutritional support, particularly TPN and PPN, has stimulated research that has lead to an important understanding of the relationship between nutrition, physiology, and outcome. A patient that cannot eat should not suffer from complications of malnutrition because these techniques are available. On the other hand, a patient should not suffer from complications secondary to nutritional support. Virtually anything that can go wrong with TPN or PPN has been evaluated, elucidated, and, therefore, can be avoided. Should these complications occur, a nutritional support team will be able to move early to adjust the protocol to correct aberrancy. Discontinuing TPN is usually not necessary if the specific problem is understood, focused on, and measures taken to avoid or correct it.

REFERENCES

1. Dudrick SJ, Wilmore DW, Vars HM, et al. Long-term total parenteral nutrition with growth, development, and positive nitrogen balance. *Surgery* 1968; 64:134–142.

2. Hansen LM, Hardie WR, Hidalgo J. Fat emulsion for intravenous administration: clinical experience with Intralipid 10%. *Ann Surg* 1976; 184:80–88.
3. Aubaniac R. L'injection intraveineuse sous-claviculaire: avantage et technique. *Presse Med* 1952; 60:1456.
4. Sitzman JV, Townsend TR, Siler MC, et al. Septic and technical complications of central venous catheterization. *Ann Surg* 1985; 202:766–770.
5. Coppa GF, Gouge TH, Hofstetter SR. Air embolism: a lethal but preventable complication of subclavian vein catheterization. *J Parent Ent Nutr* 1980; 5:166.
6. Stillman RM, Soliman F, Garcia L, et al. Etiology of catheter-associated sepsis. *Arch Surg* 1977; 112:1497–1499.
7. Bozzetti F, Scarpa D, Terno G, et al. Subclavian thrombosis due to indwelling catheters: A prospective study on 52 patients. *J Parent Ent Nutr* 1983; 7:560–562.
8. Crocker KS, Devereaux GB, Ashmore DL, et al. Clinical evaluation of elastomeric hydrogel peripheral catheters during home infusion therapy. *J Intraven Nurs* 1990; 13:89–97.
9. McKee JM, Shell JA, Warent TA, et al. Complications of intravenous therapy: a randomized prospective study—Vialon vs Teflon. *J Intraven Nurs* 1989; 12:288–295.
10. Fabri PJ, Mirtallo JM, Ruberg RL, et al. Incidence and prevention of thrombosis of the subclavian vein during total parenteral nutrition. *Surg Gynecol Obstet* 1982; 115:238–240.
11. Kaminski MV Jr, Harris DH. Prolonged uncomplicated intravascular catheterization. *Am J IV Ther* 1976; 19–24.
12. Hurtubise MR, Bottino JC, Lawson M, et al. Restoring patency of occluded central venous catheters. *Arch Surg* 1980; 115:212–213.
13. Monturo CA, Dickerson RN, Mullen JL. Efficacy of thrombolytic therapy for occlusion of long-term catheters. *J Parent Ent Nutr* 1990; 14:312–314.
14. Kaminski Jr, MV, Harris DH, Collin CF, et al. Electrolyte compatibility in a synthetic amino acid hyperalimentation solution. *Am J Hosp Pharm* 1974; 31:244–246.
15. Shulman RJ, Reed T, Pitre D, et al. Use of hydrochloric acid to clear obstructed central venous catheters. *J Parent Ent Nutr* 1988; 12:509–510.
16. Pennington CR, Pithie AD. Ethanol lock in the management of catheter occlusion. *J Parent Ent Nutr* 1987; 11:507–508.
17. Hoppe MC, Kaminski MV, Aquinas SM. Zero catheter sepsis rate during hyperalimentation in an urban community hospital. *J Parent Ent Nutr* 1977; 1(4):34A(abstract).
18. Kaminski MV, Haase TJ. Reimbursement for nutritional support. In CW Vanway, ed., *Handbook of surgical nutrition*. Philadelphia: J.B. Lippincott, 1992. 293–303.
19. McCormick DC, Knutsen CV, Griffen RE, et al. Pharmaceutical aspects of parenteral nutrition. In Kaminski MV, ed. *Hyperalimentation: a guide for clinicians*. Marcel Dekker, New York: 1985. 537–590.
20. Rypins EB, Johnson BH, Reder B, et al. Three-phase study of phlebitis in patients receiving peripheral intravenous hyperalimentation. *Am J Surg* 1990; 159:222–225.

21. Recommendations to pharmacists for solving problems with large volume parenterals. *Am J Hosp Pharm* 1980; 37:663–667.
22. Harrigan C. Care and cost justification of final filtration. *Nat Intraven Ther Assoc* 1985; 8:426–430.
23. Mershon J, Nogami W, William J, et al. Bacterial/fungal growth in a combined parenteral nutrition solution. *J Parent Ent Nutr* 1986; 10:498–502.
24. Hoppe M. Role of the nurse on the metabolic support team In Kaminski MV, ed. Hyperalimentation: a guide for clinicians. New York: Marcel Dekker, 1985. 465–492.
25. Maki DG, McCormack KN. Defatting catheter insertion sites in total parenteral nutrition is of no value as an infection control measure. *Am J Med* 1987; 83:833–840.
26. Larson E. APIC guidelines for infection control practice: guideline for use of topical antimicrobial agents. *Am J Infect Control* 1988; 16:253–266.
27. Maki DG, Ringer M, Alverado CJ. Prospective randomized trial of providone-iodine, alcohol, and chlorhexidine for prevention of infection associated with central venous and arterial catheters. *Lancet* 1991; 338:339–343.
28. Palidar PJ, Simonowitz DA, Oreskovich MR, et al. Use of Op Site as an occlusive dressing for total parenteral nutrition catheters. *J Parent Ent Nutr* 1982; 6:150–151.
29. Porter KA, Bistrian BR, Blackburn GL. Guidewire catheter exchange with triple culture technique in the management of catheter sepsis. *J Parent Ent Nutr* 1988; 12:628–632.
30. Lee RB, Buckner M, Sharp KW. Do multilumen catheters increase central venous catheter sepsis compared to single lumen catheters? *J Trauma* 1988; 28:1472–1475.
31. Sitges-Serra A, Puig P, Linares J, et al. 1984. Hub colonization as the initial step in an outbreak of catheter-related sepsis due to coagulase negative staphylococci during parenteral nutrition. *J Parent Ent Nutr* 1984; 8:668–672.
32. Stotter A, Ward H, Waterfield AH. Junctional care: The key to prevention of catheter sepsis in intravenous feeding. *J Parent Ent Nutr* 1987; 11:159–162.
33. Goode CJ, Titler M, Rakel B, et al. A meta-analysis of effects of heparin flush and saline flush: quality and cost implications. *Nurs Res* 1991; 40:324–330.
34. Wolfe BM, Walsh I, Kaminski MV Jr. Outcome of central venous catheters for home infusion therapy. *J Parent Ent Nutr* 1994;134s.18:1(Suppl), (abstract).
35. Kaminski MV. Hyperosmolar, hyperglycemic, non-ketotic dehydration—etiology, pathophysiology and prevention during total parenteral alimentation. *Excerpta Medica Int Cong* 1975; 367:290–305.
36. Freed BA, Pinchcofsky G, Nasr NJ, Kaminski MV. Normalization of serum glucose levels and decreasing insulin requirements by the addition of chromium to TPN formulas. *J Parent Ent Nutr* 1981; 5:568. 5(6):(abstract).
37. Nirgoitis M, Hennessey B, Black C, et al. 1991. Hyperglycemia eliminated immunoglobulins ability to enhance survival in septic asplenic infant rats. *J Parent Ent Nutr* 1991;15(1):215(abstract).
38. Hostetter M: Handicaps to host defense. Effects of hyperglycemia on C_3 and *candida albicans*. *Diabetes* 1990; 39:271–275.

39. Frankel W, Evans N, Rombeau J. 1990. Scientific rationale and clinical application of parenteral nutrition in critically ill patients In Rombeau JL, Caldwell MD *Clinical nutrition: parenteral nutrition* 2nd ed. Philadelphia: W.B. Saunders 1990: 597–616.
40. Booth T, Kaminski MV, Haase T. 1991. A comparison of blood glucose levels in diabetic and nondiabetic patients with tapering vs. abrupt discontinuation of TPN. *J Am Coll Nutr* 1991;10(5):551(abstract).
41. Askanazi J, Rosenbaum SH, Hyman AL, et al. Respiratory changes induced by the large glucose loads of total parenteral nutrition. *JAMA* 1980;243:1444–1447.
42. Messing B, Bories C, Kunstlinger F, et al. Does total parenteral nutrition induce gallbladder sludge formation and cholelithiasis? *Gastroenterology* 1983; 84:1012–1019.
43. Fisher RL. Hepatobiliary abnormalities associated with total parenteral nutrition. *Gastroenterol Clin North Am* 1989; 18:645–667.
44. Martins FM, Wennberg A, Meurling S, et al. Serum lipids and fatty acids and the composition of tissues in rats on total parenteral nutrition. *Lipids* 1984; 19:728–737.
45. Latham PS, Menkes E, Phillips MJ, et al. Hyperalimentation-associated jaundice: An example of serum factor inducing cholestasis in rats. *Am J Clin Nutr* 1985; 41:61–65.
46. Adams PR, Copeland EM, Dudrick SJ, et al. Maintenance of gut mass in by-passed bowel of orally vs parenterally nourished rats. *J Surg Res* 1978; 24:421–427.
47. Johnson LR, Copeland EM, Dudrick SJ, et al. Structural and hormonal alterations in the gastrointestinal tract of parenterally fed rats. *Gastroenterology* 1975; 68:1177–1183.
48. Hughes CA, Bates T, Dowling RH. Cholecystokinin and secretin prevent the intestinal mucosal hypoplasia of total parenteral nutrition in the dog. *Gastroenterology* 1978; 75:34–41.
49. Lickley HLA, Track NS, Vranic M, et al. Metabolic responses to enteral and parenteral nutrition. *Am J Surg* 1978; 135:172–175.
50. Sagor GR, Ghatei MA, AL-Mukhtar MYT, et al. Evidence for a humoral mechanism after small intestinal resection. *Gastroenterology* 1983; 84:902–906.
51. Aynsley-Green A, Lucas A, Lawson GR, et al. Gut hormones and regulatory peptides in relation to enteral feeding, gastroenteritis, and necrotizing enterocolitis in infancy. *J Pediatr* 1990; 117(suppl):24–32.
52. Bragg LE, Hollingsed TC, Thompson JS. Urogastrone reduces gut atrophy during parenteral alimentation. *J Parent Ent Nutr* 1990; 14:283–286.
53. Souba WW, Klimberg SV, Plumley DA, et al. The role of glutamine in maintaining a healthy gut and supporting the metabolic response to injury and infection. *J Surg Res* 1990;48:383–391.
54. Lee A, Gemmell E. Changes in the mouse intestinal microflora during weaning: Role of volatile fatty acids. *Infect. Immun* 1972; 5:1.
55. Kripke SA, Fox AD, Berman JM, et al. Stimulation of intestinal mucosal growth with intracolonic infusion of short-chain fatty acids. *J Parent Ent Nutr* 1989; 13:109–116.
56. Lawrence HL, Hagen TH, Jones DP, et al. Exogenous glutamine protects intestinal epithelial cells from oxidative injury. *Proc Natl Acad Sci USA* 1986; 83:4641–4646.

57. Pockley AG, Montgomery PC, In vitro adjuvant effects of interleukin 5 and 6 on rat tear IgA antibody response. *Immunology* 1991; 73:19–23.
58. Harmatz PR, Carter EA, Sullivan D, et al. Effect of thermal injury in the rat on transfer of IgA proteins into bile. *Ann Surg* 1989; 210:203–207
59. Wira CR, Sandoe CP, Steele MG. Glucocorticoid regulation of the humoral immune system. *J Immunol* 1990; 144:142–146.
60. Alverdy JC, Aoys E. The effect of dexamethasone and endotoxin administration on biliary IgA and bacterial adherence. *J Surg Res* 1992; 53:450–454.
61. Haase T, Berger AW, Kaminski Jr, MV. Gut adaptation in HPN patient. *J Am Coll Nutr* 1990; 9(5):550.
62. Kaminski Jr, MV, Williams SD. Review of the rapid normalization of serum albumin with modified total parenteral nutrition solutions. *Crit Care Med* 1990; 18(3):327–335.
63. Kaminski MV Jr, Blumeyer TJ. Albumin supplementation: Starling's law as a guide to therapy and literature review. In G.P. Zaloga, ed. Nutrition in critical care. Chicago: Mosby, 143–156.

ENTERAL NUTRITION

The Benefits of Early Enteral Nutrition

CHAPTER 14

Peter C. Muskat, M.D.

INTRODUCTION

Currently, nutritional support is recognized as a major element in the care of the critically ill or injured patient with the gastrointestinal tract considered the ideal route for administration. Since the introduction of modern nutritional support, total parenteral nutrition (TPN) has been the most widely used means of delivering calories and protein in the early postoperative or postinjury period. Central venous infusion of nutrients has been the gold standard for nutritional support. It is well tolerated by most patients and is associated with relatively minimal complications, although central line placement complications, line sepsis, and overfeeding remain common problems. The major drawback to TPN is that its use is usually in conjunction with the avoidance of nutrient administration into the intestinal tract. This condition has recently been shown to result in gut mucosal atrophy and related complications. Thus, the relationship between TPN and gut luminal nutrient deprivation, makes parenteral nutrition a less than ideal final solution for aggressive nutritional support.

The last ten years has seen a number of important studies that demonstrate the efficacy and benefits of early enteral feedings in the postoperative or postinjured patient. This chapter will review these studies and describe the many benefits achieved when nutrients are delivered into the gastrointestinal tract shortly after surgery, trauma, or illness.

HISTORICAL PERSPECTIVE

Enteral nutrition has been used only sporadically since the time of the Egyptians. Nutrients were usually delivered via enema until 1890 when Edsal and Miller demonstrated that rectally administered protein was eliminated essentially unabsorbed.[1] Surgical access to the upper gastrointestinal tract was first reported in the late 1800s with Surmay in 1878 performing the first surgical jejunostomy.[2] This was followed by Witzel's introduction of the gastrostomy tube in 1891.[3] Unfortunately, these tubes were associated with significant morbidity, specifically high leakage and obstruction rates. The early part of this century saw the popularization of transnasal feeding tube placement, first reported by Andressen in 1918.[4] Shortly thereafter, Abbott and Rawsen in the late 1930s described the double lumen tube which allowed gastric decompression and jejunal feeding simultaneously.[5] These large bore tubes were associated with a high incidence of sinusitis, otitis, reflux, and aspiration. It was not until MacDonald in 1954 reported the development of a small bore polyethylene catheter, that enteric feeding, either by the nasoenteric or direct transjejunal route began to be used as a serious means of supplemental nutrition.[6] In 1973, Delaney et al. designed the needle catheter jejunostomy tube currently used in many hospitals.[7] This endeavor was an attempt to develop an easier technique for placing operatively inserted jejunostomy tubes.

The 1960's marked the beginning of the transition from an enteral route of nutritional supplementation to a parenteral route via the central veins. Continuous central vein infusion was first used in patients in 1961 based on the work of Rhode in 1949.[8] Dudrick et al. in 1968[9] demonstrated the viability of this route of nutrition and the following 30 years has seen the development of TPN as the primary means of feeding the surgical patient in the early postoperative period. As TPN was shown to be relatively safe and easy to administer, the interest in enteral feeding (often more difficult to establish and maintain) markedly decreased. However, despite TPN's popularity, the high costs of administration and the not inconsiderable complication rate have led to renewed interest in developing better enteral solutions, tubes, and techniques for establishing reliable access into the gastrointestinal tract.

In addition to major improvements in enteral nutrient delivery systems, the last 10 years have also seen several studies that support the notion that early postinjury nutritional care of the surgical patient need not be confined to the parenteral route. Early enteral feeding has been shown to offer significant benefits and may in fact be superior to parenteral feedings.

An explosion of interest in enteral feedings has occurred, particularly in the ever-expanding variety of solutions being marketed. In addition to the development of elemental and disease-specific solutions, there has been recent interest for the inclusion of specific nutrients that are thought to modulate the metabolic processes or stimulate the immune system. These nutrients include amino acids such as glutamine and arginine, omega-3 fatty acids (fish oil), and nucleotides. Recent work has suggested that the addition of these nutrients may improve overall outcome for the critically ill or injured.

BENEFITS OF ENTERAL NUTRITION

Preservation of Gut Mucosal Integrity

Overall, enteral nutrition is a more physiologic means of maintaining adequate nutrition and should be used whenever possible as the primary route for the administration of nutrition. The lack of luminal nutrients (despite the adequate provision of nutrition with TPN) leads to gut mucosal atrophy and immunological dysfunction of the intestinal immune system.

The gastrointestinal barrier is a complex system designed to prevent the invasion of intestinal bacteria and pathogens into the host. The defense of the gut is a coordinated effort with chemical, enzymatic, physical, and immunologic components. Healthy enterocytes and the tight junctions between them form the primary physical barrier. Luminal nutrients are critical for maintaining a normal pattern of regeneration, migration, maturation, and sloughing of enterocytes. A gastric pH of 4.5 or less kills most bacteria, making the stomach a sterile place normally, with pepsin, salivary lysozyme, and lactoferrin further inhibiting the growth of bacteria. Mucin is produced throughout the gut and prevents bacteria from interacting with the enterocyte by both binding nonspecifically with the bacteria and creating a pH gradient with secreted bicarbonate. Pancreatic enzymes, intestinal motility, and even the normal bacterial flora reduce the likelihood of pathogen invasion. Impairment of the gut barrier has been shown to result in the translocation of intraluminal bacteria and toxins into the intestinal lymphatics and bloodstream.[10] Furthermore, translocation is believed to be a major factor in the development of nosocomial sepsis and multiple organ dysfunction syndrome, both significant contributors to the mortality seen in the critically injured patient. Therefore, it is of great importance that the intestinal barrier be preserved.

Experimental data suggests that maintaining animals on TPN results in mucosal atrophy and immunologic dysfunction.[11–13] It is not clear whether this effect is due to a lack of enteral stimulation or the absence of a specific nutrient. Interestingly, recent evidence suggests that if glutamine is added to TPN, enhanced cellularity of both small bowel[14] and colonic mucosa[15] is seen. Thus, it appears to be a combination of enteral stimulation and specific nutrients which is responsible for maintaining the physical barrier.

IMMUNOLOGIC BENEFITS OF ENTERAL NUTRITION

The immunologic benefits of a healthy gut are several. The wall of the intestinal tract contains a rich bed of immune cells, such as lymphocytes, macrophages, neutrophils, eosinophils, and mast cells, which are commonly referred to as the gut associated lymphoid tissue (GALT). This system is located mainly in the submucosa and mucosa. These cells secrete immunoglobulin A (IgA) into the saliva, bile, and succus entericus to assist in the bacteriolysis and prevention of adherence of microbes to the mucosa.[16] Alverdy and colleagues have shown that rats fed enterally maintained secretory IgA levels better than rats fed the same nutrients intravenously.[11] Other studies have shown decreased neutrophil chemotaxis and phagocytosis and a reversal of the T-cell helper to suppressor ratio following trauma or hemorrhagic shock.[17] Thus, it appears that an impaired immune system is a predisposing factor to bacterial translocation with evidence suggesting preservation of immunologic function with enteral feedings.

Effects on the Stress Response

The administration of intraluminal nutrients has also been shown to possibly attenuate the stress response. Lowry demonstrated lower counterregulatory hormone and C-reactive protein levels in subjects fed enterally after endotoxin exposure versus those fed intravenously.[18] Peterson and coworkers found improved synthetic protein synthesis (albumin, transferrin, retinol-binding protein) in trauma patients randomized to enteral feeding versus those given TPN.[19] However, in contrast, Eyer et al. in a prospective randomized trial of 52 patients given either early or late enteral feedings found no difference in metabolic responses as measured by plasma lactate, total urinary nitrogen, catecholamines, and cortisol.[20] Poret et al. studying immediate enteral and parenteral feedings in patients following trauma laparotomy also showed no difference in urinary catecholamine responses in the first 4 days.[21] Thus, it remains unclear whether or not the route of administration of nutrients plays a significant role in muting the stress response.

Effect on Septic Morbidity

Multiple animal studies have demonstrated significant benefits with enteral nutrition but it has been only in the last 10 years that several large prospective studies have shown the advantages of enterally feeding postinjury patients. Moore and Jones in 1985[22] reported a randomized prospective trial of 75 trauma patients who were given either no supplemental nutrition following emergent celiotomy or a constant infusion of an elemental diet through a needle catheter jejunostomy. Enteral feedings were started within 18 hr of surgery and were well tolerated in 63% of the patients. Significant improvement was seen in the enterally fed patients' nitrogen balance at 4 and 7 days. Septic morbidity was also shown to be significantly greater in the control group (29%) than in the early fed group (9%).[22] In 1985, Adams et al. compared multiple trauma patients prospectively randomized to either a TPN group or an enterally fed group with nutritional support begun on the first postoperative day. They found no differences between the groups in average daily caloric intakes, nitrogen balance, and complication rates, suggesting that early jejunostomy feeding is safe and well tolerated.[23] Moore et al followed up their earlier study with a prospective clinical trial comparing patients with major abdominal trauma fed either with TPN or total enteral nutrition (TEN). Feeding was started within 12 hr in both groups and was well tolerated by 86% in the enteral group. The incidence of major septic morbidity was 3% in the TEN group versus 20% in the TPN group. Nitrogen balance remained equivalent between the two groups. Of interest, levels of prealbumin and α_2-macroglobulin returned to normal faster in the TEN group than in the TPN group, whereas levels of haptoglobulin and α-antitrypsin increased to a greater degree in the TPN group.[24] This suggests a reprioritization of hepatic protein synthesis defined as the relative balance of acute-phase proteins compared to constitutive proteins.[25] The authors speculated that TEN reduced bacterial translocation by preserving gut mucosal integrity, which may provide an explanation for the reduced septic morbidity in the enterally fed group.[1]

Further evidence that the use of early enteral feeding decreases septic morbidity in trauma patients was reported by Kudsk et al. in 1992. Patients (98) were prospectively randomized to either enteral or parenteral feedings begun within 24 hr of injury. The enteral group had significantly fewer pneumonias, intraabdominal abscesses, episodes of line sepsis, and total number of infections per patient.[26]

Further Benefits of Enteral Nutrition

Although a low gastric pH protects the gastrointestinal tract by eliminating pathogens, it has also been shown to increase the risk of upper gastrointestinal bleeding in critically ill patients. Layon et al. showed that patients fed into the jejunum with an enteral feeding solution versus a saline solution were able to maintain a gastric pH greater than 4.[27] This was thought to be regulated by the secretion of gut hormones stimulated by the luminal nutrients. These patients required less antacids than the controls. Enteral nutrition, therefore, could protect the gastric lining from bleeding without the cost and potential complications of antacids or histamine receptor blockers.

In 1992, McClave et al. demonstrated that enterally fed critically ill patients had improved gallbladder contraction, increased pancreatic stimulation, and improved healing of gastrointestinal anastomoses.[28] These findings also suggest that the use of the gut for the delivery of nutrition is more physiologic than the parenteral route and may improve wound healing. Although no studies to date have shown a reduction in mortality, a recent study was able to show a reduction in hospital days for patients fed enterally. Bower et al. reported a multicenter prospective, randomized trial with 296 intensive care patients comparing a commonly used formula (Osmolite HN, Ross Laboratories, Columbus, OH) with a formula supplemented with arginine, dietary nucleotides, and fish oil (Impact, Sandoz Nutrition, Minneapolis, MN). All patients were started on nutritional supplements within 48 hr of admission to the intensive care unit. The mortality rate was significantly lower than predicted in both groups but not statistically different between the two groups. Of great significance, however, was that the hospital median length of stay was reduced by 8 days in the Impact group. A subgroup of septic patients benefited most from the Impact feedings with a reduction in the median number of hospital days and in the frequency of acquired infections being noted.[29]

Another major advantage of enteral nutrition is the cost benefit seen over parenteral nutrition. Even today, parenteral nutrition remains five to ten times more expensive to administer to the patient when the cost of materials and preparation are calculated. Bower et al. found parenterally fed patients following abdominal surgery had nutrition support charges of $2312 versus $849 for the enterally fed patients.[30] Moore and coworkers[22] were able also to show decreased hospital costs for patients fed enterally compared with those fed parenterally ($16,000 versus $19,000, respectively).

HOW EARLY IS EARLY?

While most surgical nutritionists would advocate the early institution of enteral nutrition in the critically ill or injured patients, it is currently not known how early it must be initiated to obtain the benefits. Although jejunal feeding can be safely begun shortly after injury or laparotomy, does it need to be started that soon? Currently, there are few reports in the literature that have looked at this question. Grahm et al. compared early jejunal feeding (within 36 hr of injury) to late gastric feeding in head-injured patients and demonstrated a reduced incidence of bacterial infections and hospital stay in the early fed group.[31] However, as previously described, Eyer and coworkers[20] could find no difference in the stress response or improvement in outcome between early enteral feeding (within 24 hr of injury) and late (after 72 hr of injury). Despite the lack of strong clinical evidence to support very early enteral nutrition, many have advocated initiating gut feeding via jejunostomy within 48 hr of injury or surgery in a stable patient, because it is well-tolerated and may improve outcome.[32] In any case, it is probably not wise to initiate feeding in any patient who is hypotensive or hemodynamically unstable.

DIFFICULTIES WITH CLINICAL TRIALS

Despite the increasing evidence that enteral nutritional supplementation is a more physiologic method of maintaining the critically injured patient in the intensive care unit, it remains, nevertheless, the second choice for many patients and physicians. Several factors have contributed to this, particularly the limitations inherent to the clinical trials. The small size of most sample populations and the heterogeneity of the trial groups remain significant problems in definitively demonstrating the efficacy of early enteral nutrition. In an effort to answer these questions, Moore et al.[33] published a meta-analysis in 1991 that incorporated data from eight prospective randomized trials designed to compare enteral and parenteral supplements in the early postoperative care of high-risk surgical patients. The analysis revealed homogeneous groups and a significant difference in septic morbidity between the TPN and TEN patients with the TEN group having half the number of septic complications. In addition to the important finding of improved septic morbidity, this study also reported what many clinicians have believed for years: enteral nutrition can lead to increased rates of abdominal distension, diarrhea, and abdominal discomfort. In the meta-analysis, gastrointestinal discomfort was significantly greater in the enterally fed group with abdominal distention (TEN 46%, TPN, 24%), and diarrhea (TEN 34%, TPN 9%) being the major differences.[33] These findings suggest that early enteral nutrition is of value in reducing septic complications; however, tube feedings can be difficult and must be administered with great care, particularly in the early postoperative period. Special attention must be given to the type of supplement, the timing of when to start and when to advance the rate, proper patient selection, and the proper placement of the feeding tube.

CONCLUSION

Enteral nutrition offers several advantages over parenteral supplementation. Decreased septic morbidity, preserved gut mucosal integrity, a strengthened immune system and significant cost savings all can be shown clearly as benefits of early enteral feeding in the acutely injured surgical patient. However, parenteral and enteral nutrition are not mutually exclusive. There is still an important role for parenteral nutrition in patients with inadequate gut function, gastrointestinal injury or failure, severe fluid restrictions, or when diarrhea or abdominal distention suggest intolerance to tube feeds. It is important to note that these issues do not always necessitate the cessation of TEN. Often an adjustment of the rate and/or solution can succeed. Additionally, it is important to note that early enteral feedings are best tolerated if the feeding tube is placed postpyloric and preferably beyond the ligament of Treitz as gastric dysmotility is common after injury, surgery, or illness. Continued studies are needed to definitively demonstrate the advantages of enteral nutrition, better formula composition, and better methods of delivery with the multicenter, prospective, randomized trial remaining the gold standard for answering these questions.

REFERENCES

1. Jones TN, Moore EE, Moore FA. Early postoperative feeding. In Borlase BC, Bell SJ, Blackburn GL, Forse RA, eds. *Enteral nutrition.* New York: Chapman and Hall, 1994:78–80.
2. Shackelford RT. *Surgery of the alimentary tract.* Philadelphia: W.B. Saunders, 1986.
3. Torosian MH, Rombleau JL. Feeding by tube enterostomy. *Surg Gynecol Obstet* 1980;150:918–927.
4. Andresen AFR. Immediate jejunal feeding after gastroenterostomy. *Ann Surg* 1918;67:565–566.
5. Abbott WO, Rawson AJ. A tube for use in the postoperative care of gastroenterostomy patients. *JAMA* 1937;108:1873–1874.
6. MacDonald HA. Intrajejunal drip in gastric surgery. *Lancet* 1954;1:1007.
7. Delany HM, Carnevale NJ, Garvey JW. Jejunostomy by needle catheter technique. *Surgery* 1973;73:786–790.
8. Rhode CM, Parkins W, Tourtelotte D, et al. Method for continuous intravenous administration of nutritive solutions suitable for prolonged metabolic studies in dogs. *Am J Physiol* 1949;159:409.
9. Dudrick SJ, Wilmore DW, Vars HM, et al. Long term parenteral nutrition with growth, development and positive nitrogen balance. *Surgery* 1968;64:134–142.
10. Deitch EA, Ma WJ, Ma L, et al. Endotoxin-induced bacterial translocation: a study of mechanisms. *Surgery* 1989;106:292–300.
11. Alverdy JC, Aoys E, Moss GS. Total parenteral nutrition promotes translocation of bacteria from the gut. *Surgery* 1988;104:185–190.
12. Eastwood GL. Small bowel morphology and epithelial proliferation in intravenously fed rabbits. *Surgery* 1977;82:613–620.
13. Alverdy JC, Chi HS, Sheldon GF. The effect of parenteral nutrition on gastrointestinal immunity: the importance of enteral stimulation. *Ann Surg* 1985;202:681–684.
14. Hwang TL, O'Dwyer ST, Smith RJ, et al. Preservation of the small bowel mucosa using glutamine-enriched parenteral nutrition. *Surg Forum* 1986;37:56–58.

15. Jacobs DO, Evans DA, O'Dwyer ST, et al. Trophic effects of glutamine-enriched parenteral nutrition on colonic mucosa. *J Parent Ent Nutr* 1988;12(1, suppl):6s.
16. Alverdy JC. The GI tract as an immunologic organ. *Contemp Surg* 1989;35 (suppl):14–19.
17. Christou NV, McLean APH, Meakins JL. Host defense in blunt trauma: interrelationships of kinetics of anergy and depressed neutrophil function, nutritional status and sepsis. *J Trauma* 1980;20:833–841.
18. Lowry SF. The route of feeding influences injury responses. *J Trauma* 1990;30:s10–s15.
19. Peterson VM, Moore EE, Jones TN, et al. Total enteral nutrition versus total parenteral nutrition after major torso injury: attenuation of hepatic protein reprioritization. *Surgery* 1988;104:199–207.
20. Eyer DE, Micon LT, Konstantinides FN, et al. Early enteral feeding does not attenuate metabolic response after blunt trauma. *J Trauma* 1993;34:639–644.
21. Poret HA, Kudsk KA, Croce MA, et al. The effect of enteral feeding on catecholamine response following trauma. *Surg Forum* 1991;62:11.
22. Moore EE, Jones TN. Benefits of immediate jejeunostomy feeding after major abdominal trauma-a prospective, randomized study. *J Trauma* 1986;26:874–881.
23. Adams S, Dellinger EP, Wertz MJ, et al. Enteral versus parenteral nutritional support following laparotomy for trauma: a randomized prospective trial. *J Trauma* 1986;26:882–891.
24. Moore FA, Moore EE, Jones TN, et al. TEN versus TPN following major abdominal trauma-reduced septic morbidity. *J Trauma* 1989;29:916–923.
25. Sganga G, Siegel JH, Brown G, et al. Reprioritization of hepatic plasma protein release in trauma and sepsis. *Arch Surg* 1985;120:187–199.
26. Kudsk KA, Croce MA, Fabian TC, et al. Enteral versus parenteral feeding: effects on septic morbidity after blunt and penetrating abdominal trauma. *Ann Surg* 1992;215:503–513.
27. Layon AJ, Florete OG, Day AL, et al. The effect of duodenojejunal alimentation on gastric pH and hormones in intensive care patients. *Chest* 1991;99:695–702.
28. McClave SA, Lowen CC, Snider HL. Immunonutrition and enteral hyperalimentation of critically ill patients. *Digest Dis Sci* 1992;37:1153–1161.
29. Bower RH, Cerra FB, Bershadsky B, et al. Early enteral administration of a formula (Impact) supplemented with arginine, nucleotides, and fish oil in intensive care unit patients: results of a multicenter, prospective, randomized, clinical trial. *Crit Care Med* 1995;23:436–449.
30. Bower RH, Talamini MA, Sax HC, et al. Postoperative enteral vs parenteral nutrition. A randomized controlled trial. *Arch Surg* 1986;121:1040–1045.
31. Grahm TW, Zadrozny DB, Harrington T. The benefits of early jejunal hyperalimentation in the head-injured patient. *Neurosurgery* 1989;25:729–735.

32. Minard G, Kudsk KA. Is early feeding beneficial? How early is early? *N Horizons* 1994;2:156–163.

33. Moore FA, Feliciano DV, Andrassy RJ, et al. Early enteral feeding, compared with parenteral, reduces postoperative septic complications: the results of a meta-analysis. *Ann Surg* 1992;216:172–183.

Enteral Access Options

CHAPTER 15

Robert C. Gorman, M.D.,
Jon B. Morris, M.D., and
James L. Mullen, M.D.

INTRODUCTION

Adequate nutrition is essential for all hospitalized patients. This is particularly true for critically ill patients who, for any number of reasons, are unable to maintain their nutritional status. Recent advances in techniques for long-term vascular and enteral access provide the clinician with many options for providing nutritional support to patients who are unable to fully maintain themselves via normal oral alimentation.

In this chapter, the currently available techniques for obtaining access to the gastrointestinal tract to provide enteral feedings will be presented. Surgical, endoscopic, laparoscopic, and radiologic-assisted techniques will be presented. The indications, contraindications, complications, and relative benefits of each procedure will also be described.

HISTORICAL PERSPECTIVE

Advances in techniques for providing adequate enteral alimentation to patients unable to maintain nutrition orally have paralleled advances in gastrointestinal surgery. Attempts to provide enteral nutrition via mechanical means date back to the ancient Egyptians, who used emetics and nutrient enemas to preserve health.[1] The use of nutrient enemas continued, and in the late nineteenth century were considered by some the method of choice for providing nutrition to patients unable to eat.[2]

Technical problems with rectal alimentation included the tendency to induce evacuation, poor absorption, and mucosal irritation. These problems combined with increasing knowledge about normal digestion caused Einhorn in 1910 to propose gastric and duodenal feedings as an alternative to nutrient enemas.[1] A similar technique had been employed by John Hunter who in 1790 treated a 50-year-old patient who was rendered unable to swallow by a recent stroke. Hunter recommended that a hollow flexible tube be passed into the patient's stomach so that medicines and nourishment could be provided to sustain the patient. Hunter's patient was treated in this manner for 5 weeks until the ability to swallow normally was recovered.[1]

Success with nasoenteric feedings continued throughout the early twentieth century. In 1939, Stengel and Ravdin (3) used a nasogastric-jejunal tube placed at the time of surgery to provide jejunal feedings as well as gastric decompression.

Surgical access to the gastrointestinal tract for tube feedings was a logical extension of previously described nasoenteric methods for nutrient delivery. Advances in surgical techniques stimulated development of this type of enteral feeding access. The forerunners of surgical enteral access were gastrocutaneous fistulas resulting from trauma. The best known case of traumatic gastric fistula is that of Alexis St. Martin, who sustained a gunshot wound to the abdomen in 1822. Beaumont's detailed study of this patient led to further understanding of gastric physiology and demonstrated the feasibility of feeding via surgical gastrostomy.[4]

Gastrostomy as a planned surgical procedure was proposed first by Egeberg[5,6] in 1837 and performed initially by Sedillot[5,7] in 1849 and 1853. The first gastrostomy resulting in survival was performed by Jones[5,8] in 1875. The chronological development of surgical gastrostomies is presented in Figure 15.1. Gauderer et al. described an endoscopic assisted technique for gastrostomy without laparotomy.[9] This technique offered the advantage of decreased procedure time, local anesthesia, lack of wound complications and a decrease in the incidence of postoperative ileus. Since its initial description, the percutaneous endoscopic gastrostomy has been widely adopted and has largely replaced surgical gastrostomies at most institutions.

The introduction of the laparoscopic cholecystectomy in the late 1980s has

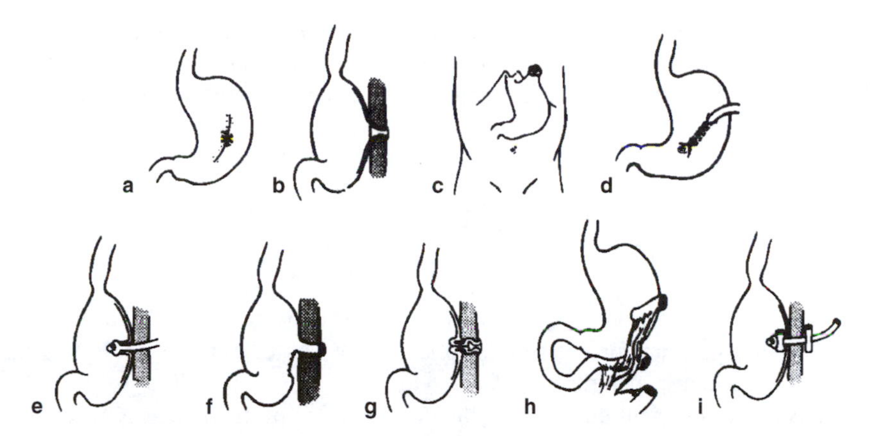

Figure 15.1. The six basic types of gastrostomies arranged chronologically in order of development. Type 1: (a) gastric fistula secondary to trauma (1635). Type 2: formation of a gastric cone, (b) through the incision (1846) and (c) through a counterincision (1890). Type 3: formation of a channel from the anterior gastric wall, (d) catheter parallel to stomach (1891) and (e) catheter perpendicular to stomach (1894). Type 4: formation of a tube from the gastric wall, (f) without valve (1899) and (g) with a valve (1901). Type 5: (h) formation of a tube from small or large bowel (1906). Type 6: (i) gastrostomy without celiotomy (percutaneous endoscopic) (1980). From Gauderer et al.[9]

stimulated the use of laparoscopic modalities in many aspects of gastrointestinal surgery. Innovative surgeons have developed laparoscopic techniques for gaining access to the upper gastrointestinal tract to provide enteral nutrition. These procedures, although still evolving, may ultimately provide alternatives to standard open surgical procedures in patients who are not candidates for endoscopic procedures.

Surgical jejunostomies for feeding were initially constructed to palliate patients with esophageal and gastric cancers in the late nineteenth century. In 1912, W. T. Mayo proposed the extensive use of surgical jejunostomy in patients with esophageal obstruction and severe pelvic ulcer disease.[10] Intraoperative placement of feeding jejunostomies in patients with gastric outlet obstruction was described by Andersen in 1918.[4] In 1952, Bowles and Zollinger reported on 103 patients that had been fed by a surgically placed jejunostomy at the time of abdominal exploration. A Stamm technique was employed, and feedings were typically begun within 12 h of surgery.[11]

Recent studies have demonstrated the efficacy of jejunal feedings in aspiration risk patients and as an adjunct to major hepatobiliary, pancreatic, and gastric surgery.[12,13] Alternatives to standard surgical options for jejunal access have paralleled advances in endoscopic and laparoscopic procedures. In the late 1980s and 1990s various techniques for laparoscopic jejunostomies and endoscopic conversion of gastrostomies to jejunostomies were developed and described. These minimally invasive options, as well as the more conventional techniques for providing safe and effective enteral feeding access, are discussed in depth in this chapter.

PATIENT SELECTION FOR ENTERAL NUTRITIONAL SUPPORT

Presented in this section is a brief review of the indication for enteral feeding access. An algorithm for determining the best means of enteral access for the individual patient is also discussed.

Intake of a sufficient diet to meet the caloric needs of the patient is necessary to maintain normal body composition and function. This equilibrium may be disturbed by decreased intake, increased requirements, or alteration in the body's normal milieu that prevents nutrients from being utilized effectively for tissue repair. A detailed presentation on nutritional assessment can be found elsewhere in this text; however, to establish the need for supplementary feedings, a careful evaluation of the patients' nutritional status should demonstrate that the patient's current dietary intake is insufficient to meet nutrient needs. Once volitional oral intake is found to be inadequate, the route for forced feeding must be selected.

It is generally accepted that in patients with a functional gastrointestinal tract, enteral feedings are superior to the parenteral route. Increased safety is a major factor in support of enteral feeding. Although complications of enteral feeding do occur, they are usually more easily managed and usually are not life threatening. Common complications are related to the catheter (occlusion, migration, granulation tissue, persistent fistula) or the enteral infusions (nausea, vomiting, and diarrhea). The potentially life-threatening complications of pneumothorax, hemothorax, arterial puncture, catheter or air embolism, electrolyte abnormalities, and line sepsis associated with central venous cannulation and feedings are avoided.

Another benefit of enteral feedings over venous feedings is reduced cost. Most of this benefit is a result of a lower cost for enteral feeding formulas compared with comparable parenteral formulas. Parenteral formulas can be two to ten times more expensive than enteral formulas that provide similar amounts of nutrients.[14] Finally, there is evidence that enteral feedings are superior to parenteral nutrition in sustaining intestinal structure and function.[15]

As mentioned previously, patients who require nutritional support and have a functioning gastrointestinal tract should receive feedings enterally. Common indications for enteral feedings include neurologic disorders, trauma, critical illness, head or neck surgery, upper gastrointestinal surgery, and prolonged respiratory failure (Table 15.1).

Enteral nutrients may be delivered intragastrically (prepylorically) or postpylo-

Table 15.1. Indications for Enteral Nutritional Support

Neurological and Muscular Diseases
Cerebrovascular accidents
Dementia
Head trauma
Brain neoplasms
Parkinson's disease
Myopathy
Critical illness/trauma
Cancer
Head and neck
Gastrointestinal
Respiratory failure with prolonged intubation
Other (variable)
Inflammatory bowel disease
Enterocutaneous fistula
Anorexia nervosa

rically. There is considerable debate as to which of these two basic yet physiologically different techniques is most appropriate. A primary concern is the risk of aspiration of enteral contents, particularly in neurologically impaired patients who frequently require enteral nutrition. The authors favor postpyloric feedings for patients judged to be at risk for aspiration, an approach that substantially reduces episodes of feeding-related aspiration.[12]

Risk factors for aspiration of enteral feeds are many (Table 15.2). After a thorough history and physical examination, which sometimes necessitates direct and indirect laryngoscopy, tests may be necessary to assess aspiration risk, including barium studies, fluoroscopy, manometry, and scintigraphy.[16] Evaluation of the patient by a neurologist and/or speech pathologist may also help to assess the patient at risk for aspiration.[17]

Once it is determined that enteral supplementation is indicated, the clinician

Table 15.2. Risk Factors for Aspiration

Altered mental status
Swallowing dysfunction
Central (cerebrovascular accident)
Local (vagal disruption, trauma)
History of aspiration
Severe gastroesophageal reflux
Gastric outlet obstruction
Gastroparesis

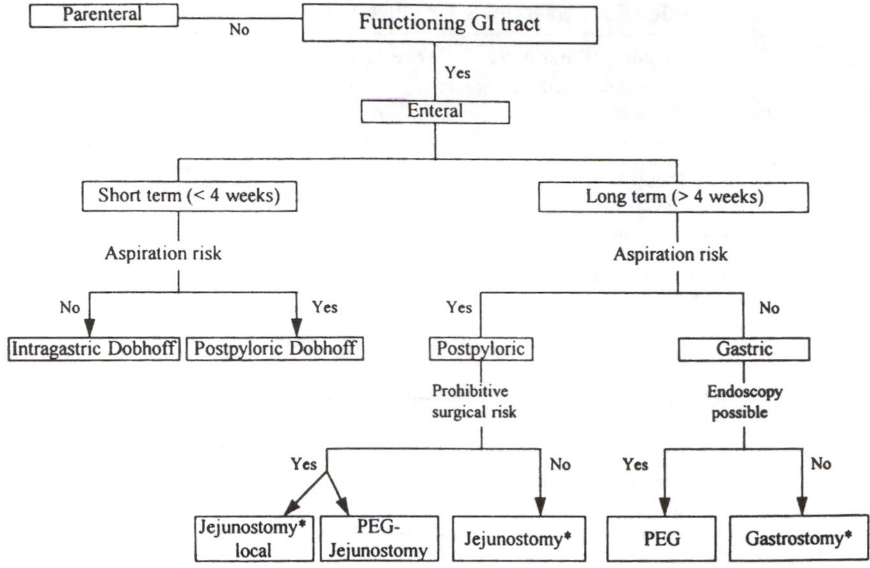

Figure 15.2. Enteral access algorithm for selection of the most appropriate approach for an individual patient.

must choose among the many available delivery techniques for enteral feedings. Possible access techniques include surgical (open or laparoscopic), endoscopic, and radiologic intervention, which may be performed with either local or general anesthesia. The type of technique employed often varies among institutions and clinicians, but each choice should be based on the nutritional goals of the individual patient, the patient's ability to tolerate the procedure, and the expected duration of therapy. An algorithm for determining the most appropriate means of enteral access is presented in Figure 15-2.

NASOENTERIC TUBES

Nasoenteric tubes are an attractive means for the delivery of enteral feedings because they are relatively easy to use, cause relatively few complications, entail limited intervention, and can be used in the aspiration-risk patient if the tip of the tube is placed postpylorically. Nasoenteric tubes are not recommended for enteral access for more than 4 weeks, however, an exception may be the terminal patient in whom only minimal intervention is desired. Nasoenteric tubes are most appropriate as an initial technique until a more definitive technique is selected.

Small-bore silicone rubber or polyurethane tubes are preferred over large-bore, rubber or plastic tubes, since the former are less likely to be associated with the complications of dislodgment, pulmonary compromise, tissue erosion, sinusitis, and tracheobronchial injury.[5]

Placement of nasoenteric tubes is straightforward and usually uncomplicated. The lubricated tip is placed through a nostril into the nasopharynx and swallowed. Use of viscous lidocaine as a lubricant and local topical anesthetic often makes insertion easier and less traumatic for the patient. Most devices come packaged with a metal stylet, which temporarily increases tube stiffness and allows easier insertion. After the tip of the tube is confirmed to be within the stomach by abdominal plain film, the patient may be placed on his or her right side to facilitate transit through the pylorus in aspiration-risk patients. Any number of prokinetic agents, such as metoclopramide or erythromycin, may be administered to promote transpyloric passage of the tube if patient positioning is ineffective. At times, fluoroscopic or endoscopic means are necessary to achieve postpyloric placement of nasoenteric tubes.

OPEN SURGICAL TECHNIQUES

Open Gastrostomy

The open gastrostomy is now more commonly performed with another intraabdominal procedure. It is indicated as an isolated procedure only when the endoscopic technique is contraindicated or not technically possible.

An open gastrostomy may be performed through a small upper midline or left upper quadrant transverse incision. These procedures may be done under local, regional, or general anesthesia.

Stamm method. Upon entering the peritoneal cavity, the stomach is identified and the anterior wall grasped with noncrushing clamps, ensuring that the posterior wall is not included. Two concentric pursestring sutures are placed. An 18 to 22-gauge French foley catheter is introduced through the gastrostomy, and the pursestring sutures are tied sequentially to invert the seromuscular gastric wall around the tube. The gastrostomy tube is then pulled through a separate stab wound. The tube should exit the peritoneal cavity at least 2 cm below the costal margin to avoid the pain produced by pressure from the tube on the lower ribs. The anterior gastric wall is then sutured to the abdominal wall with four equally spaced sutures. The abdomen is closed and the tube is secured to the skin with a nonabsorbable monofilament suture (Figure 15.3). Use of the foley catheter allows for easy and painless tube changes, once the tract has matured.

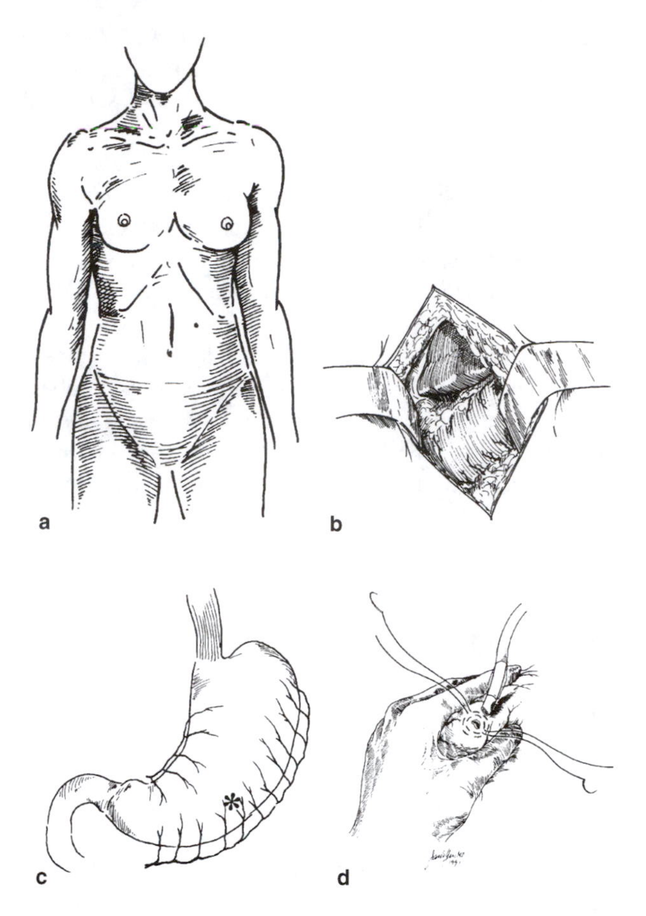

Figure 15.3. Operative gastrostomy: (a) Assuming there are no previous abdominal incisions, a 5-cm vertical midline incision is made above the umbilicus. The tube exit site is designed to be 5 cm lateral to the midpoint of the incision on the left side or at least 2 cm inferior to the costal margin. If the tube exit site is made either more cephalad or caudad, the intraperiotoneal tacking of the stomach to the parietal peritoneum is made more difficult. (b) The peritoneal cavity is entered through this incision, and minimal exploration is performed to decrease any postoperative gastric ileus. The procedure is done on the left side of the round ligament. One must be sure that the stomach exhibits no dilatation or thickened wall, possibly signaling partial obstruction downstream. (c) The proposed gastrostomy site is identified on the anterior lateral wall, leaving just enough room for two pursestring sutures between the site and the gastroepiploic vessels. The site should not be placed down in the narrowed antrum where a balloon or tube could partially obstruct the gastric outlet. (d) We use a nonabsorbable 00 silk suture for the pursestring. After the pursestring sutures are placed, the seal is checked with infusion of water through the tube. The tube is clamped to prevent any spillage of enteric contents. The tube is then pulled up tightly and the anterior gastric wall pulled against the anterior abdominal wall. Sutures are used to oppose the visceral peritoneum on the stomach to the parietal peritoneum on the abdominal wall.

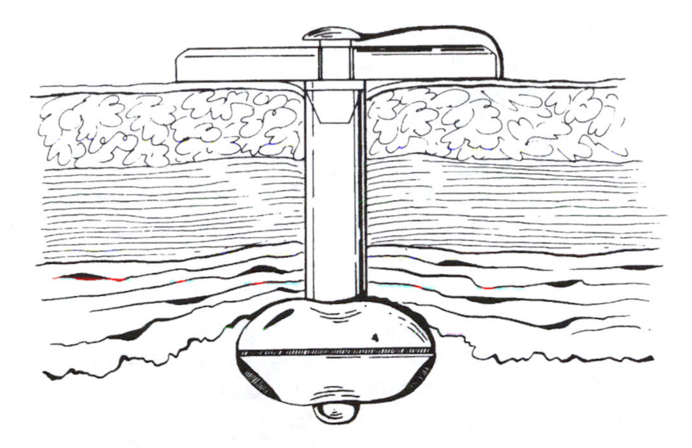

Figure 15.4. Skin-level stoma device—the button.

Gastric buttons. Skin-level intestinal conduits known as buttons (Figure 15.4) are silastic devices that replace the often unsightly, vulnerable external tubes typically used for gastrostomies. Popularized in 1984 by Gauderer et al.[18] and originally described for use in children, the button can be placed percutaneously in a mature gastrostomy stoma at least 8 weeks postoperatively. The potential advantages of this device over conventional tubes are decreased pivoting motion, fewer inadvertent extractions, improved patient self-image, and no external fixation sutures. Exchanging a catheter for a button occurs in the office without anesthesia, making it an excellent option for home-care patients. These devices may also be inserted at the time of open gastrostomy using a technique similar to Stamm gastrostomy and button jejunostomy (to be described subsequently).

Open Jejunostomy

Jejunostomy for enteral feeding access is most commonly performed as an adjuvant to major upper abdominal surgery. There are few absolute contraindications to jejunostomy in patients who would otherwise tolerate a small laparotomy incision under local or general anesthesia; however, in cases where primary small bowel disorders such as radiation enteritis or Crohn's disease exist, it is usually wise to avoid this type of enteral access due to the potential for enterocutaneous fistula. Distal obstruction is also an absolute contraindication to jejunal feedings.

Modified Witzel jejunostomy. This procedure represents the gold standard for chronic, postpyloric feeding. A review of the Hospital of the University of Pennsylvania (HUP) experience with 100 consecutive open jejunostomy tube placements in aspiration-risk patients demonstrated that the technique substantially reduced feeding-related aspiration pneumonia.[12] In addition, it can be performed with an acceptable rate of morbidity and mortality in a population that typically has significant underlying disease.

Under general endotracheal anesthesia, an upper midline incision is made. The jejunum, 25 cm distal to the ligament of Treitz, is carefully identified and brought up into the wound. An enterotomy is made on the antimesenteric border large enough to admit insertion of a 12- or 14-gauge French red rubber catheter. Additional holes are cut in the distal end of the red rubber feeding tube prior to insertion to decrease the risk of tube obstruction. The tube is inserted into the lumen in the aboral direction until only 3 to 4 cm of catheter are exposed on the skin surface once the bowel is approximated to the anterior abdominal wall. A 2 to 3 cm seromyotomy is made in the jejunal wall proximally from the point of insertion, and the tube is set into the seromyotomy, after which the edges are reapproximated over the tube, using interrupted, inverting 3-0 silk sutures (Figure 15.5). The tube is brought externally through a separate incision in the left upper quadrant at least 2 cm below the costal margin. The bowel is carefully secured to the under surface of the anterior abdominal wall by a series of interrupted 3-0 silk sutures, such that the tube and previous suture line are encircled. The midline incision is closed in the standard fashion. The tube is secured to the skin with two sutures placed on opposite sides. The tube is left to straight drainage until adequate bowel function returns, at which time feedings are initiated.

Button jejunostomy. Surgical jejunostomy has been demonstrated to be a safe and effective technique for providing chronic postpyloric nutritional support in aspiration-risk patients; however, the long-term presence of an externally protruding catheter is associated with several commonly occurring problems. Leakage around the catheter due to the pivoting action of the tube about the exit site, inadvertent tube dislodgment, and the need for long-term fixation to the skin are all problems associated with the typical catheter-type access devices. A randomized, prospective trial comparing a button jejunostomy with the standard tube jejunostomy at HUP demonstrated that both techniques were equally effective for delivery of nutrients, but the button was associated with fewer device-related complications, including inadvertent tube dislodgment, leakage around the tube, and device occlusion.[19]

Under general anesthesia, an upper midline incision is made and the jejunum, 20–25 cm distal to the ligament of Treitz, is identified and brought into the wound. A 3-0 PDS® suture is placed in a pursestring fashion around the enterotomy site (Figure 15.6). Additional holes are cut in the button tip to minimize the potential for obstruction of the intestinal lumen. An enterotomy is made within the pursestring, and the device is inserted. The device is brought through the abdominal wall through a separate stab wound lateral to the incision. The jejunum is fixed to the anterior abdominal wall using a second PDS® pursestring suture. The abdomen is closed in the standard fashion. The button remains capped until

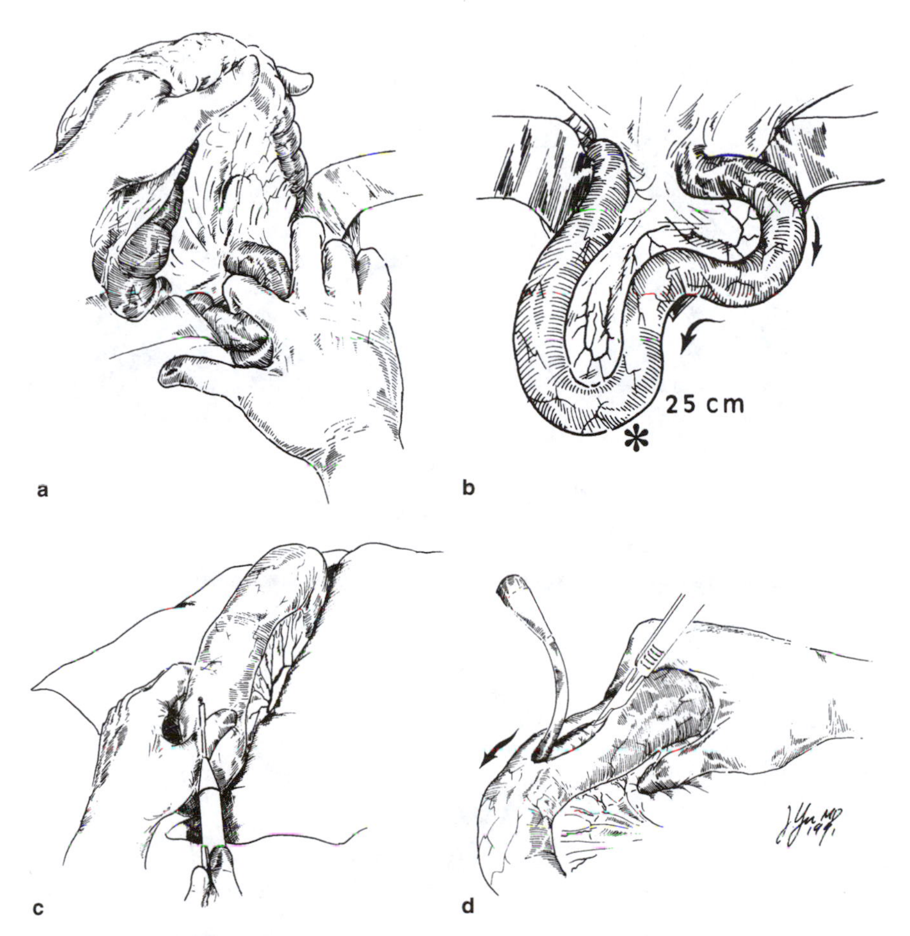

Figure 15.5. Modified Witzel jejunostomy. (a) The ligament of Treitz is identified. (b) The jejunostomy site is selected 25 cm distally. The jejunum is examined for any signs of ongoing obstruction, such as dilation or thickened muscular wall. The jejunostomy site must allow its attachment to the skin exit site without any excessive tension on the jejunum between the ligament of Treitz and the proposed abdominal wall exit site. (c) The jejunostomy is prepared by making a tiny hole, usually using cautery, on the antimesenteric side. A 12- or 14-gauge French red rubber Silastic catheter is inserted in the jejunostomy with the tip directed downstream. It is important to cut the tip off the catheter to allow exchange of a wire in the future. (d) A 3-cm seromuscular incision is then made proximal to the jejunostomy site. One can estimate the proper depth by identifying protruding mucosa and bleeding submucosa.

continued on next page

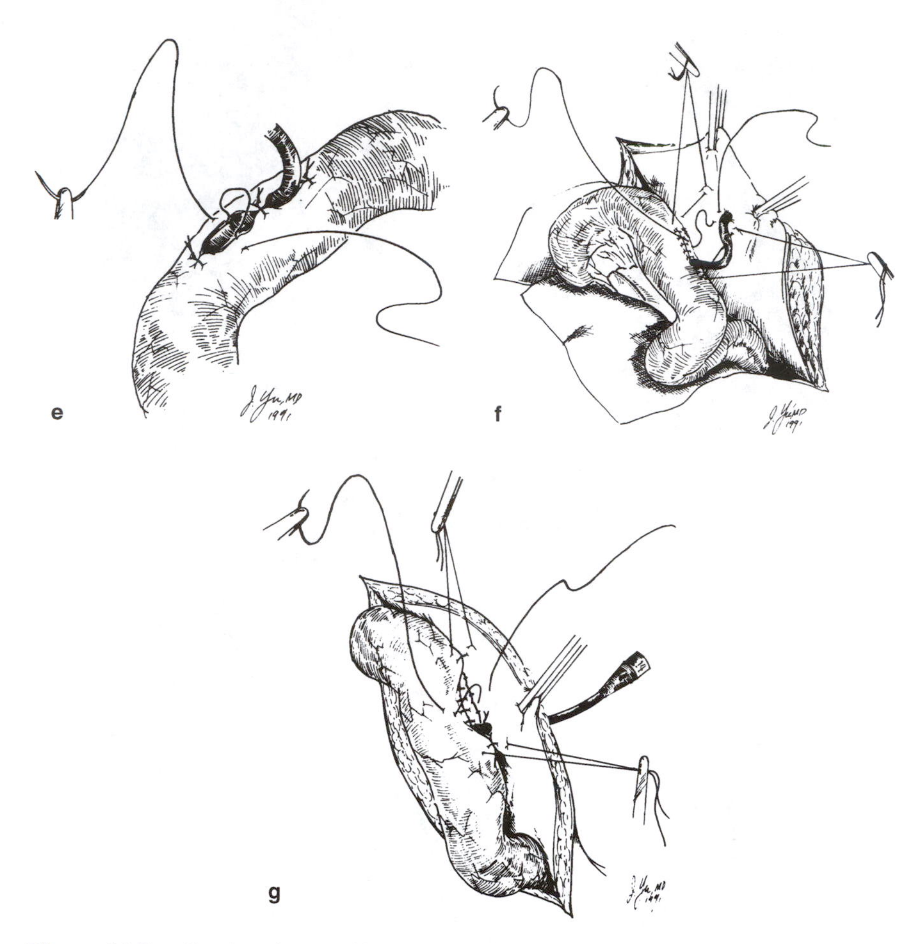

Figure 15.5. *Continued* (e) This seromuscular incision needs to be carried up to and including the jejunostomy site. The jejunum is then closed over the tube with interrupted seromuscular sutures creating a subserosal tunnel for approximately 3 cm. The tube is fixed in place by tightening the sutures around the catheter. No pursestring suture is used because the strangulation effects lead to an excessively large jejunostomy in the future. (f) The jejunum is then aligned appropriately to avoid the risk of volvulus. The goal is to use a broad-based attachment to the parietal peritoneum to again guard against volvulus. The objective is to create an effective peritoneal seal completely surrounding the jejunal suture line. (g) The seal is performed to completely oppose the visceral peritoneum of the jejunum to the parietal peritoneum of the abdominal wall. The lateral row is done first and the medial row last using 000 silk suture.

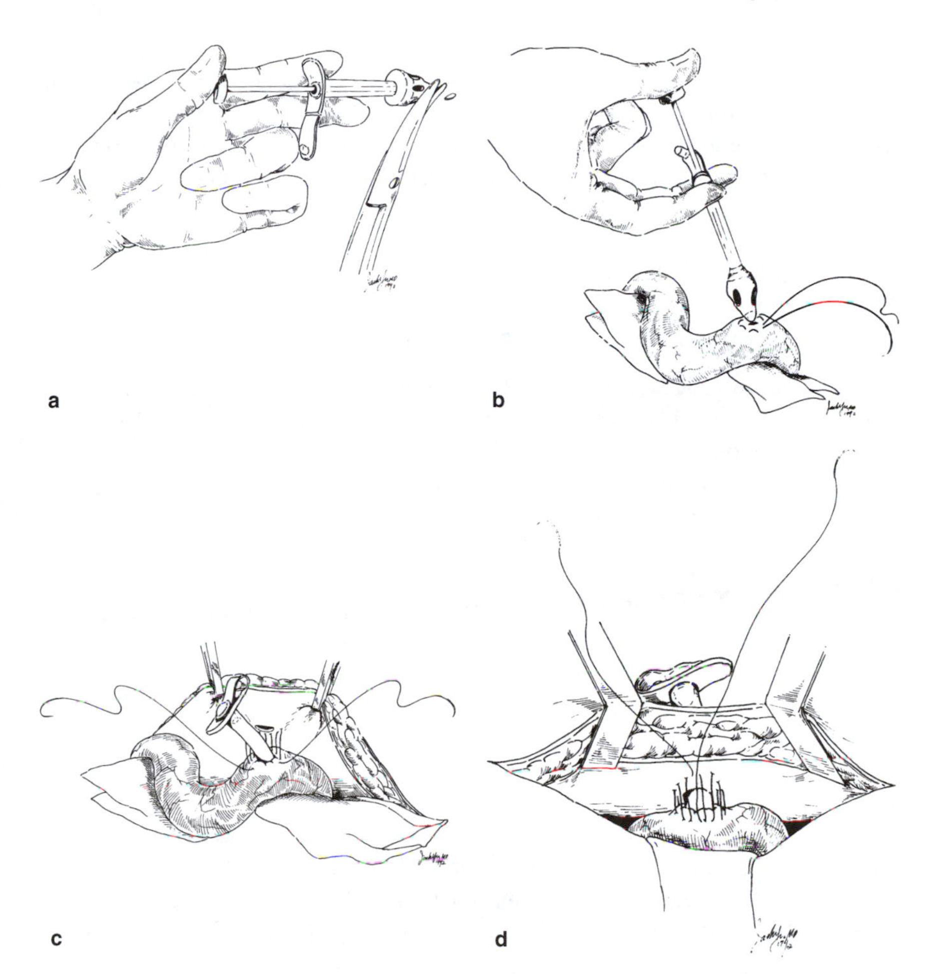

Figure 15.6. (a) Additional holes are cut from a standard gastrostomy button to allow better flow of nutrients and to prevent bowel obstruction. (b) The button is deformed with an obturator and placed through a jejunotomy made within a previously placed pursestring suture. (c, d) A running 3-0 double-armed prolene suture is used to secure the jejunum to the anterior abdominal wall.

adequate bowel function permits enteral feeding. The button devices are manufactured in a variety of shaft lengths. The appropriate size should be chosen such that the other bumper rests at skin level without tension.

Needle jejunostomy. A technique that has become popular because of the ease and speed with which it can reportedly be performed is the needle catheter jejunostomy. Several commercial kits are now available to the surgeon. A disadvantage of the technique has been the high rate of tubal occlusion related to the small bore of the catheter. For this reason, the authors do not generally recommend this technique for long-term use. The previously described techniques of Witzel and button jejunostomies are safe, effective, durable and easy to perform. We, therefore, believe the indications for this method of enteral access to be limited.

ENDOSCOPIC TECHNIQUES

Percutaneous Endoscopic Gastrostomy

Percutaneous endoscopic gastrostomy (PEG), which Gauderer et al. first described in 1980,[9] has been a major advance in long-term nutritional support and has nearly supplanted the open gastrostomy. Its rapid acceptance and popularity as a means of enteral access are largely due to its time and cost effectiveness. Many variations upon Gauderer's initial technique have been described. The initial steps are, however, similar for all methods and will be described prior to the various options for completion of the procedure.

The patient is kept NPO for at least 8 hr prior to the procedure. Prophylactic antibiotics, usually a cephalosporin, are recommended preoperatively. The patient lies supine on the operating room table and intravenous sedation as well as local pharyngeal anesthesia is administered. The gastroscope is then introduced into the stomach and air is insufflated to distend the stomach. Adequate air insufflation opposes the stomach to the anterior abdominal wall and displaces the transverse colon caudally to avoid injury to this organ. The abdominal wall is then transilluminated with the gastroscope. The endoscopist watches as the assistant indents the abdominal wall at the point of the proposed gastrostomy. This position should be at least 2 cm below the costal margin to avoid painful contact between the tube and the rib. It is at this point that the various techniques differ (Figure 15.7).

Pull technique of Gauderer and Ponsky. This classic technique was originally described by Gauderer, Ponsky, and Izant[9] and is our preferred method for performing the PEG procedure. At the point where the assistant's finger indents the abdominal wall and stomach, a small skin incision is made after the administration of local anesthesia. An angiocatheter is introduced through the incision. Entrance of the needle into the stomach is verified by the endoscopist. A biopsy snare is passed through the endoscope and maneuvered around the catheter. The needle is removed and a long suture is placed through the catheter into the stomach. The biopsy snare grasps the suture; the endoscope and suture are withdrawn through the mouth. The suture is secured to the gastrostomy tube, and, with gentle but firm traction, the tube is pulled through the esophagus into the stomach. The gastroscope is then passed again as the tube is pulled against the anterior stomach wall. The endoscopist verifies tube position and the scope

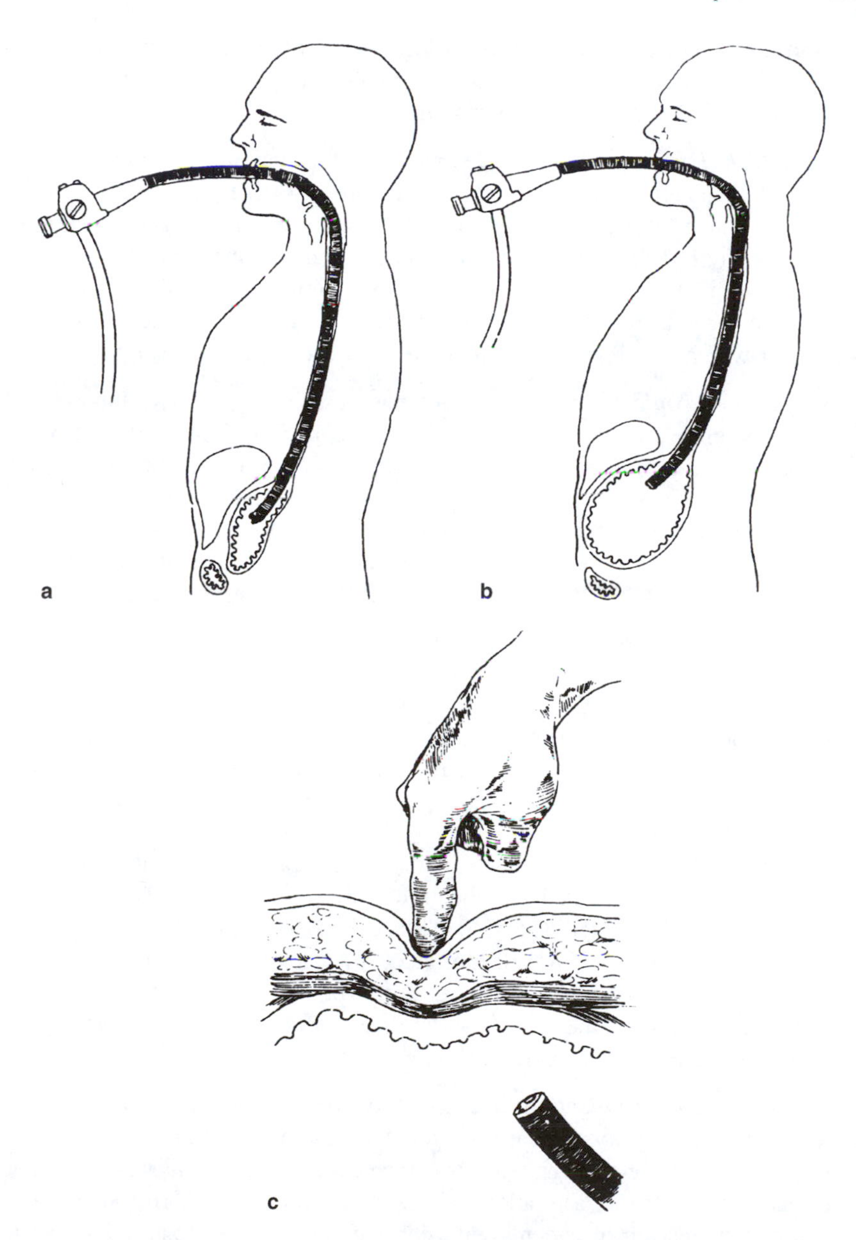

Figure 15.7. Initial steps common to all PEG techniques. (a) The gastroscope is inserted into the esophagus and the stomach is intubated. (b) Air insufflation distends the stomach to ensure opposition with the anterior abdominal wall and caudad displacement of the transverse colon. (c) Endoscopic gastric translumination at the proposed incision site is confirmed by digital palpation.

is removed. A bolster is applied to the outside of the tube and sutured to the abdominal wall. Excessive traction must be avoided to prevent ischemic necrosis of the gastric and abdominal walls (Figure 15.8).

Push technique of Sachs and Levine. This technique is identical to the method of Gauderer and Ponsky, except that a long flexible guide wire is placed into the stomach through the angiocatheter. Using the gastroscope and biopsy snare, the wire is pulled through the esophagus and out of the patient's mouth. The gastrostomy tube is then pushed over the wire into position. The gastroscope is passed a second time to verify catheter location against the anterior stomach wall, after which the procedure is completed as described above (Figure 15.9).[20]

Single endoscopic technique. Grant has described a PEG technique that does not require a second passing of the gastroscope.[21] This method is identical to the original pull technique of Gauderer and Ponsky except that the gastrostomy tube has markings at 2-cm intervals that allows proper positioning of the tube by estimating the thickness of the abdominal wall without a second endoscopy. Using this technique in 598 patients, Grant reported a major complication rate of 1.3% and one death.

Introducer technique of Russell. This technique differs from the standard pull and push methods in that it requires only one endoscopy and the gastrostomy tube is not pulled through the mouth. This has the theoretical advantage of reducing wound infection. Using the previously described procedure, the gastroscope is passed into the stomach and the sight of the proposed gastrostomy is identified. A small skin incision is made and the angiocatheter is placed in the stomach through the abdominal wall. A flexible guide wire is passed through the catheter and, using the Seldinger method, a 16-gauge French dilator with a peel-away introducer is passed over the guide wire (Figure 15.10). The endoscopist must verify the entrance of the dilator and introducer into the stomach. The dilator is then removed and a well-lubricated 14-gauge French Foley catheter is placed through the introducer as it is peeled away. The Foley catheter balloon is inflated and secured to the anterior abdominal wall, and the endoscopist verifies adequate device position (Figure 15.11).[22]

Indication, contraindication, and complications. The indications for PEG are the same as those for standard surgical gastrostomy. The procedure is most often performed to provide enteral feeding access in the patient who is unable to eat and is not at risk for aspiration. Occasionally a PEG is performed to provide gastric decompression in patients with chronic small bowel obstruction due to malignancy and in patients with proximal enterocutaneous fistulae.[23]

The only absolute contraindication to PEG are an inability to pass the endoscope, usually because of a malignant obstruction of the pharynx or esophagus, and failure to adequately transilluminate the abdominal wall because of obesity.

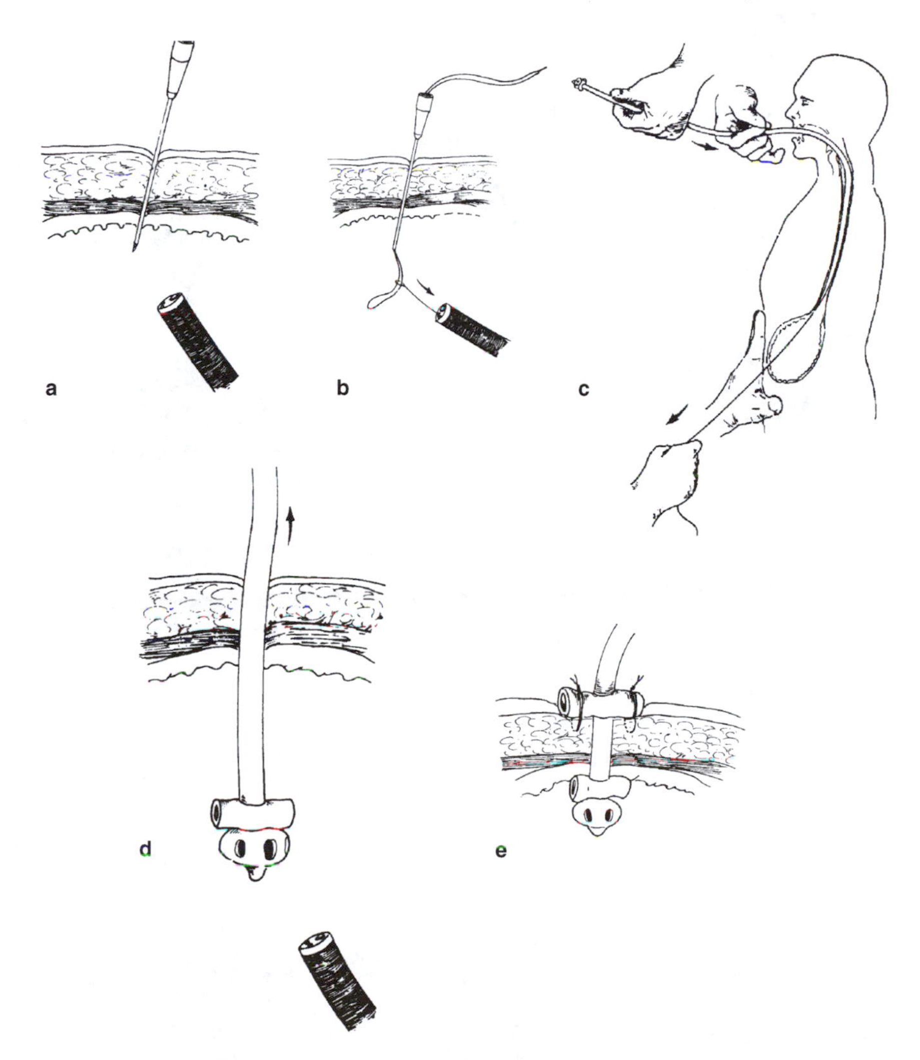

Figure 15.8. Gauderer-Ponsky PEG technique (a) After installation of local anesthesia, a 10-mm transverse incision is made, through which a tapered cannula needle is introduced under direct endoscopic vision. (b) A looped heavy suture is directed through the catheter into the stomach, secured with a polypectomy snare, and withdrawn from the patient's mouth. (c) The well-lubricated PEG catheter is then secured to the suture and with steady traction directed down the posterior pharynx into the esophagus. (d) The endoscope is reintroduced, and under direct vision the catheter is pulled across the gastroesophageal junction and then approximated to the anterior gastric wall. It is imperative that the inner crossbar gently approximate the mucosa without excess tension to avoid ischemic necrosis. The stomach is decompressed by aspiration and the gastroscope withdrawn. (e) The outer crossbar is gently approximated to the skin level and secured with two 00 Prolene sutures.

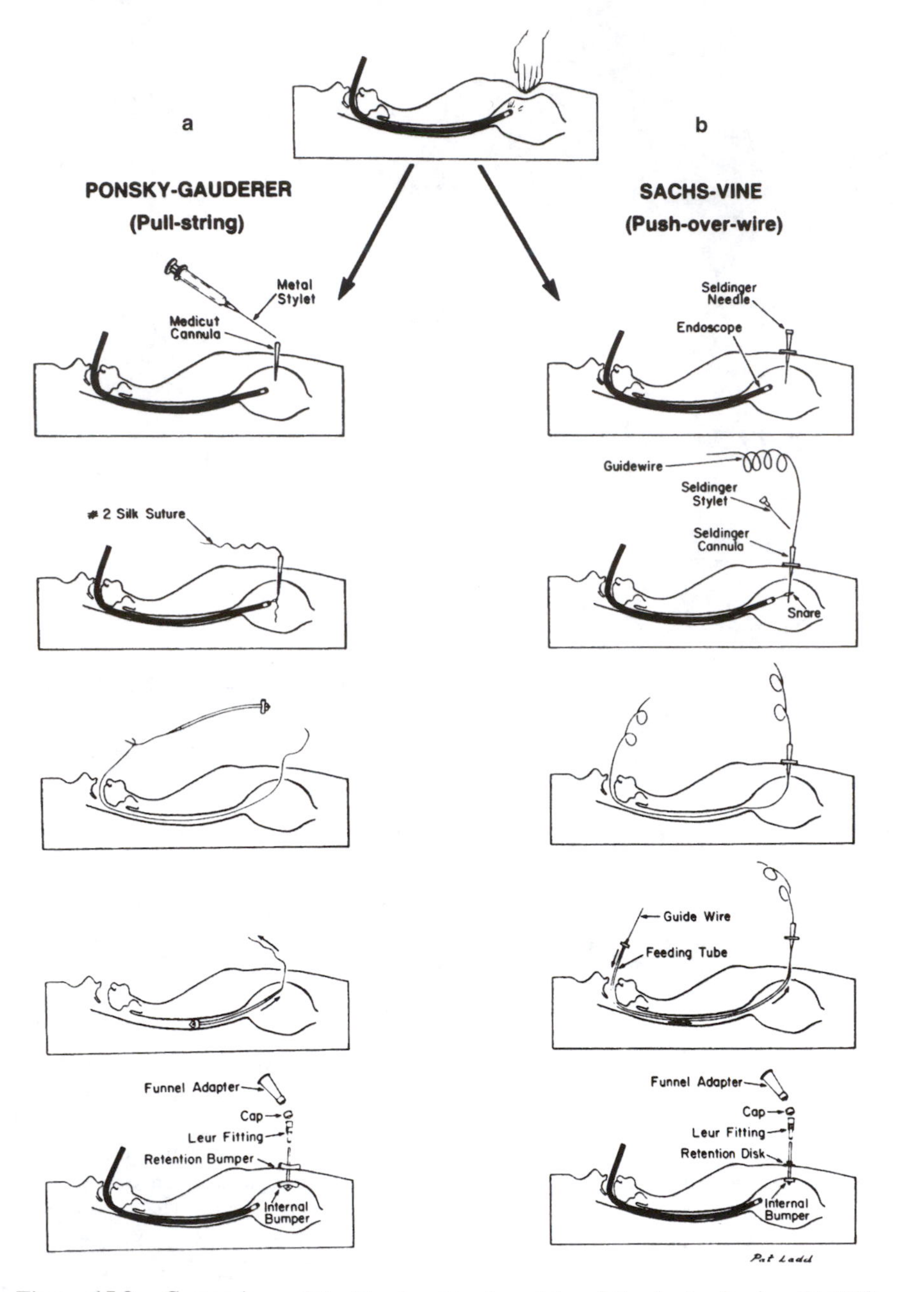

Figure 15.9. Comparison of the Ponsky–Gauderer (a) and the Sachs–Levine (b) PEG techniques. From Hogan et al.[20]

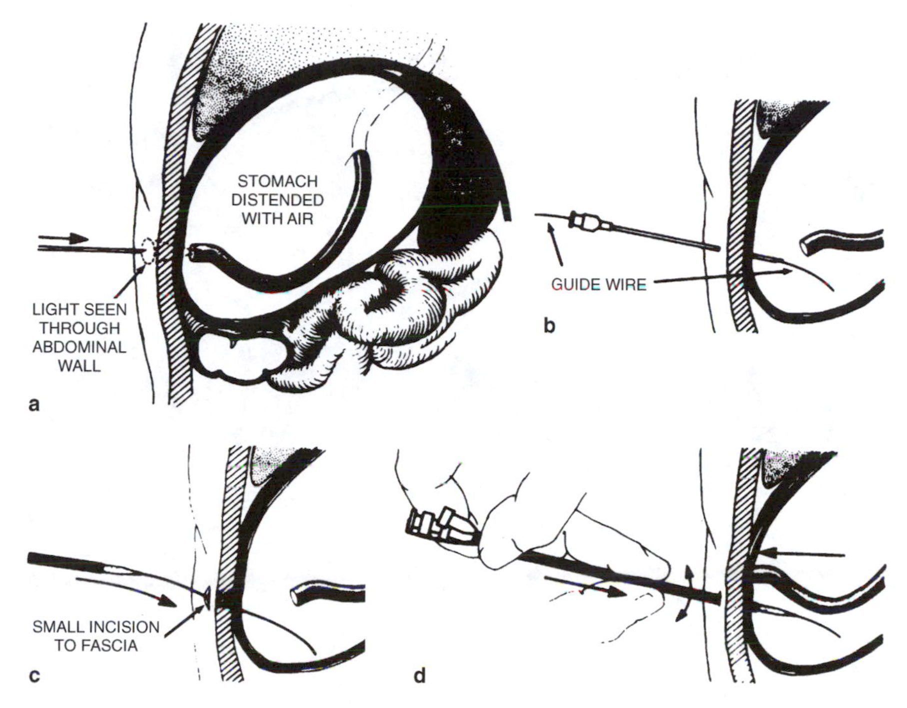

Figure 15.10. Russell techniques for PEG. (a) Gastric dissension with needle directed at light source. (b) The guide wire is passed through the needle into the stomach. (c, d) The introducer is passed over the wire into the stomach. From Russell et al.[22]

Ascites and previous abdominal surgery at one time were considered contraindications to PEG; however, as experience with the procedure has grown, several authors have demonstrated acceptable results in these two groups of patients.[24,25] Ascites remains a relative contraindication with ascitic leak rate reported to be approximately 25% in some studies.[25] The incidence of ascitic leak is reportedly greatly reduced by preoperative paracentesis.[25]

Several large studies have demonstrated the relative ease and safety with which the PEG procedure can be performed,[23,24,26,27] major complications are rare (<10%) and are often the result of aspiration which could seemingly be minimized by proper patient selection. Extrusion of the gastrostomy is a serious complication that occurs in 1% to 11% of patients and is the result of excessive tightness between the inner and outer bolster which leads to pressure necrosis of the abdominal wall. Minor complications occur in 24% to 43% of patients[23,24,26,28] and typically include wound infections, peristomal leaks, cellulitis, and clogged tubes.

Studies comparing the PEG to open gastrostomy have demonstrated similar morbidity and mortality for both techniques. Most studies indicate that PEG

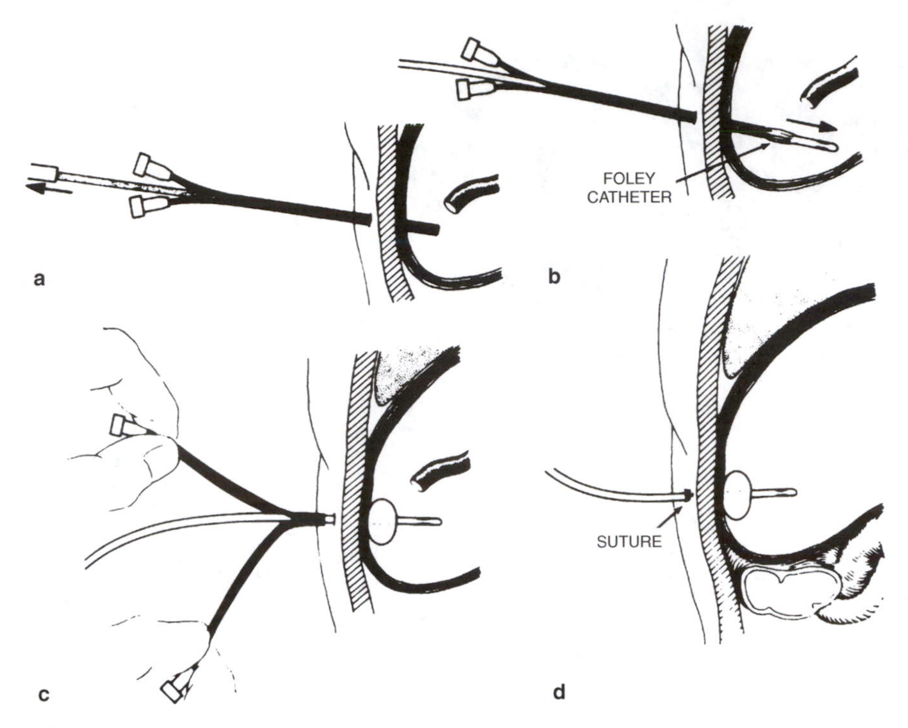

Figure 15.11. Additional Russell techniques for PEG. (a) The dilator is removed from the introducer. (b) A Foley catheter is passed through the introducer into the stomach. (c) The introducer is peeled away. (d) The Foley balloon is inflated and secured against the abdominal wall. From Russell et al.[22]

procedures can be formed slightly faster than open gastrostomy at 50% to 75% of the cost.[28–30]

Comparison between the various techniques for performing PEG's have demonstrated similar safety and efficacy for all available methods. There is no clear advantage of one percutaneous technique over another.[20,31]

Percutaneous Endoscopic Jejunostomy

A significant number of patients who require enteral access for feeding have substantial risk factors for aspiration and are best served by a feeding jejunostomy.[12] This fact combined with the ease and efficacy with which percutaneous access to the stomach can be performed has stimulated several innovative techniques for obtaining postpyloric enteral access via endoscopically assisted percutaneous methods.[26,32–37]

Two basic techniques for percutaneous endoscopic jejunostomy have been described. The most common method involves the placement of a postpyloric feeding tube through a previously placed PEG using endoscopic guidance (transpyloric percutaneous endoscopic jejunostomy—PEJ)[26,32–34]. More recently techniques for direct PEJ have been reported[35–37]. Both procedures will be described in the following sections.

Jejunal Access via transpyloric percutaneous endoscopic jejunostomy. The principle underlying this technique is simple and is illustrated in Figure 15.12. A small feeding tube with a heavy suture tied to the weighted tip is passed via previously placed gastrostomy. A gastroscope is then passed and the suture is grasped with a biopsy forceps. The tube is then guided as far as possible into the duodenum. The biopsy forcep is then held in place as the gastroscope is withdrawn alone to prevent inadvertent dislodgment of the feeding tube. An

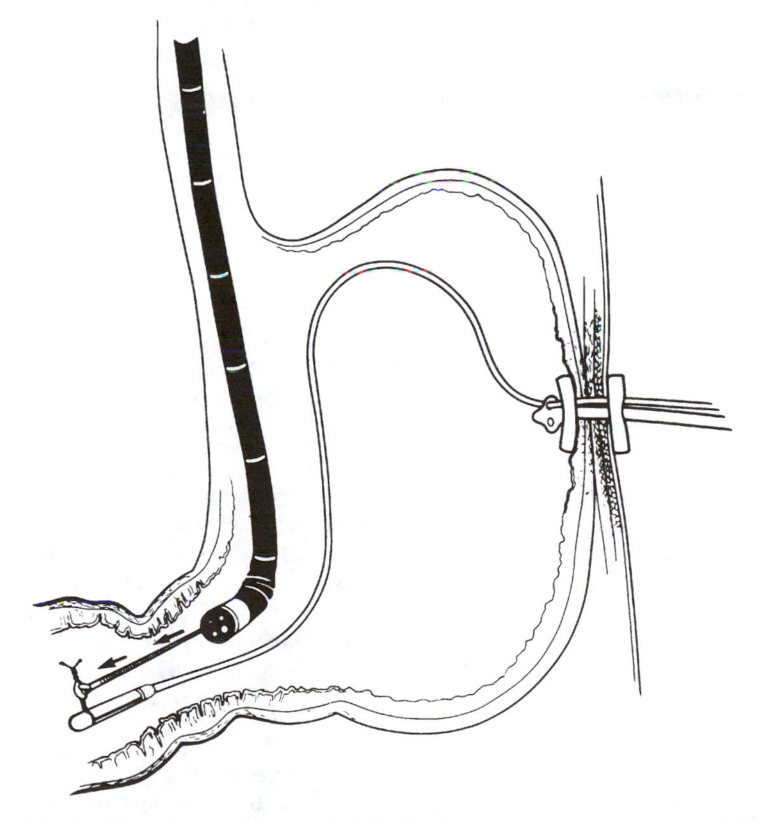

Figure 15.12. Technique for transpyloric PEJ. The endoscope guides the weighted tip of the feeding tube into the duodenum. From Ponsky et al.[26]

excess amount of tubing is left within the stomach to allow peristalsis to pull the tip of the feeding tube past the ligament of Trietz.

Several studies have documented the relative ease of performing PEG with a jejunal extension as well as the short-term efficacy of the procedure.[26,32–34] Most authors report that the addition of a transpyloric PEJ to a PEG procedure only increases the procedure time from 10 to 15 min.[26,32–34,38]

Studies evaluating the long-term efficacy of transpyloric PEJ have not been as positive.[39,40] Kaplan et al. followed 23 patients undergoing transpyloric PEJ over a 2-year period. They found that 84% of the tubes failed and were functional for an average of only 39.5 days. The reason for tube failures were (a) separation of the inner PEJ tube from the outer gastrostomy tubes resulting in tube dislodgment (59%); (b) clogging due to small PEJ diameter (32%); and (c) kinking and knotting of the tube (9%). In addition, the authors did not demonstrate a significant decrease in the risk of aspiration pneumonia in patients fed via transpyloric PEJ. The failure of transpyloric PEJ to prevent aspiration is thought to be due to frequent retrograde migration of the tube into the stomach. The transpyloric nature of this type of enteral access also decreases the ability of the pyloric sphincter mechanisms to perform effectively, therefore predisposing to duodeno-gastric reflux and subsequent aspiration.

Direct percutaneous endoscopic jejunostomy. To avoid the mechanical and physiologic complications associated with transpyloric PEJ, methods for direct PEJ have been described recently by several authors.[35–37] These techniques all require advancement of an endoscope into the lumen of the small bowel to direct placement of the feeding tube. Initial reports of direct PEJ procedures were confined to patients who had undergone previous partial gastrectomies with Billroth II reconstruction and had had previous jejunostomies. Success with such procedures lead to the development of direct PEJ procedures for patients with intact foreguts. This section will present the available techniques for direct PEJ.

1. *Direct PEJ after Gastrojejunostomy and Feeding Jejunostomy.* In the patient who has had a previous Billroth II procedure with a concomitant feeding jejunostomy, an endoscope is passed through the gastric remnant into the jejunum. The site of the previous jejunostomy is identified and the abdominal wall transilluminated. The jejunum is cannulated percutaneously with a peel-a-way plastic sheath using a Seldinger technique. Endoscopic visualization is maintained continually. A 12F feeding catheter is placed through the sheath and, using the endoscope is advanced distally into the small intestine. After removal of the sheath, catheter position is confirmed by contrast study (Figure 15.13).[37]

2. *Direct PEJ without Previous Jejunostomy.* Direct PEJ in patients without previous jejunostomy with and without intact foreguts has been described by Adams[35] and Shike et al.[36] A 160-cm colonoscope is passed into the jejunum approximately 20 cm distal to the ligament of Trietz. The abdominal wall is transilluminated marking the

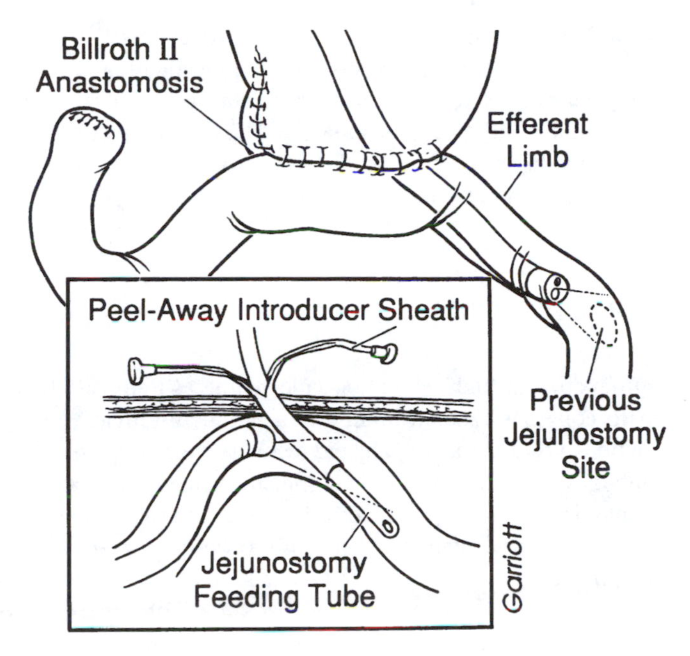

Figure 15.13. Technique for direct PEJ after Billroth II. The site of the previous jejunostomy is visualized. Inset: The jejunum is cannulated over a wire. A feeding tube is placed through a peel-away introducer. From Pritchard and Bloom.[37]

site of opposition of the small bowel to the abdominal wall. After this point in the procedure two alternate techniques have been described. The first method is a modification of the Gauderer-Ponsky PEG procedure. Under endoscopic visualization the jejunum is cannulated with a small gauge needle. Once the intraluminal portion of the needle is confirmed, a heavy thread is passed through the needle, grasped with biopsy forceps and withdrawn through the mouth. The thread is tied to a mushroom tipped feeding tube and pulled into position under endoscopic guidance.[36]

3. *Alternative Approach for Direct PEJ without Previous Jejunostomy.* An alternative approach requires a 3–4-cm incision in the left upper quadrant. Endoscopic intubation of the proximal jejunum is evident by transillumination of the intestines and palpation of the endoscope. The proximal end of the jejunum is grasped with a babcock clamp, and a short segment of small bowel 10 to 15 cm in length is eviscerated into the operative field. The feeding tube is placed using a Witzel technique. After placement of the feeding tube the jejunum is returned to the abdominal cavity, fixing the proximal and distal ends of the jejunum to the abdominal wall.[35]

The techniques for direct PEJ appear to address the problems associated with transpyloric PEJ (tube migration, clogging, and aspiration risk); however, experience with these endoscopically demanding procedures is limited and the final evaluation of their usefulness awaits further investigation.

Percutaneous endoscopic techniques for obtaining postpyloric enteral access can be time consuming and difficult to master. The authors feel strongly that in most cases the patient is best served by the standard open procedures, which are easily and safely performed by most surgeons.

LAPAROSCOPIC TECHNIQUES

The recent introduction of high-resolution video cameras to the established techniques of laparoscopy has ushered in an era of minimally invasive surgery.[41] A dramatic reduction in postoperative pain and recovery time coupled with excellent cosmetic results has led to the rapid development of many new and innovative procedures employing laparoscopy. Several techniques for obtaining enteral feeding access by laparoscopic means have recently been described.[42–49] Experience with these new procedures is limited, and indications for their use continue to evolve. It appears that these techniques are best suited for patients who cannot undergo endoscopically assisted procedures but who would be best served by a minimally invasive approach. Presented in this section are several techniques for performing both laparoscopic gastrostomies and jejunostomies.

Laparoscopic Gastrostomies

Laparoscopic gastrostomy is indicated when gastric enteral access is required and PEG cannot be performed. This may be particularly true when the patient is obese and a large incision would be required to perform a safe surgical gastrostomy. The most straightforward approach to laparoscopic gastrostomy is a modification of the Russell introducer technique for PEG placement.[42,43]

Pneumoperitoneum is created in the usual manner via a Verres needle inserted through a small supraumbilical incision. The camera port is placed through this incision and a 5-mm port is placed in the epigastrum. The stomach is grasped with an atraumatic instrument, and the site of the proposed gastrostomy is identified in the left upper quadrant. A 7-cm 18-gauge angiocatheter is placed through the anterior abdominal wall into the stomach at the site chosen for the gastrostomy. The needle is removed and a soft J-wire is passed through the catheter into the stomach. A 12-gauge French dilator followed by a 14-gauge French dilator are passed into the stomach. A 16-gauge French peel-a-way catheter over a dilator is passed into the stomach. A 16-gauge French Foley catheter is then placed through the sheath as it is peeled away. The balloon is inflated and secured snugly to the anterior abdominal wall. This procedure may be performed using either local or general anesthesia.

Laparoscopic Jejunostomy

Several methods for performing laparoscopic feeding jejunostomy have been described.[45–50] The indication for these procedures is currently evolving; however, laparoscopic jejunostomies may represent a minimally invasive alternative to PEJ techniques. Further experiences with both PEJ and laparoscopic jejunostomy will be needed to determine their usefulness for providing safe, effective, and reliable postpyloric feeding access. The principal advantage of the minimally invasive laparoscopic approaches is a reduction in postoperative pain and rehabilitation time while achieving cosmetic results. Other theoretical advantages include reduction in wound infection rates and subsequent incisional herniation, as well as reduced contact between the surgeon and the patients blood.[51] The cost of the procedures remains a drawback. In one recent study,[48] comparing the cost of various forms of postpyloric feeding access, the laparoscopic jejunostomy was found to be almost three times more expensive than PEJ and two times more expensive than standard surgical jejunostomy. We have recently reported a simple technique for performing laparoscopically assisted feeding jejunostomies and describe it again below. Exactly where this operation fits into the overall armamentarium for achieving enteral access remains unclear. Further experience will better define its indications and contraindications.

Laparoscopic jejunostomy and the extracorporeal tube placement. After induction of general endotracheal anesthesia, pneumoperitoneum is established using a Verres needle placed through a small incision just above the umbilicus. A 12-mm trocar is inserted and the laparoscope is introduced. Under laparoscopic guidance, an accessory 10-mm trocar is placed in the left anterior axillary line approximately 5 cm above the anterior superior iliac spine. The operating camera is then moved to the accessory port, the abdomen is inspected, and the ligament of Treitz is identified. The jejunum is grasped on the antimesenteric border at a point 25 cm distal to the ligament of Treitz and withdrawn through the umbilical wound, incising the fascial edges slightly if necessary. A 12-gauge French straight red rubber catheter is introduced through an enterotomy within two concentric silk purse string sutures, or a Witzel tunnel may be used. The bowel is secured to the fascial edge around the tube with four seromuscular 3-0 silk sutures, then returned to the abdominal cavity. The fascia and skin are closed in a standard fashion, and pneumoperitoneum is reestablished. Under direct vision, the bowel is distended with saline injected through the catheter to confirm the absence of leak and aboral positioning of the tube. The red rubber catheter is tunneled subcutaneously and brought externally through the lateral trocar site (Figure 5.14).[45]

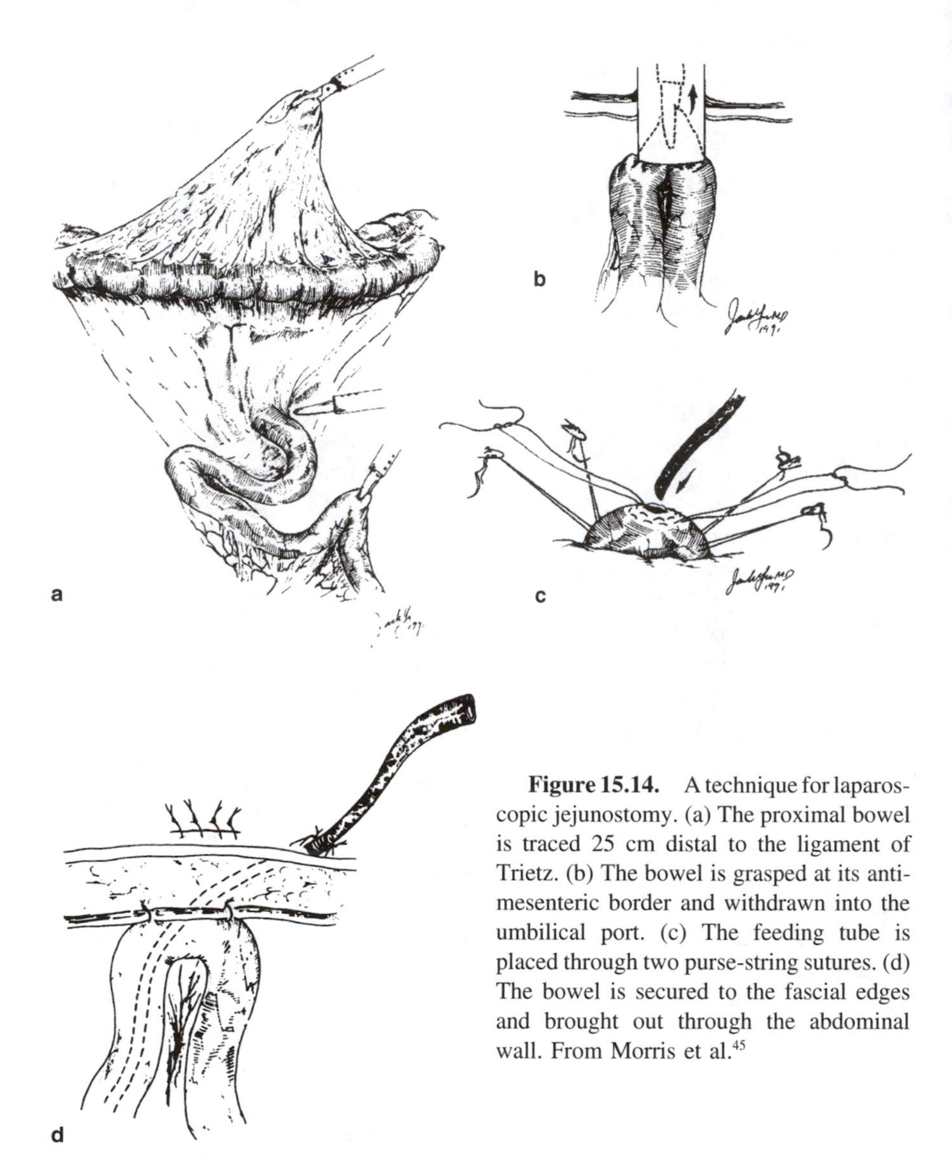

Figure 15.14. A technique for laparoscopic jejunostomy. (a) The proximal bowel is traced 25 cm distal to the ligament of Trietz. (b) The bowel is grasped at its antimesenteric border and withdrawn into the umbilical port. (c) The feeding tube is placed through two purse-string sutures. (d) The bowel is secured to the fascial edges and brought out through the abdominal wall. From Morris et al.[45]

RADIOLOGICALLY ASSISTED TECHNIQUES

Recent advances in the field of interventional radiology have led to the development of methods for percutaneous radiologically assisted gastrostomies and jejunostomies. Although these techniques are seldom considered a first-line approach for enteral access, they do provide attractive alternatives for patients for whom surgery is a prohibitive risk and in whom endoscopic procedures are impossible for anatomic reasons.

Percutaneous Radiologically Assisted Gastrostomy

An option for achieving access to the stomach using fluoroscopy has been described by Willis and Oglesby.[52] Prior to the procedure, an ultrasound of the left upper quadrant is obtained to evaluate the relationship of the left lobe of the liver to the stomach. This ensures that the gastrostomy will not pass through the liver before entering the stomach. The patient is placed supine and the stomach is distended via insufflation of air into the stomach via a nasogastric tube. The position of the stomach and transverse colon are determined via fluoroscopy. The distance from the anterior stomach wall to the abdominal wall is determined via a cross-table view. After placement of local anesthesia and under fluoroscopic guidance, a long 18-gauge needle with a Teflon sheath is inserted to the desired depth, and the needle is removed. Intragastric location is verified by the injection of water-soluble contrast. A guide wire is then passed, and the tract is dilated using progressive large dilators until a 10- to 12-gauge French pigtail catheter can be passed. The catheter is secured to the skin. The pigtail catheter prevents inadvertent removal.

Percutaneous Radiologically Assisted Jejunostomy

Techniques similar to those described for radiologically assisted access to the stomach have been described for the jejunum.[53] Experience with these techniques is limited, but they may provide a viable option for patients who are prohibitive surgical risks.

Radiologically assisted percutaneous access to the stomach is first achieved as described previously. A guide wire is then passed through the duodenum into the proximal jejunum. A 8-gauge French balloon occluder catheter is passed over a wire into the jejunum, and the balloon is inflated with 30 cc of air and 10 cc of water-soluble contrast. The balloon's intrajejunal location and position adjacent to the abdominal wall is confirmed by fluoroscopy. In thin patients, the balloon may also be palpated. Under fluoroscopic guidance, an 18-gauge needle is passed into the jejunum. Successful jejunal puncture is confirmed by both palpable and fluoroscopic evidence of balloon puncture. A guide wire is then passed and the tract dilated to approximately 10-gauge French. A pigtail catheter is then placed with or without an introducer. The balloon occluder catheter is removed and the feeding catheter is secured to the skin.

SUMMARY

Many techniques have been described for obtaining access to the gastrointestinal tract for forced feedings. Gastrostomy feeds are most physiologic and easiest for caregivers to administer in patients who are not at risk for aspiration. A PEG procedure is now the method of choice for gastrostomy placement. When a PEG is contraindicated, an open gastrostomy is a safe, effective, and quick operation for gastric feeding access. In patients at risk for aspiration, the modified Witzel jejunostomy is the method of choice for forced enteral feeding access. Laparoscopic enteral access techniques and the various PEJ procedures are specialized operations whose place in the surgeon's armamentarium for providing enteral feeding access requires further elucidation.

REFERENCES

1. Randall HT. Enteral nutrition: tube feedings in acute and chronic illness. *J Parent Ent Nutr* 1984;8:113–136.
2. Brown-Seguard CE. Feedings per rectum in nervous affections. *Lancet* 1978;1:144.
3. Stengel A, Ravdin IS. Maintenance of nutrition in surgical patients with description of an orojejunal method of feeding. *Surgery* 1939;6:511–519.
4. Andersen AFR. Immediate jejunal feedings after gastroenterostomy. *Ann Surg* 1918;67:565–566.
5. Gauderer MW, Stellato TA. Gastrostomies: evolution, techniques, indications and complications. *Curr Prob Surg* 1986;23:661–719.
6. Egeberg CA, Christianana med gessellsch. Om behandlingen af impenetrable strecturer: madroevect. *Norsk Mag Laegevidensk* 1841;2:97–105.
7. Sedillot O. De la gastrostomie fistulenge. *Compt Mend Acad Sci* 1846;23:222–230.
8. Jones S. Gastrostomy for stricture of the esophagus; death from bronchitis four days after operation. *Lancet* 1875;1:678–681.
9. Gauderer MW, Ponsky JL, Izant RJ. Gastrostomy without laparotomy: a percutaneous endoscopic technique. *J Pediatr Surg* 1980;15:872–875.
10. Torosian MH, Rombeau JL. Feeding tube enterostomy. *Surg Gynecol Obstet* 1980;150:918–927.
11. Bowles T, Zollinger RM. Critical evaluation of jejunostomy. *Arch Surg* 1952;65:358–366.
12. Weltz CR, Morris JB, Mullen JL. Surgical jejunostomy in aspiration risk patients. *Ann Surg* 1992;215:140–145.
13. Ryan JA, Page CP. Intrajejunal feeding: development and current status. *J Parent Ent Nutr* 1984;8:187–198.

14. Steinberg EP, Andersons GF. Implication of medicare's prospective payment system for special nutrition services. *Nutr Clin Pract* 1986;1:12–28.
15. Eastwood GL. Small bowel morphology and epithelial proliferation in intravenously alimented rabbits. *Surgery* 1977;82:612–620.
16. Sonies BC, Baum BJ. Evaluation of swallowing pathology. *Otolaryngol Clin North Am* 1988;21:637–648.
17. Nance ML, Morris JB. Enteral nutritional support: techniques and common complications. *Hosp Phys* 1992;28:24–29.
18. Gauderer MW, Picha GJ, Inzant RJ. The gastrostomy button—a simple, skin-level, non-refluxing device for long term enteral feedings. *J Pediatr Surg* 1984;19:803–805.
19. Gorman RC, Morris JB, Metz C, Mullen JL. The button jejunostomy for long term jejunal feeding. *J Parent Ent Nutr* 17:428–431.
20. Hogan RB, DeMarco DC, Hamilton JK, et al. Percutaneous endoscopic gastrostomy: to push or pull. *Gastrointest Endosc* 1986;32:253–258.
21. Grant JP. Percutaneous endoscopic gastrostomy. *Ann Surg* 1993;217:168–174.
22. Russell TR, Brotman M, Norris F. Percutaneous gastrostomy: a new simplified and cost effective technique. *Am J Surg* 1984;142:132–137.
23. Stellato TA, Gauderer MWL. Percutaneous endoscopic gastrostomy for gastrointestinal decompression. *Ann Surg* 1987;205:115–126.
24. Stellato TA, Gauderer MWL, Ponsky JL. Percutaneous endoscopic gastrostomy following previous abdominal surgery. *Ann Surg* 1984;200:46–51.
25. Lee MJ, Saini S, Brink JA, et al. Malignant small bowel obstruction and ascite: not a contraindication to percutaneous endoscopic gastrostomy. *Clin Radiol* 1991; 44:332–374.
26. Ponksy JL, Gauderer MWL, Stellato TA, et al. Percutaneous approaches to enteral nutrition. *Am J Surg* 1985;149:102–105.
27. Miller RE, Kummer BA, Kotler DP, et al. Percutaneous endoscopic gastrostomy: procedure of choice. *Ann Surg* 1986;204:543–547.
28. Kirby DE, Craig RM, Tsang TK, et al. Percutaneous endoscopic gastrostomy: a prospective evaluation and review of the literature. *J Parent Ent Nutr* 1986;10:155–159.
29. Tanker MS, Scheinfeldt BD, Steerman PH, et al. A prospective randomized study comparing surgical gastrostomy and percutaneous endoscopic gastrostomy. *Gastrointest Endoscop* 1986;32:144–148.
30. Stiegmann G, Goff J, VanWay C, et al. Operative versus endoscopic gastrostomy: preliminary results of a prospective randomized trial. *Am J Surg* 1988;155:89–92.
31. Kozarek RA, Ball TJ, Ryan RJ. When push comes to shove: a comparison between two methods of percutaneous endoscopic gastrostomy. *Am J Gastroenterol* 1986;81:642–647.
32. Ponsky JL, Agzodi A. Percutaneous endoscopic jejunostomy. *Am J Gastroenterol* 1984;79:113–116.

33. McFadyen BV, Catalno MF, Raijman I, et al. Percutaneous gastrostomy with jejunal extension: a new technique. *Am J Gastroenterol* 1992;87:725–728.
34. Bumpers HL, Luchette CA, Doerr RJ, Hoover EL. A simple technique for insertion of PEJ via PEG. *Surg Endosc* 1994;8:121–123.
35. Adams DB. Feeding jejunostomy with endoscopic guidance. *Surg Gynecol Obstet* 1991;172:239–241.
36. Shike M, Wallach C, Likier H. Direct percutaneous jejunostomies. *Gastrointest Endosc* 1991;37:62–65.
37. Pritchard TJ, Bloom AD. A technique of direct percutaneous jejunostomy tube placement. *Surg Gynecol Obstet* 1994;178:173–174.
38. Patterson DJ, Kozavek RA, Ball TJ, Botoman VA. Comparison of percutaneous endoscopic gastrostomy alone versus PEG with jejunal extension. Gastrointest Endosc 1987;33:176.
39. Kaplan DS, Murthy UK, Linscheer WH. Percutaneous endoscopic jejunostomy: long-term follow-up of 23 patients. *Gastrointest Endosc* 1989;35:403–406.
40. Disario JA, Foutch PG, Sanowski PA. Poor results with percutaneous endoscopic jejunostomy. *Gastrointest Endosc* 1990;36:257–260.
41. Dent TL, Ponsky JL, Berci G. Minimal access general surgery: the dawn of a new era. *Am J Surg* 1991;161:323–326.
42. Edelman DS, Unger SW. Laparoscopic gastrostomy. *Surg Gynecol Obstet* 1991; 173:401–402.
43. Edelman DS, Unber SW, Russin DR. Laparoscopic gastrostomy. *Surg Lapar Endosc* 1991;1:251–253.
44. Duh Q-Y, Way LW. Laparoscopic gastrostomy using T-fasteners as retraction and anchors. *Surg Endosc* 1993;7:60–63.
45. Morris JB, Mullen JL, Yu JC, Rosato EF. Laparoscopic-guided jejunostomy. *Surgery* 1992;112:96–99.
46. Duh QY, Way LW. Laparoscopic jejunostomy using T-fasteners as retractors and anchors. *Arch Surg* 1993;128:105–108.
47. Reed DN. Percutaneous peritonendoscopic jejunostomy. *Surg Gynecol Obstet* 1992;174:527–529.
48. Sangster W, Swanstrom L. Laparoscopic-guided feeding jejunostomy. *Surg Endosc* 1993;7:308–310.
49. O'Regan PJ, Scarrow GD. Laparoscopic jejunostomy. *Endoscopy* 1990;22:39–40.
50. Eltringham WK, Roe AM, Galloway SW, et al. A laparoscopic technique for full thickness intestinal biopsy and feeding jejunostomy. *Gut* 1993;34:122–124.
51. Cuschieria A. Minimal access surgery and the future of interventional laparoscopy. *Am J Surg* 1991;161:404–407.
52. Willis JS, Oglesby DF. Percutaneous gastrostomy. *Radiology* 1987;149:449–453.
53. Rosenblum J, Taylor FC, Lu CT, Martich V. A new technique for direct percutaneous jejunostomy tube placement. *Am J Gastroenterol* 1990;85:1165–1167.

Selection of Enteral Formulas

CHAPTER 16

Elaine B. Trujillo, M.S., R.D., and Stacey J. Bell, D.Sc., R.D.

INTRODUCTION

Enteral nutrition is the provision of nutrients via the gastrointestinal tract. This can be accomplished through oral intake, or when used in the context of nutritional support, through tube feedings. Enteral nutrition is the preferred method for alimentation. It is more physiologic, safer, and cheaper than parenteral nutrition.

The provision of nutrients through the gastrointestinal tract maintains normal gut function.[1] This has been demonstrated in animals who have improved gut weights, mucosal thickness, protein, and deoxyribonucleic acid (DNA) contents, disaccharidase activity, and epithelial cell proliferation when fed orally as opposed to intravenously.[2,3] Bowel rest from starvation or administration of parenteral nutrition leads to villous atrophy.[3] This atrophy not only affects nutrient assimilation, but affects the intestinal barrier.[4] The disruption of the intestinal barrier and alteration of the bacterial microflora contained within the intestinal barrier allow greater translocation of bacteria and absorption of endotoxins from the gut lumen.[5] Bacterial translocation is the process of bacterial migration or invasion across the intestinal mucosa into mesenteric lymph nodes and the portal bloodstream.[4]

Bowel rest with parenteral nutrition produced bacterial translocation in rats by increasing the cecal bacterial counts and impairing the intestinal barrier.[6]

Enteral nutrition is safer than parenteral nutrition, because it is not associated with the more severe metabolic, mechanical, and infectious complications that can occur with parenteral nutrition. However, enteral nutrition is not risk free and care should be taken in its administration to avoid problems such as pulmonary aspiration of the tube-feeding formula, which is one of the most serious and life-threatening complications of enteral nutrition.[7]

INDICATIONS AND CONTRAINDICATIONS

The indications for enteral nutrition include malnutrition, insufficient oral alimentation, and a functioning or at least partially functioning gastrointestinal tract. The contraindications as described by the American Society for Parenteral and Enteral Nutrition (ASPEN) are listed in Table 16.1.[7] These contraindications are not an absolute. Oftentimes, enteral nutrition may be beneficial for one patient, but deleterious for another with the same disorder. For instance, ileus and diarrhea are not necessarily contraindications. Patients with ileus can be fed enterally as long as the diagnosis of mechanical obstruction has been ruled out and there is access to the small bowel for feeding with simultaneous gastric decompression. Enterally feeding patients with diarrhea can be difficult. If the diarrhea is from prolonged bowel rest, administration of tube feedings would be warranted, but care should be taken to advance the formula slowly. However, if the diarrhea is due to gut failure from underlying ischemia or bacterial overgrowth, enteral nutrition would be contraindicated.[4]

Table 16.1. Contraindications to Enteral Nutrition

Contraindications
Diffuse peritonitis
Intestinal obstruction that prohibits use of the bowel
Intractable vomiting
Paralytic ileus
Severe diarrhea
Potential contraindications
Severe pancreatitis
Enterocutaneous fistulae
Gastrointestinal ischemia
Not recommended
Early stages of short bowel syndrome
Severe malabsorption

Source: From ASPEN Board of Directors.[7]

It is not always certain that enteral nutrition will be well tolerated. It is sometimes necessary to use both enteral and parenteral nutrition concomitantly. Parenteral nutrition can provide the majority of the nutrients, all of the fluid, and all of the electrolytes, and act as a vehicle for medications and acid–base management, while enteral nutrition can provide progressive stimulation to the gastrointestinal tract.

Formulas. There are over 100 enteral formulas available for oral and/or tube feeding use. They have evolved from blenderized foodstuffs to very specific nutrient compositions for various disease states. They all contain protein, carbohydrate, and fat of varying quantities and composition.

Enteral formulas can be categorized into three major groups: (a) standard formulas, which is the largest group and includes concentrated formulas, high protein formulas, and fiber-supplemented formulas. (b) predigested formulas for patients with some degree of gastrointestinal dysfunction and are more easily absorbed than intact nutrients. (c) disease-specific formulas, in which nutrient profiles have been altered for specific disease states or immune-enhancement.

Standard

Standard formulas are most widely used and are generally well tolerated in tube-fed patients with functioning gastrointestinal tracts. They are nutritionally complete in that they can provide most of a patient's macro- and micronutrient needs in a volume to meet their caloric needs. Generally, they provide 1.0 cal/mL of formula and contain 50% to 55% carbohydrate, 15% to 20% protein, and 30% fat. They are usually isotonic, with an osmolality between 280 and 350 mOsm/L.

Carbohydrate is the primary energy source of standard formulas. The form of carbohydrate contained in the formulas ranges from starch to simple sugars. Starch is well tolerated; however, it is not often in enteral formulas because of its insolubility in water.[8] Oligosaccharides and polysaccharides are the most predominant forms of carbohydrate found in enteral formulas. They are usually well tolerated, but they require pancreatic enzymes for digestion.[8] Maltodextrins, hydrolyzed cornstarch, and corn syrup are examples. Disaccharides, such as sucrose, lactose, and maltose require specific disaccharidases in the small bowel mucosa for hydrolysis and absorption.[8,9] Lactase deficiency is the most prevalent disaccharidase deficiency and most enteral formulas are therefore lactose-free. Monosaccharides such as glucose and fructose do not require hydrolysis for digestion, but tolerance may be limited by the absorptive capacity of the small bowel.[8,9]

Protein is considered to be the most critical component of enteral formulas. In standard formulas, it is provided mostly as intact proteins, which are whole protein from food and protein isolates. Protein isolates are intact proteins that have been separated from their original source.[8] For example, lactalbumin is a protein isolate derived from whey. Other examples of intact proteins and protein isolates in enteral formulas include meat, soy protein isolate, and calcium, sodium, and potassium caseinates.[10] Intact protein and protein isolates require normal levels of pancreatic enzymes to catabolize large proteins to small polypeptides and free amino acids.[10]

Fat is provided in enteral formulas as a calorically dense energy source. It provides a vehicle to absorb fat-soluble vitamins and supplies essential fatty acids. Vegetable oils, rich in long-chain triglycerides (LCTs), are the most predominant sources of fat in enteral formulas. A mix of LCTs and medium-chain triglycerides (MCTs) are often used as well. Medium-chain triglycerides offer advantages over LCTs in terms of digestion. They can be absorbed intact without appreciable pancreatic or biliary function and are subject to more rapid clearance from the bloodstream than LCTs.[11] Other lipid sources that are emerging in enteral formulas include fish oil, structured lipids, and short-chain fatty acids.

Concentrated formulas. Concentrated formulas provide more nutrients and calories per milliliter of formula (i.e., 1.5 or 2.0 cal/mL). They are used most often when fluid restriction is necessary, such as in renal failure, liver failure, or heart disease. They are not intended for the patient who is not volume restricted, as their use can result in dehydration. Renal solute load, which refers to the constituents within the formula that must be excreted by the kidneys (protein, sodium, potassium, and chloride), is higher with more concentrated formulas.[8] As the renal solute load increases, so does the patient's fluid needs.

High protein formulas. Protein in the diet provides amino acids and nitrogen for various physiologic compounds, such as neurotransmitters, creatinine, glutathione, and nucleic acids. Proteins function as enzymes, hormones, and antibodies and are used for the structural formation of cells and the synthesis of organ and muscle tissues. Inadequate protein intake leads to diminished protein content in the cells and organs and a deterioration in cellular function.[12] The protein needs of a healthy adult are approximately 0.82 $g/kg \cdot day^{-1}$.[13] In severe stress, protein needs can increase to 15% to 20% of total calories (1.5 to 2.0 $g/kg \cdot day^{-1}$) as compared to 8% to 10% in the absence of stress.[12,14] This increase in protein requirement is related to not only an increase in body protein breakdown, but also less efficient utilization of both endogenous and exogenous protein for synthesis of body proteins.[12]

High protein formulas are widely used in the acute care setting. They generally provide greater than 20% of the total calories as protein and often can meet the protein needs of the patient without overfeeding. Protein modules in the form of casein or whey can be added to a formula to increase the protein content.

Fiber-supplemented formulas. Fiber-supplemented enteral formulas are often used to alleviate the symptoms of diarrhea and constipation, which are common complaints of tube-fed patients. Although there are studies that demonstrate improved bowel function with fiber-supplemented formulas,[15,16] there are others that show little effect on bowel function.[17,18] The use of fiber-supplemented enteral formulas does not appear to be harmful. The potential for reduced mineral bioavailability has been investigated, and no clinically adverse effects have been demonstrated.[19] There are numerous benefits in consuming a diet high in fiber, including decreased risk of different types of cancer, improved blood sugar control, and sterol binding. However, during the relatively short period of time that a tube feeding formula is administered, fiber need not be included. For patients who cannot eat and require only tube feedings, fiber may be included in the formula.

The average American consumes approximately 10 to 15 g of fiber/day. It is recommended that we consume 20 to 25 g/day.[20,21] Fiber-supplemented enteral formulas contain anywhere from 4 to 20 g of dietary fiber/L. The predominant fiber source contained in enteral formulas is soy polysaccharide. Soy polysaccharide is primarily an insoluble fiber (94%) derived from dehulled, defatted soybeans.[21] Most of the fiber sources contained in enteral formulas are from insoluble fiber. Insoluble fiber has the physiologic effects of decreasing transit time and increasing stool weight by its water holding ability. Soluble fiber increases stool weight by increasing bacterial mass.[21]

Soluble fiber has been found to decrease the incidence of diarrhea in tube-fed and enterally supplemented patients.[22,23] Soluble fibers, such as pectin, guar, and gum arabic are fermented in the colon to produce short-chain fatty acids, which are the preferred respiratory fuels for colonocytes and stimulate their proliferation. In animals, the addition of pectin to an elemental diet has been shown to significantly enhance intestinal adaptation.[24] Soluble fiber in the form of gum arabic, guar gum, and pectin are being used in enteral formulas, usually as a mix with soy fiber.

Predigested

Predigested or elemental formulas refer to enteral formulas that are in easily digestible form. This specifically refers to the protein form, as carbohydrate is generally well tolerated and the fat content is usually very low or contains a large portion of MCTs. Predigested formulas are indicated in patients with impaired digestion and absorption, such as pancreatitis, acquired immunodeficiency syndrome, gastrointestinal cancer, critical illness, short gut, enterocutaneous fistulas, and diarrhea.[25]

The protein source is what designates a predigested formula. Historically, protein in the form of free amino acids was thought to be superior to intact or other hydrolyzed protein as it was directly absorbed into the bloodstream.[26] This was expanded when a peptide carrier system was described.[27,28] Small peptides in the form of dipeptides and tripeptides have been found to be absorbed more readily into the bloodstream and result in improved nitrogen balance.[29,30] Subsequently, no appreciable differences were found in the tolerance of severely hypoalbuminemic geriatric patients to a diet containing free amino acids and one containing small chain peptides.[31] It remains unclear whether small-chain peptides are superior to free amino acids. It has also been suggested that whole protein or protein isolates are just as good in certain clinical situations. One study has shown no advantage of a peptide enteral formula over a standard enteral formula in selected measurements of tolerance and nutritional outcome in acutely injured, hypoalbuminemic patients.[32] Despite the uncertainties, it is reasonable to give predigested formulas to all patients in whom intolerance may be expected, such as in the critically ill and early postgastrointestinal surgery patients.

As predigested formulas are more expensive than standard formulas, transition should begin once tolerance to the goal rate of the predigested formula has been established. Transitioning should be gradual and can be accomplished by mixing the predigested formula with the standard formula as follows:

Step 1. ¾ elemental, ¼ standard
Step 2. ½ elemental, ½ standard
Step 3. ¼ elemental, ¾ standard
Step 4. full-strength standard

If tolerance is good and the goal rate achieved, the full transition can be made in 4 to 5 days.[33]

Disease-Specific

Immunity-enhancing. The most interesting recent advances in enteral nutrition have been in the area of immunity-enhancing formulas. Specific nutrients (i.e., arginine, glutamine, nucleotides, omega-3 fatty acids) have been shown to possess immunity-enhancing properties.[34–38] At present, there are at least three immunity-enhancing formulas that are commercially available. They are enriched with various combinations of these specific nutrients.

In a study of postoperative gastrointestinal cancer patients from a single institution, patients received either a formula supplemented with arginine, ribonucleic acid (RNA), and omega-3 fatty acids or a standard enteral formula. Results showed that the patients fed the supplemental formula had significantly fewer infectious and wound complications and a significantly shorter hospital stay.[37] The study was repeated in a multicenter trial with 296 critically ill patients.[38] Results were similar. The hospital median length of stay was reduced by 8 days in patients receiving the supplemental formula. There was also a significant reduction in the frequency of acquired infections in those patients who were stratified as septic and received the supplemental formula. In another multicenter trial, trauma patients were randomized to receive early enteral nutrition with a different immunity-enhancing diet or a standard stress enteral formula. Patients receiving the immunity-enhancing diet experienced significantly fewer intraabdominal abscesses and significantly less multiple organ failure.[34]

The major criticism of these trials has been that the immunity-enhancing formulas and the control formulas were not isonitrogenous; the study formulas usually provided more protein. Therefore, it is difficult to discern whether the beneficial effects were because of immunity-enhancing active nutrients or because of the higher protein load. However, when a modular tube feeding recipe enriched with omega-3 fatty acids, arginine, and other selected micronutrients was compared to two standard enteral formulas in burn patients, there was a significant reduction of wound infection and length of stay. These formulas were isonitrogenous.[35]

Taken together, the results of these clinical trials are promising. They support the idea of nutritional pharmacology where nutrients provided in pharmacologic doses can induce a desired outcome.

Renal. Traditionally, formulas designed for renal failure contained predominantly essential amino acids (EAAs). Although the theory may be reasonable for short-term use (i.e., 5–7 days), over the long term, protein synthesis most likely will be impaired. The use of branch-chain amino acids (BCAAs) has been proposed as chronic renal failure is characterized by subnormal concentrations of plasma BCAAs and their keto acids.[39] The BCAAs are more efficiently utilized and thereby, less total protein may be given to patients when the protein is enriched with BCAAs. Standard enteral formulas contain approximately 22% BCAA. Enteral formulas with higher amounts (approximately 45% BCAA) are commercially available.

There are also formulas that contain balanced amino acids but modest electrolytes. In some cases, patients with renal insufficiency may benefit from formulas devoid of potassium, magnesium, and phosphorus. Some of these formulas are free of these minerals. It is as important to avoid micronutrient toxicity as it is to avoid protein toxicity.

In the nutritional management of patients with renal insufficiency, it is important that they are not underfed protein, because this will lead to body cell mass catabolism and malnutrition. For critically ill patients, it is best to use dialysis and feed an adequate protein diet, than to underfeed protein.

Hepatic. Hepatic enteral formulas contain large amounts of the BCAA's valine, leucine, and isoleucine, and low amounts of the aromatic amino acids (AAAs) phenylalanine and tryptophan. These products were formulated for patients with hepatic encephalopathy with the rationale that the infusion of BCAA corrects the increased AAA/BCAA ratio in plasma and may thereby lower the penetration of AAA across the blood–brain barrier with normalization of neurotransmitter synthesis.[39,40] Results of various clinical trials are conflicting. As with renal formulas, their use over the short term may be beneficial, by improving nitrogen balance and lessening encephalopathy. However, as with the renal formulas, their use for longer periods may limit protein synthesis, as these hepatic formulas have a poor protein or chemical score.[41] Briefly, the chemical score is a numeric indicator of protein quality in which the proportion of each EAA in the protein component of a food or formula is compared with the corresponding amount in a reference profile of EAAs, which is usually egg or milk protein. The EAA with the lowest individual score is designated as the limiting EAA, and its score is assigned as the chemical score for the protein.[42] Hepatic formulas have low amounts of aromatic amino acids, and, because some of them are also EAAs, the chemical score is low.

The more important point in patients with liver disease may be the provision of adequate nutrition. Studies have demonstrated improved liver function[43] and lower mortality rates[44] in patients who received supplemental tube feedings than in patients on oral diets alone.

Pulmonary. Pulmonary formulas are designed to be high in fat (approximately 50% of calories) and low in carbohydrate to reduce carbon dioxide production and minimize carbon dioxide retention in patients with chronic obstructive pulmonary disease, cystic fibrosis, or respiratory failure. This is based on the premise that carbohydrate provided in excess of energy needs promotes lipogenesis. This causes increased carbon dioxide production and a concurrent rise in the respiratory quotient (RQ).[45] The respiratory quotient is the ratio of carbon dioxide production to oxygen consumption. Each substrate has its own RQ (carbohydrate=1.0, fat=0.7, protein=0.82). An RQ>1.0 implies lipogenesis, or the synthesis of fat from glucose[45].

Low-carbohydrate low-fat diets and enteral formulas are not ideal. Fats high in omega-6 fatty acids are immunosuppressive. Omega-6 fatty acids are precursors to arachidonic acid, which gives rise to the two-series prostanoids and leukotrienes of the 4-series. These are eicosanoids and mediate potent immune and inflammatory effects.[46,47] On the other hand, omega-3 fatty acids from fish oils are known to interfere in the production of eicosanoids from arachidonic acid and thereby may be beneficial to the immune system. The fat source of most pulmonary formulas is predominantly from omega-6 fatty acids.

In addition, the carbohydrate content of pulmonary formulas may be insufficient to meet endogenous needs of 2.5 $mg/kg \cdot min^{-1}$. A minimum of 100 g/day of carbohydrate is needed to avoid ketosis. Glucose is required by the brain (120 g/day), red blood cells (30 to 40 g/day) and wound healing (20–60 grams/day). Their requirements will be covered by gluconeogenesis from endogenous protein stores or amino acids from the diet if an adequate glucose supply is not maintained.[48] It is therefore important to determine whether the prescribed rate of formula administered will supply enough carbohydrate to prevent this.

LONG-TERM FEEDINGS AND HOMECARE

For a patient to be discharged to chronic care or homecare, a permanent feeding access is usually required. There are instances when a patient may be discharged with a nasal tube for feeding. It is often when shorter-term feedings are expected (i.e., hyperemesis gravidarum) or when a patient is able to self-intubate. It is imperative that whatever the access is, it is well secured, and kept patent by frequent flushes and avoidance of medication infusion.[49] Dislocation and clogging of the tubing are some of the most common complications of home tube-fed patients.[50]

As soon as it is established that a patient will be discharged on tube feedings, a nutritional plan should be initiated. Begin by choosing an appropriate formula. Many patients benefit from fiber-supplemented formulas for long-term feedings. Fiber-supplemented formulas should be initiated cautiously, as an abrupt increase in fiber can cause abdominal discomfort, including gas and cramping. Such discomfort occurs more commonly in patients who have not been receiving fiber or who have had a minimal fiber intake for an extended period of time. The transition to a fiber-supplemented formula can be done gradually and, if not completed in the hospital, can be continued at home. Often this is accomplished by instructing the patient to substitute some of the current formula for a fiber formula (i.e. replace one can of the fiber formula for one can of the current formula) and gradually increase the number of cans of the fiber formula until the patient is infusing all of the fiber-supplemented formula.

Another concern in formula selection for long-term use is the free water content. Oftentimes, practitioners recommend the use of a concentrated formula to reduce the infusion time while providing adequate nutrition. This practice should be limited to patients who require fluid restrictions or can drink sufficiently to meet their fluid needs. Even with standard 1.0 cal/mL formulas, it is necessary to provide additional free water flushes or boluses to meet the patient's fluid needs. See Table 16.2 for a suggested flushing schedule.

Finally cost and availability should be considered when choosing a formula for discharge. This is especially important when the patient is responsible for purchasing his or her own formula.

Once the formula has been selected and tolerance has been established, cycling can begin. Cyclic feeding is the provision of tube feeding formula over a designated period. It is often provided at night and it is thought to increase the patient's appetite and allow for increased ambulating during the day.[51] If possible, cycling

Table 16.2. Suggested Flushing Schedule

60–120 ml of water at the following times:
before starting tube feeding
at bedtime
once during the night, if awake
first thing in the morning
after stopping tube feeding
30 mL of water before and after administration of medications via the tube

should be initiated in the hospital, but can also be accomplished after discharge. A suggested cyclic regimen is as follows:

Step 1. 20 hr cycle
Step 2. 16 hr cycle
Step 3. 14 hr cycle
Step 4. 10–12 hr cycle

As the cycle time is decreased, the tube feeding rate is increased to provide the desired volume.[51]

If the patient is to be discharged home on tube feedings, it is necessary to evaluate the home situation for safety.[52] A home health care agency needs to be contacted if the patient requires a feeding pump and or nursing care. The patient or caregiver should be instructed on how to administer the feedings prior to discharge. This will ensure that the patient or caregiver is willing and able to perform home tube feeding procedures.[52] Clearly written instructions for both routine procedures and how to deal with common problems should be available to the patient or caregiver. Problems that are of little concern to the health professional may be of major concern to some patients. Arrangements for follow-up with a dietitian should be made and telephone contacts should be provided so that the patient feels secure and can obtain appropriate advice when necessary.[53] The steps to be carried out in discharging a patient home on tube feedings are summarized in Table 16.3.

Table 16.3. Steps to Discharge a Patient Home on Tube Feedings

1. Assess nutrient and fluid needs
2. Select appropriate formula
3. Cycle feedings
4. Psychosocial assessment of the patient and home environment for safety
5. Contact home health care agency
6. Arrange for nutrition follow-up

CONCLUSION

Enteral nutrition provides nutrients via the gastrointestinal tract and can be accomplished through oral intake or tube feedings. This method is more physiologic, safer, and cheaper than parenteral nutrition. Enteral nutrition should begin as early as possible in patients who require nutritional support. Care should be taken in its administration to avoid complications or intolerances. Appropriate formula selection is an integral part of successful enteral therapy.

There are numerous enteral formulas that are commercially available. The large number of formulas can make the selection process difficult. This process can be simplified by categorizing the formulas into groups, such as standard; predigested; and disease-specific. Once a formula group has been chosen, specific macro- and micronutrient contents should be evaluated more closely. Formulas for long-term use should take into consideration fiber content, free water content, and cost.

REFERENCES

1. Steele KW, Seidner DL, Jensen GL. 1994. Gastrointestinal physiology and the gastrointestinal tract as a central organ in the hypermetabolic response to injury. In Borlase BC, Bell SJ, Blackburn GL, Forse RA, eds. *Enteral Nutrition.* New York: Chapman and Hall, 1994:3–14.
2. Eastwood GL. Small bowel morphology and epithelial proliferation in intravenously alimented rabbits. *Surgery* 1977;82:613–620.
3. Levine GM, Denen JJ, Steiger ET, et al. Role of oral intake in maintenance of gut mass and disaccharide activity. *Gastroenterology* 1974;67:975–982.
4. Rombeau JL. Enteral nutrition and critical illness. In Borlase BC, Bell SJ, Blackburn GL, Forse RA eds. *Enteral Nutrition.* New York: Chapman and Hall, 1994:25–36.
5. Berg RD. Promotion of the translocation of enteric bacteria from the gastrointestinal tracts of mice by oral treatment with penicillin, clindamycin, or metronidazole. *Infect Immun* 1981;33:854–861.
6. Alverdy JC, Aoys E, Moss GS. Total parenteral nutrition promotes bacterial translocation from the gut. *Surgery* 1988;104:185–190.
7. ASPEN Board of Directors. Guidelines for the use of parenteral and enteral nutrition in adult and pediatric patients. *J Parent Ent Nutr* 1993;17(4):8SA.
8. MacBurney MM, Russell C, Young LS. Formulas. In Rombeau JL, Caldwell MD, eds. *Clinical Nutrition: Enteral and Tube Feeding,* 2nd ed. Philadelphia: W.B. Saunders, 1990:149–170.
9. Mayes PA. Carbohydrates. In Martin DW, Mayes PA, Radwell VW, Granner DK, ed. *Harper's Review of Biochemistry,* 20th ed. Los Altos, CA: Lange Medical Publications, 1985:147–157.

10. Eisenberg P. Enteral nutrition. Indications, formulas, and delivery techniques. *Nurs Clin North Am* 1989;24:315–338.
11. Adam S, Yeh YY, Jensen GL. Changes in plasma and erythrocyte fatty acids in patients fed enteral formulas containing different fats. *J Parent Ent Nutr* 1993;17:30–34.
12. Heimburger DC, Young VR, Bistrian BR, et al. The role of protein in nutrition, with particular reference to the composition and use of enteral feeding formulas. A consensus report. *J Parent Ent Nutr* 1986;10:425–430.
13. *Energy and Protein Requirements.* Report of a joint FAO/WHO/UNU Consultation. WHO Tech Rep Series. Geneva: World Health Organization, 1985.
14. Duke JH, Jorgensen SB, Broel JR. Contribution of protein to caloric expenditure following injury. *Surgery* 1970;68:168–174.
15. Zarling EJ, Edison T, Berger S, et al. Effect of dietary oat and soy fiber on bowel function and clinical tolerance in a tube feeding dependent population. *J Am Coll Nutr* 1994;13:565–568.
16. Liebl BH, Fischer MH, Van Calcar SC, et al. Dietary fiber and long-term large bowel response in enterally nourished nonambulatory profoundly retarded youth. *J Parent Ent Nutr* 1990;14:371–375.
17. Kapadia SA, Raimundo AH, Grimble GK, et al. Influence of three different fiber-supplemented enteral diets on bowel function and short-chain fatty acid production. *J Parent Ent Nutr* 10:63–68.
18. Frankenfield DC, Beyer PL. Soy-polysaccharide fiber: effect on diarrhea in tube-fed, head-injured patients. *J Clin Nutr* 1989;50:533–538.
19. Heymsfield SB, Roongsopisuthipong C, Evert M, et al. Fiber supplementation of enteral formulas: effects on the bioavailability of major nutrients and gastrointestinal tolerance. *J Parent Ent Nutr* 1988;12:265–273.
20. Slavin J. Commercially available enteral formulas with fiber and bowel function measures. *Nutr Clin Pract* 1990;5:247–250.
21. Fredstrom SB, Baglien KS, Lampe JW, et al. Determination of the fiber content of enteral feedings. *J Parent Ent Nutr* 1991;15:450–453.
22. Homann HH, Kemen M, Fuessenich C, et al. Reduction in diarrhea incidence by soluble fiber in patients receiving total or supplemental enteral nutrition. *J Parent Ent Nutr* 1994;18:486–490.
23. Zimmaro DM, Rolandelli RH, Koruda MJ, et al. Isotonic tube feeding formula induces liquid stool in normal subjects: reversal by pectin. *J Parent Ent Nutr* 1989;13:117–123.
24. Koruda MJ, Rolandelli RH, Settle RG, et al. The effect of a pectin-supplemented elemental diet on intestinal adaptation to massive small bowel resection. *J Parent Ent Nutr* 1986;10:343–350.
25. Martindale RG, Andrassy RJ. Elemental nutrition overview. *Nutritional Support Strategies for the Catabolic Patient,* proceedings from a symposium held at the American Dietetic Association Meeting (Denver, Colorado), Oct. 16–18, 1990.

26. Winitz M, Graff J, Gallagher N, et al. Evaluation of chemical diets as nutrition for man in space. *Nature* 1965;205:741–743.

27. Silk DBA, Perrett D, Clark ML. Intestinal transport of two dipeptides containing the same two neutral amino acids in man. *Clin Med* 1973;45:291–299.

28. Silk DBA, Perrett D, Webb JPW, et al. Tripeptide absorption in man. *Gut* 1973;14:427–428.

29. Albina JE, Jacobs DO, Melnik G, et al. Nitrogen utilization from elemental diets. *J Parent Ent Nutr* 1985;9:189–195.

30. Meguid MM, Landel AM, Terz JJ, et al. Effect of an elemental diet on albumin and urea synthesis: comparison with partially hydrolyzed protein diet. *J Surg Res* 1984;37:16–24.

31. Borlase BC, Bell SJ, Lewis EJ, et al. Tolerance to enteral tube feeding diets in hypoalbuminemic critically ill, geriatric patients. *Surg Gynec Obstet* 1992;174:181–188.

32. Mowatt-Larssen CA, Brown RO, Wojtysiak SL, et al. Comparison of tolerance and nutritional outcome between a peptide and a standard enteral formula in critically ill, hypoalbuminemic patients. *J Parent Ent Nutr* 1992;16:20–24.

33. Trujillo EB. Transitional feeding strategies for discharge. In Borlase BC, Bell SJ, Blackburn GL, et al. eds. *Enteral Nutrition,* New York: Chapman and Hall, 1994:199–205.

34. Moore FA, Moore EE, Kudsk KA, et al. Clinical benefits of an immune-enhancing diet for early postinjury enteral feeding. *J Trauma* 1994;37:607–615.

35. Gottschlich MM, Jenkins M, Warden GD, et al. Differential effects of three enteral dietary regimens on selected outcome variables in burn patients. *J Parent Ent Nutr* 1990;14:225–236.

36. Cerra FB, Lehmann S, Konstantinides N, et al. Improvement in immune function in ICU patients by enteral nutrition supplemented with arginine, RNA and menhaden oil independent of nitrogen balance. *Nutrition* 1991;7:193–199.

37. Daly JM, Lieberman MD, Goldine J, et al. Enteral nutrition with supplemental arginine, RNA and omega-3 fatty acids in patients after operation: immunologic, metabolic and clinical outcome. *Surgery* 1992;112:56–67.

38. Bower RH, Cerra FB, Bershadsky B, et al. Early enteral administration of a formula (Impact) supplemented with arginine, nucleotides, and fish oil in intensive care unit patients: results of a multicenter, prospective, randomized, clinical trial. *Crit Care Med* 1995;23:436–449.

39. Skeie B, Kvetan V, Gil KM. Branch-chain amino acids: their metabolism and clinical utility. *Crit Care Med* 1990;18:549–571.

40. Elia M. Changing concepts of nutrient requirements in disease: implications for artificial nutritional support. *Lancet* 1995;345:1279–1284.

41. Bell SJ, Bistrian BR, Ainsley BM, et al. A chemical score to evaluate the protein quality of commercial parenteral and enteral formulas: emphasis on formulas for patients with liver failure. *J Am Diet Assoc* 1991;91:586–589.

42. Dubin S, McKee K, Battish S. Essential amino acid reference profile affects the evaluation of enteral feeding products. *J Am Diet Assoc* 1994;94:884–887.
43. Kearns PJ, Young H, Garcia G, et al. Accelerated improvement of alcoholic liver disease with enteral nutrition. *Gastroenterology* 1992;102:200–205.
44. Cabre E, Gonzalez Huix F, Abad-Lacruz A, et al. Effect of total enteral nutrition on the short-term outcome of severely malnourished cirrhotics. *Gastroenterology* 1990;98:715–720.
45. Spector N. Nutritional support of the ventilator-dependent patient. *Nurs Clin North Am* 1989;24:407–415.
46. Adams S, Yeh YY, Jensen GL. Changes in plasma and erythrocyte fatty acids in patients fed enteral formulas containing different fats. *J Parent Ent Nutr* 1993;17:30–34.
47. Mascioli EA, Iwasa Y, Trimbo S, et al. Endotoxin challenge after menhaden oil diet: effects on survival of guinea pigs. *Am J Clin Nutr* 1989;49:277–282.
48. Askanazi J, Weissman C, Rosenbaum SH, et al. Nutrition and the respiratory system. *Crit Care Med* 1982;10:163–172.
49. Thibault A. Care of feeding tubes. In Borlase BC, Bell SJ, Blackburn GL, et al. eds. *Enteral Nutrition.* New York: Chapman and Hall, 1994:197–198.
50. Nelson JK, Palumbo PJ, O'Brien PC. Home enteral nutrition: observations of a newly established program. *Nutr Clin Pract* 1986;1:193–199.
51. Trujillo EB, Queen PM. Transition feeding. In Borlase BC, Bell SJ, Blackburn GL, et al. eds. *Enteral Nutrition.* New York: Chapman and Hall, 1994:107–111.
52. ASPEN Board of Directors. Standards for home nutrition support. *Nutr Clin Pract* 1992;7:65–69.
53. Elia M. Home enteral nutrition: general aspects and a comparison between the United States and Britain. *Nutrition* 1994;10:115–123.

Special Nutrients for Gut Feeding

CHAPTER 17

Scott A. Shikora, M.D.

INTRODUCTION

The intestinal tract or "gut" is a tubular structure partitioned into two functionally and morphologically distinct organs known as the small and large intestine. The large intestine or colon is approximately 120 to 200 cm in length and functions primarily for water reabsorption and stool storage. The small intestine is the principle organ of nutrient preparation and absorption. It is measured to be approximately 6 to 8 m in length at autopsy but is about half that long *in vivo.* To accomplish the complex tasks of digestion and absorption of nutrients, minerals, electrolytes and water, the small intestine has a luminal surface area of about 22 m^2. Traditionally, the small intestine has been thought to function solely for nutrient absorption. Over the last decade, it has become increasingly clear that the small bowel has other major functions. It produces and secretes numerous hormones that have both local and distant effects such as gastrin, somatostatin, secretin, and motilin. In addition, the small bowel contains an abundant amount of lymphoid tissue and is a major component of the body's defense system. With such a large interface with the external environment, it is therefore no surprise that the gut is the first line of defense to invasion.

To prevent the invasion of microbes into the systemic circulation, the small intestine relies on both a physical barrier and an immunologic shield (Table 17.1). The physical barrier is created by the tight junctions between intact epithelial cells, gastric acid, digestive enzymes, mucus production, intestinal motility, and a normal bacterial flora. The intestinal immune system, often referred to as the gut associated lymphoid tissue (GALT), consists of cells such as lymphocytes and macrophages situated throughout the intestinal wall and the production of the antibody immunoglobulin A (IgA) which is secreted into the lumen and prevents the adherence of microbes to the mucosa, a step necessary for invasion.[1]

The spread of pathogens or their products through the intestinal wall and into the systemic circulation is referred to as translocation.[2] Factors thought to promote translocation include luminal bacterial overgrowth, impaired host defense mechanisms, protein-calorie malnutrition, trauma or critical illness, or any process that leads to mucosal atrophy, such as interruption of the luminal nutrient stream.[3] Translocation is believed to be associated with, and even may be the underlying cause, of the nosocomial sepsis and multisystem organ dysfunction seen with critical illness.[2,4]

While present-day parenteral nutrition can substitute nutritionally for the interruption of oral nutrition, gut mucosal atrophy and immune dysfunction leading to translocation is not prevented.[5–7] A stream of luminal nutrients seems to be essential. It has also become apparent that certain nutrients are indispensable for maintaining the intestinal barrier to microbes (Table 17.2). Both the epithelial cells of the mucosa as well as the lymphoid cells of the GALT system require specific nutrients for growth and normal function.

Recent investigations have resulted in conflicting data as to whether the supplementation of standard enteral and parenteral solutions with these nutrients improve the integrity of the small bowel mucosa and GALT systems and/or improve

Table 17.1. Intestinal Barrier to Microbial Invasion

Structural
Intact mucosa
Mucus layer
Gastric acid
Digestive enzymes
Normal gut motility
Normal gut bacterial flora
Immunologic (Gut Associated Lymphoid Tissue)
Cellular immunity
Secretory IgA

Table 17.2. Nutrients Thought to be Important for Maintaining the Integrity of the Gut Mucosa and GALT

Glutamine
Arginine
Nucleotides
Omega-3 fatty acids (fish oil)
Dietary fiber (short-chain fatty acids)

outcome. In addition, most of these studies were performed on animals. Unfortunately, the results obtained from animal models may not always apply to humans. This chapter will review that literature.

GLUTAMINE

The amino acid glutamine has been recognized as one of the most important amino acids for normal metabolism. It has the highest concentration of any amino acid in the plasma (0.5 to 0.8 mmol/L), the largest pool size (61% of the intracellular amino acids in muscle), and the greatest interorgan flux (>25 g/day). Glutamine is the preferred respiratory fuel of rapidly dividing cells, such as those of the intestinal mucosa and the immune system.[8,9] It is a regulator of acid–base balance via the production of ammonia from one of its nitrogen groups and it is a precursor of nucleic acids, nucleotides, amino sugars, and proteins.[8,10] While the body normally can produce adequate glutamine to meet its needs, during stress from injury, sepsis or critical illness, the need may outstrip production, thus making glutamine conditionally essential.[10] During stress, muscle liberates greater than normal quantities of glutamine into the circulation, the majority of which is taken up by the intestinal mucosa.[11,12] These cells in fact have a dual source of glutamine. They can extract it from either the circulation or the luminal nutrient stream. The gut is the principle organ of glutamine utilization.[13] Its metabolism of glutamine may function as a regulator of nitrogen metabolism in normal and catabolic states.[14] Therefore, the gut may modulate the body's protein catabolic response to stress.[7] In settings such as the interruption of oral nutrition or the depletion of muscle-liberated glutamine, which may be seen with critical illness, mucosal atrophy, GALT dysfunction, and translocation can occur.

At present, because of its instability, none of the commercial parenteral nutrition solutions contain glutamine.[15] Furthermore, the majority of the enteral formulas contain only small quantities of this important amino acid. Currently, there are only a few studies in humans that have evaluated the benefits of supplemental glutamine. A number of animal studies have been able to demonstrate beneficial effects of supplementing parenteral and enteral diets with glutamine. These effects include improving the gut structural barrier by increasing the cellularity and mass of the mucosal layer and by improving the immune response of the GALT system. In similar studies in rats, both O'Dywer et al.[5] and Grant et al.[6] demonstrated that the supplementation of parenteral nutrition with glutamine could prevent the mucosal atrophy associated with the interruption of oral feeding and reliance on parenteral nutrition. In both investigations, the glutamine-enriched total parenteral nutrition (TPN) was nearly as effective as the oral rat chow diet. Burke et al.[16] studied the effect of intravenous glutamine administration on gut immune function. They found that parenterally administered glutamine could improve gut immune function in TPN-fed rats with normal intestinal tracts. The incidence of bacterial translocation and biliary levels of IgA were significantly better than the group given a standard parenteral solution devoid of glutamine. The improved immune parameters seen in the glutamine-fed rats were only slightly inferior to the control rats fed standard rat chow.

Human studies also found intravenous glutamine to be beneficial (Table 17.3). In a group of bone marrow transplant patients, Scheltinga et al.[17] demonstrated a significantly smaller increase in extracellular water and lower infection rate in patients given a glutamine-supplemented parenteral solution compared with a similar group of patients given a standard TPN formula. Tremel and colleagues found that the addition of glutamine to TPN resulted in an attenuation of mucosal atrophy and decrease in the atrophy-associated gut permeability in twelve critically ill patients.[18] In a similar report, van der Hulst and coworkers documented preservation of mucosal structure and permeability in postoperative TPN patients.[19] In contrast, Hornsby-Lewis et al. found no benefit to intestinal permeability in stable home TPN patients when given glutamine.[20] Conflicting data also exists for the effect of intravenous glutamine on immune function. Ogle and coworkers demonstrated improved neutrophil bactericidal function when cells from burned children were cultured in glutamine-enriched baths.[21] O'Riordain and colleagues found that surgical patients given a glutamine-supplemented TPN had increased T-cell deoxyribonucleic acid (DNA) synthesis after only 5 days of parenteral nutrition.[22] In contrast, van der Hulst et al. found no improvement in GALT function in 20 patients after 2 weeks of TPN supplemented with glutamine.[23]

In certain circumstances, orally administered glutamine may be preferable to intravenous administration. In animals subjected to methotrexate, a chemotherapeutic agent known to damage intestinal mucosa, Fox et al. found that those animals fed a glutamine-enriched oral diet demonstrated decreased intestinal injury, less bacterial translocation, and improved survival.[24] In a similar investigation, the oral administration of glutamine caused a 10-fold improvement in survival and a significantly lower incidence of bacterial translocation to the spleen when compared with the same amount of glutamine intravenously.[25] In the rat

Table 17.3. Trials in Humans Evaluating the Effects of Glutamine Supplementation[a]

Study	Methodology	Population	Intervention	Outcome
Scheltinga et al.[17] 1991	Randomized, blinded	BMT patients (n=20)	(1) TPN-glutamine 0.57 g/kg/day (2) Standard TPN	Clinical infections reduced 0 vs 5[b] Decreased expansion of ECF 20% vs 0%
Tremel et al.[18] 1994	Randomized	ICU patients (n=12)	(1) TPN-Ala-Gln 0.3 g/kg/day (2) Standard TPN	Decreased mucosal atrophy Decreased permeability Outcome not evaluated
van der Hulst et al.[19] 1993	Randomized, blinded	TPN patients (n=20)	(1) TPN-Gly-Gln 0.23 g/kg/day (2) Standard TPN	Decreased mucosal atrophy Decreased permeability Outcome not evaluated
Hornsby-Lewis et al.[20] 1994	Nonrandomized, baseline compared with 4-week results	Home TPN patients (n=7)	TPN-glutamine 0.285 g/kg/day	No improvement in permeability Increase in liver enzymes
Ogle et al.[21] 1994	Consecutive patients, PMNs isolated, unburned adult controls	Pediatric burn patients (n=12)	(1) 20–30 mmol/L glutamine (2) standard broth	Increased PMN bactericidal function in both burn patients and normal, unburned controls
O'Riordain et al.[22] 1994	Randomized	Surgical patients (n=20)	(1) TPN-Gly-Gln 0.18 g/kg/day (2) Standard TPN	Enhanced T-cell DNA synthesis No improvement in outcome found
van der Hulst et al.[23] 1994	Randomized	TPN patients (n=20)	(1) TPN-glutamine (2) Standard TPN	No effect on GALT Outcome not evaluated

[a]BMT, bone marrow transplant; ECF, extracellular fluid; Ala-Gln, L-alanyl-L-glutamine dipeptide; Gly-Gln, Glycyl-L-Glutamine dipeptide; PMNs, polymorphonuclear cells; GALT, gut associated lymphoid tissue

[b]$p<0.05$

sarcoma model of cancer, oral glutamine was even shown to enhance the tumoricidal effectiveness of methotrexate while reducing its morbidity and mortality.[26]

There have been few reported studies investigating the benefits of glutamine when administered via the gastrointestinal tract. In animals, enterally delivered glutamine has been shown to improve intestinal morphologic parameters and decrease translocation after abdominal radiation[14,27] and after massive resection.[28] In contrast, Scott and coworkers[29] found that glutamine was ineffective when given intravenously to animals subjected to abdominal radiation. Although they could demonstrate attenuated weight loss with the glutamine-enriched parenteral solution, no improvement was found in mucosal DNA content or villous height. It is not entirely clear why intravenous glutamine was ineffective. Souba has demonstrated that intestinal glutamine uptake can vary with stress (Figure 17.1). Plasma glutamine uptake decreased after the administration of endotoxin which resulted in loss of mucosal integrity and barrier function.[30] Similar work by Dudrick and colleagues confirmed a drop in glutamine consumption from the bloodstream after an endotoxin challenge, whereas luminal glutamine extraction increased shortly after exposure.[24] Therefore, in conditions such as sepsis, the intravenous administration of glutamine may be insufficient to support metabolism. This may explain the lack of benefit when glutamine is given parenterally.

Additional factors may be involved to explain the inconsistency of glutamine supplementation, particularly in human studies. At present, it is not known what dose of glutamine is necessary for effect. In a recent editorial, Wilmore suggested 0.37 to 0.5 $g/kg \cdot day^{-1}$, which would equate to approximately 19 to 35 g daily for averaged-sized patients.[31] Hornsby-Lewis et al. gave only 0.285 $g/kg \cdot day^{-1}$ to their patients and demonstrated no effect.[20] Some studies, such as that of Tremel and coworkers,[18] gave glutamine as a dipeptide for which the appropriate dosing is not known. van der Hulst et al. demonstrated improved gut permeability with 0.23 $g/kg \cdot day^{-1}$.[19] Additionally, patient condition may influence the likelihood of demonstrating a benefit from glutamine supplementation. Studies in stable patients such as that by Hornsby-Lewis and colleagues might be less likely to show effect because those patients were not stressed and may not have developed glutamine deficiencies.[20] Investigations done in postoperative or critically ill patients were successful.[18,19,21]

ARGININE

Arginine is another amino acid considered to be conditionally essential.[32] It possesses a number of important metabolic properties (Table 17.4). It enhances nitrogen retention, improves wound healing, is essential for collagen synthesis, acts as a precursor for nitric oxide synthesis, enhances secretion of growth hormone, prolactin, insulin, and glucagon, and is required for nucleotide synthe-

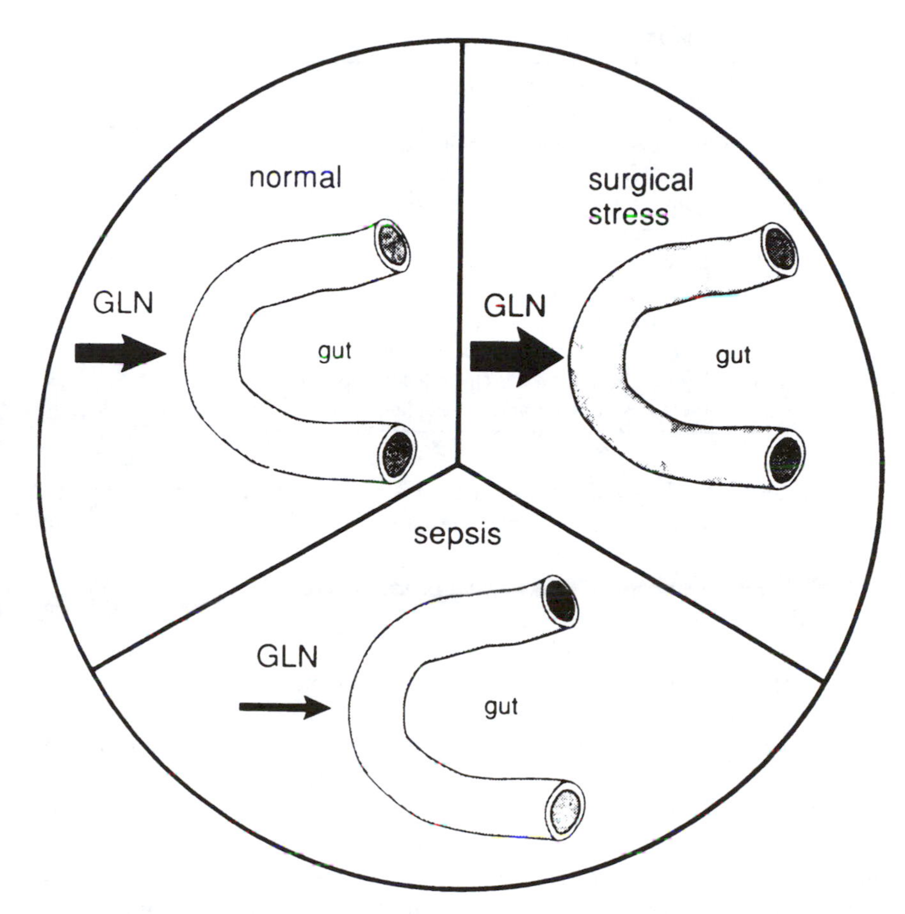

Figure 17.1. Glutamine uptake from the bloodstream in normal, postoperative and septic states. Reprinted from Souba et al.[8]

sis.[33] The most significant characteristic of arginine relates to its immunostimulatory effects. These effects include increased thymic lymphocyte blastogenesis, increased responsiveness to mitogens, increased interleukin-2 (IL-2) production and up-regulation of IL-2 receptors, increased natural killer (NK) cell and macrophage cytotoxicity to tumor cells and bacteria and overall augmentation of cellular immunity.[34]

Previous work by Barbul et al. supported the assumption that the wound healing and thymotrophic effects of supplemental arginine may be indirect and secondary to its secretagogue properties on pituitary and pancreatic hormones.[35] In their study, an intact hypothalamic–pituitary axis was essential. More recent work has demonstrated that arginine also has direct effects on macrophages and

Table 17.4 Metabolic Effects of Arginine

Enhances nitrogen retention
Improves wound healing
Stimulates the release of numerous hormones
Acts as a nitric oxide precursor
Required for nucleotide synthesis
Has antitumor properties
Enhances immune response even in normal conditions
Stimulates thymic lymphocyte blastogenesis
Increases the lymphocyte responsiveness to mitogens
Increases interleukin-2 production
Up-regulates the interleukin-2 receptors
Increases NK cell and macrophage cytotoxicity

lymphocytes that enhance their responsiveness.[34,35] These effects are thought to be mediated by nitric oxide, an intermediate metabolite generated during the conversion of arginine to citrulline.

Although it has no known direct effect on gut mucosa, arginine may still have trophic influences on the gut mucosa via its stimulation of the release of pancreatic insulin, somatostatin, pancreatic polypeptide, and various other polyamines.[37] More significantly are the profound effects of supplemental arginine on immune response in both healthy human volunteers with intact immune systems[38] and in stressed patients.[39] In a randomized prospective study involving patients with gastrointestinal malignancies undergoing major surgery, Daly and coworkers[39] found that patients given arginine-enriched enteral nutrition demonstrated significantly improved T-lymphocyte response to antigenic stimulation, a greater expression of CD_4 phenotype (T-helper cells) and a modest improvement in nitrogen balance. However, they did not find an improvement in outcome for patients given the supplemental arginine.

Outcome was demonstrated to be improved in mice fed an enteral diet supplemented with arginine in a study done by Gianotti and colleagues.[37] Mice given such a diet displayed significantly better survival after cecal puncture peritonitis and after major thermal burn. The arginine-fed animals could better kill translocated bacteria. Because the administration of a nitric oxide inhibitor negated this enhanced killing mechanism, it was postulated that the production of nitric oxide was responsible for the improved bactericidal effects. Improved survival after major thermal burn was also achieved by the administration of an arginine-enriched diet by Saito et al.[40] In that study, groups of guinea pigs were given different concentrations of arginine (0%, 2%, 3%, and 4% of total energy intake). Of these concentrations, 2% had the most benefit and 4% had no effect over control.

In the setting of established sepsis, enterally administered supplemental arginine did not improve outcome. In studies involving animals subjected to cecal puncture peritonitis prior to the initiation of the diet, no enhancement of immune response or improvement in outcome was noted.[41,42] Impaired intestinal absorption or markedly increased arginine requirements in the setting of sepsis may explain this phenomena. In addition, the provision of arginine at even higher levels of administration (6% of total calories) appeared to be detrimental.[41]

NUCLEOTIDES

Nucleotides, the precursors of ribonucleic acid (RNA) and DNA, are involved in many biochemical processes. Synthesis of DNA and RNA is necessary for most cellular functions including protein synthesis and cell division. Organs with rapid cell turnover such as lymphoid tissue and the gut mucosa are particularly sensitive to nucleotide availability. In health, the requirement for nucleotides was assumed to be satisfied by *de novo* synthesis in the liver.[43] The significance of dietary sources is not clear. In stress, the requirement for nucleotides may outstrip production.

Standard parenteral and enteral diets are nucleotide-free. Studies have demonstrated decreased cellular immunity and resistance to infection in animals provided a nucleotide-free diet then inoculated with either bacteria[44] or fungus.[45] When these same animals were provided a nucleotide-enriched oral diet, improved cellular immunity and survival were achieved. Nucleotide supplementation may also have a significant role in attenuating the gut mucosal atrophy seen with parenteral feeding. Tsujinaka et al.[46] demonstrated improved intestinal mucosal growth and maturity in mice given a nucleotide-enriched diet compared to a group of mice fed a standard nucleotide-free TPN. However, these animals still fared worse than a control group fed an oral diet of chow. One possible explanation is that parenteral nutrition enriched with nucleotides still lacked some of the gut trophic effects induced by oral nutrition.

OMEGA-3 FATTY ACIDS

Presently, nearly all enteral and parenteral formulations contain omega-6 fatty acids as their lipid source. When given in excess, omega-6 fatty acids may depress cell-mediated immunity and promote the production of the inflammatory mediators prostaglandin E_2 (PGE_2) and leukotriene B_4.[47] Omega-3 fatty acids, derived mainly from fish oil, seem to attenuate inflammatory responses and enhance cellular immunity.[47] In a burn model, Alexander and coworkers[48] demonstrated attenuated weight loss, lower resting metabolic expenditure, and improved cell-mediated immunity in animals given an enteral diet containing omega-3 fatty acids.[48] Studies have also been able to demonstrate improved immune response and survival in fish oil-fed animals given bacterial peritonitis.[49,50]

Protein-calorie malnutrition is known to be a major cause of depressed cellular immune function. Standard nutritional support devoid of immunostimulatory nutrients such as arginine, fish oil, and nucleotides may not restore immunocompetence. This was demonstrated by Van Buren and coworkers.[51] After subjecting mice to a protein-free diet, they demonstrated that a standard protein and calorie oral diet restored weight loss but not cell-mediated immunity. Mice given diets enriched with either RNA, or omega-3 fatty acids recovered normal immune response.

Recently, a few randomized prospective trials in humans have produced some exciting results with specialized enteral formulas that are supplemented with arginine, nucleotides, and fish oil (Table 17.5). These preliminary studies have all demonstrated decreased incidence of sepsis and improved outcome for the critically ill with the use of these formulations. Gottschlich et al. reported significantly lower rates of wound infection and length of hospitalization for burn patients given a specialized arginine and fish oil formula.[52] Moore et al. reported the results of a multicenter prospective trial in trauma patients using a specialized formula containing glutamine, arginine, fish oil, nucleotides, and branched-chain amino acids.[53] They found that the patients given this "designer" formula had enhanced immune function, less intraabdominal sepsis, and significantly less multiple organ failure. Cerra and coworkers demonstrated improved immune function in septic ICU patients fed with an arginine, fish oil, and nucleotide product.[54] Kemen et al. also demonstrated enhanced immune responses in postoperative patients.[55] Daly and coworkers found reduced septic complications and hospital length of stay in critically ill cancer patients after major surgery given this formula compared to those patients given a standard enteral product.[56] In addition, a recent multicenter trial in critically ill patients by Bower and colleagues also demonstrated a decreased incidence of acquired infections and shortened hospital stay with the same enteral formula.[57]

Despite the optimistic results reported from these investigations, it is not advisable at this time to widely use these products. As with most disease-specific formulas, these tend to be more expensive. At present, a number of questions still need to be answered; for example, which patient populations should receive these enteral products? It is unlikely all patients or even all critically ill patients will benefit from them. Since glutamine, arginine, and nucleotides are conditionally essential, most patients would have adequate reserves of these nutrients. Therefore, it is plausible that only patients who have developed deficiencies would actually benefit from supplementation. This would narrow the indications to patients with severe malnutrition and wasting, prolonged interruption of oral nutritional intake, and those with severe critical illness. In addition, there are no human data that demonstrate improved outcome from critical illness with only

Table 17.5. Outcome-Based Trials in Humans Evaluating the Effects of Enteral Formulas Supplemented With Arginine, Nucleotides, and Omega-3 Fatty Acids

Study	Methodology	Population	Intervention[a]	Outcome[b]
Gottschlich et al[52] 1990	Randomized, blinded	Burn patients (n=50)	(1) Study formula (2) Osmolyte/ Promix (3) Traumacal	Wound infections reduced 2 vs 8 vs 8, $p<0.03$ LOS/% burn 0.83 vs 1.21 vs 1.21, $p<0.02$
Moore et al[53] 1994	Randomized, multicenter	Trauma patients (n=98)	(1) Immum-Aid (2) Vivonex TEN	Increased TLC 2.15 vs 1.73, $p=0.014$ Decreased intraabdominal sepsis and MOF 0% vs 11%, $p=0.023$
Cerra et al[54] 1991	Randomized, blinded	Trauma patients (n=20)	(1) Impact (2) Osmolyte HN	Improved immune function Outcome not evaluated
Kemen et al[55] 1995	Randomized, blinded	Surgical patients with GI cancers (n=42)	(1) Impact (2) Placebo	Improved immune function Outcome not evaluated
Daly et al[56] 1992	Randomized, complications evaluated blindly, retrospectively	Surgical patients with GI cancers (n=85)	(1) Impact (2) Osmolyte HN	Infections/wound complications 5 vs 21, p=0.02 LOS 15.8 vs 20.2, $p=0.01$ Improved immune function
Bower et al[57] 1995	Randomized, blinded, multicenter	Adult ICU patients (n=326)	(1) Impact (2) Osmolyte HN	Infections reduced in septic subgroups bacteremias 0 vs 7, $p<0.05$ LOS reduced in septic subgroup 18 vs 28, $p<0.05$

[a]Study formula is a specially-prepared modular mixture supplemented with arginine and fish oil. Osmolyte and Osmolyte HN are standard formulas (Ross Laboratories, Columbus, Ohio). Promix is a whey protein additive (ProMix R.D., Navaco Laboratories, Phoenix, Arizona). Traumacal is a standard formula used for trauma patients (Mead Johnson, Evansville, Indiana). Immun-Aid is a specialized formula supplemented with glutamine, arginine, nucleotides, fish oil, and branched chain amino acids (McGaw, Inc, Irvine California). Vivonex TEN is a standard elemental formula (Sandoz Nutrition, Minneapolis, Minn). Impact is a specialized formula supplemented with arginine, nucleotides, and fish oil (Sandoz Nutrition, Minneapolis, Minn).

[b]LOS, length of stay (days); TLC, total lymphocyte count (k/mm^3); MOF, multiple organ failure.

fish oil supplementation. Other issues to address include the duration of use for these products and whether they are effective in patients with organ dysfunction. Until further research better defines the role of specialized enteral products, their higher cost mandates very selective use.

FIBER AND SHORT-CHAIN FATTY ACIDS

Dietary fiber is plant-origin nonstarch polysaccharides that resist enzymatic breakdown in the small intestine and reaches the colon intact. Although there is growing support for including fiber in the diet, all parenteral solutions and most enteral formulas are fiber free. A number of beneficial effects have been attributed to fiber. Undigested fibers rich in cellulose and lignin (i.e., wheat bran) decrease stool transit time and increase stool bulk. This characteristic is thought to be advantageous for diseases such as diverticulitis and may be helpful for decreasing the likelihood for the development of colon cancer. Other types of fibers such as pectin are fermented by colonic anaerobic bacteria into short-chain fatty acids (SCFA). Other presumed benefits of dietary fiber include improved colonic sodium and water absorption, maintenance of the normal colonic bacteria, enhanced colonocyte proliferation, improved anastomic healing, delayed small intestinal glucose absorption, and a decrease in the incidence of tube feed diarrhea.[58,59]

The SCFAs are the bacterial breakdown products of dietary fibers, the most common of which are acetate, propionate, and butyrate.[59,60] Although SCFA exert most of their effects on the colon, they also seem to have significant impact on the mucosa of the small intestine. SCFA are the preferred respiratory fuel of the colonocyte. In fact, over 5% to 30% of the patient's daily energy expenditure could be satisfied through this route.[61] In patients with short bowel syndrome, the colonic absorption of these energy-rich compounds may be sufficient to enable enteral nutrition to fully meet total energy requirements.[59] Some SCFA are not metabolized by the colonocyte but instead are transported through the portal circulation to the liver. They are then metabolized into ketone bodies and glutamine, both of which are important for small bowel mucosal integrity.[59] Although the enterocyte can absorb SCFA as readily as the colonocyte, this rarely occurs, because their concentration is normally quite low in the small intestine.[60] The effects on the small bowel are therefore likely to be indirect. The addition of SCFA to parenteral solutions has been shown to be as effective in preventing gut mucosal atrophy as the addition of dietary fiber to an enteral diet, and in fact, the effects were more pronounced in the jejunum than the ileum.[62] The mechanisms are not well-understood. Possible explanations include increased intestinal blood flow, increased pancreatic secretions, stimulated release of insulin, and the conversion of SCFA by the liver into glutamine and ketone bodies.[62]

The addition of SCFA directly into parenteral solutions, or indirectly through the addition of dietary fiber to the enteral formulations may prevent or reverse the gut mucosal atrophy associated with TPN and may be useful in the management of the patient with massive small bowel resection.[63]

CONCLUSION

It is becoming increasingly clear that the gastrointestinal tract has a major role in protection from microbial invasion. In the setting of illness and luminal nutrient deprivation, this defense weakens and translocation becomes more likely to occur. In the critically ill, this may be the pivotal event in the development of multiple organ system failure. Although nutritional support has been very successful for preventing protein-calorie malnutrition, it appears that the provision of specific nutrients such as glutamine, arginine, nucleotides, fish oil, and SCFAs are indispensible for maintaining the integrity of the gut barrier by supporting the intestinal mucosa and/or stimulating the immune system.

At present, there is a body of literature that has reported on the physiologic benefits of supplementing standard parenteral and enteral formulations with one or more of these nutrients. However, most of these studies have been performed in animals and the results of the human research have not been conclusive. Although addition of these nutrients to standard formulas may offer many important theoretical benefits, further research is needed to better define their indications, and appropriate dosing before their use can justify their higher costs.

REFERENCES

1. Alverdy JC. Effects of glutamine-supplemented diets on immunology of the gut. *J Parent Ent Nutr* 1990;14(suppl);109S–113S.
2. Deitch EA. Multiple organ failure. Pathophysiology and potential future therapy. *Ann Surg* 1992;216;117–134.
3. Johnson LR, Copeland EM, Dudrick JT, et al. Structural and hormonal alterations in the gastrointestinal tract of parenterally fed rats. *Gastroenterology* 1975;68;1170–1183.
4. Carrico J, Meakin JL. Multiple organ failure syndrome. *Arch Surg* 1986;121;196–208.
5. O'Dwyer ST, Smith RJ, Hwang TL, et al. Maintenance of small bowel mucosa with glutamine-enriched parenteral nutrition. *J Parent Ent Nutr* 1989;13;579–585.
6. Grant JP, Snyder BA. Use of L-glutamine in total parenteral nutrition. *J Surg Res* 1988;44;506–513.

7. Wilmore DW, Smith RJ, O'Dwyer ST, et al. The gut: a central organ after surgical stress. *Surgery* 1988;104;917–923.
8. Souba WW, Herskowitz K, Salloum RM, et al. Gut glutamine metabolism. *J Parent Ent Nutr* 1990;14(suppl);45S–50S.
9. Windmueller HG, Spaeth AE. Respiratory fuels and nitrogen metabolism in vivo in small intestine of fed rats: quantitative importance of glutamine, glutamate and aspartate. *J Biol Chem* 1980;255;107–112.
10. Lacey JM, Wilmore DW. Is glutamine a conditionally essential amino acid? *Nutr Rev* 1990;48;297–309.
11. McAnena OJ, Moore FA, Moore EE, et al. Selective uptake of glutamine in the gastrointestinal tract: confirmation in a human study. *Br J Surg* 1991;78;480–482.
12. Souba WW, Wilmore DW. Postoperative alterations of arteriovenous exchange of amino acids across the gastrointestinal tract. *Surgery* 1983;94;342–350.
13. Souba WW, Smith RJ, Wilmore DW. Glutamine metabolism by the intestinal tract. *J Parent Ent Nutr* 1985;9;608–617.
14. Souba WW, Klimberg VS, Hautamaki RD, et al. Oral glutamine reduces bacterial translocation following abdominal radiation. *J Surg Res* 1990;48;1–5.
15. Khan K, Hardy G, McElroy B, et al. The stability of glutamine in total parenteral nutrition solutions. *Clin Nutr* 1991;10;193–198.
16. Burke DJ, Alverdy JC, Aoys E, et al. Glutamine-supplemented total parenteral nutrition improves gut immune function. *Arch Surg* 1989;124;1396–1398.
17. Scheltinga MR, Young LS, Benfell K, et al. Glutamine-enriched intravenous feedings attenuate extracellular fluid expansion after a standard stress. *Ann Surg* 1991;214;385–395.
18. Tremel H, Kienle B, Weilemann LS, et al. Glutamine dipeptide-supplemented parenteral nutrition maintains intestinal function in the critically ill. *Gastroenterology* 1994;107;1595–1601.
19. van der Hulst RRWJ, van Kreel BK, von Meyenfeldt MF, et al. Glutamine and the preservation of gut integrity. *Lancet* 1993;341;1363–1365.
20. Hornsby-Lewis L, Shike M, Brown P, et al. L-Glutamine supplementation in home total parenteral nutrition patients: stability, safety, and effects on intestinal absorption. *J Parent Ent Nutr* 1994;18;268–273.
21. Ogle CK, Ogle JD, Mao J-X, et al. Effect of glutamine on phagocytosis and bacterial killing by normal and pediatric burn patient neutrophils. *J Parent Ent Nutr* 1994;18;128–133.
22. O'Riordain MG, Fearon KCH, Ross JA, et al. Glutamine-supplemented total parenteral nutrition enhances T-lymphocyte response in surgical patients undergoing colorectal resection. *Ann Surg* 1994;220;212–221.
23. van der Hulst RRWJ, von Meyenfeldt MF, Tiebosch A, et al. The effect of glutamine enriched parenteral nutrition on intestinal mucosa immune cells. *Gastroenterology* 1994;106;A636 (abstract).
24. Fox AD, Kripke SA, De Paula J, et al. Effect of a glutamine-supplemented enteral diet on methotrexate-induced enterocolitis. *J Parent Ent Nutr* 1988;12;325–331.

25. Burke DJ, Alverdy JC, Aoys E, et al. Effect of route of glutamine administration on mortality following experimental enterocolitis. *J Parent Ent Nutr* 1990;14(suppl);8S (abstract).
26. Klimberg VS, Nwokedi E, Hutchins LF, et al. Glutamine facilitates chemotherapy while reducing toxicity. *J Parent Ent Nutr* 1992; 16(suppl);83S–87S.
27. Klimberg VS, Dolson DJ, Hautamaki RD, et al. Prophylactic glutamine protects the intestinal mucosa from radiation injury. *Cancer* 1990;66;62–68.
28. Wang X-D, Jacobs DO, O'Dwyer ST, et al. Glutamine-enriched parenteral nutrition prevents mucosal atrophy following massive bowel resection. *Surg Forum* 1988;39;44–46.
29. Scott TE, Moellman JR. Intravenous glutamine fails to improve gut morphology after radiation injury. *J Parent Ent Nutr* 1992;16;440–444.
30. Dudrick PS, Salloum RM, Copeland EM, et al. The early response of the jejunal brush border glutamine transporter to endotoxemia. *J Surg Res* 1992;52;372–377.
31. Wilmore DW. Glutamine and the gut. *Gastroenterology* 1994;107;1885–1886.
32. Siefter E, Rettura G, Barbul A, et al. Arginine: an essential amino acid for injured rats. *Surgery* 1978;84;224–230.
33. Barbul A, Lazarou SA, Efron G, et al. Arginine enhances wound healing and lymphocyte immune responses in humans. *Surgery* 1990;108;331–337.
34. Reynolds JV, Daly JM, Zhang S, et al. Immunomodulatory mechanisms of arginine. *Surgery* 1988;104;142–151.
35. Barbul A, Rettura G, Prior E, et al. Supplemental arginine, wound healing and thymus: arginine-pituitary interaction. *Surg Forum* 1978;29;93–95.
36. Hibbs JB, Vavrin Z, Taintor RR. L-Arginine is required for expression of the activated macrophage effector mechanism causing selective metabolic inhibition in target cells. *J Immunol* 1987;138;550–565.
37. Gianotti L, Alexander JW, Pyles T, et al. Arginine-supplemented diets improve survival in gut-derived sepsis and peritonitis by modulating bacterial clearance. The role of nitric oxide. *Ann Surg* 1993;217;644–654.
38. Barbul A, Sisto DA, Wasserkrug HL, et al. Arginine stimulates lymphocyte immune response in healthy human beings. *Surgery* 1981;90;244–251.
39. Daly JM, Reynolds J, Thom A, et al. Immune and metabolic effects of arginine in the surgical patient. *Ann Surg* 1988;208;512–523.
40. Saito H, Trocki O, Wang S, et al. Metabolic and immune effects of dietary arginine supplementation after burn. *Arch Surg* 1987;122;784–789.
41. Gonce SJ, Peck MD, Alexander JW, et al. Arginine supplementation and its effect on established peritonitis in guinea pigs. *J Parent Ent Nutr* 1990;14;237–244.
42. Madden HP, Brelin RJ, Wasserkrug HL, et al. Stimulation of T cell immunity by arginine enhances survival in peritonitis. *J Surg Res* 1988;44;658–663.
43. Rudolph FB, Kulkarni AD, Fanslow WC, et al. Role of RNA as a dietary source of pyrimidines and purines in immune function. *Nutrition* 1990;6(suppl);45–52.

44. Kulkarni AD, Fanslow WC, Rudolph FB, et al. Effect of dietary nucleotides on bacterial infections. *J Parent Ent Nutr* 1986;10;169–171.
45. Fanslow WC, Kulkarni AD, Van Buren CT, et al. Effect of nucleotide restriction and supplementation on resistance to experimental murine candidiasis. *J Parent Ent Nutr* 1988;12;49–52.
46. Tsujinaka T, Iijima S, Kido Y, et al. Role of nucleosides and nucleotide mixture in intestinal mucosal growth under total parenteral nutrition. *Nutrition* 1993;9;532–535.
47. Wan JM-F, Teo TC, Babayan VK, et al. Invited comment. Lipids and the development of immune dysfunction and infection. *J Parent Ent Nutr* 1988;12(suppl);43S–52S.
48. Alexander JW, Saito H, Trocki O, et al. The importance of lipid type in the diet after burn injury. *Ann Surg* 1986;204;1–8.
49. Peck MD, Ogle CK, Alexander JW. Composition of fat in enteral diets can influence outcome in experimental peritonitis. *Ann Surg* 1991;214;74–82.
50. Mascioli EA, Leader L, Flores E, et al. Enhanced survival to endotoxin in guinea pigs fed IV fish oil emulsion. *Lipids* 1988;23;623–625.
51. Van Buren CT, Rudolph FB, Kulkarni AD, et al. Reversal of immunosuppression induced by a protein-free diet: comparison of nucleotides, fish oil, and arginine. *Crit Care Med* 1990;18(suppl):S114–S117.
52. Gottschlich MM, Jenkins, Warden GD, et al. Differential effects of three enteral dietary regimens on selected outcome variables in burn patients. *J Parent Ent Nutr* 1990;14;225–236.
53. Moore FA, Moore EE, Kudsk KA, et al. Clinical benefits of an immune-enhancing diet for early postinjury enteral feeding. *J Trauma* 1994;37;607–615.
54. Cerra FB, Lehmann S, Konstantinides N, et al. Improvement in immune function in ICU patients by enteral nutrition supplemented with arginine, RNA, and menhaden oil is independent of nitrogen balance. *Nutrition* 1991;7;193–199.
55. Kemen M, Senkal M, Homann HH, et al. Early postoperative enteral nutrition with arginine-omega-3 fatty acids and ribonucleic acid-supplemented diet versus placebo in cancer patients: an immunologic evaluation of Impact®. *Crit Care Med* 1995;23;652–659.
56. Daly JM, Lieberman MD, Goldfine J, et al. Enteral nutrition with supplemental argnine, RNA, and omega-3 fatty acids in patients after operation: Immunologic, metabolic, and clinical outcome. *Surgery* 1992;112;56–67.
57. Bower RH, Cerra FB, Bershadsky B, et al. Early enteral administration of a formula (Impact®) supplemented with arginine, nucleotides, and fish oil in intensive care unit patients: results of a multicenter, prospective, randomized, clinical trial. *Crit Care Med* 1995;23;436–449.
58. Scheppach W, Burghardt W, Bartram P, et al. Addition of dietary fiber to liquid formula diets: the pros and cons. *J Parent Ent Nutr* 1990;14;204–209.
59. Settle RG. Invited comment: Short-chain fatty acids and their potential role in nutritional support. *J Parent Ent Nutr* 1988;12(suppl);104S–107S.
60. Rombeau JL, Kripke SA. Metabolic and intestinal effects of short-chain fatty acids. *J Parent Ent Nutr* 14(suppl);181S–185S.

61. Ruppin H, Bar-Meir S, Soergel KH, et al. Absorption of short chain fatty acids by the colon. *Gastroenterology* 1980;78;1500–1507.
62. Koruda MJ, Rolandelli RH, Bliss DZ, et al. Parenteral nutrition supplemented with short-chain fatty acids: effect on the small-bowel mucosa in normal rats. *Am J Clin Nutr* 1990;51;685–689.
63. Koruda MJ, Rolandelli RH, Settle RG, et al. The effect of parenteral nutrition supplemented with short chain fatty acids on adaption to massive small bowel resection. *Gastroenterology* 1988;95;715–720.

CHAPTER

18 Prevention and Management of Enteral Feeding Complications

Peter G. Janu, M.D.,
Gayle Minard, M.D., and
Kenneth A. Kudsk, M.D.

INTRODUCTION

The goals of specialized nutritional support are to preserve host defenses, replete lean body mass, supplement nutrient deficiencies, and most important, improve clinical outcome. Over the last 15 years, evidence demonstrating the benefits of nutrition to the high-risk patient has greatly increased. Although parenteral nutrition is indicated in a variety of clinical conditions, such as short-gut syndrome, enterocutaneous fistula, inflammatory bowel disease, radiation or chemotherapy-induced enteritis,[1] cost, safety, and simplicity favor the enteral route.[2] In addition, enteral nutrition provides a variety of physiologic benefits to the critically ill patient and decreases septic morbidity.[3–7] Although enteral nutrition provides many benefits, complications can occur and can be grouped into four major categories: infectious, gastrointestinal, mechanical, and metabolic. Close attention to detail and risk management can substantially limit them, however. This chapter serves to discuss each of these complications, with particular attention to preventive strategies and suggestions for management.

INFECTIOUS COMPLICATIONS

The use of enteral compared with parenteral feeding has been shown to decrease infectious morbidity in several patient populations.[3–10] However, a few infectious complications, including aspiration, gastric colonization, and contamination of feedings, are specifically associated with enteral feeding.

Aspiration

Aspiration pneumonia is one of the most serious complications of enteral feeding and is reported to occur in 1% to 70% of enterally fed patients. The wide range in incidence is primarily due to the absence of a consensus definition, inadequate study designs, and inadequate follow-up.[11–17] The diagnostic difficulty is exemplified by a study by Pellegrini et al. in which the incidence of aspiration based on historical criteria such as nocturnal cough, recurrent pneumonia, wheezing spells, or morning hoarseness was 48%, but the incidence using objective documentation such as decreases in esophageal pH followed by an acid taste in the mouth and respiratory symptoms was only 8%.[18] Addition of blue food coloring to the formula or testing of pulmonary aspirate for glucose may aid in diagnosis but may not be reliable.[11,14] Depending on the volume and the extent of pulmonary involvement, the associated mortality can be significant, ranging from 41% to 100%.[12–14]

The physiologic response to aspiration is related to the quantity. A small volume of aspirate may initially go unnoticed, as shown by a study of healthy volunteers in which 45% of subjects aspirated during sleep without clinical sequelae.[17] If the acid aspirate exceeds 0.3 to 0.4 mL/kg body weight,[12] however, injury to the pulmonary capillary bed will occur with resultant exudation of protein-rich extracellular fluid. This is manifested clinically by reduced ventilation and shunting from destruction of surfactant and accumulation of edema fluid. Immediately after aspiration, chest X-rays usually show no abnormalities, and the full radiographic extent of pulmonary injury may not be evident until 24 hr after the event.[11,12] If massive volumes are aspirated, however, pulmonary edema may develop within minutes.

The initial step in preventing aspiration is identification of those patients at risk. Patients who are paralyzed and mechanically ventilated, those with large bore tubes in place, and those who are conscious but have diminished gag, cough, or swallowing reflexes are at increased risk,[11–13] as well as those of advanced age and without intensive nursing care.[15] Aspiration can occur despite an endotracheal tube or tracheostomy tube with an inflated cuff.[12–19] Brain-injured patients bear special mention, since dysphagia is found in 25% to 50%.[13] In addition, they frequently have delayed gastric emptying and prolonged ileus (up to 2 weeks), which increases aspiration risk.[14,6] Delayed gastric emptying is also seen in patients with intraabdominal pathology such as peritonitis, pancreatitis, or retroperitoneal hematoma; medication use such as anticholinergics or narcotics; abdominal trauma; diabetic neuropathy; myocardial infarction; hepatic coma; hypercalcemia; myxedema; and malnutrition.[14,20]

The site of enteral access may influence the chance of aspiration, although comparisons of the risks of intragastric or jejunal feeding shows conflicting results in the literature.[12–15,21,22] Gastric access is the easiest to obtain, but may be associated with a higher incidence of aspiration. Unfortunately, nasogastric or nasointestinal access is an unacceptable long-term option because of patient discomfort, frequent tube dislodgement, and the risk of sinusitis.[11,23] The small bowel is the route of choice when surgical exploration is already indicated for patients with documented recurrent aspiration or for patients with gastric dysmotility. Access can be obtained via small-bore nasoenteric tubes or with surgical, laparoscopic, or endoscopic jejunostomies.[24] Theoretically, feeding distal to the stomach is safer because the pylorus, gastroesophageal, and cricopharyngeal sphincters serve as protective gateways. Animal data supports the duodenal route as optimal for nutrient absorption and reducing gastroesophageal reflux.[25] Aspiration from jejunal tube feedings does occur, however.[13,14,16,22]

Careful attention to tube position, patient position, and gastric residuals are the cornerstones of complication prevention. A variety of methods are available to verify feeding tube position, although each has its limitations.[14] Checking gastric tube aspirates for pH is inexpensive and easy. An acid pH implies gastric placement, whereas an alkaline pH implies misplacement into the pulmonary tree. This method is not foolproof, because conditions such as pulmonary malignancy or empyema may cause an acidic pH, and the use of antacids or histamine antagonists can elevate gastric pH. Alkaline aspirates will also be obtained if the feeding tube migrates into the duodenum. The color and consistency of the aspirate is frequently used to confirm tube placement; however, straw-colored pleural fluid may be mistaken for gastric fluid. Patients should be observed for respiratory difficulty, which occurs with tracheal tube placement. Some patients may be asymptomatic or even able to talk with a soft small-bore tube in the bronchial tree. Auscultation over the stomach for air infused through the tube can be very misleading, even for the experienced clinician. Bubbling may be seen from the end of the feeding tube if it has been positioned in the lungs, but if the tube is wedged in a bronchiole, this may not occur. Although radiography is the most reliable method of determining tube location, it may require multiple exposures and is expensive. The most cost-effective method to confirm tube position is to combine pH monitoring with auscultation, aspiration, and observation for respiratory symptoms while maintaining a high index of suspicion for malposition. The exit site should be marked on the tube and monitored to help detect changes in tube position.

Both scientific evidence and common sense suggest that elevation of the head of the bed can reduce the chance of regurgitation and aspiration, since it is more difficult for fluid to travel against gravity. Although elevation may not prevent aspiration, it may reduce its frequency and severity.[14,26] Patients receiving enteral nutrition should have the head of the bed elevated 30° to 45°, and the amount of time spent in the supine position should be kept at a minimum.

Monitoring of gastric residuals is an important method to detect feeding intolerance. Increased intragastric volume may lead to reduced lower esophageal sphincter tone and thus increased risk for aspiration.[27] Recent work with stable patients, healthy volunteers, and critically ill patients has led to the recommendation that 200-mL residuals in patients with nasogastric tubes and 100-mL residuals in patients with gastrostomy tubes should raise concerns for impending intolerance.[28] For a single high residual, feedings should be held for a short period of time and resumed after the residual volumes decrease to acceptable limits. Persistently high residuals require reduction or cessation of gastric feeds because of the high aspiration risk. Elemental diets and formulas with high osmolality or high carbohydrate content can delay gastric emptying and should be monitored closely.[11,29] Monitoring should be more frequent in the early phases of feeding than after tolerance has been established. Aspirates from residual volume checks should be reinfused because doing so allows better evaluation of transit as well as replaces electrolytes, medications, and nutrients.

Management of aspiration pneumonia involves elimination of the source and removal of as much aspirate as possible from the bronchopulmonary tree. This can be accomplished with nasogastric decompression, along with aggressive tracheal suctioning or bronchoscopy. Even if the volume of tracheal aspirate is minimal, the induced coughing promotes expulsion of small particulate matter. Controversy surrounds the use of bronchial lavage, as evidence suggests that lavage may disseminate the aspirate into the distal pulmonary tree, thus enlarging the affected area. Unless aspiration involves solid matter, tracheal suctioning is felt to be as efficacious as bronchoscopy.[12] Many animal trials have shown benefits of steroid administration in the management of aspiration, but unless given before or within minutes of the actual event, steroids provide little or no benefit and may in fact be harmful.[11,12,30,31] Ventilator support with positive end-expiratory pressure (PEEP) may be required in severe cases.[12] Finally, treatment of any subsequent infection following aspiration should be tailored to the cultured organisms and should include coverage of anaerobes.[32]

Gastric Colonization

Another infectious complication of enteral nutrition is gastric colonization. The pH of the normally acidic stomach increases with the institution of tube feedings,which have a pH range between 6.4 and 7.0. This increases the risk of bacterial colonization, particularly with gram-negative organisms.[19] The common use of histamine antagonists or antacids adds to the neutralization of gastric acid. The consequence is tracheal colonization with gastric organisms, presumably due to aspiration or reflux mechanisms, and possible progression to pneumonia.[19,33,34] Use of sucralfate in place of histamine antagonists or antacids, as well as intermittent versus continuous feeds, have been suggested as having clinical advantages.[33] Also, acidification of the formula can reduce bacterial colony counts.[35] Whether these maneuvers can reduce pneumonia rates and improve patient outcome is still unclear.

Bacterial Contamination

Bacterial contamination of enteral formulas is an unusual but preventable complication of enteral feeding.[11,35–38] Commercially available products are sterile and can be given safely. Jejunal feeding without either a pump or filter has been shown to become contaminated via an ascending route, particularly with large-bore tubes.[36] Exogenous bacteria can contaminate any kitchen-prepared diet, despite close attention to clean technique, so this practice is discouraged. If the commercially available product is handled aseptically, using a closed-system infusion with a pump or filter, sterility can be maintained. Using this technique, infusion times up to 24 hrs are safe.[11,35]

GASTROINTESTINAL COMPLICATIONS

Gastrointestinal complications are relatively frequent and may indicate some degree of feeding intolerance. They encompass such conditions as diarrhea, nausea, abdominal distention, elevated gastric residuals, and, less commonly, pneumatosis intestinalis and bowel infarction.

Diarrhea

The incidence of diarrhea in enterally fed patients varies widely; this can be based partially on the difficulty in its documentation and definition.[11,39–41] From a physiologic standpoint, it is defined as passage of more than 200 g of stool/day.[11] Practically, an adequate clinical definition is three or more loose or liquid stools per day.[39] The etiology can include osmolality of the formula, rate of administration, hypoalbuminemia, or medications. It is generally agreed that diarrhea due to enteral feeding is osmotic in nature, rather than secretory.[11,41,42] To objectively determine this, one can calculate the stool osmotic gap: stool osmotic gap = (stool osmolality) – 2 × (conc. stool sodium + conc. stool potassium).[41,42] Osmotic diarrhea will have a value greater than 140 mmol/L, whereas a secretory diarrhea will have a low or negative gap. The formula is usually not the source of the diarrhea, however, because most standard formulas have osmolalities in the range of 300 to 500 mOsm/L. Osmolality up to 700 mOsm/L has little effect on tolerance and diarrhea when fed gastrically,[43] so there is no need for diluted starter regimens. In addition, the use of diluted feedings also causes inadequate nutrient delivery.[11,43] On the other hand, jejunal feeds may need to be diluted initially, although this is unusual.

The rate of formula administration can contribute to diarrhea, particularly in patients with dysmotility or malabsorption syndromes. When feeding postpylorus, the natural regulatory mechanism is bypassed. Decreasing the rate to half of the desired goal and slower advancement to goal (e.g., increasing the rate by 25 mL/ hr $\cdot$ day^{-1}) often results in reduced diarrhea and successful delivery. In unusual cases, diluting the formula may be helpful.

Hypoalbuminemia may be a contributory factor to diarrhea,[40,44] but is probably not clinically significant unless serum albumin is less than 1.5 g/dL. Administration of albumin to improve tolerance is usually not justified, unless doing so will prevent switching to TPN.

Concurrent pharmacotherapy is the major etiology of diarrhea associated with tube feeding. In a study of tube-fed patients with diarrhea, over 75% of patients had diarrhea attributed to sorbitol and/or magnesium or antibiotic-induced pseudomembranous colitis.[42] Large amounts of sorbitol are found in aminophylline solution or theophylline elixir, but it can also be found in acetaminophen elixir, cough preparations, codeine, cimetidine, isoniazid, lithium and vitamins. Prokinetic drugs, such as metoclopramide, can also cause diarrhea. Concomitant antibiotics are significantly related to diarrhea; in a study of hospitalized patients receiving enteral nutrition, over 50% of patients with diarrhea had positive assays for *Clostridium difficile*.[40]

Management of diarrhea requires identification of the etiology. The initial steps should include a careful review of the medication profile, diet and rate of administration, along with a rectal examination to rule out fecal impaction or blood. If the cause is not apparent or it does not resolve with simple measures, such as changing the rate of administration, stool osmotic gap should be determined. If the diarrhea is osmotic in nature, further review of medications is warranted. If it appears to be secretory, further investigation including stool cultures should be pursued. Most formulas are lactose free, so lactose intolerance is an unlikely cause. Continuous infusions can help minimize intolerance even with proven malabsorption.[11,41] A change in formula may be necessary to reduce osmolality, caloric density, nutrient content, or residue content. Addition of fiber may lead to symptomatic improvement.[40] Fiber is reduced to volatile fatty acids in the colon via bacterial fermentation, which promote sodium and water absorption.[40,41] This process may be disrupted by altered colonic flora resulting from antibiotic use; in these cases addition of fiber may have variable results. These patients may benefit from this addition of *Lactobacillus acidophilus*, which can help restore normal gastrointestinal flora.[41] Use of antidiarrheal medications may be beneficial, but caution should be used in face of an infectious etiology because they may prolong the clinical course.[1,41] Opiates, such as codeine or paregoric, may also be helpful. In some patients, for example, those with severe pseudomembranous colitis, active inflammatory bowel disease, or acquired immunodeficiency syndrome (AIDS) enteropathy, conversion to parenteral nutrition may be indicated.

Nausea

Nausea and vomiting occur in 10% to 20% of patients receiving enteral nutrition.[11] A variety of factors including medications, infections, and abdominal surgery can be implicated, but particular attention should be paid to nutrient administration rate and osmolality. In a study of healthy volunteers, tolerance to bolus feeding was maximized at a rate of 30 mL/min, independent of the volume administered; however, gastric motility was decreased with higher volumes.[45] Intragastric feeding is dependent on adequate gastric emptying, so any disorder that delays gastric transit time may result in nausea.[20] High osmolality has also been shown to decrease gastric emptying,[29] but this is not much of a problem with most currently used formulas. Treatment is directed toward initial slow, continuous delivery of formula with gradual advancement of the rate. Transpyloric or jejunal access may be required for refractory patients.

Abdominal Bloating

Abdominal bloating is a common problem that is rarely well documented. It occurs as a result of a delay in gastric emptying and other forms of dysmotility. In a study of trauma patients, 8% of enterally fed patients complained of severe distention and required alteration of their feeding schedule, and several required conversion to parenteral nutrition.[46] However, a similar incidence of bloating was found in the patients fed parenterally, implying that the disease process itself, and not the feeding, was to blame. Again, as with the treatment of nausea, slow continuous delivery of formula with gradual advancement of rate may be preventative. Manipulation of the feeding schedule, particularly the rate, may alleviate symptoms. Early ambulation can help promote peristalsis. Prokinetic agents may also be of some benefit.

Elevated Gastric Residuals

Delayed gastric emptying is common in patients with head trauma, sepsis, multiple trauma, and a variety of other conditions, as mentioned previously.[6,14,20] A strict protocol of checking residuals regularly, for example every 4 hr, is recommended for those receiving intragastric nutrition. Most reviews cite gastric residuals greater than 100 or 200 mL as excessive.[11,14,28] Sudden intolerance in a patient who has previously tolerated enteral feedings is suggestive of an occult process, such as sepsis or a pyloric channel ulcer. If a high residual is obtained, feedings are held for a short period of time and resumed if subsequent values are within acceptable limits. If residuals are persistently elevated, prokinetic agents, such as metoclopramide, cisapride, or low-dose erythromycin, may be used to increase gastric emptying. On a cautionary note, residual volume alone can be unreliable. Tubes may be coiled, kinked, or malpositioned. Excessive back pressure can cause the tubing to collapse and give false low levels. Volumes are best checked with a 60-mL syringe from tubes 10-gauge French or greater.[28] Patients receiving jejunostomy feedings do not need residuals checked.

Pneumatosis Intestinalis

One of the most devastating complications associated with enteral feeding is pneumatosis intestinalis, which can lead to bowel necrosis or infarction. The incidence is as high as 5% in some series[47] and is associated with a significant mortality rate. The pathophysiologic mechanism is unknown, but theories include defective carbohydrate metabolism, bacterial contamination of the bowel wall with gas-forming organisms, disruption in mucosal integrity, and distention pressure from obstruction.[47–51] Pneumatosis arises primarily in patients who are critically ill and hemodynamically unstable, suggesting a mesenteric low-flow state as a possible etiology. Hyperosmolar feeds may increase the risk of pneumatosis by promoting increased blood flow to an already taxed blood supply in the hemodynamically compromised patient. This is illustrated by one case report in which a patient who underwent radical cystectomy along with placement of a needle catheter jejunostomy had necrosis of the entire small bowel except for the ileal conduit.[52]

Most patients evidence abdominal distention prior to developing pneumatosis.[47,48,52–54] Despite this, clinical findings may not be a prerequisite. A retrospective review of trauma patients receiving needle-catheter jejunostomy feedings identified seven patients with radiographic evidence of pneumatosis who exhibited no symptoms or signs of intolerance.[48] Careful physical examination and discontinuation of enteral feedings during periods of hemodynamic instability are key steps in prevention of pneumatosis. A high index of suspicion needs to be maintained for any jejunostomy-fed patient exhibiting signs or symptoms of intolerance. If pneumatosis is identified early, cessation of feedings can lead to prompt resolution. The benefits of antibiotics are unproven but recommended. Anecdotal evidence suggests resolution can occur with the jejunostomy tube intact,[54] and asymptomatic patients without metabolic acidosis may be safely observed.[50] Surgical intervention is indicated in life-threatening cases and has a high mortality rate, but resection of any necrotic intestine is mandatory. Use of high-flow oxygen (70% to 100% FiO_2), hyperbaric oxygen, or use of pure elemental diet may be of benefit.[49]

MECHANICAL COMPLICATIONS

Mechanical complications associated with enteral nutrition are related to the placement of the tube, dislodgement once placed, and obstruction during use of the feeding tube.

Tube Placement Complications

Complications associated with tube placement are specific to the route of administration and the techniques involved in obtaining access. Route is determined by the clinical situation and length of time that enteral support is required, function of the gastrointestinal tract, and availability of access techniques. Nasogastric feeding is best for the patient requiring short-term support who has a fully functional gastrointestinal tract and intact regurgitation reflexes, based on cost and ease of placement. Complications specific to nasoenteral tubes occur particularly with large bore tubes, including esophageal erosion, ruptured esophageal varices, nasal erosion, sinusitis, and tracheoesophageal fistula.[11] More pliable small-bore nasogastric tubes are better tolerated by patients but are more difficult to insert and more easily dislodged. Placement complications are related to blind passage of the tube and include transesophageal and transbronchial intrapleural placement, pneumothorax, hydrothorax, nasopharyngeal perforation, gastric perforation, and even intracranial placement in patients with facial trauma.[55] Several methods to aid in placement have been developed, including pH probes and capnometers,[14,56] but have not met with overwhelming acceptance.

For longer term intragastric access, a gastrostomy tube can be inserted at laparotomy, laparoscopically, under fluoroscopic guidance, or percutaneously with an endoscope. Benefits include improved cosmesis and patient comfort, less risk of dislodgement, and easier replacement. Disadvantages include infection, intraabdominal leakage, bleeding, and fistula formation. Laparoscopic placement theoretically reduces postoperative morbidity by eliminating a laparotomy incision. Fluoroscopic placement also eliminates an incision, but intraabdominal or intragastric pathology cannot be assessed and perforation of other visceral organs has occurred.[24] Percutaneous endoscopic gastrostomy (PEG) tube placement has become more widespread, but as with fluoroscopic placement, perforation of visceral organs has also occurred. As operator experience increases, complications can be reduced.

For patients with delayed gastric emptying, tube placement beyond the pylorus can facilitate enteral delivery. Nasoduodenal or nasojejunal tubes can be associated with similar complications as nasogastric tubes, although direct endoscopic placement distal to the pylorus may limit complications associated with blind passage. The literature supports intraoperative placement of jejunostomy tubes during laparotomy as being safe and having a positive effect on outcome.[3,4,6–8,10] The benefit of jejunal tubes compared with gastric tubes with regard to aspiration, are debated.[13,16,21,22,57] Nonoperative jejunal access can be accomplished via a variety of endoscopic or radiologic techniques, or placement can be achieved laparoscopically. The risks and benefits of the particular techniques are similar to those mentioned for gastric access.[24] One rare complication particular to jejunal access is pneumatosis intestinalis as discussed earlier.

Dislodgement

The most common complication of all access routes is inadvertent dislodgement, and its incidence ranges from 5% to 70%.[11,16,58] If it is a nasoenteric tube, it may be viewed as a minor inconvenience; however, dislodgement results in cessation of feeding, patient discomfort during reinsertion, and the risk of errant placement. The reason for almost 90% of all incidents of dislodgement is accidental or purposeful patient manipulation.[14,16] Patients with retching, vomiting, or coughing are at high risk for inadvertent spontaneous migration of the tube. For recurrent dislodgement, use of a "bridle" is recommended. This involves placement of a 5-gauge French polyvinylchloride tube loosely around the nasal septum to which the feeding tube is secured.[58,59] Refractory patients should be considered for other access routes. If a new gastrostomy or jejunostomy tube is displaced, it may result in leaking of intestinal contents and intraabdominal sepsis; therefore, dislodgement needs to be addressed immediately. Replacement of tubes with well-formed tracts should also be done immediately, particularly jejunostomy tubes, because the tracts close rapidly. If there is any question of correct placement, radiologic studies should be obtained.

Tube Obstruction

Small-bore feeding tubes can become clogged if certain viscous medications are administered via the tube or if feedings are stopped for several hours. It is prudent to flush the tube with a small amount of water (e.g., 30 mL) prior to and following administration of each medicine or bolus feed. Certain medications, such as carbamazepine, have a propensity for adherence to tubing and increase the risk of obstruction.[60] Reinsertion of a guide wire is generally unsuccessful, can lead to perforation of the tube and/or intestine, and is not recommended. Many declogging agents have been recommended, including carbonated beverages, cranberry juice, and a slurry of pancreatic enzyme, sodium bicarbonate, and water. The best prevention of clogging is frequent flushing and using the small-bore tubes for nothing other than formula.

METABOLIC COMPLICATIONS

Although it may be difficult to meet patient nutritional needs via the gastrointestinal tract, the metabolic and physiologic benefits compared with those from parenteral nutrition are well documented.[61–64] Enteral administration is more labor intensive than parenteral and has a higher likelihood of interruptions from gastrointestinal intolerance, medical or surgical procedures, and mechanical problems with feeding tubes. A study of hospitalized adults receiving nasoenteric feeding found that patients received an average of only 61% of their goal rate for many of the above reasons,[65] and current formulas can compound the problem in the pediatric population.[66] However, many of the benefits of enteral nutrition can still be provided without achieving goal rates.[4] Nevertheless, strict attention to electrolytes, nitrogen balance, and visceral proteins, such as prealbumin, is mandatory for safe and effective enteral nutrition.

Electrolyte abnormalities are the most frequently encountered metabolic complications associated with enteral feedings and arise predominantly from poor choice of formula, misguided administration, and inadequate monitoring. The incidence and severity vary, depending on the condition of the patient. Metabolic abnormalities are more frequent in patients with critical illnesses and organ dysfunction.

Hypokalemia

Hypokalemia is the most commonly encountered electrolyte abnormality.[11] Metabolic alkalosis, administration of dextrose, and various medications can cause an intracellular shift of potassium or cause wasting at the level of the renal tubule. It also is seen as a part of the "refeeding" syndrome in severely malnourished patients.[67] Many formulas have insufficient potassium to meet patient needs; therefore, supplementation may be required. Potassium chloride elixir is frequently used but may cause diarrhea, presumably from the high osmotic load, and is incompatible with some formulas.[60] Injectable potassium salts can be given either intravenously or diluted in 500–1000 mL of enteral formula. Chloride and acetate salts, at dosages of 20 mEq/L, or phosphate salts, at dosages of 15 to 22.5 mM/L, can be used effectively.

Hyponatremia

Hyponatremia tends not to occur as a result of enteral feeding, but rather as a result of the patient's clinical condition. Evaluation of the patient's fluid balance is likely to reveal the etiology. Many intravenous medications are administered in dextrose and water and, depending on the volumes of medications received, can have a dramatic effect on sodium balance. In many cases, changing the medication solution from dextrose and water to saline will correct the problem.

Alterations in formula composition are uncommonly required to treat hyponatremia, except for select clinical situations such as renal failure. Patients with head injuries may develop inappropriate antidiuretic hormone secretion (SIADH). Calorically dense formulas are helpful in meeting the fluid restriction requirements of these patients, as well as those patients with other conditions needing fluid and salt restriction.

Hyperglycemia

Hyperglycemia can occur from high carbohydrate formulas, but less commonly than with parenteral administration.[61] It tends to occur in patients with insulin resistance related to their illness or undiagnosed diabetes.[11] Formulas with higher fat and lower carbohydrate content can help combat this problem. Insulin may be administered subcutaneously or by intravenous infusion if necessary. Hypoglycemia can occur if enteral feedings are abruptly discontinued. This can usually be avoided with maintenance intravenous dextrose while the feeds are stopped.

Hypophosphatemia

Hypophosphatemia can be a dangerous metabolic complication. Periods of inadequate nutrition lead to total body depletion of phosphorus, but serum levels remain normal. Upon repletion of carbohydrate, along with its concomitant rise in insulin, phosphorus is shifted intracellularly leading to profound extracellular hypophosphatemia. The consequences are cardiac or neuromuscular dysfunction, pulmonary insufficiency, reduced oxygen delivery to tissue, and depressed leukocyte function. Patients at greatest risk are those with alcoholism, chronic weight loss, hyperglycemia, exogenous insulin requirements, or those receiving chronic antacid or diuretic therapy.[67] Standard formulas fail to meet the phosphorus requirements for patients with a high metabolic demand.[68] Intravenous sodium or potassium phosphate at dosages between 15 mM/L and 30 mM/L can be used to treat patients with moderate to severe hypophosphatemia. Less severe cases should have phosphorus added to the enteral formula. Addition of 5 to 10 mL of Fleets™ Phosphosoda (C. B. Fleet Co., Lynchburg, VA) per liter of feeding is usually an adequate supplement. Higher doses may cause diarrhea and are not recommended. Alternatively, injectable sodium or potassium salts at doses of 15 mM/L to 22.5 mM/L can be used.

Medication Interactions

The administration and metabolism of various medications can be affected by concomitant enteral feeding. If pills cannot be swallowed, tablets must be crushed or capsules opened to give some medications. This can lead to mechanical problems such as obstructed tubes if they are not flushed properly. For products in sustained-release form, this practice is hazardous and should be avoided; therefore, non-sustained-release products should be substituted. The interaction of enteral feeding with phenytoin causing inadequate drug levels are well documented.[60,69,70] To avoid this problem, feedings should be withheld for 1 to 2 hr prior to phenytoin administration and resumed 1 to 2 hours afterward. Warfarin resistance may be encountered if the enteral formula contains supplemental vitamin K, so coagulation studies should be followed closely. In some situations, a formula change will be required. Although all drugs may clog feeding tubes, carbamazepine needs to be administered in a very dilute form because of adherence to the tubing,[60] causing inadequate delivery and potential mechanical obstruction.

CONCLUSION

Enteral delivery of nutrients provides many advantages to patients compared with the parenteral route. Benefits include decreased cost, increased safety, and improvement in patient outcome. Unfortunately, enteral nutrition is associated with several complications including infectious, gastrointestinal, mechanical, and metabolic problems. With close attention to detail and patient management, complications can be minimized.

REFERENCES

1. Sax HC, Souba WW. Enteral and parenteral feedings: guidelines and recommendations. *Clin Nutr* 1993; 77(4):863–880.
2. ASPEN Board of Directors. Guidelines for the use of parenteral and enteral nutrition in adult and pediatric patients. *J Parent Ent Nutr* 1993; 17(4):18A–265A.
3. Moore FA, Feliciano DV, Andrassy RJ, et al. Early enteral feeding, compared with parenteral, reduces postoperative septic complications. *Ann Surg* 1992; 216(2):172–183.
4. Kudsk KA, Croce MA, Fabian TC, et al. Enteral versus parenteral feeding: effects on septic morbidity after blunt and penetrating abdominal trauma. *Ann Surg* 1992; 215(5):165–173.
5. Dent D, Kudsk KA, Minard G, et al. Risk of abdominal septic complications after feeding jejunostomy placement in patients undergoing splenectomy for trauma. *Am J Surg* 1993; 166:686–691.
6. Grahm TW, Zadrozny DB, Harrington T. The benefits of early jejunal hyperalimentation in the head-injured patient. *Neurosurgery* 1989; 25(5):729–735.

7. Alexander JW, MacMillan BG, Stinnett JD, et al. Beneficial effects of aggressive protein feeding in severely burned children. *Ann Surg* 1980; 29:916–923.
8. Moore EE, Jones TN. Benefits of immediate jejunostomy feeding after major abdominal trauma—a prospective, randomized study. *J Trauma* 1986; 26(10):874–881.
9. Veterans Affairs TPN Co-op Study Group. Perioperative total parenteral nutrition in surgical patients. *N Engl J Med* 1991; 325(8):525–532.
10. Moore EE, Jones TN. Benefits of immediate jejunostomy feeding after major abdominal trauma—a prospective randomized study. *J Trauma* 1986; 26(10):874–880.
11. Silk DB, Payne-James JJ. 1990. Complications of enteral nutrition. In Rombeau JL, Caldwell MD, eds. *Clinical Nutrition: Enteral and Tube Feeding*, 2nd ed., Philadelphia: W. B. Saunders, 1990:510–531.
12. Broe PJ, Toung TJK, Cameron JL. Aspiration pneumonia. *Surg Clin North Am* 1990; 60(6):1551–1564.
13. Lazurus B, Murphy JB, Culpepper L. Aspiration associated with long-term gastric versus jejunal feeding: a critical analysis of the literature. *Arch Phys Med Rehab* 1990; 71:46–53.
14. Metheny N. Nasogastric tube feedings: minimizing respiratory complications of nasoenteric tube feedings: state of the science. *Heart Lung* 1993; 22(2):13–23.
15. Mullan H, Roubenoff RA, Roubenoff R. Risk of pulmonary aspiration among patients receiving enteral nutrition support. *J Parent Ent Nutr* 1992; 16(2): 160–164.
16. Cogen R, Weinryb J, Pomerantz C, et al. Complications of jejunostomy tube feedings in nursing facility patients. *Am J Gastroenterol* 1991; 86(11):1610–1613.
17. Huxley EJ, Viroslav J, Gray WR, et al. Pharyngeal aspiration in normal adults and patients with depressed consciousness *Am J Med* 1978; 64:564–568.
18. Pellegrini CA, DeMeester TR, Johnson LF, et al. Gastroesophageal reflux and pulmonary aspiration: incidence, functional abnormality, and results of surgical therapy. *Surgery* 1979; 86(1):110–119.
19. Pingleton SK, Hinthorn DR, Liu C. Enteral nutrition in patients receiving mechanical ventilation: multiple sources of tracheal colonization include the stomach. *Am J Med* 1986; 80:827–832.
20. Melnik G, Wright K. Pharmacologic aspects of enteral nutrition. In Rombeau JL, Caldwell MD, eds. *Clinical Nutrition: Enteral and Tube Feeding*, 2nd ed. Philadelphia: W. B. Saunders, 1990:472–509.
21. Montecalvo MA, Steger KA, Farber HW, et al. Nutritional outcome and pneumonia in critical care patients randomized to gastric versus jejunal tube feeds. *Crit Care Med* 1992; 20(10):1377–1387.
22. Strong RM, Condon SC, Solinger MR, et al. Equal aspiration rates from postpylorus and intragastric-placed small-bore nasoenteric feeding tubes: a randomized, prospective study. *J Parent Ent Nutr* 1992; 16(1):59–63.
23. Alessi DM, Berci G. Aspiration and nasogastric intubation. *Otolaryngol Head Neck Surg* 1986; 94(4):486–489.

24. Minard G. Enteral access. *Nutr Clin Pract* 1994; 9:172–182.
25. Curet-Scott MJ, Meller JL, Shermeta DW. Transduodenal feedings: a superior route of enteral nutrition. *J Pediatr Surg* 1987; 22(6):516–518.
26. Torres A, Serra-Batlles J, Ros E, et al. Pulmonary aspiration of gastric contents in patients receiving mechanical ventilation: the effect of body position. *Ann Intern Med* 1992; 116:540–543.
27. Coben R, Weintraub A, DiMarino AJ. Gastroesophageal reflux during gastrostomy feeding. *Gastroenterology* 1994; 106:13–18.
28. McClave SA, Snider HL, Lowen CC, et al. Use of residual volume as a marker for enteral feeding intolerance: prospective blinded comparison with physical examination and radiographic findings. *J Parent Ent Nutr* 1992; 16(2):99–105.
29. Bury KD, Jambunathan G. Effects of elemental diets on gastric emptying and gastric secretion in man. *Am J Surg* 1974; 127:59–64.
30. Wolfe JE, Bone RC, Ruth WE. Effects of corticosteroids in the treatment of patients with gastric aspiration. *Am J Med* 1977; 63:719–722.
31. Wynne JW, DeMarco FJ, Hood CI. Physiological effects of corticosteroids in foodstuff aspiration. *Arch Surg* 1981; 116:46–49.
32. Lorber B, Swenson RM. Bacteriology of aspiration pneumonia: a prospective study of community- and hospital-acquired cases. *Ann Intern Med* 1974; 81:329–331.
33. Jacobs S, Chang RWS, Lee B, et al. Continuous enteral feeding: a major cause of pneumonia among ventilated intensive care unit patients. *J Parent Ent Nutr* 1990; 14(4):353–356.
34. Driks MR, Craven DE, Celli BR, et al. Nosocomial pneumonia in intubated patients given sucralfate as compared with antacids or histamine type 2 blockers. *N Eng J Med* 1987; 317(22):1376–1882.
35. Heyland DK, Cook DJ, Guyatt GH. Enteral nutrition in the critically ill patient: a critical review of the evidence. *Intens Care Med* 1993; 19:435–442.
36. De Leeuw IH, Vandewoude MF. Bacterial contamination of enteral diets. *Gut* 1986; 27(suppl 1):56–57.
37. Paauw JD, Fagerman KE, McCamish MA, et al. Enteral nutrient solutions; limiting bacterial growth. *Am Surg* 1984; 50(6):308–316.
38. Schroeder P, Fisher D, Volz M, et al. Microbial contamination of enteral feeding solutions in a community hospital. *J Parent Ent Nutr* 1983; 7(4):364–368.
39. Bliss DZ, Guenter PA, Settle RG. Defining and reporting diarrhea in tube-fed patients-what a mess. *Am J Clin Nutr* 1992; 55:753–759.
40. Guenter PA, Settle RG, Perlmutter S, et al. Tube feeding-related diarrhea in acutely ill patients. *J Parent Ent Nutr* 1991; 15(3):277–280.
41. Eisenberg PG. Causes of diarrhea in tube-fed patients: a comprehensive approach to diagnosis and management. *Nutr Clin Pract* 1993; 8:119–123.
42. Edes TE, Walk BE, Austin JL. Diarrhea in tube-fed patients: feeding formula not necessarily the cause. *Am J Med* 1990; 88:91–93.

43. Keohane PP, Attrill H, Love M, et al. Relation between osmolality of diet and gastrointestinal side effects in enteral nutrition. *Br Med J* 1984; 288:678–680.
44. Ford EG, Jennings LM, Andrassy RJ. Serum albumin (oncotic pressure) correlates with enteral feeding tolerance in the pediatric surgical patient. *J Pediatr Surg* 1987; 7:597–599.
45. Heitkemper ME, Martin DL, Hansen BC, et al. Rate and volume of intermittent enteral feeding. *J Parent Ent Nutr* 1981; 5(1):125–129.
46. Jones TN, Moore FA, Moore EE, et al. Gastrointestinal symptoms attributed to jejunostomy feedings after major abdominal trauma; a critical analysis. *Crit Care Med* 17(11):1146–1150.
47. Smith-Choban P, Max MH. Feeding jejunostomy: a small bowel stress test? *Am J Surg* 1988; 155:112–117.
48. Cogbill TH, Wolfson RH, Moore EE, et al. Massive pneumatosis intestinalis and subcutaneous emphysema: complication of needle catheter jejunostomy. *J Parent Ent Nutr* 1983; 7(2):171–175.
49. Galandiuk S, Fazio VW. Pneumatosis cystoides intestinalis: a review of the literature. *Dis Colon Rectum* 1986; 29:358–363.
50. Knechtle SJ, Davidoff AM, Rice RP. Pneumatosis intestinalis: surgical management and clinical outcome. *Ann Surg* 1990; 212(2):160–165.
51. Suarez V, Chesner IM, Price AB, et al. Pneumatosis cystoides intestinalis: histological mucosal changes mimicking inflammatory bowel disease. *Arch Pathol Lab Med* 1989; 113:898–901.
52. Brenner DW, Schellhammer PF. Mortality associated with feeding catheter jejunostomy after radical cystectomy. *Urology* 1987; 30(4):337–340.
53. Gaddy MC, Max MH, Schwab CW, et al. Small bowel ischemia: a consequence of feeding jejunostomy. *So Med J* 1986; 79(2):180–182.
54. Smith CD, Sarr MG. Clinically significant pneumatosis intestinalis with postoperative enteral feedings by needle catheter jejunostomy: an unusual complication. *J Parent Ent Nutr* 1991; 15(3):328–331.
55. Bohnker BK, Artman LE, Hoskins WJ. Narrow bone nasogastric feeding tube complications. *Nutr Clin Pract* 1987; 2(5):203–209.
56. D'Souza CR, Kilam SA, D'Sousa U, et al. Pulmonary complications of feeding tubes: a new technique of insertion and monitoring malposition. *Can J Surg* 1994; 37(5):404–408.
57. Montecalvo MA, Steger KA, Farber HW, et al. Nutritional outcome and pneumonia in critical care patients randomized to gastric versus jejunal tube feeding. *Crit Care Med* 1992; 20:1377–1387.
58. Meer JA. Inadvertent dislodgement of nasoenteral feeding tubes: incidence and prevention. *J Parent Ent Nutr* 1987; 11(2):187–189.
59. Armstrong C, Luther W, Sykes T. A technique for preventing extubation of feeding tubes: the bridle. *J Parent Ent Nutr* 1980; 4(6):603.
60. Estoup M. Approaches and limitations of medication delivery in patients with enteral feeding tubes. *Crit Care Nurse* 1994; 14(1):68–79.

61. McArdle AH, Palmason C, Morency I, et al. A rationale for enteral feeding as the preferable route for hyperalimentation. *Surgery* 1981; 90(4):616–621.
62. Lickley HLA, Track NS, Vranic M, et al. Metabolic responses to enteral and parenteral nutrition. *Am J Surg* 1978; 135:172–175.
63. Enrione EB, Gelfand MJ, Morgan D, et al. The effects of rate and route of nutrient intake on protein metabolism. *J Surg Res* 1986; 40:320–325.
64. Saito J, Trocki O, Alexander JW, et al. The effect of route of nutrient administration on the nutritional state, catabolic hormone secretion, and gut mucosal integrity after burn injury. *J Parent Ent Nutr* 1987; 11(1):1–7.
65. Abernathy GB, Heizer WD, Holcombe BJ, et al. Efficacy of tube feeding in supplying energy requirements of hospitalized patients. *J Parent Ent Nutr* 1989; 13(4):387–391.
66. Brammer EM. Shortcomings of current formulae for long-term enteral feeding in pediatrics. *Nutr Clin Pract* 1990; 5:160–162.
67. Solomon SM, Kirby DF. The refeeding syndrome: a review. *J Parent Ent Nutr* 1990; 14(1):90–97.
68. Hayek ME, Eisenberg PG. Severe hypophosphatemia following the institution of enteral feedings. *Arch Surg* 1989; 124:1325–1328.
69. Saklad JJ, Graves RH, Sharp WP. Interaction of oral phenytoin with enteral feeding. *J Parent Ent Nutr* 1986; 10(3):322–323.
70. Maynard GA, Jones KM, Guidry JR. Phenytoin absorption from tube feedings. *Arch Intern Med* 1987; 147:1821.

SPECIAL SITUATIONS IN NUTRITIONAL SUPPORT

Total Parenteral Nutrition in the Diabetic Patient

CHAPTER 19

John K. Siepler, Pharm.D., B.C.N.S.P., and Stephen Phinney, M.D., Ph.D.

Total parenteral nutrition (TPN) is provided by infusing hypertonic dextrose, amino acids, and lipid emulsion in a central line. Traditionally, the majority of TPN energy is supplied by the dextrose component.[1] Therefore, the diabetic patient often presents the physician with difficulty in maintaining glucose control. Resolving this dilemma is the primary focus of this chapter, along with the additional nutritional concerns of the diabetic patient.

This chapter covers the physiology of glucose tolerance including glucose metabolism and the effects of insulin, discussing why glucose control is necessary and what level of glucose control should be achieved. Special consideration is given to the differences in glucose metabolism in the insulin-dependent compared to the non-insulin-dependent diabetic, as well as stressed and septic patients. Also included will be a discussion of the consequences that occur when the delivery of glucose in TPN exceeds the ability of the body's metabolic ability.

Because the essence of good TPN delivery involves anticipation and prevention of problems, patients at risk for hyperglycemia accompanying TPN should be identified. These include not only patients who have insulin-dependent and non-insulin-dependent diabetes but also those who have hyperglycemia associated with clinical and pharmacologic stress. In addition, patients who develop glucose intolerance from excessive carbohydrate administration (both from the dextrose in TPN and from dextrose given by non TPN fluids) is discussed.

Lastly, effective management of hyperglycemia in this patient population is discussed. This management can include not only administration of insulin but also alteration of energy provision by decreasing the glucose load. This can occur by increasing the lipid calories while keeping the level of energy provision the same. Another option is to decrease the total energy provision by decreasing dextrose calories.

CARBOHYDRATE METABOLISM IN DIABETES

Although the two general categories of diabetes share the common manifestation of hyperglycemia, type I diabetes and type 2 diabetes differ fundamentally in the mechanisms of altered glucose control. Type I, or juvenile-onset diabetes, is generally a disease of impaired insulin production resulting from pancreatic beta-cell destruction. Given adequate insulin, glucose control in type I patients is facilitated by the fact that peripheral insulin-mediated glucose disposal remains relatively normal. Type 2, or maturity-onset diabetes, on the other hand, is characterized by peripheral insulin resistance. That is, skeletal muscle, the primary site of glucose disposal, is relatively unresponsive to glucose uptake, requiring a high dose of insulin to accomplish glucose clearance. Because increasing insulin does not resolve insulin resistance, there is increasing hepatic glucose disposal via lipogenesis as the glucose load is increased in type II diabetics, contributing to steatosis and hypertriglyceridemia. Thus the insulin resistance of type II diabetes is poorly managed by incremental increases in insulin therapy.

In the acute care setting, management of diabetes is exacerbated by a number of clinical variables. Both the type I and type 2 diabetic patient respond to infection, surgical stress and trauma, and glucocorticoid medications by increased resistance to insulin's action. The additional result of stress induced by trauma, surgery, or infection is increased hepatic glucose output. Whereas the normal fasting state is associated with hepatic glucose production of 1 $mg/kg \cdot min^{-1}$, this can increase fivefold with stress, and it is ineffectively suppressed by exogenous glucose provision.[2] This provides a third source of metabolically significant glucose appearance in addition to that delivered via the gastrointestinal tract plus TPN and other intravenous fluids, further exacerbating glucose disposal in many diabetic inpatients receiving TPN.

THE IMPORTANCE OF GLUCOSE CONTROL

The long held view that tight diabetic control is desirable has recently been validated in the DCC trial.[3] However, although this study showed reduced progression of diabetic complications over a multiple year period, it probably has little direct relevance to TPN in the inpatient setting, which is given for periods of weeks rather than years. Nonetheless, there are other reasons to support effective glucose control in the short term, including avoidance of osmotic diuresis induced when the kidney threshold for glucose recovery is surpassed and also considerable evidence that hyperglycemia is deleterious to leukocyte function and host defense against infection.[4]

An important fact to keep in mind when setting standards for glucose control is that a patient receiving continuous TPN is constantly in the fed state. In this setting, normal values for fasting glucose have little meaning. Thus a desirable range for glucose control during TPN is probably in the range of 150 to 200 mg/dL.[4]

On the other side of the spectrum, hypoglycemia is less of an immediate threat in the patient receiving continuous TPN than in an orally fed patient, as the ongoing provision of dextrose calories during hypoglycemia meets the body's ongoing metabolic demand for glucose despite the low serum level. Thus a patient with a blood glucose in the 30 to 40-mg/dL range may have absolutely no clinical symptoms as long as TPN provision is not interrupted. Nonetheless, this represents a fragile state that should be avoided because of the potential for immediate hypoglycemic manifestations should the TPN be stopped for even a short interval.

PATIENTS AT RISK FOR GLUCOSE INTOLERANCE WITH TOTAL PARENTERAL NUTRITION

Although all diabetic patients are at increased risk for development of hyperglycemia, some have more risk than others. Patients with a history of type II diabetes mellitus, and also type I patients with a high insulin requirement prior to the initiation of TPN (i.e., greater than 50 units daily) are usually difficult to manage. Thus, the insulin requirement of the type I patient at home prior to illness may provide a clue to their ability to tolerate the dextrose loads required of TPN. Type II patients with poor glucose control on oral hypoglycemics may have similar difficulties. In addition, diabetic patients who are experiencing complications of their disease such as renal failure or retinopathy often have difficulties tolerating carbohydrate calories necessary in TPN. Lastly, while chromium deficiency is rare, glucose tolerance can be impaired if it is present.[5]

During hospitalization, both groups of patients often experience stress that produces marked elevations in their serum glucose.[6] Thus, a review of the insulin requirement in the 24 to 48 hr prior to initiation of TPN will provide a preview to the difficulty in maintaining serum glucose in a reasonable range. When determining this pre-TPN insulin requirement, it is often helpful to take into account the grams of dextrose delivered by isotonic intravenous fluids. Even a moderate insulin requirement such as 15 to 25 units over the prior 24 hr can be significant if the intravenous fluids were primarily saline with less than 50 g of dextrose infused. In this setting, this moderate insulin requirement may need to be increased considerably when the dextrose load is increased to 200 to 300 g with the initiation of TPN. Anticipation of this increased insulin requirement can help prevent severe hyperglycemia.

When the diabetic patient becomes septic, glucose tolerance almost always worsens. This effect is usually additive in those patients who already have insulin resistance.[7,8] In part, the mechanism for this effect is elevated levels of glucagon, cortisol, and growth hormone. This increases hepatic glucose production and decreases peripheral glucose uptake. In these patients, control of hyperglycemia with insulin is often difficult, requiring very high insulin doses even with the dextrose loads of normal intravenous fluids. The same problem occurs in the diabetic patient who is receiving corticosteroids. Careful attention when corticosteroids are given to these patients will be helpful in anticipating difficulty in controlling serum glucose.

Lastly, some otherwise well-controlled diabetic patients become hyperglycemic because of inadvertent excessive carbohydrate administration. This can occur either because the TPN dextrose content is excessive or because of additional provision of dextrose from fluids or nutrients separate from the TPN, such as when attempting to provide additional free water to a dehydrated patient with 5% dextrose and water. Additional glucose-control problems often emerge when a patient stabilized on TPN begins receiving meaningful calories during transition to gastrointestinal feeding.

MINERAL METABOLISM IN THE DIABETIC PATIENT

Evaluating and correcting abnormalities in major mineral and trace mineral metabolism are part of the usual provision of TPN to patients, particularly when the patient is malnourished with significant mineral deficits. In the diabetic patient, mineral deficits may be further accentuated beyond the patient's degree of malnutrition. For example, the osmotic diuresis that occurs in poorly controlled type 1 and type 2 diabetics results in chronic increases in mineral and trace mineral loss, accentuating the likelihood for major tissue deficits. In addition, with inadequate insulin provision or insulin resistance, muscle glycogen stores tend to be depleted, and, when these limitations are overcome the rapid accumulation of muscle glycogen mandates muscle uptake of all intracellular minerals.[9] Because deficits in intracellular minerals markedly impede protein anabolism,[10] close attention to and prompt correction of any mineral deficits observed during initiation of TPN is of great importance.

Potassium

The well-known effect of insulin and glucose on cellular uptake of potassium results from the above-noted intracellular glycogen accretion, during which each gram of cellular glycogen is associated with 3 mL of intracellular water. Although this is a transient effect when running 5% dextrose, the higher carbohydrate loads and the ability of liver and muscle to accumulate as much as 500 grams of glycogen mean that this effect is often sustained for multiple days during initiation of TPN. This effect can also be seen in midcourse of TPN when a previously infected or stressed patient resolves the stress, reducing insulin resistance, and allowing cellular glycogen to accumulate.

Phosphorous

Reversal of diabetic ketoacidosis is commonly associated with transient hypophosphatemia, and opinion remains divided as to whether this transient hypophosphatemia requires aggressive replacement. With sustained hypophosphatemia lasting longer than 24 hr, however, cellular regeneration of adenosine triphosphate (ATP) can be impaired, resulting in reduced function of multiple organ systems and markedly increased risk of complications or mortality.[11–13] An exacerbating factor in inducing phosphate depletion can be osmotic diuresis as noted above. In the setting of initiating TPN, close attention to serum phosphate levels is mandatory with aggressive replacement for any serum phosphorus level that falls even slightly below the lower limit of normal due to the major and sustained intracellular uptake known to occur with initiation of TPN.[12]

Magnesium

The diagnosis and treatment of magnesium depletion is often neglected because in part of the lack of an accurate, noninvasive method of assessment. The serum level is significant if low, but has a notoriously high rate of false-negative results. As many as 90% of cases of magnesium depletion yield normal serum levels. Thus, the clinical history, particularly diuretic and alcohol use, physical exam signs such as hyperreflexia, and associated biochemical signs such as persistent hypokalemia are important in the initial evaluation. Replacement of significant tissue deficit can take many days, although the serum level is usually normalized with the first day of therapy. Because adequate renal function (i.e., a GFR > 40) can clear any excess administered, it is often appropriate to give a modest extra dose of magnesium (20 to 40 mEq/day) to patients with adequate renal function, especially in diabetics with the problems of osmotic losses and intracellular shifts noted above.

Zinc

Although zinc deficiency is uncommon in most healthy humans, it is increasingly recognized as a significant clinical problem in many disease states. Along with alcohol, diuretic medications, and chronic infections, the osmotic diuresis associated with poorly controlled diabetes can lead to accelerated urinary zinc loses and eventually cause depletion. Thus, the clinician should have a high level of suspicion for zinc depletion in poorly controlled diabetic patients of both type 1 and type 2 classifications and include the usual specific history of change in taste and smell and the physical findings of dry skin and impaired wound healing in their normal workup.

PRACTICAL APPLICATIONS

The principal limitations to providing TPN in diabetic patients are the result of difficulties in their disposal of the carbohydrate calories necessary to provide adequate energy. This is complicated by the limitations in using soybean oil emulsion as a major fuel, as diabetics are frequently hypertriglyceridemic, and current oil emulsions may be immunosuppressive.[14] In addition, these patients often present with mineral and electrolyte deficiencies that make their initial management a challenge.

The major goal of the safe provision of TPN in diabetic patients is to prevent the adverse effects of hyperglycemia that may result from the increased carbohydrate delivery (see Table 19.1). It is generally desirable to keep the blood sugar under 220 mg/dL, as it has been shown that immune function diminishes when that level is exceeded for even modest time periods.[4] This is monitored either by using the hospital laboratory or by using finger stick blood glucose determinations (FSBG) with a system, such as Accucheck® (FSBG). Monitoring is usually accompanied by an alteration of the dose of insulin, often by intermittent subcutaneous or intravenous injections. The specific insulin dose increases with higher

Table 19.1. Adverse Effects of Hyperglycemia in Patients on TPN

Immune function depression
Direct effect of hyperglycemia
Indirect due to zinc losses with osmotic diuresis
Electrolyte disturbances
Hypophosphatemia
Hypomagnesemia
Hypokalemia
Osmotic diuresis (dehydration, mineral losses)
Hypertriglyceridemia

blood sugar readings using a sliding scale. Another option is to use an insulin infusion with the dose based on the blood sugar results.

Significant glycosuria accompanied by osmotic diuresis can also occur when the blood sugar is elevated,[15] causing both dehydration and electrolyte aberrations that complicate the safe delivery of TPN. Monitoring the urine sugar (in addition to blood glucose determinations) is useful in these patients, because the real threshold for glucosuria can vary considerably with individuals. Keeping the urine glucose under 2+ will usually ensure that significant glycosuria and osmotic diuresis are avoided.

When elevated blood sugar is treated with insulin, anticipate that electrolyte shifts will result as the insulin mobilizes glucose into cells. The most common electrolyte disturbances are hypophosphatemia, hypomagnesemia, and hypokalemia. It may thus be necessary to monitor the phosphorous, magnesium, and potassium level as often as two or three times daily in the first few days in diabetic patients on TPN. Keeping these values in the normal range may require that high doses of minerals be added to the TPN and/or given as separate intravenous infusions.

INITIATING TOTAL PARENTERAL NUTRITION

The major challenge in managing TPN in the diabetic patient is to provide the patient with a solution that contains a significant amount of carbohydrate calories without developing severe hyperglycemia.

The first 24 hr of TPN can be the most challenging, as the delivery of dextrose calories is usually increased above that provided with maintenance intravenous fluids in order to advance toward the patient's eventual caloric goals. At the same time, the physician and pharmacist must take anticipatory steps to control the patient's blood sugar and electrolytes. Although this is a goal in the initiation of TPN in any patient, the diabetic will require more care and management than most others.

There are two general principles used for achieving this goal. The first involves initiating the carbohydrate calorie load at a more modest level than would be used in the nondiabetic patient. The second involves using insulin to control the expected hyperglycemia. In practice, a combination of these two principles is often the best course of action.

The level of difficulty in initiating TPN in the diabetic patient can be predicted by reviewing the patient's history and hospital course to assess the level of control of hyperglycemia prior to the initiation of TPN. A careful review will reveal that most patients fall into one of several categories (see Table 19.2). In addition, other conditions can occur in the diabetic patient receiving TPN. Within the three risk categories described, the addition of these confounding conditions increases the risk of hyperglycemia further.

Table 19.2. Difficulty in Initiating TPN in the Diabetic Patient

Minimal risk for hyperglycemia

History of diabetes (type 1 or 2) with blood sugar values below 180mg/dL in hospital without insulin requirement

Moderate risk for hyperglycemia

History of diabetes (type 1 or 2) with blood sugar values in Hospital below 180 mg/dL with insulin required (risk increases: A<B<C)

INSULIN DOSE

Low	A	Insulin dose is less than 25 U daily by drip or by sliding scale using urine or blood sugars
Moderate	B	Insulin required is more than 25 U but less than 50 U daily by drip or by sliding scale using urine or blood sugars
High	C	Insulin dose is more than 50 U daily by drip or by sliding scale using urine or blood sugars

Significant risk for hyperglycemia

History of diabetes (type 1 or 2) with poor control of blood sugar values in Hospital with insulin required (poor control = blood sugars consistently above 250 mg/dL) (risk increases: A<B<C)

INSULIN DOSE

Low	A	Insulin dose is less than 25 units daily by drip or by sliding scale using urine or blood sugars.
Moderate	B	Insulin required is more than 25 units but less than 50 units daily by drip or by sliding scale using urine or blood sugars.
High	C	Insulin dose is more than 50 units daily by drip or by sliding scale using urine or blood sugars.

OTHER VARIABLES THAT INCREASE RISK OF HYPERGLYCEMIA

Presence of any of these variables increase the risk of hyperglycemia one category

1. Insulin resistance
   a. Sepsis
   b. Concurrent corticosteroid administration
2. Presence of complications of diabetes (e.g., renal failure)
3. Hypertriglyceridemia accompanying hyperglycemia.

The diabetic patient (either type I or type II) whose disease is well controlled at home without any complications of diabetes and has normal blood sugar values in hospital without requiring insulin prior to initiation of TPN (considered a minimal risk in Table 19.2) should have a low risk of hyperglycemia when TPN is initiated. These patients also will not have the presence of any of the additional confounding factors that affect glucose tolerance summarized in the bottom of Table 19.2. In this patient (see Table 19.3), initiation of TPN using a similar

Table 19.3. Initiating TPN in the Diabetic Patient: Specific Recommendations by Level of Risk for Hyperglycemia

Risk	Recommendation
Minimal risk for hyperglycemia	
Initial dextrose dose	200–250 g
Use insulin sliding scale	
K, PO_4, Mg in normal range	use normal doses
Normal doses: K, 60 mEq; Mg, 16 mEq; PO_4, 30mM	
K, PO_4, Mg low or in low-normal range	use increased doses or extra supplementation
Increased doses: K, 80–120 mEq; Mg, 24–40 mEq; PO_4, 45 mM	
Repeat K, PO_4, Mg determination in 6–12 hr	
Repeat supplementation if repeat values are low.	
Moderate risk for hyperglycemia	
Initial Dextrose dose	150 g
Use insulin sliding scale using FSBG.	
Add 50% of insulin requirement to TPN if requirement is greater than 25 U	
K, PO_4, Mg in normal range	use normal doses
Normal doses: K, 60 mEq; Mg, 16 mEq; PO_4, 30 mM	
K, PO_4, Mg low or in low-normal range	use increased doses or extra supplementation
Increased doses: K, 80–120 mEq; Mg, 24–40 mEq; PO_4, 45 mM	
Repeat K, PO_4, Mg determination in 6–12 hr	
Repeat supplementation if repeat values are low	
Significant risk for hyperglycemia	
Initial Dextrose dose	100–150 g
Use insulin sliding scale using FSBG	
Add 75% of insulin requirement to TPN if requirement is greater than 25 U	
Alternative: supplemental insulin drip (in addition to TPN)	
Start at 1–2 U/hr, increase dose based on q 1–2 hr FSBG determinations	
K, PO_4, Mg in normal range	use normal doses in TPN
Normal doses: K, 60 mEq; Mg, 16 mEq; PO_4, 30 mM	
K, PO_4, Mg low or in low-normal range	use increased doses or extra supplementation
Increased doses: K, 80–120 mEq; Mg, 24–40 mEq; PO_4, 45mM	
Repeat K, PO_4, Mg determination in 6–12 hours	
Repeat extra supplementation if repeat values are low.	

amount of dextrose as a nondiabetic patient is indicated. Using an initial dose of 200 to 250 g of dextrose should provide the patient with a significant bridge to the TPN caloric goal. A supplemental insulin sliding scale is indicated using laboratory measurements or FSBG to ensure maintenance of glucose control. Careful attention to the patient's electrolytes is indicated, and extra supplementation of phosphate, potassium and magnesium may be necessary if the laboratory values of these electrolytes are below the normal range or in the low-normal range. This can be provided in the initial TPN, but additional oral or intravenous supplementation separate from the TPN provides greater flexibility of dosing. Repeating these mineral determinations are indicated 6 to 12 hr after the initiation of TPN if these values are in the low-normal range or if supplementation was required. The goal is to have the patient tolerate the first 24 hr of TPN without developing hyperglycemia, or hypophosphatemia, hypokalemia, or hypomagnesemia.

The diabetic patient who has good control of blood glucose with exogenous insulin administration has a moderate risk of hyperglycemia when TPN is initiated. In addition, this risk rises with higher insulin doses (Table 19.2). Also a patient who falls in the minimal-risk category in Table 19.3 but has any of the confounding variables that affect glucose tolerance in the bottom of the table should also be considered to have a moderate risk of hyperglycemia. In these patients, the initial dose of TPN should be no larger than 150 g. If the patient is already requiring insulin, the total insulin dose for the previous day should be calculated, and if that is greater than 25 U, 50% of that insulin requirement should be added to the initial TPN bag. An insulin sliding scale using FSBG should also be employed. Careful attention to electrolytes as with the previous risk group is also indicated (see Table 19.3) for summary of recommendation).

The diabetic patient who has poorly controlled blood sugar presents a significant risk when initiating TPN. In addition, even patients who have moderate control of blood glucose but require large doses of supplemental insulin have a high risk of hyperglycemia in this setting. Within this general category, the risk of serious hyperglycemia rises with the amount of exogenous insulin required (Table 19.2). In this class of patients, it is often best to start with an initial dose of TPN dextrose no larger than 100 g. If the patient already requires insulin, the insulin requirement for the previous day should be calculated, and if that is greater than 25 U, 75% of that insulin requirement should be added to the TPN. Frequent monitoring of blood glucose using either the laboratory or FSBG is indicated. An insulin sliding scale providing supplemental insulin given either intravenously or subcutaneously using FSBG should also be employed. An insulin infusion that supplements the insulin in the TPN may be required. A starting dose of 1 U/hr and increased to keep the blood sugar under about 200 to 250 mg/dL is indicated. This method allows specific dosing of insulin and greater flexibility and prevents the chance of overdosing with the addition of a very large amount of insulin to the initial TPN.

As in the previous group, the chance of experiencing severe hypokalemia, hypomagnesemia, and hypophosphatemia is quite high, and this is particularly true when good glucose control is actually achieved. These patients are thus quite labile in their response to TPN and require the greatest attention to electrolytes.

The patient who is described in the significant risk group but has sepsis, is receiving corticosteroids, or has end organ damage from diabetes (e.g., renal failure) is perhaps the most difficult to manage on TPN. The risk of hyperglycemia is higher than any previous group described, and initiation of TPN in these patients must be accompanied with the utmost care. On occasion, an initial dextrose dose below 100 g will be necessary to prevent severe hyperglycemia. A supplemental insulin infusion with hourly determination of FSBG may be necessary to control the hyperglycemia, using insulin doses of 10 to 20 U/hr. In addition, electrolyte shifts can be severe, requiring additional phosphate, magnesium, and potassium with renal dysfunction an added complication by limiting clearance if excess minerals are administered.

SETTING CALORIE GOALS

The main challenge after initiating TPN in this patient population is advancing TPN to appropriate energy goals without inducing hyperglycemia. In the more difficult patients, it may not be realistic to give the patient full adult anabolic goals of 30 to 35 kcal/kg $\cdot$ day^{-1}. Thus, a hypocaloric goal may be a temporary option for many of these patients.

Since the original intravenous protein-sparing therapy of Blackburn et al.,[16] there have been a number of subsequent studies reporting excellent nitrogen homeostasis and clinical outcome data for overweight patients given varying degrees of calorie-restricted TPN. This is not advocated as a mechanism for clinical weight loss; rather it is an effective means of dealing with poorly controlled levels of energy-yielding substrates such as glucose and triglycerides during feeding regimens in diabetic and stressed patients.

Because all overweight patients do not have excess lean body mass, particularly those with chronic wasting illnesses where the loss of lean body mass may have exceeded the loss of adipose tissue, hypocaloric TPN in not appropriate for all overweight patients. Thus, its use should be reserved for those patients in whom the clinician has determined that there is reasonably normal underlying lean body mass. In this setting, hypocaloric TPN can be used for days or weeks at a time as long as adequate nitrogen conservation is demonstrated following initiation of reduced calorie feeding.

If determined appropriate for a specific patient, the published literature supports TPN goals in the range of 15 to 25 cal/kg reference body weight.[17–20] As adipose tissue in obese patients contains adequate linoleic acid for essential needs, the amount mobilized to cover even a modest energy deficit will yield adequate essential fatty acids such that parenteral lipid emulsion is not required during such hypocaloric regimens. In most such patients, following resolution of the stress of illness and increased physical activity (which itself enhances muscle response to insulin), a hypocaloric regimen can be advanced to a eucaloric level to achieve optimum clinical results should longer term parenteral feeding (i.e., months or greater) be required.

ADVANCING THE PATIENT TO THE CALORIE GOAL

The patient who falls in the minimal-risk category (see tables 19.2 and 19.3) can be advanced in a similar fashion to a nondiabetic patient. It must be remembered that advancing this patient does not need to be done in one day and may be carried out over several days. On day two, the protein dose can be advanced to the goal, and the dextrose can be advanced by 100 to 150 g daily (see Table 19.4). Intravenous fat emulsion can be started as necessary at any time. The insulin sliding scale should be continued, and, if the patient's blood sugar is not consistently under 200 to 250 mg/dL, the patient should be treated as a moderate-risk patient (see Table 19.4). Care should be taken during this period to ensure that phosphorus, magnesium, and potassium remain in the normal range. Extra doses may be necessary.

For the moderate-risk patient, the TPN protein and intravenous fat emulsion can be treated like a minimal-risk patient. The dextrose dose should be advanced by no more than 50 to 100 g daily. If the blood sugar is kept under 200 to 250 mg/dL and the insulin requirement is more than 20 U, half of that (10 U) should be added to the TPN for the next 24 hr. The additional insulin should be continued. In these cases, where there is poor control of the blood sugar, it may be necessary to initiate a supplemental insulin infusion. If blood glucose control becomes very poor, it may be necessary to stop or interrupt the infusion of the TPN and continue the insulin infusion to regain control. As in the minimal risk category, care should be taken with patients in this risk group to ensure that phosphate, magnesium, and potassium remain in the normal range.

For the significant-risk patient, the TPN protein dose and intravenous fat emulsion can be advanced to goal as necessary on day 2. The dextrose dose should be advanced to goal with extreme care to avoid hyperglycemia. This should be done by increasing the dextrose dose by no more than 50 g daily. An insulin sliding scale or supplemental insulin infusion using FSBG should be used. Starting doses and titration are described in the moderate-risk section. If the

Table 19.4. Advancing TPN in the Diabetic Patient: Specific Recommendations by Level of Risk for Hyperglycemia

Risk	Recommendation
Minimal risk for hyperglycemia	
Protein	Advance to goal on day 2
Dextrose	Advance 100–150g daily to goal
Use insulin sliding scale	
Fat emulsion:	Add if necessary
K, PO_4, Mg	See table 3
Moderate risk for hyperglycemia	
Protein	Advance to goal on day 2
Dextrose dose	Advance 50–100g daily to goal
Use insulin sliding scale using FSBG	
Add 50% of insulin requirement to TPN if requirement is greater than 20 U	
Fat Emulsion	Add if necessary
K, PO_4, Mg	See table 3
Significant risk for hyperglycemia	
Protein	Advance to goal on day 2
Dextrose dose:	Advance no more than 50g daily to goal
Use insulin sliding scale using FSBG	
Add 75% of insulin requirement to TPN if requirement is greater than 20 U	
Alternative: supplemental insulin drip (in addition to TPN)	
Start at 1–2 U/hr, increase dose based on q 1–2 hr FSBG determinations	
If insulin drip is unable to control blood glucose, interrupt TPN and continue insulin	
Fat Emulsion	Add if necessary
K, PO_4, Mg	See table 3

insulin requirement exceeds 20 U daily, 75% of that should be added to the next day's TPN, and the additional insulin continued. Extreme care should be taken to ensure that phosphate, magnesium, and potassium remain in the normal range. Measurements of serum levels may be necessary several times daily, and additional doses are often required.

ADDITIONAL TIPS

When the diabetic patient on TPN is stable with good glucose tolerance, loss of glucose control can occur when transitioning the patient to gastrointestinal feedings. Thus, when adding enteral or oral feedings in these patients, additional care to the insulin coverage is necessary. Glucose tolerance in the diabetic septic patient can improve quickly as the sepsis is controlled. Thus, in these patients, it may be desirable to have a significant portion of the insulin coverage outside of the TPN to allow the flexibility to reduce dosage if necessary.

REFERENCES

1. Duke DH, Jorgensen SB, Broel JR, et al. Contributions of protein to calorie expenditure following surgery. *Surgery* 1970; 78:168–174.
2. Wolfe RR, Herndon DN, Jahoor F, et al. Effect of severe burn injury on substrate cycling by glucose and fatty acids. *N Engl J Med* 1987; 317:403–408.
3. The Diabetes Control and Complications Trial Research Group. The effect of intensive treatment of diabetes on the development and progression of insulin dependent diabetes mellitus. *N Engl J Med* 1993; 329:977–986.
4. Baxter RK, Babineau TJ, Aprovian CM, et al. Perioperative glucose control predicts increased nosocomial infection in diabetics. *Crit Care Med* 1992; 18 (4):S207.
5. Jeejeebhoy KN, Chu RC, Marliss EB, et al. Chromium deficiency, glucose intolerance and neuropathy reversed by chromium supplementation in a patient receiving long-term TPN. *Ann J Clin Nutr* 1977; 30:531–538.
6. Wolfe RR, Alsop JR, Burke JF. Glucose metabolism in man: response to intravenous glucose infusions. *Metabolism* 1979; 28: 210–220.
7. Filkins JP. Monokines and the metabolic pathways of septic shock. *Fed Proc* 1985; 44:300–304.
8. Shearer JD, Amal JF, Caldwell MD. Glucose metabolism of injured muscle: the contribution of inflammatory cells. *Circ Shock* 1988; 28:131.
9. Bergstrom J. Muscle electrolytes in man. *Scand J Clin Lab Med* 1962; 14:S11–S13.
10. Rudman D, Millikan WJ, Richardson TJ, et al. Elemental balances during intravenous hyperalimentation of underweight adult subjects. *J Clin Invest* 1975; 55:94–104.
11. Jacob HS, Amsden P. Acute hemolytic anemia with rigid red cells in hypophosphatemia. *N Engl J Med* 1971; 285:1446–1450.
12. Knochel JP. The clinical status of hypophosphatemia. *N Engl J Med* 1985; 313:447–449.
13. Halevy J, Bulvick S. Severe hypophosphatemia in hospitalized patients. *Arch Intern Med* 1988; 148:153–155.
14. Phinney SD, Siepler JK, Bach HT. Is there a role for parenteral feeding in clinical medicine? *Western J Med* 1995; (in press).
15. Van Way CW, Buekk CA, Peterson R, et al. Nitrogen balance and electrolyte requirement in intralipid based hyperalimentation. *J Parent Ent Nutr* 1979; 3:174–176.

16. Blackburn GL, Flatt JP, Clowes GHA, Jr., et al. Protein sparing therapy during periods of starvation with sepsis or trauma. *Ann Surg* 1973; 177:588–594.
17. Greenburg GR, Jeejeebhoy KN. Intravenous protein sparing therapy with gastrointestinal disease. *J Parent Ent Nutr* 1979; 3:427–432.
18. Dickerson RN, Rosato EF, Mullen JL. Net protein anabolism with hypocaloric parenteral nutrition in obese stressed patients. 1986; *Am J Clin Nutr* 44:747–755.
19. Burge JC, Goon A, Chopan PS, et al. Efficacy of hypocaloric total parenteral nutrition in hospitalized obese patients: a prospective double-blind trial. *J Parent Ent Nutr* 1994; 18:203–207.
20. Parnes HL, Mascioli EA, La Civita CL, et al. Parenteral nutrition in overweight patients: are intravenous lipids necessary to prevent essential fatty acid deficiency? *J Nutr Biochem* 1994; 5:243–247.

Nutrition Support in Liver, Pulmonary, and Renal Disease

Douglas L. Seidner, M.D., Raed Dweik, M.D., and Bharat Gupta, M.D.

NUTRITION AND THE LIVER

The liver is involved in orchestrating a vast array of biochemical reactions that are responsible for the synthesis and degradation of a variety of nutrients and chemicals. Central to these activities is the role that the liver plays in the metabolism of nitrogen, carbohydrate and lipid and maintenance of the body cell mass. Significant perturbation of these processes can occur in both acute and chronic liver disease, and so it is important to understand the impact of these illnesses on the nutritional status of the patient and that timely and appropriate intervention can often help optimize liver function.

Patients with parenchymal disease of the liver quite frequently have associated malnutrition with a prevalence of 20% to 80% being noted in patients with alcoholic liver disease admitted to tertiary care hospitals.[1] In patients with established cirrhosis, this number approaches 100%. These facts along with the known complications associated with protein-calorie malnutrition have compelled clinicians to provide specialized nutrition support to patients with liver disease in an attempt to improve their outcome. In this section, we first review the etiology and assessment of malnutrition in patients with liver disease. Then we discuss the use of enteral and parenteral nutrition and branched-chain amino acids (BCAAs) in acute and chronic liver disease. Hopefully, this will provide a rational strategy for providing nonvolitional feedings to these patients.

Pathogenesis of Malnutrition in Liver Disease

Impaired dietary intake is the primary cause for the development of nutritional deficiencies with documented protein and calorie intakes of 47 g/day and 1320 kcal/day in patients with chronic liver disease.[2] Exocrine pancreas insufficiency and malabsorption have also been demonstrated.[3] The synthesis and activation of certain nutrients, such as choline and vitamin D, can also be impaired in patients with liver disease.[4] Lastly, although increased protein catabolism has been thought to play a major role in the development of a negative nitrogen balance, it appears that an impaired ability to store adequate quantities of glycogen in the liver results in a rapid transition from the fed to the fasted state and therefore a greater reliance on amino acids for gluconeogenesis.[5] This observation stresses the importance of providing an adequate amount of protein in patients with chronic liver disease.

Nutrition Assessment in Liver Disease

Identifying patients with malnutrition may be difficult because of the progressive retention of sodium and water in chronic liver disease. Weight loss may not be a sensitive indicator of malnutrition because of fluid retention, which is manifest as ascites and pedal edema. Hypoalbuminemia has often been attributed to decreased synthesis by the liver; however, it has been shown that synthetic rates are at the lower end of the normal range in advanced liver disease.[6] More often, low albumin concentrations reflect a delusional effect associated with intravascular volume expansion along with an increased protein catabolic rate. A compensatory increase in albumin synthesis does not occur because of inadequate protein intake, inadequate synthetic reserve, or both. Despite these limitations, albumin and other visceral proteins may still be good markers of protein-calorie malnutrition in this population. In fact, albumin remains an integral part of the Child's score, which categorizes the severity of liver disease. Finally, upper arm anthropometry, which is an estimate of fat reserves and muscle mass, is reasonably accurate, because edema tends to accumulate in the lower extremities.

Energy Requirements

The Harris–Benedict equation can be used to determine the resting energy expenditure (REE) in most patients with mild to moderately severe liver disease. Hypercatabolism appears to be present in patients with advanced liver disease relative to lean body mass[7]; however, in most of these patients, lean body mass tends to be overestimated because of fluid retention. Therefore, when compared to actual body weight, these patients have calculated energy expenditures that approximate measured values. Despite the presence of these offsetting errors, it may be useful to determine REE by indirect calorimetry in select patients with severe liver disease. In general, most patients require 25 to 35 kcal/kg of body weight or an REE $\times$ 1.2 to 1.4.

Caution should be exercised when providing nonprotein calories to patients with severe liver disease. Hepatic steatosis, nonspecific triaditis, and intrahepatic cholestasis have all been described in patients requiring nutrition support, especially parenteral nutrition.[8] In many of these cases, patients were provided nutrition support that was in excess of their requirements. These patients also had other factors that may have been responsible for these abnormalities including sepsis and medications. In general, dextrose should not be infused above its oxidative capacity (5 $mg/kg \cdot min^{-1}$), because it may increase hepatic acetyl coenzyme A carboxylase and fatty acid synthetase activity[9] and suppress hepatic triglyceride secretion.[10] Excessive amounts of intravenous fat emulsion should also be avoided since liver phospholipidosis has also been reported.[11] We advise that the proportion of energy supplied as protein, carbohydrate, and fat be approximately 15%, 55%, and 30%, respectively.

Enteral Nutrition in Chronic Liver Disease

Two randomized controlled trials comparing supplemental tube feeding to a standard diet have shown improvement in hepatocellular function as measured by serum albumin, bilirubin, antipyrine clearance, and Child's score with the tube feeding.[12,13] In one of these trials, encephalopathy improved even though the formula was casein-based.[12] Positive nitrogen balance did not exacerbate encephalopathy, azotemia, edema, or ascites.

Other randomized controlled trials comparing BCAA to casein-based formulas provided as a diet supplement have shown no clear advantage for the BCAA group.[14–16] In general, these trials were performed in patients with latent hepatic encephalopathy. The use of BCAAs will be discussed in more detail below.

Peripheral Parenteral Nutrition for Alcoholic Hepatitis

The rationale for providing nutritional supplementation to patients with moderate-to-severe alcoholic hepatitis is that these patients are uniformly malnourished and that nutrition support may hasten recovery of the damaged liver and prevent the development of complications such as infections.[17] This theory, which is based primarily on epidemiologic and basic research, led to the use of peripheral parenteral nutrition (PPN) as a part of the treatment of this disease. Nasrallah and Galambos were the first to demonstrate significant improvement in serum albumin, bilirubin, and survival in patients who received PPN compared with controls who were given a regular diet.[18] Although several subsequent studies have shown biochemical and histologic improvement with a similar solution, none have been able to duplicate the improvement in clinical outcome.[19–21] Bonkovsky et al. compared PPN, the anabolic steroid oxandrolone, and PPN with oxandrolone to standard therapy and showed that PPN and/or oxandrolone led to more marked improvement in liver function (albumin, transferrin, prothrombin time, antipyrine, and galactose metabolism).[17] Interestingly, the groups that received PPN had a significant increase in the concentration of BCAAs despite the fact that the formula was not BCAA enriched.

These studies emphasize the importance of nutritional supplementation in patients with modest-to-severe alcoholic hepatitis; however, they do not provide convincing proof that PPN should be used in all patients in this clinical setting. Enteral supplementation using small-bore feeding tubes is probably the treatment of choice, because the aforementioned studies showed similar results and the cost of enteral nutrition is less than PPN. Further investigation is needed to clarify the role of nutrition support with regards to other therapies used for this condition, such as corticosteroids and propylthiouracil.

Branched-Chain Amino Acids for Hepatic Encephalopathy

The accumulation of aromatic amino acids (AAA) is caused by an increase in the breakdown of endogenous proteins along with an increase in the oxidation of BCAA. This accumulation leads to an imbalance in the plasma concentration of these amino acids. This imbalance in the ratio of AAA to BCAA is believed to play a role in the development of hepatic encephalopathy secondary to the accumulation of false neurotransmitters within the central nervous system. Because the accumulation of false neurotransmitters and other toxins has been implicated in the development of hepatic encephalopathy, this theory led to the development of a BCAA-enriched and AAA-restricted parenteral solution for its treatment by Fischer.[22] Enteral formulas with BCAA enrichment have also been developed for patients with chronic and latent hepatic encephalopathy.

The parenteral solutions, which have been shown to improve survival in encephalopathic animals, have been studied extensively in patients with acute hepatic encephalopathy. Overall, the number of randomized controlled clinical trials have been few. In only one of these studies was the encephalopathy and mortality clearly shown to improve.[23] However, only one of the randomized controlled trials comparing parenteral BCAA-enriched TPN used standard amino acids (SAA) in the control solution[24]; all of the other trials used lactulose or neomycin along with intravenous solutions containing large amounts of dextrose as the control arm.[25] In that study, there was no difference in outcome between the BCAA and the SAA group. Only one other study provided isonitrogenous quantities BCAA-enriched and SAA TPN; this study of patients with cirrhosis who were given postoperative TPN found there was no clinical advantage for the BCAAs over the SAAs.[26]

It is difficult to compare the results of these trials because of the varying degrees of hepatic encephalopathy upon entry, the differences in the parenteral solutions used, and the dissimilar end points examined. Despite these confusing results, we still believe there are situations where BCAA-enriched hepatic solutions should be used. The first line of action in the management of hepatic encephalopathy should be to treat the precipitating event, such as dehydration, electrolyte and acid–base imbalances, and infection. In most cases, mental status will improve with this strategy. The SAAs can be used for these patients if they require nutrition support and can usually be dosed at 1 g/kg·day^{-1} without difficulty. When it appears that an SAA formula cannot be tolerated, even at a restricted dose, then a BCAA formula may be used so that further deterioration in mental status can be prevented, while providing adequate quantities of amino acids to meet nitrogen requirements.

Goals of Nutrition Support in Liver Disease

Dietary counseling is advised in all ambulatory patients with advanced liver disease to assure that an adequate amount of protein and a properly balanced diet is consumed. Protein restriction should only be imposed when hepatic encephalopathy cannot be managed with conventional medications. In general, a protein intake of 1.0 to 1.2 g/kg is tolerated well in most patients. A bedtime snack should also be provided to avoid depletion of intrahepatic glycogen stores which can promote breakdown of skeletal muscle protein.

Patients who are unable to eat, as may occur in hospital settings, may require supplementation with either enteral or parenteral nutrition. Both enteral and parenteral nutrition have been shown to improve hepatocellular function; but these therapies should be directed only toward repleting patients with moderate-to-severe malnutrition. Enteral nutrition is preferred because it is more physiologic, safer, and less expensive than the parenteral route. The SAA formulas should be used in most patients except when encephalopathy cannot be controlled with conventional medical therapy. As previously noted, BCAA-enriched hepatic solutions should be limited to a select group of patients with hepatic encephalopathy refractory to standard therapy who require nutrition support.

PULMONARY DISEASE

The primary function of the respiratory system is to allow oxygen and carbon dioxide to move between the gaseous phase of the atmosphere and the liquid phase of the blood, so that protein, carbohydrate, and fat metabolism can occur. Given this central role in metabolism, it is no surprise that malnutrition can adversely effect gas exchange. The prevalence of malnutrition in patients with stable chronic obstructive pulmonary disease (COPD) is as high as 19% to 24%.[27,28] Studies have shown that inadequate intake of protein and energy can adversely effect central respiratory drive, respiratory muscle function, and structural and functional integrity of the lung parenchyma. Conversely, pulmonary diseases can result in diminished appetite, decreased oral intake of food and an increase in energy demand resulting in malnutrition. This complex interaction underscores the importance of providing timely nutrition support to patients with pulmonary disease. We will now discuss the interaction of malnutrition and pulmonary function, the impact of malnutrition and specific nutrients in lung function, and recommendations on how to provide effective and safe nutrition support to these patients.

Pathogenesis of Malnutrition in Pulmonary Disease

Anorexia is a frequent complaint of COPD patients and in some cases is believed to be due to hyperinflation of the lung, which results in the feeling of abdominal fullness and early satiety. Alternatively, inflammatory mediators such as tumor necrosis factor may play a role in the development of anorexia in these patients.[29] Energy requirements may be increased as a result of an increase in the work of breathing.[30] Whereas the daily energy cost of breathing ranges from 36 to 72 kcal in normal subjects, this may increase tenfold to 430 to 720 kcal in patients with significant COPD. Patients with cystic fibrosis frequently develop malabsorption due to pancreatic and biliary insufficiency.[31] These factors contribute to the development of malnutrition, which adversely effects morbidity and mortality in these patients.[32]

Impact of Malnutrition on Pulmonary Function

Protein-calorie malnutrition (PCM) impairs several structural and functional components of the respiratory system. Ventilatory drive to hypoxemia has been shown to decrease in healthy volunteers subject to brief courses of starvation.[33] The diaphragm is composed of a mixture of slow-twitch (type II) and fast-twitch (type I) muscle fibers, which despite their continued use, are not spared from the effects of malnutrition. An autopsy study in patients with COPD who lost 31% of their usual weight demonstrated a loss of diaphragmatic muscle mass of 43%.[34] This loss appears to predominantly affect the fast-twitch muscle fibers, which rely on glycogen for their primary source of fuel.[35] Starvation has also been shown to diminish the production of surfactant and to decrease the number and size of alveoli.[36] Functionally, these changes result in respiratory muscle weakness with a decrease in vital capacity, resting minute ventilation, and maximal voluntary ventilation along with inefficient gas exchange.[33,37,38] Furthermore, PCM has been shown to have a direct impact on several components of the defense mechanisms aimed against respiratory infections. These include a decrease in the production of secretory immunoglobulin A,[39] an impairment of alveolar macrophage recruitment and function,[40] and an increase in bacterial adhesion to respiratory epithelium.[41] Collectively, these pathophysiologic events lead to an increased risk of nosocomial pneumonia, prolonged mechanical ventilation, and an increase in mortality in patients who suffer from acute and chronic respiratory failure.[42–44]

Nutrient Impact on Pulmonary Function

Nutrition support has two opposing effects on ventilation for patients with marginal respiratory reserve capacity. The initiation of nonvolitional feeding increases carbon dioxide production ($\dot{V}CO_2$) and ventilatory drive because of diet-induced thermogenesis and thus may put a substantial load on the respiratory system, which may make it difficult to wean a patient from mechanical ventilation. Prolonged nutritional therapy, on the other hand, increases serum albumin, reduces extravascular water, and improves respiratory function, leading to a favorable effect on lung parenchyma and surfactant metabolism, which may make weaning easier. We will now consider some of the physiologic effects of macro- and micronutrients on respiratory function.

Carbohydrate. The provision of carbohydrates are essential to provide oxidative fuel for certain tissues, including the central nervous system, the renal medulla, and red and white blood cells. The average adult requires 125 g of glucose each day to meet this requirement.[45] The upper limit of glucose that can be infused over a 24-hr period is 600 g. This has been demonstrated in injured patients where the maximal rate of glucose infusion to allow complete oxidation ranges from 4 to 7 $mg/kg \cdot min^{-1}$ (20–34 $kcal/kg \cdot day^{-1}$) with the lower end of this range being the oxidative limit for the patients with more severe stress.[46] Higher infusion rates of glucose exceeds the oxidative capacity of the body and results in fat synthesis and an increase in carbon dioxide production and the respiratory quotient (RQ). This increase in carbon dioxide can cause respiratory distress in patients with compromised pulmonary function and may prolong the need for mechanical ventilation.[47,48] Where past studies have focused on the impact of carbohydrate infusion on $\dot{V}CO_2$, more recent data underscores the importance of excessive total calories on increases in $\dot{V}CO_2$. Talpers et al. demonstrated that an increase in energy supply from 1.5 to 2.0 $\times$ REE had a significant effect on $\dot{V}CO_2$, whereas little change was noted when the percentage of calories as carbohydrate was increased for a fixed energy intake of 1.3 $\times$ REE.[49] Ambulatory patients with COPD also appear to be adversely effected by diets high in carbohydrate.[50]

Fat emulsions. Intravenous fat emulsion (IVFE) may be used as a valuable source of calories in patients with respiratory failure. The substitution of fat for glucose lowers the RQ and can result in a decreased minute ventilation and ventilatory demand.[51] Along with these benefits, the adverse effects of IVFE, which include pulmonary and immune dysfunction, must be considered. Rapid infusion of IVFE can result in vasoconstriction of the pulmonary circulation with ventilatory profusion mismatching.[52,53] When these emulsions are administered slowly, there is a tendency for vasodilatation of the pulmonary vasculature.[52] Although the mechanism of these effects on the pulmonary circulation is not entirely known, it appears that the rate of clearance of IVFE is limited and that lipid particles may coalesce and impair gas exchange at the alveolar-capillary level. Excessive quantities of *n*-6 polyunsaturated fatty acids may also favor the production of vasoconstrictive prostaglandins. Furthermore, rapid infusion of IVFE can impair reticuloendothelial system function of the liver and limit the clearance of bacteria from the systemic circulation, thereby enhancing pulmonary uptake of bacteria and the development of pneumonia.[54] The provision of a conservative dose of fat emulsion in the range of 20% to 40% of nonprotein calories should be infused over a period of 12 to 24 hr to avoid these adverse effects.

Protein. Adequate protein is necessary to enhance protein synthesis and maintain lean body mass. Infusion of standard amino acids results in an increase in central ventilatory drive and minute ventilation.[55] Part of this effect seems to be the result of enhanced central chemosensitivity to carbon dioxide. The ventilatory response to carbon dioxide inhalation is greater after the infusion of a BCAA formula, demonstrating that the composition of amino acids may effect the ventilatory response.[56] Even though protein may adversely effect ventilation in patients with compromised respiratory reserve, the development of protein malnutrition is far more grave and so protein restriction is not advised. We currently provide a generous protein dose of 1.5 $g/kg{\cdot}day^{-1}$ in patients with a moderately severe injury response. This dose may also be given to patients who are unstressed and nutritionally depleted. Stable unstressed patients should receive 1 $g/kg{\cdot}day^{-1}$.

Indirect calorimetry is a technique whereby the rate of carbon dioxide production and oxygen consumption can be measured and used to determine energy requirements and macronutrient utilization for a given subject. A portable device, often referred to as a metabolic cart, can be used to easily make these determinations at the patient's bedside. Resting energy expenditure (REE), which is determined with a patient at rest in a thermoneutral environment after an overnight fast, is calculated by the metabolic cart using the formula derived by Weir: $\text{REE} + (3.94 \times \dot{V}O_2) + (1.11 \times \dot{V}CO_2) - (2.17 \times \text{urinary nitrogen})$.[57] The RQ, which describes the ratio of carbon dioxide production to oxygen consumption for the metabolism of a given macronutrient, can be used to adjust the mixture of fuels in a nutrient prescription to avoid the detrimental effects of overfeeding. The RQ for the oxidation of fat, protein, and carbohydrate is 0.7, 0.8, and 1.0, respectively. The synthesis of fat from carbohydrate given in excess of energy requirements has a theoretic value of 8.0, but in the clinical setting is usually seen as an RQ slightly greater than 1.0. When energy is provided as a mixed-fuel system, the RQ approximates 0.85. Mechanically ventilated patients who are difficult to wean from the ventilator or those who are ventilator dependent who show no improvement in their nutritional status are good candidates for indirect calorimetric measurement. Indirect calorimetry may also be useful where over- and underfeeding leads to undesirable effects. Clinical conditions that fit these parameters include the following: persistently elevated liver associated enzymes, increased blood sugars that are difficult to manage, and where further weight loss in the severely marasmic or weight gain in the morbidly obese patient would be detrimental.

Micronutrients. Patients who have respiratory disease or respiratory infections appear to be particularly susceptible to hypophosphatemia.[58] Ficcadori demonstrated that muscle phosphorous content is usually reduced in patients with COPD who have hypophosphatemia.[59] Gravelyn reported a high prevalence of respiratory muscle weakness in hospitalized patients with hypophosphatemia and that maximal inspiratory and expiratory pressures improved significantly after oral repletion of phosphorous.[60] Oxygen transport is also affected by hypophosphatemia, because the oxygen–hemoglobin dissociation curve is shifted to the left with low levels of 2,3-diphosphoglycerate. Hypophosphatemia can result under a variety of circumstances in patients with lung disease including initiation of nutrition support, increased urinary losses from corticosteroids and theophylline, and in alkalosis. Therefore, serum phosphorous concentrations should be monitored regularly in patients with respiratory disease, especially those who require mechanical ventilation and specialized nutrition support.

Potassium, calcium, and magnesium must also be monitored since depletion of these electrolytes can also impact on respiratory muscle function.[61–63]

Specific Lung Diseases

Chronic obstructive pulmonary disease. The overall incidence of malnutrition in patients with COPD is approximately 25%. Changes in body weight closely correlates with morbidity and mortality.[30,32] Nutritional depletion is more common in patients with emphysema as compared with bronchitis.

Cystic fibrosis. It seems that malnutrition and gross retardation in cystic fibrosis patients are due to unfavorable energy balance rather than due to inherent factors of the disease itself.[64] Excess energy consumption from the increased work of breathing, together with inadequate oral intake and malabsorption all contribute to malnutrition in these patients. The prognosis of patients with cystic fibrosis seems to be linked primarily to their nutritional status.[65,66] In addition to the adverse impact on muscle strength and functioning of pulmonary parenchyma, malnutrition impairs immunologic function. Patients with cystic fibrosis may also have central fatty acid deficiencies because of the accompanying biliary tract disease. It has been suggested that the provision of adequate quantities of polyunsaturated fatty acids may be beneficial in patients with cystic fibrosis by providing sufficient substrate for prostaglandins, which may improve underlying pulmonary inflammation.[67] The energy and protein needs of patients with cystic fibrosis depend on the age of the patient and specific complications of the disease.[68]

ARDS. Aggressive treatment of Adult Respiratory Distress Syndrome (ARDS) with mechanical ventilation may lead to alterations in gastrointestinal motility and ileus.[69] The data in the literature regarding nutrition support in patients with ARDS is scant. A small prospective study[53] demonstrated that lung function worsened in patients with ARDS following a 4-hr infusion of 500 mL of 10% intralipid. As patients with other respiratory diseases who are critically ill, one should maintain patients with ARDS on moderate calorie intakes.

Lung transplantation. Lung transplantation has become an accepted treatment for selected patients with end-stage lung disease. Following lung transplantation, malnourished patients with end-stage lung disease were able to achieve normal nutrition within 1 year posttransplantation.[70] There are currently no specific dietary recommendations for patients following lung transplantation.

RENAL DISEASE

Acute renal failure occurs in approximately 5% of all hospitalized patients and is associated with a mortality rate of 50% to 90%.[71,72] The incidence of acute renal failure is certainly greater than this in the intensive care setting where factors such as hemodynamic instability, sepsis, and exposure to nephrotoxic drugs are common. In addition, patients with chronic renal insufficiency may develop acute reduction in renal function as a result of intercurrent infections, such as pneumonia or peritonitis, or complications of their underlying illness, such as atherosclerotic heart disease and diabetes mellitus. Protein-calorie malnutrition is associated with many of the conditions that lead to acute renal failure and, without timely treatment, can adversely effect morbidity and mortality. Aggressive nutrition support has been shown to improve survival in these patients.[73,74] Recent advances in continuous renal replacement therapy (CRRT) have made it easier to provide adequate nutrition to these patients.

Chronic renal failure currently effects more than 200,000 patients in the United States. Despite advances in technology over the past two decades, mortality rates continue to be as high as 23.4%/year.[75] Chronic renal failure is also commonly associated malnutrition. The prevalence of mild-to-moderate and severe malnutrition in patients who require hemodialysis has been reported to be 30% and 8%, respectively.[76] In patients on chronic ambulatory peritoneal dialysis, the prevalence of moderate-to-severe malnutrition is as high as 40%.[77] Malnutrition may contribute to several aspects of the uremic syndrome, including decreased resistance to infection, impaired wound healing, and suboptimal rehabilitation and quality of life. A broad range of nutritional interventions are available to these patients ranging from oral supplements to intradialytic parenteral nutrition.

In this section, we review the pathogenesis of malnutrition in patients with renal disease, discuss the impact of renal replacement therapies on nutritional status and therapies, and provide our general approach to these patients.

Pathogenesis of Malnutrition in Renal Disease

As with other patients who experience the metabolic response to injury, patients with acute renal failure also demonstrate altered protein metabolism with a marked increase in skeletal muscle protein degradation, a decrease in skeletal muscle protein synthesis and an increase in the synthesis of acute-phase proteins. The diseased kidney has a diminished capacity to metabolize some of the polypeptide hormones that direct the injury response including insulin, glucagon, and parathyroid hormone.[78] The metabolic acidosis associated with acute renal failure can also hasten skeletal muscle protein breakdown. These factors contribute to the marked rate of protein catabolism seen in patients with acute renal failure and can result in a protein loss of 150 to 200 g/day.[79]

Several factors contribute to the development of malnutrition in end-stage renal disease. Poor oral intake secondary to uremia and its associated gastrointestinal symptoms is one of the major reasons. A marked spontaneous decrease in dietary protein intake has been shown to correlate with a decline in creatinine clearance.[80] Uremia-induced gastroparesis, medications with gastrointestinal side effects, diets that taste poorly because they are restricted in protein, sodium, potassium, and phosphate, economic hardship, and depression all contribute to diminished oral intake.

The adequacy of dialysis also has an influence on the development of malnutrition. In the past, the standard measurement of predialysis blood urea nitrogen (BUN) less than 100 mg/dL was judged as adequate. The advantage of this measure is that it is readily available, inexpensive, and easy to interpret. A second measure is the urea reduction ratio (URR) which is defined as the percentage reduction of BUN concentration during a single dialysis treatment. In a retrospective study of 13,473 patients with end-stage renal disease treated with hemodialysis, a high risk of death was seen in patients with an URR of less than 60%.[81] This study also points out the important interaction between dialysis dose and protein intake, because the only other factor found to adversely effect mortality was a serum albumin concentration of less than 4.0 g/dL. Unfortunately these measures do not account for protein intake and tissue catabolism that occurs between dialysis sessions.

The impact of dialysis dose and adequate nutrition on survival was appreciated by the authors of the National Cooperative Dialysis Study who advised that the time-averaged BUN concentration be at or below 50 mg/dL while maintaining a protein intake of at least 0.8 g/kg·day^{-1}.[82,83] In this study, the daily protein intake was determined by calculating the protein catabolic rate (PCR).[84] As one might expect, the PCR has also been shown to correlate with morbidity and mortality. Patients with estimated protein intakes, as measured by PCR, of less than 0.8 g/kg·day^{-1} have a greater number of hospitalizations, a longer length of stay, and a greater likelihood of mortality than patients who consume more protein.[85] Most recently, urea kinetic modeling (UKM), which measures urea movement from the patient during and between dialyses, has been used to assess adequacy of dialysis. It has been expressed using the fraction Kt/V, where K is the dialyzer urea clearance, t is time on dialysis in minutes, and V is the volume in liters, of the body pool that contains urea of the patient being dialyzed. A linear relationship has been demonstrated between Kt/V and PCR, suggesting that an adequate dose of dialysis is an important contributing factor to adequate oral intake and malnutrition in chronic dialysis patients.[86] Currently, it is recommended that Kt/V be measured monthly to assess adequacy of dialysis and that a level of 1.2 to 1.4 be attained. It has also been suggested that PCR and serum albumin be used as independent measures of nutritional status.[87]

Nutrition Assessment

Serum albumin has been shown to predict morbidity and mortality for hospitalized patients with renal failure.[88] The relative risk of death is two times greater for patients with serum albumin between 3.5 g/dL and 3.9 g/dL, and five times greater for patients with values between 3.0 g/dL and 3.4 g/dL.[89] Transferrin is another visceral protein that correlates with nutritional status; however, its level may be falsely elevated in patients with iron-deficient anemia, chronic blood loss, and dehydration, all of which are common in patients with renal failure. Transthyretin (thyroxin-binding prealbumin) and retinal-binding protein have also been used to measure the visceral protein compartment, but, because they are filtered and metabolized by the kidney, their concentrations are not reliable markers of malnutrition in patients with kidney disease. The anthropometry of dialysis patients should be compared with standards derived from a stable dialysis population. It should be noted that anthropometric variables in diabetic dialysis patients are higher than nondiabetic patients because of a high prevalence of obesity.[90]

Protein and Energy Requirements in Renal Failure

Critically ill patients with acute renal failure have the same protein and energy requirements as any other patient in intensive care. It must be realized that the metabolic rate of a given patient depends on the underlying cause of the renal disease and on coexisting conditions. Indirect calorimetry has shown that the presence of renal disease and its severity has no direct impact on energy metabolism.[91] Therefore, patients with acute renal failure and burns, trauma, or sepsis are far more catabolic than patients with renal failure secondary to nephrotoxic drugs. The Harris–Benedict equation should be used to calculate basal energy requirements, which are then multiplied by an activity and stress factor to determine total energy expenditure (TEE). This calculation usually estimates TEE to range from 25 to 35 kcal/kg·day^{-1}. Indirect calorimetry may be performed in patients with a prolonged intensive care unit stay to make certain that energy requirements are being met. Protein requirements may range from 1.5 to 1.8 g/kg·day^{-1} to achieve a nitrogen balance.[92] Murray has outlined a formula where protein requirements can be established using UKM.[93] However, because these methods are quite cumbersome, we use them only in the most challenging cases.

Patients with chronic renal failure have energy requirements that are only slightly greater than basal needs and range from 25 to 30 kcal/kg·day^{-1}.[94] As mentioned previously, protein intake must be sufficient so that adequate protein synthesis occurs. In patients with stable chronic renal failure, we provide 0.8 g/kg·day^{-1}.

Renal replacement therapy can also alter protein and energy requirements. The dialysis membrane used in hemodialysis can activate cytokines, the complement system, and leukocytes that induce tissue catabolism.[95] In hemodialysis, 9 to 13 g of amino acids may be lost into the dialysis solution.[96] Chronic ambulatory peritoneal dialysis (CAPD) patients can lose between 5 and 12 g protein/day.[97] This protein loss consists largely of albumin, immune globulins, and free amino acids. This loss may be increased further during episodes of peritonitis. We generally attend to these loses by providing an adequate amount of protein in the nutritional prescription for the patient.

Provision of full support often requires dialytic therapy to avoid the complication of uremia and fluid overload. Besides protein and calorie needs, fluid status, electrolyte balance, vitamins and trace elements are important aspects of nutritional therapy. It is of utmost importance that food and protein restriction not be imposed primarily to delay or to lessen the intensity of dialysis.

Continuous Renal Replacement Therapy in Acute Renal Failure

Conventional hemodialysis and peritoneal dialysis are often not possible in critically ill patients with acute renal failure as a result of hemodynamic instability.

Continuous renal replacement therapy (CRRT) can offer adequate control of azotemia and fluid balance so that nutrition support can be given. The loss of nutrients, including glucose, protein, vitamins, and trace minerals, is small relative to conventional forms of dialysis.[98] The goals of CRRT are often the same as conventional hemodialysis. Various modalities of CRRT are available and include continuous arteriovenous hemofiltration (CAVH), continuous veno-venous hemofiltration (CVVH), continuous arteriovenous or veno-venous hemodiafiltration (CAVHD and CVVHD), and slow continuous ultrafiltration (SCUF). The choice of which form of CRRT is to be used is mainly influenced by personal preference and availability of equipment.

Essential Amino Acids in Renal Disease

Specialized renal disease solutions were initially designed to improve nitrogen balance and retard or reverse the accumulation of nitrogenous waste. It was believed that urea would accumulate in the gastrointestinal lumen in patients with renal failure and that bacterial ureases would produce carbon and ammonia fragments that would be absorbed and passed via the portal circulation to the liver where they would be converted to nonessential amino acids (NEAA). Preliminary studies by Wilmore[99] and Dudrick[100] showed that parenteral essential amino acid (EAA) solutions reduce the adverse metabolic and clinical consequences of uremia and diminish the requirement for dialysis in patients with acute renal failure.[101] Abel performed a prospective randomized controlled study comparing an EAA solution with hypertonic dextrose to hypertonic dextrose alone in 53 patients with acute renal failure. The survival rate from the initial episode of renal failure was improved (75% vs 44%, $p = 0.02$); however, overall hospital survival was no different. Subsequent studies that compared EAA to standard amino acid solutions failed to show improvement in nitrogen balance or survival.[102,103] Therefore, it appears that EAA solutions offer no benefit over standard amino acid solutions for the management of acute renal failure.

Ambulatory patients with chronic renal failure have been maintained in nitrogen balance while decreasing the endogenous breakdown of protein when provided with only 2 g EAA nitrogen along with adequate calories, vitamins, and minerals.[104,105] Why then do these EAA solutions fail to show benefit in patients with acute rather than chronic renal failure? The first reason may be that stable patients with chronic renal failure who are placed on low-protein diets undergo an adaptive phase where there is initially a net loss of nitrogen followed by the development of nitrogen balance. Patients with acute renal failure often experience an acute injury response and do not undergo this adaptative response. Secondly, urea cycling may not occur to a sufficient degree to provide substrate for NEAA or urea cycling may not occur at all.[106] Lastly, the amino acid solutions in these earlier studies may not have provided an adequate amount of nitrogen and histidine, which has now been shown to be essential in patients with renal failure.[107,108]

Although protein restriction may be tolerated in patients with chronic renal failure, prolonged restriction in stressed patients with renal insufficiency will ultimately lead to significant muscle wasting and a kwashiorkor-like state with immune dysfunction and poor wound healing. In general, we use standard parenteral amino acid solutions in patients with acute renal failure and do not endorse the use of EAA. In instances where severe intolerance to standard amino acids occurs despite intensive dialysis, we provide a mixture of equal amounts of EAA with standard amino acids in a dose of 1 g/kg·day^{-1}. Our rationale is to limit azotemia and to potentially provide a variety of amino acids, some of which may be conditionally essential for the critically ill. In addition, we give most patients with chronic renal failure standard amino acid formulas and diets in amounts described below. Enteral tube feeding for renal failure is used on occasion following the criteria as stated for the use of parenteral EAA.

Intradialytic Parenteral Nutrition

Smolle et al.[109] studied the use of intradialytic parenteral nutrition in malnourished patients with chronic renal failure. They provided an intradialytic parenteral nutrition (IDPN) in a dose of 0.8 g amino acid/kg three times a week for 16 weeks and showed improvement in visceral protein concentration and immune competents. The plasma amino acid profile did not change as a result of this intervention.[109] Although the use of IDPN seems attractive for malnourished patients with renal failure and poor gastrointestinal function, no prospective studies of this therapy have been performed, and so its general use can not be endorsed at this time.

Diet Therapy

Diet is an important part of the total rehabilitation of end-stage renal disease and dialysis patients. The goal of nutritional therapy is to promote and maintain adequate nutritional status without causing electrolyte and fluid imbalance or increasing uremic symptoms. The Modification of Diet in Renal Disease study (MDRD) has shown that low-protein diets have no demonstrable benefit in patients with moderate degrees of chronic renal insufficiency and that these patients should receive standard minimal protein allowances of 0.8 g/kg·day^{-1}.[110,111] Patients with Glomerular Filtration Rates (GFR) of less than 25 may benefit slightly from very-low-protein intake, but this is difficult to achieve and requires monthly monitoring of nutritional safety, including measurement of body weight, serum albumin, and transferrin concentrations. The MDRD experience supports the safety of low-protein diets only within the achieved level rather than the prescribed levels, and only for limited durations of time which average 2.2 years.

We currently provide three oral diets for patients with renal insufficiency. The first is a hemodialysis diet, which is restricted in protein, phosphorous, and potassium. The protein allowance in this diet is between 1.0 and 1.2 g/kg. The second diet is a protein restricted diet that provides between 0.6 and 0.8 g/kg. This is intended for patients with renal failure who are not yet on renal replacement therapy. A CAPD diet is restricted in phosphorous but high in protein to account for the modest loss of protein and amino acids seen in patients with this therapy.

REFERENCES

1. McCullough AJ, Mullen KD, Smanik EJ, et al. Nutritional therapy and liver disease. *Gastroenterol Clin North Am* 1989;18:619–643.
2. Munro HN, Fernstrom JD, Wurtman RJ. Insulin, plasma amino acid imbalance and hepatic comas. *Lancet* 1975;1:722–724.
3. Mezey G. Intestinal function in chronic alcoholism. *Ann NY Acad Sci* 1975;252:215–227.
4. Chowla RK, Wolf DC, Kutner MH, et al. Choline may be an essential nutrient in malnourished patients with cirrhosis. *Gastroenterology* 1989;97:1514–1520.
5. Swart GR, Zillikens MC, van Vuure JK, et al. Effect of a late evening meal on nitrogen balance in patients with cirrhosis of the liver. *Br Med J* 1989;299:1202–1203.
6. Hehir DJ, Jenkins RL, Bistrian BR, et al. Nutrition in patients under-going orthotopic liver transplantation. *J Parent Enteral Nutr* 1985;9:695–700.
7. Vitale GC, Neill GD, Fenwick MK, et al. Body composition in the cirrhotic patient with ascites: assessment of total exchangeable sodium and potassium with simultaneous serum electrolyte determination. *Am Surg* 1985;15:675–681.
8. Sheldon GF, Petersen SR, Sanders R. Hepatic dysfunction during hyperalimentation. *Arch Surg* 1978;113:504–508.
9. Kaminski DL, Adams A, Jellinek M. The effect of hyperalimentation on hepatic lipid content and lipogenic enzyme activity in rats and man. *Surgery* 1980;88:93–100.
10. Hall RI, Grant JP, Ross LH, et al. Pathogenesis of hepatic steatosis in the parenterally fed rat. *J Clin Investigation* 1984;74:1658–1668.
11. Degott C, Messing B, Moreau D, et al. Liver phospholipidosis induced by parenteral nutrition: histologic histochemical, and ultrastructural investigations. *Gastroenterology* 1988;95:183–191.
12. Kearns PJ, Young H, Garcia G, et al. Accelerated improvement of alcoholic liver disease with enteral nutrition. *Gastroenterology* 1992;102:200–205.
13. Cabre E, Gonzalez-Huix F, Abad-Lacruz A, et al. Effect of total enteral nutrition on the short-term outcome of severely malnourished cirrhotics. *Gastroenterology* 1990;98:715–720.

14. Christie ML, Sack DM, Pomposelli J, Horst D. Enriched branched-chain amino acid formula versus a casein-based supplement in the treatment of cirrhosis. *J Parent Ent Nutr* 1985;9:671–678.

15. Marchesini G, Dioguardi FS, Bianchi GP, et al. Long-term branched-chain amino acid treatment in chronic hepatic encephalopathy. *J Hepatol* 1991;11:92–101.

16. McGhee RD, Henderson JM, Millikan Jr WJ, et al. Comparison of the effects of hepatic-aid and a casein modular diet on encephalopathy, plasma amino acids and nitrogen balance in cirrhotic patients. *Ann Surg* 1983;197:288–293.

17. Bonkovsky HL, Fiellin DA, Smith GS, et al. A randomized, controlled trial of treatment of alcoholic hepatitis with parenteral nutrition and oxandrolone. I. Short-term effects on liver function. *Am J Gastroenterol* 1991;86:1200–1208.

18. Nasrallah SM, Galambos JT. Amino acid therapy of alcoholic hepatitis. *Lancet* 1980;2:1276–1277.

19. Diehl AM, Boitnott JK, Herlong HF, et al. Effect of parenteral amino acid supplementation in alcoholic hepatitis. *Hepatology* 1985;5:57–63.

20. Achord JL. A prospective clinical trial of peripheral amino acid-glucose supplementation in acute alcoholic hepatitis. *Am J Gastroenterol* 1987;82:871–875.

21. Simon D, Galambos JT. A randomized controlled trial of peripheral parenteral nutrition in moderate and severe alcoholic hepatitis. *J Hepatology* 1988;7:200–207.

22. Fischer JE, Funovics JM, Aquirre A, et al. The role of plasma amino acids in hepatic encephalopathy. *Surgery* 1975;78:276–290.

23. Cerra FB, Cheung NK, Fischer JE, et al. Disease-specific amino acid infusion (F080) in hepatic encephalopathy: a prospective, randomized, double-blind, controlled trial. *J Parent Ent Nutr* 1985;9:288–295.

24. Michel H, Pomier-Layrargues G, Aubin JP, et al. Treatment of hepatic encephalopathy by infusion of a modified amino-acid solution: results of a controlled study in 47 cirrhotic patients. In: Capocaccia L, Fischer JE, Rossi-Fanelli F, eds. *Hepatic encephalopathy in chronic liver failure.* New York: Plenum 1984:323–333.

25. Naylor CD, O'Rourke K, Detsky AS, et al. Parenteral nutrition with branched-chain amino acids in hepatic encephalopathy. *Gastroenterology* 1989;97:1033–1042.

26. Kanematsu T, Koyanagi N, Matsumata T, et al. Lack of preventive effect of branched-chain amino acid solution on postoperative hepatic encephalopathy in patients with cirrhosis: a randomized, prospective trial. *Surgery* 1988;104:482–88.

27. Wilson DO, Rogers RM, Wright EC, et al. Body weight in chronic obstructive pulmonary disease; the National Institute of Health intermittent positive-pressure breathing trial. *Am Rev Respir Dis* 1989;139:1435–1438.

28. Schols A, Mostert R, Soeters P, et al. Inventory of nutritional status in patients with COPD. *Chest* 1982;82:586–571.

29. Swedlund AP, Gorelick FS. Cachectin, tumor necrosis factor and anorexia. *J Clin Gastroenterol* 1987;9:256–257.

30. Vandenbergh E, van de Woestijue KP, Gyselen A. Weight changes in the terminal stages of chronic obstructive pulmonary disease. *Am Rev Respir Dis* 1967;95:556–566.
31. Ramsey BW, Farell P, Pencharz PB. Nutritional assessment and management in cystic fibrosis: a consensus report. *Am J Clin Nutr* 1992;55:108–116.
32. Openbrier DR, Irwin MM, Rogers RM, et al. Nutritional status and lung function in patients with emphysema and chronic bronchitis. *Chest* 1983;83:17–22.
33. Doekel RC Jr, Zwillich CW, Scoggin CH, et al. Clinical semi-starvation: depression of hypoxic ventilatory response. *N Eng J Med* 1979;295:358–361.
34. Arora NS, Rochester DF. Effect of body weight and muscularity on human diaphragm muscle mass, thickness and area. *J Appl Physiol* 1982;52:64–70.
35. Arora NS, Rochester DF. Respiratory muscle strength and maximal voluntary ventilation in under nourished patients. *Am Rev Respir Dis* 1982;126:5–8.
36. Sahebjami H, MacGee J. Effects of starvation on lung mechanics and biochemistry in young and old rats. *J Appl Physiol* 1985;58:778–784.
37. Kerr JS, Riley DJ, Lanza-Jacobi S, et al. Nutritional emphysema in the rat: influence of protein depletion and impaired lung growth. *Am Rev Resp Dis* 1985;131:644–650.
38. Rubin JW, Clowes Jr GHA, Macnicol MF, et al. Impaired pulmonary surfactant synthesis in starvation and severe non thoracic sepsis. *Am J Surg* 1972;123:461–467.
39. Martin TR, Altman LC, Alvares OF. The effects of severe protein-calorie malnutrition on anti-bacterial defense mechanism in the rat lung. *Am Rev Resp Dis* 1983;128:1013–1019.
40. Shennib H, Chiu RCJ, Muldar DS, et al. Depression and delayed recovery of alveolar macrophage function during starvation and re-feeding. *Surg Gynecol Obstet* 1984;158:535–540.
41. Niederman MS, Merill WW, Ferranti RD, et al. Nutritional status and bacterial binding in the lower respiratory tract in patients with chronic trachiostomies. *Ann Intern Med* 100:795–800.
42. Driver AG, McAlvey MT, Smith JL. Nutritional assessment of patients with chronic obstructive pulmonary disease and acute respiratory failure. *Chest* 1982;82:568–571.
43. Bassilli HR, Deitel M. Effect of nutritional support on weaning patients off mechanical ventilation. *J Parent Ent Nutr* 1981;5:161–163.
44. Larca L, Greenbaum DM. Effectiveness of intensive nutritional regimens in patients who fail to wean from mechanical bentilation. *Crit Care Med* 1982;10:297–300.
45. Cahill Jr GF. Starvation in man. *N Engl J Med* 1970;282:668–675.
46. Wolfe RR, O'Donnell TF Jr, Stone MD, et al. Investigation of factors determining the optimal glucose infusion rate in total parenteral nutrition. *Metabolism* 1980;29:892–900.
47. Askanazi J, Rosenbaum SH, Hyman AL, et al. Respiratory changes induced by the large glucose loads of total parenteral nutrition. *JAMA* 1980;245:1444–1446.
48. Dark DS, Pingleton SK, Kerby GR. Hypercapnia during weaning: a complication of nutritional support. *Chest* 1985;88:141–143.

49. Talpers SS, Romberger DJ, Bunce SB, et al. Nutritionally associated increased carbon dioxide production: excess total calories vs high proportion of carbohydrate calories. *Chest* 1992;102:551–555.
50. Angelillo VA, Sukhdarshan B, Durfee D, et al. Effects of low and high carbohydrate feedings in ambulatory patients with chronic obstructive pulmonary disease and chronic hypercapnia. *Ann Intern Med* 1985;103:883–885.
51. Askanazi J, Nordenstrom J, Rosenbaum SH, et al. Nutrition for the patient with respiratory failure. *Anesthesiology* 1981;54:373–377.
52. Skeie B, Ankanazi J, Rothkopf MM, et al. Intravenous fat emulsions and lung function: a review. *Crit Care Med* 1988;16:183–194.
53. Hwang TL, Huang SL, Chen MF. Effects of intravenous fat emulsion on respiratory failure. *Chest* 1990;97:934–938.
54. Hamawy KJ, Moldawer LL, Georgieff M, et al. The effect of lipid emulsions on reticuloendothelial system function in the injured animal. *J Parent Ent Nutr* 1985;9:559–565.
55. Weissman C, Askanazi J, Rosenbaum S, et al. Amino acids and respiration. *Ann Intern Med* 1983;98:41–44.
56. Takala J, Askanazi J, Weissman C, et al. Changes in respiratory control induced by amino acid infusions. *Crit Care Med* 1988;16:465.
57. Weir JB. New methods for calculating metabolic rate with special reference to protein metabolism. *J Physiol* 1949;109:1–9.
58. Fischer J, Magie N, Kallman C, et al. Respiratory illness and hypophosphatemia. *Chest* 1983;83:504–508.
59. Fiaccadori E, Coffrini E, Ronda N, et al. Hypophosphatemia in course of chronic obstructive pulmonary disease: prevalence, mechanisms, and relationships with skeletal muscle phosphorus content. *Chest* 1990;97:857–868.
60. Gravelyn TR, Brophy N, Siegert C, et al. Hypophosphatemia-associated respiratory muscle weakness in a general inpatient population. *Am J Med* 1988;84:870–876.
61. Bilbrey G, Herbin L, Carter N, et al. Skeletal muscle resting membrane potential in potassium deficiency. *J Clin Invest* 1973;52:3011–3018.
62. Aubier M, Vires N, Piquet J, et al. Effects of hypocalcemia on diaphragmatic strength generation. *J Appl Physiol* 1985;58:2054–2061.
63. Benotti PN, Bistrian B. Metabolic and nutritional aspects of weaning from mechanical ventilation. *Crit Care Med* 1989;17:181–185.
64. Vaisman N, Pencharz PB, Corey M, et al. Energy expenditure of patients with cystic fibrosis. *J Pediatr* 1987;111:496–500.
65. Huang NN, Schidlow DV, Szatrowski TH, et al. Clinical features, survival rate, and prognostic factors in young adults with cystic fibrosis. *Am J Med* 1987;82:871–879.
66. Corey M, McLaughlin FJ, Williams M, et al. A comparison of survival, growth, and pulmonary function in patients with cystic fibrosis. Toronto and Boston. *J Clin Epidemiol* 1988;41:583–591.
67. Askanazi J, Rothkoph M, Rosenbaum SH, et al. Treatment of cystic fibrosis with long term home parenteral nutrition. *Nutrition* 1987;3:277–279.

68. Pencharz PB, Durie PR. Nutritional management of cystic fibrosis. *Annu Rev Nutr* 1993;13:111–136.
69. Golden GT, Chandler JG. Colonic ileus and cecal perforation in patients requiring mechanical ventilatory support. *Chest* 1975;68:661–664.
70. Madill J, Mauer Jr, de Hoyos A. A comparison of preoperative and postoperative nutritional states of lung transplant recipients. *Transplantation* 1993;56:347–350.
71. Goldstein DJ. Nutrition for acute renal failure patients on continuous hemofiltration. *Nutr Clin Pract* 1988;3:238–241.
72. Hou SH, Bushinsky DA, Wish JB, et al. Hospital acquired renal insufficiency: a prospective study. *Am J Med* 1983;74:243–248.
73. Bartlett R, Mault J, Dechert R, et al. Continuous arteriovenous hemofiltration: improved survival in surgical acute renal failure. *Surgery* 1986;100:400–408.
74. Cerra F. Hypermetabolism, organ failure and metabolic support. *Surgery* 1987; 101:1–14.
75. Proceedings from the Morbidity, Mortality, and Prescription of Dialysis Symposium. Dallas, TX. *Am J Kidney Dis 13* 1989;375–383.
76. Steinman TI, Mitch WE. Nutrition in dialysis patients. In Maher JF, ed. *Replacement of renal function by dialysis.* Dordrecht: *Kluwer Academic* 1989:1088–1106.
77. Young GA, Kopple JD, Lindholm B, et al. Nutritional assessment of continuous ambulatory peritoneal dialysis patients: an international study. *Am J Kidney Dis* 1991;17:462–471.
78. Seidner DL, Matarese LE, Steiger E. Nutritional care of the critically ill patient with renal failure. *Sem Nephrol* 1994;14:(1):53–63.
79. Feinstein EI, Blumenkrantz MJ, Healy M, et al. Clinical and metabolic responses to parenteral nutrition in acute renal failure. *Medicine* 1981;60:124–137.
80. Ikizler TA, Greene J, Wingard RL, et al. Effects of progression of chronic renal failure on spontaneous dietary protein intake in chronic renal failure patients. *J Am Soc Nephrol* 1994;5:332A.
81. Owen WF, Lew NL, Liu Y, et al. The urea reduction ratio and serum albumin concentration as predictors of mortality in patients undergoing hemodialysis. *N Engl J Med* 1993;329:1001–1008.
82. Harter H. Review of findings from the National Cooperative Dialysis Study and recommendations. *Kidney Int* 1983;23(suppl 13):S107–S112.
83. Gotch FA, Sargent JA. A mechanistic analysis of the National Cooperative Dialysis Study (NCDS). *Kidney Int* 1985;28:526–534.
84. Borah MF, Schoenfeld PY, Gotch FA, et al. Nitrogen balance during intermittent dialysis therapy of uremia. *Kidney Int* 1978;14:491–500.
85. Acchiardo S, Moore L, LaTour P. Malnutrition as the main factor in morbidity and mortality of hemodialysis patients. *Kidney Int* 1983;24(Suppl. 16):S199–S203.
86. Lindsay RM, Spanner E. A hypothesis: the protein catabolic rate is dependent upon the type and amount of treatment in dialyzed uremic patients. *Am J Kidney Dis* 1989;13:382–389.

87. Spinowitz BS, Gupta BK, Kulogowski J. Dialysis adequacy in hypoalbuminemic CAPD patients. *Peritoneal Dialysis Int* 1993;13(Suppl 2):S221–S223.
88. Murry MJ, Marsh M, Wocho DN, et al. Nutritional assessment of intensive care unit patients. *Mayo Clinic Proc* 1988;63:1106–1115.
89. Lowrie EG, Lew NL. Death risk in hemodialysis patients: the predictive value of commonly measured variables and an evaluation of death rate difference between facilities. *Am J Kidney Dis* 1990;15:458–482.
90. Gupta BK, Plotner E, Spinowitz BS, et al. Severe malnutrition. *Sem Dialysis* 1993;6(16):366–369.
91. Schneeweiss B, Graninger W, Stockenhuber F, et al. Energy metabolism in acute and chronic renal failure. *Am J Clin Nutr* 1990;55:468–472.
92. Macias WL, Alaka KJ, Murphy MH, et al. Impact of the nutritional regimen on protein catabolism and nitrogen balance in patients with acute renal failure. *J Parent Ent Nutr* 1996;20(1):56–62.
93. Murray R. Protein and energy requirements. In Krey SH, Murray R, eds. *Dynamics of Nutrition Support.* Norwalk, CT: Appleton-Century-Crofts, 1986:185–217.
94. Kopple J, Monteon F, Shaib J. Effect of energy intake on nitrogen metabolism in non-dialyzed patients with chronic renal failure. *Kidney Int* 1986;29:734–742.
95. Gutierrez A, Bergstrom J. Alvestrand A, et al. Protein catabolism in sham-hemodialysis: the effect of different membranes. *Clin Nephrol* 1992;38:20–29.
96. Ikizler TA, Wingard RL, Haking R. Intervening to beat malnutrition in dialysis patients: the role of the dose of dialysis, intradialytic parenteral nutrition, and growth hormone. *Am J Kidney Dis* 1995;26(1):256–265.
97. Blumenkrantz MJ, Kopple JD, Moran JK, et al. Metabolic balance studies and dietary protein requirements in patients undergoing continuous ambulatory peritoneal dialysis. *Kidney Int* 1982;21:849–861.
98. Bommel EFM, Leunissen KML, Weimar W. Continuous renal replacement therapy for critically ill patients: an update. *J Intens Care Med* 1994;9:265–280.
99. Wilmore DW, Dudrick SF. Treatment of acute renal failure with venous essential L-amino acids. *Arch Surg* 1969;99:669–673.
100. Dudrick SF, Steiger D, Long JM. Renal failure in surgical patients: treatment with intravenous essential amino acids and hypertonic glucose. *Surgery* 1970;68:180–186.
101. Abel RM, Beck CH, Abbot WM, et al. Improved survival from acute renal failure after treatment with intravenous essential L-amino acids and glucose: results of a prospective, double-blind study. *N Engl J Med* 1973;288:695–699.
102. Freund J, Hoover H, Atamian S, et al. Infusion of the branched-chain amino acids in postoperative patients. *Ann Surg* 1979;190:18–23.
103. Mirtallo JM, Schneider PJ, Mauko K, et al. A comparison of essential and general amino acid infusions in the nutritional support of patients with compromised renal function. *J Parent Ent Nutr* 1982;6:109–113.
104. Giordano C. Use of exogenous and endogenous urea for protein synthesis in normal and uremic subjects. *J Lab Clin Med* 1963;62:231–246.

105. Givannetti S, Maggiore Q. A low-nitrogen diet with proteins of high biological value for severe chronic uraemia. *Lancet* 1964;1:1000–1003.
106. Varcoe R, Halliday D., Carson ER, et al. Efficiency of utilization of urea nitrogen for albumin synthesis by chronically uremic and normal man. *Clin Sci Mol Med* 1975;48:379–390.
107. Furst P, Ahlberg M, Alvestrand A, et al. Principles of essential amino acid therapy in uremia. *Am J Clin Nutr* 1978;31:1744–1755.
108. Alvestrand A, Bergstrom J, Furst P. Plasma and muscle free amino acids in uremia: influence of nutrition with amino acids. *Clin Nephrol* 1982;18:297–305.
109. Smolle KH, Kaufmann P, Holzer H, et al. Intradialytic parenteral nutrition in malnourished patients on chronic haemodialysis therapy. *Nephrol Dial Transplant* 1995;10:1411–1416.
110. Klahr S, Levey AS, Beck GJ, et al. The effects of dietary protein restriction and blood-pressure control on the progression of chronic renal disease. *N Engl J Med* 1994;330(13)877–884.
111. Dennis VW. Decoding the modification of diet in renal disease study. *Cleveland Clin J Med* 1994;61(4):254–257.

CHAPTER 21

Nutrition Support and Pancreatitis

Virginia M. Herrmann, M.D., and
M. Patricia Fuhrman, R.D., C.N.S.D.

The pancreas gland is located in the upper epigastrium in the retroperitoneal area and has both exocrine and endocrine functions. In adults, the pancreas can secrete from 1 L to greater than 2.5 L of exocrine fluid/day, containing bicarbonate, electrolytes, protein, and various enzymes. The active enzymes secreted by the pancreas gland include amylase, lipase, and phospholipase. Other proteins are secreted in their inactive form (zymogens) and become activated in the small intestine by an enterokinase in the lumen of the small intestine. The endocrine function of the pancreas relates primarily to release of insulin from the beta islet cells. The pancreas serves as an important source of the enzymes needed for digestion. Pancreatic secretion of these enzymes is altered with malnutrition, and this may lead to worsening malabsorption in an already compromised patient. When the pancreas becomes inflamed or diseased as in pancreatitis, there is an increased secretion of enzymes with altered feedback mechanisms leading to adverse sequela.

The pancreas gland is traversed by various ducts, the dorsal duct (duct of Santorini), which is located in the body and the tail of the pancreas, and the ventral duct (duct of Wirsung), which is located in the head of the pancreas. The duct of Wirsung empties into the duodenum along with the common bile duct at the ampulla of Vater. In rare instances, the pancreas drains through the dorsal duct and directly into the duodenum through an accessory ampulla.

The endocrine cells in the pancreas produce peptide hormones, which have both systemic and local or regional effects. Hormones released into the circulation act systemically. The pancreas also has *paracrine* function, whereby a hormone is released locally and has most of its effect in a local or regional area. The endocrine and exocrine function of the pancreas, although closely related, operate independently of one another.

The pancreas is innervated by the vagus nerves and also receives sympathetic innervation from T_5 through T_{10}. Stimulation of the vagus nerves lead to increased pancreatic secretion. Sympathetic stimulation leads to decreased blood flow affecting both the exocrine and endocrine pancreatic function and can actually inhibit pancreatic secretion. Pancreatic juice is generally isotonic and varies in overall volume, as well as the concentration of bicarbonate and enzymes. During periods of bowel rest, pancreatic secretory volume may be negligible, whereas during digestion pancreatic secretion increases dramatically. Pancreatic secretion is generally divided into three distinct but closely interrelated phases. The first phase is known as the *cephalic phase*. The vagus nerves provide direct cholinergic stimulation to the pancreas. Indirect stimulation occurs through release of gastrin from the antrum of the stomach and vasoactive inhibitory proteins (VIP) from the small intestine. The release of gastrin during the cephalic phase stimulates pancreatic enzyme secretion. The cephalic phase is followed by the *gastric phase*. The presence of food within the stomach causes gastric dilation, further stimulating the pancreas and release of gastrin. Gastric distention is mediated by both vagus nerves. Gastrin also acts directly on pancreatic acini to produce other enzymes needed for digestion. The third phase, the *intestinal phase,* is closely related to the gastric phase. Gastrin release from the antrum causes the contents of the stomach to become acidic by stimulating hydrochloric acid release. The presence of acid or acidic content in the duodenum signals the release of secretin from the small intestine, primarily the duodenum.[1] Secretin acts directly on the pancrease to increase fluid and bicarbonate production.

Other gastrointestinal enzymes and hormones impact pancreatic function. Substrates from the digestion of fat and protein in the duodenum cause release of the enzyme pancreozymin, which further stimulates pancreatic secretion. Vasoactive inhibitory protein is a homologue that further stimulates pancreatic bicarbonate secretion. Undigested protein has minimal, if any, stimulatory effect on pancreatic secretion, but polypeptides, oligopeptides, and amino acids have a marked stimulatory effect on the gland. Ingestion of combined amino acids has a greater stimulatory effect on the pancreas than any single amino acid alone. Amino acids that have the most potent effect on pancreatic stimulation include phenylalanine, valine, and tryptophan. The presence of calcium in the duodenum stimulates pancreatic enzyme release and bile salts stimulate both pancreatic enzyme and bicarbonate secretion. Gastrointestinal hormones that stimulate or inhibit the pancreas are interrelated in their function and are often synergistic.

Pancreatic enzymes (amylase, lipase, phospholipase A, trypsinogen, chymotrypsinogen, aminopeptidases, and cholesterol esterase) are stored in their precursor forms, which prevent autodigestion of the pancreas by these enzymes. When the enzymes are released, they are acted upon by enterokinases, which convert them into their active forms. Knowledge of pancreatic secretory function has been the main impetus for attempting bowel rest in patients. Pancreatic secretory function can be markedly reduced by altering the amount of intake and content of the diet. Pancreatic rest can be achieved by eliminating oral intake, or by the administration of inhibitors to pancreatic secretion. The sympathetic nervous system inhibits pancreatic secretion through the splanchnic nerves and circulating catecholamines. Gastrointestinal hormones that inhibit pancreatic secretion include glucagon, somatostatin and pancreatic polypeptide. Pancreatic secretion is inhibited by the presence of trypsin or chymotrypsin in the duodenum, and this effect is mediated by gastrointestinal hormones.[2]

Pancreatic secretory response varies with different substrates. A diet rich in casein, peptides, or amino acids increases proteolytic enzymes, whereas a high-carbohydrate diet increases amylase secretion. High-fat diets are associated with an increase in the production and secretion of lipase. The pancreas can adapt to various dietary intakes. Adaptation is associated with an increase in both the production and the release of various enzymes.[3] Cholecystokinase (CCK) may play a role in this adaptive mechanism. Amylase content of the pancreas adapts to either oral or intravenous glucose. This effect is most likely mediated by insulin. The concentration of lipase in pancreatic secretions similarly responds to either oral fat or intravenous fat emulsions.

ETIOLOGY OF PANCREATITIS

The etiology of pancreatitis is extremely varied. Cholelithiasis is the most common cause of acute pancreatitis in the United States. Chronic pancreatitis is most often related to ethanol abuse and may be further exacerbated by diets high in protein or fat.[3] The association between pancreatitis and hyperlipidemia is well established. Patients with underlying hyperlipidemia are predisposed to pancreatitis, and this effect is aggravated by alcohol intake or abuse. Patients who do not have underlying hyperlipidemia, may develop this condition during an episode of pancreatitis. This is most often observed when the etiology of the pancreatitis is ethanol intake or abuse, as hyperlipidemia is rarely seen with other forms of pancreatitis.[4] The relationship of alcohol and pancreatitis seems well documented in men, but is less clear in women.[5] There does appear to be a dose–response curve that explains the association, demonstrating a direct relationship between the mean daily ethanol intake and the risk of chronic pancreatitis. Alcohol stimulates pancreatic secretion, and this response is directly proportional to the amount and type of alcohol consumed. Sarles et al.[6] identified a particular protein secreted by the pancreas, labeled pancreatic stone protein (PSP). Present in pancreatic secretions, PSP inhibits the precipitation of calcium carbonate within the duct. There is a decreased amount of PSP produced in patients with chronic ethanol abuse and a corresponding increase in the total protein content of pancreatic juice. This facilitates the precipitation of calcium carbonate, or stones, which can obstruct small pancreatic ducts. Pancreatic duct obstruction leads to further acute and chronic changes, and repeated bouts of chronic pancreatitis.

Cholelithiasis is a common cause of acute pancreatitis in the United States. Pancreatitis secondary to cholelithiasis is usually mild or self-limiting.

Acute pancreatitis may be related to a number of pharmacologic agents or medications, including steroids, antibiotics (Tetracycline), furosemide, and thiazide diuretics, as well as some chemotherapeutic agents. Patients who are on chronic medications associated with pancreatitis may suffer repeated bouts and develop a chronic form of this disease. Hyperlipidemia is both a causative factor in pancreatitis, and a sequela of the disease. Other causes of acute pancreatitis include direct trauma to the pancreas gland, viral infections, peptic ulcer disease with posterior penetration, and iatrogenic pancreatitis most often seen today with endoscopic retrograde cholangiopancreatography (ERCP).

In approximately 10% of cases of pancreatitis, the cause is idiopathic. Chronic pancreatitis is most often associated with chronic ethanol abuse and biliary tract disease. Type I and V hyperlipoproteinemia has also been associated with chronic pancreatitis.

Pancreatitis is a general term that encompasses a wide variation in etiology, clinical course, and treatment. Mild pancreatitis is associated with mild edema, minimal enzyme elevation, and relatively low mortality and morbidity. Serum amylase and lipase levels in mild pancreatitis may be minimally elevated and patients experience mild bouts of abdominal pain, nausea, vomiting, and anorexia. Approximately 10% of patients with pancreatitis will develop frank pancreatic necrosis or hemorrhage, and both necrosis and hemorrhage are associated with a high mortality and morbidity.[7,8] Most patients present with a mild form of pancreatitis and edema that resolves in several days with conservative management.[9]

Pancreatitis is determined to be mild or severe on the basis of clinical findings, biochemical or laboratory evaluations, and actual pathologic changes. Ranson and coworkers defined the severity of the disease by a set of criteria which would predict outcome.[10–13] (Table 21.1) The initial five criteria are determined on admission with assessment of the actual degree of local inflammation. The remaining six criteria are assessed within 48 hr of admission or onset of the disease

Table 21.1. Ranson's Criteria

On Admission	After 40 hr
Age > 55 yr.	Calcium < 8 mg/dL
Glucose > 200 mg/dl	Arterial pO_2 < 60 mmHg
WBC > 16,000 mm^3	Base deficit > 4 meq/L
LDH > 350 IU/L	BUN increase > 5 mg/dL
SGOT > 250 U/dl	Hct decrease > 10%
	Fluid sequestration > 6 Liters
Severity of Pancreatitis	**Number of Signs**
Mild pancreatitis	< 3 signs
Moderately severe pancreatitis	3–5 signs
Severe pancreatitis	> 5 signs

and reflect the degree of systemic response. Mortality is higher in patients who demonstrate three or more of these criteria. The patient groups who demonstrate five or more of Ranson's criteria have a mortality of 40%. Other methods to determine prognosis in pancreatitis have also been used. The acute physiology and chronic health evaluation (APACHE II) assessment has been used in other critically ill patients (trauma, sepsis, etc.) and has also been applied to assess outcome in pancreatitis.[14,15] Within the first 48 hr, patients with APACHE II scores ≤ 9 were more likely to survive. Patients with scores ≥ 13 had a significantly higher mortality. Mortality is obviously worse in patients who develop other organ system failure, such as respiratory failure, respiratory distress syndrome, acute renal failure, liver failure, and cardiovascular collapse.

DIAGNOSIS AND ASSESSMENT OF PANCREATITIS

A number of laboratory imaging tests have been used to diagnose and follow patients with pancreatitis. The laboratory data most commonly used are serum amylase and lipase values. Amylase is often elevated in acute pancreatitis, although levels of amylase can actually be normal or low in chronic pancreatitis. Generally patients with pancreatitis have amylase levels greater than 1000 u, and commonly levels are increased to several thousand units. Increased serum lipase levels have also been used as an indicator and means of following the course of pancreatitis. Phospholipase A_2 is a very accurate indicator of pancreatic necrosis.[16,17] Hypocalcemia, hyperglycemia, and hypertriglyceridemia are often noted in acute pancreatitis; therefore, serum levels of calcium, glucose, and triglycerides become important parameters in assessing and following these patients.

Radiological imaging tests are very helpful in the diagnosis and following the disease progress. Computerized tomography (CT) scans with contrast are helpful and currently the most commonly used modality to diagnose and follow patients with pancreatitis. A CT scan usually demonstrates diffuse enlargement of the pancreas with loss of the surrounding fat planes. Sequential CT scanning is used to follow the course of pancreatitis and the CT findings have been closely associated with expected prognosis and mortality. To establish the diagnosis, CT scanning is generally recommended when patients show no improvement in the first 72 hr or in any patient who demonstrates more than three of Ranson's criteria or an APACHE score ≥ 9.[18] Patients generally show clinical improvement much earlier than improvement noted on CT scanning.[19]

METABOLIC ALTERATIONS IN PANCREATITIS

Physiologic changes that occur with pancreatitis induce a state of hypermetabolism and hypercatabolism, similar to that seen in patients with other critical illness and sepsis. The hormonal alterations seen in critical illness (increased catecholamines, insulin, glucagon) affect carbohydrate, protein and fat metabolism, elevating resting metabolism and energy expenditure, and causing other metabolic derangements.

CARBOHYDRATE METABOLISM

Hyperglycemia is a common finding in patients with pancreatitis, even in the absence of underlying diabetes mellitus. The etiology of hyperglycemia is multifactorial and includes increased levels of cortisol and adrenocorticoid trophic hormone (ACTH), increased ratio of glucagon to insulin and peripheral insulin resistance, and decreased rate of glucose oxidation.[19] The administration of total parenteral nutrition (TPN) with hypertonic dextrose further complicates the management of these patients. Significant hyperglycemia requiring substantial insulin replacement has been observed in over 80% of patients with pancreatitis.[20] Outcome has been related to the severity of hyperglycemia and the insulin requirements in these patients, with a significant increase in mortality noted in patients requiring greater than 3 U/hr of insulin.[21]

PROTEIN METABOLISM

Hypercatabolism with loss of lean muscle mass, negative nitrogen balance, and increased ureagenesis, are all observed in acute pancreatitis.[22] Circulating levels of the major gluconeogenic amino acid, alanine, are decreased, and plasma levels of glutamine are markedly depressed, even though other amino acids are rapidly released by skeletal muscle. Depression in plasma alanine and glutamine levels is similar to that seen in other forms of critical illness.[23,24]

LIPID METABOLISM

Hyperlipidemia is frequently noted in acute and chronic pancreatitis, and may precipitate pancreatitis, or be a sequelae of the disease. Patients with type I and V hyperlipoproteinemia are more likely to develop recurrent bouts of pancreatitis. Ethanol abuse leading to acute pancreatitis is frequently associated with hypertriglyceridemia. Decreased clearance of lipids has been observed in this group of patients, necessitating that a triglyceride level be obtained on admission and that follow-up levels be obtained with nutrition therapy.

GOALS: RATIONALE FOR NUTRITION SUPPORT

Nutrition problems are generally observed in more severe or chronic cases of pancreatitis. Patients with severe acute or chronic pancreatitis may be undernourished secondary to the disease process itself, or other underlying diseases, such as ethanol abuse. Approximately 10% of patients with acute hemorrhagic pancreatitis will have significant morbidity and need operative intervention. Most of these patients often have a deterioration of their nutritional status mandating the need for nutrition support.[11,12,20]

ALTERATIONS IN ENERGY METABOLISM

Septic patients with pancreatitis usually are hypermetabolic.[25] However, in the absence of severe infection or sepsis, there is little evidence that pancreatitis increases energy expenditure.[26,27] The hormonal alterations associated with acute pancreatitis and sepsis (increased cortisol and catecholamines) have the overall effect of increasing the resting energy expenditure (REE). The need for caloric support varies with the patient's age, height, weight, severity of disease, and associated fever or sepsis.[25,27] Dickerson et al. evaluated 48 patients with either acute or chronic pancreatitis with or without sepsis.[27] The REE was determined by indirect calorimetry and compared with predicted energy expenditure, as determined by the Harris-Benedict equation. The REE was significantly greater in patients with pancreatitis who were septic (120 ± 11%) compared to those without sepsis (105 ± 14%). However, REE varied widely in patients with pancreatitis (77% to 139% of predicted energy expenditure) and they concluded that the Harris-Benedict equation was an unreliable means of determining caloric needs in this group of patients. Sepsis was the most critical denominator in determining elevations in REE in patients with pancreatitis.

Energy requirements can be determined by the Harris-Benedict equation and adjusted for various stress factors that correlate with the degree of illness.[28] Although determination of energy expenditure by this method is commonly done, its accuracy has been challenged.[25,27,29,30] Most data indicate that determination of caloric needs is overestimated by the Harris-Benedict equation. Resting energy expenditure is most accurately determined by indirect calorimetry (metabolic cart) which should be used when available. Elevations of REE by as much as 30% to 50% may be observed in patients with severe progressive pancreatitis, and in patients with a Ranson score of ≥ 6.[25,30]

CARBOHYDRATE METABOLISM

The goal in patients with pancreatitis is the provision of adequate energy substrate and avoidance of hyperglycemia and hypertriglyceridemia. Parenteral glucose should not be given in an amount that exceeds the glucose oxidation rate (5 mg/kg·min^{-1}). Glucose given in excess of the patient's ability to oxidize this fuel will result in hyperglycemia and fatty liver.[3,21] Glucose infusion should not exceed 3 to 4 mg/kg·min^{-1} in hyperglycemic patients, and glucose should generally provide 60% to 70% nonprotein calories each day, the balance being supplied by fat.[23] A mixed fuel substrate for energy is recommended in patients with pancreatitis.[23,25,27]

Parenteral fat emulsions are frequently used as an energy substrate in patients with pancreatitis. Because patients with pancreatitis are predisposed to hypertriglyceridemia, it is advisable to check serum triglyceride levels initially and then sequentially, particularly if the patient is receiving intravenous fat. Patients with preexisting hyperlipoproteinemia may have a decreased ability to metabolize fat.[31] Most patients with acute pancreatitis that is not secondary to hyperlipoproteinemia or hypertriglyceridemia demonstrate tolerance for fat emulsions. The safety of parenteral fat has been well documented. Patients with pancreatitis usually tolerate intravenous fat at a daily dose of 1 to 3 g/kg without significant elevations in plasma triglyceride levels.[3,32,33] There are a number of benefits associated with providing a portion of needed calories as fat. Patients with pancreatitis have decreased ability to oxidize glucose, and peripheral insulin resistance is common. The use of fat results in decreased incidence of hyperglycemia, with the resultant fatty liver. Nitrogen balance is facilitated in these patients when fat is used as part of a mixed fuel source for energy.[34,35] Patients with pancreatitis generally have a respiratory quotient (RQ) between 0.76 and 0.9, suggesting a preference for mixed fuel utilization. High-fat diets should be avoided in septic patients because of associated immunosuppression with long chain fatty acids.

PROTEIN METABOLISM AND REQUIREMENTS

Amino acid profiles in patients with pancreatitis indicate the need for additional alanine and glutamine supplementation. Studies have demonstrated that parenteral delivery of glutamine dipeptides in patients with pancreatitis helps maintain plasma and skeletal muscle levels of glutamine.[36] Other investigators have documented beneficial effects of glutamine supplementation in stressed or critically ill patients; however, these studies do not specifically address beneficial effects in patients with acute pancreatitis.[36,37]

Protein needs in this patient group approach 1.5 to 2.5 $g/kg \cdot day^{-1}$. When appropriate calories and protein are provided, the calorie-to-nitrogen ratio is 100:1. Provision of protein may be affected by renal and hepatic function, existing degree of protein depletion, presence of sepsis, and the severity of pancreatitis.

METHODS OF NUTRITIONAL SUPPORT: TOTAL PARENTERAL VERSUS ENTERAL NUTRITION

Gut rest for treating pancreatitis is directed toward reducing pancreatic secretory volume (primarily bicarbonate) and minimizing enzyme release.[38–40] The availability of total parenteral nutrition (TPN) allows maintenance of nutritional status during prolonged periods of gut rest. Several studies have shown improvement in patients with pancreatitis who received TPN.[20] Despite the decrease in pancreatic secretions with bowel rest, the clinical efficacy as measured by outcome has yet to be determined. Sax et al. conducted a prospective, randomized trial to evaluate the effect of early TPN in acute pancreatitis compared with intravenous fluids alone and analgesia.[41] This study showed that the group that received TPN had significantly longer hospitalization (16 versus 10 days), and a significantly higher rate of catheter sepsis (10.5% versus 1.5%).[41] Other studies similarly do not confirm clinical benefit of TPN in pancreatitis.[42]

Some animal studies show that intravenous administration of amino acids and fat increases pancreatic secretion, whereas others present conflicting data. Intravenous amino acids in humans does not alter pancreatic enzyme or protein secretion.[43]

The safety and efficacy of intravenous fat in acute pancreatitis has been well demonstrated. A few studies suggest that the infusion of intravenous fat stimulates pancreatic secretions.[44,45] However, these studies involve patients with inflammatory bowel disease who are also receiving other medications, including steroids, which may well have altered lipid kinetics. Most studies in both animals[38,46,47] and humans[20,33,48–51] affirm that TPN with fat emulsions is safe and does not stimulate pancreatic secretion. Controlled trials have demonstrated no increase in triglyceride levels and no change in outcome associated with use of lipid emulsions.[52,53] Even in severe acute pancreatitis, lipid emulsions seem to be well tolerated.[54] Lipid emulsions may be safely given daily to provide 25% to 30% of nonprotein calories. Fat emulsion should be cautiously used in patients with type I and V hyperlipidemia, and in children, with inflammatory bowel disease who are also receiving steroid therapy.[29,44,45] Hypertriglyceridemia has been observed in patients receiving intravenous lipid emulsions and, therefore, may need to be monitored. Serum triglycerides should be measured prior to initiating TPN and carefully monitored during the delivery of parenteral fat. Triglyceride levels should be maintained at less than 400 mg/dL during continuous lipid infusion.[55]

Despite recent trends toward early enteral feeding, some studies strongly suggest that TPN is the preferred method of nutrition support in patients who develop complications from pancreatitis (i.e., worsening severe inflammation, abscess formation, or fistulas).[56–60] Total parenteral nutrition is safe, well tolerated, and does not increase pancreatic secretion or worsen pancreatitis. However, randomized prospective trials have not shown that TPN influences mortality or morbidity despite the decrease in pancreatic secretions, and TPN is associated with certain risks. Central line sepsis associated with TPN has been shown to be significantly higher in patients with pancreatitis.[20,34,41,61] The increased incidence of catheter sepsis may be exacerbated by the hyperglycemia noted in these patients. Hennessey et al. have demonstrated that complement fixation by immunoglobulin G (IgG) is significantly impaired by even transient elevations in blood glucose.[62] Therefore, the risks associated with TPN, including metabolic complications and catheter sepsis, may not justify its use, particularly if enteral nutrition is a reasonable alternative.

Latifi and Dudrick suggest that TPN is preferred in patients during the acute phase of pancreatitis and in any patients who develop pancreatic necrosis, fistula, or abscess. These authors reason that parenteral nutrition reduces pancreatic stimulation and secretions and recommend that TPN be given until acute symptoms resolve.[57,60] This evidence has not been widely supported by other investigators. Jackson et al.[59] examined the effect of TPN in 40 patients with acute pancreatitis who developed pancreatic pseudocyst. The majority of patients (68%) showed pseudocyst regression and clinical improvement. However, 15 patients (35%) developed catheter related complications, 26% of patients developed sepsis, with one patient diagnosed with septic right atrial thrombosis.

ENTERAL NUTRITION IN PANCREATITIS

Enteral nutrition is increasingly used to support patients with pancreatitis despite previous tendencies to encourage bowel rest. Initial studies examined the efficacy of elemental diets in the treatment of acute pancreatitis.[63,64] It has been suggested that elemental formulae are less likely to stimulate pancreatic secretion than more complex formulae. Early studies in animals with pancreatitis fed an elemental diet showed conflicting data. Some investigators showed a significant decrease in the amount and enzyme concentration of pancreatic secretions with elemental feeding.[63,64] Others have shown an increase in pancreatic secretion, most likely attributed to an increase in the amino acid content of elemental diets.[38,65,66] Most early studies evaluated duodenal feeding instead of jejunal administration of the enteral formula. Jejunal administration of formulae has shown no increase in pancreatic stimulation.[67] Guan et al. recently studied animals with pancreatitis and concluded that elemental diets did not stimulate pancreatic secretion in contrast to polymeric diets which did increase pancreatic exocrine output. The differences noted were attributed to the intact protein in the polymeric diet versus the free amino acids in elemental diets.[68]

Early studies in humans have documented the efficacy of elemental diets in pancreatitis.[69,70] Feller et al. retrospectively reviewed the clinical course of 200 patients treated initially with TPN, but then converted to enteral nutrition. Nitrogen balance was achieved and mortality decreased from 21.6% to 14% in this study.[70] More recently, Bodoky et al. compared the use of parenteral with enteral nutrition in patients with pancreatitis and found no significant differences between pancreatic secretion in the two groups.[71]

Kudsk et al. studied the use of enteral support in 11 patients with acute pancreatitis, fed with an elemental diet through a feeding jejunostomy. Patients were followed for a mean of 31 days with no worsening of pancreatitis.[72] Parekh followed the clinical course of 17 patients receiving an elemental diet for a mean of 16 days with similar results.[73] Grant et al. used defined formula diets in patients with severe pancreatitis, and reported no diarrhea or worsening of the disease process.[74] However, almost half the patients in this study required exogenous insulin, to maintain blood glucose levels within acceptable range. Other studies are emerging which document the success of repleting patients using enteral feeding without exacerbating symptoms in acute pancreatitis.[75,76] Some have suggested that there is a need for supplemental pancreatic enzymes to facilitate absorption of both elemental and polymeric formulae in severe pancreatic insufficiency.[77]

Gram-negative sepsis is the most common cause of death in patients with acute pancreatitis. Efforts to alter this outcome include the use of enteral nutrition in an effort to maintain the gut barrier and reduce intestinal permeability. Current research efforts are examining the effects of specific nutrients (glutamine, arginine, short-chain fatty acids, nucleotides) in maintaining immune function and the gut mucosal barrier. The addition of trophic hormones (CCK, bombesin, neurotensin, growth hormone, and epidermoid growth factor) may also contribute to the maintenance of gut mucosal integrity.[78] Current evidence suggests that enteral feeding, preferably in the jejunum, is the optimal method of nutrition support in patients with acute pancreatitis who can tolerate this method of feeding.

REFERENCES

1. Schaffalitzky de Muckadell OB, Fahrenkung J. Role of secretin in man. In Bloom SR, ed. *Gut.* Edinburgh: Churchill Livingstone, 1978:197–201.
2. Slaff J, Jacobson P, Tieman CR, et al. Protease specific suppression of pancreatic exocrine secretion. *Gastroenterology* 1984;87:44–52.
3. McMahon MJ. Diseases of the exocrine pancreas. In Kinney JM, Jejeebhoy KN, Hill GL, Owen OE, eds. *Nutrition and metabolism in patient care.* Philadelphia: W.B. Saunders, 1988.
4. Dickson AP, O'Neill J, Imrie CW. Hyperlipidemia, alcohol abuse and acute pancreatitis. *Br J Surg* 1984;71:685–688.
5. Yen S, Hsieh CC, MacMahon B. Consumption of alcohol and tobacco and other risk factors for pancreatitis. *Am J Epidemiol* 1982;116:407–414.
6. Sarles H. Epidemiology and pathophysiology of chronic pancreatitis and the role of the pancreatic stone protein. *Clin Gastroenterol* 1984;13:895–912.
7. Robin AP, Campbell R, Palani CK. Total parenteral nutrition during acute pancreatitis: clinical experience with 156 patients. *World J Surg* 1990;14:572–579.
8. Wilson C, Imrie CW, Carter DC. Fatal acute pancreatitis. *Gut* 1988;29:782–788.
9. Levelle-Jones M, Neoptolemos JP. Recent advances in the treatment of acute pancreatitis. *Surg Annu* 1990;22:235–261.
10. Ranson JHC. Etiological and prognostic factors in human pancreatitis: a review. *Am J Gastroenterol* 1983;77:663–668.
11. Ranson JHC. Acute pancreatitis: Pathogenesis, outcome and treatment. *Clin Gastroenterol* 1984;13:843–863.
12. Ranson JHC. The role of surgery in the management of acute pancreatitis. *Ann Surg* 1990;211:382–393.
13. Ranson JHC, Rifkina KM, Roses DF, et al. Prognostic signs and the role of operative management in acute pancreatitis. *Surg Gynecol Obstet* 1974;139:69–81.
14. Larvin M, McMahon MJ. APACHE II score for assessment and monitoring of acute pancreatitis. *Lancet* 1989;2:201–204.
15. Wilson C, Heath DI, Imrie CW. Prediction of outcome in acute pancreatitis: a comparative study of APACHE II clinical assessment and multiple scoring systems. *Br J Surg* 1990;77:1260–1264.
16. Pieper-Bigelow C, Stroechi A, Levitt MO. Where does serum amylase come from and where does it go? *Gastroenterol Clin North Am* 1990;19:793–810.
17. Steinberg WM. Predictors of severity of acute pancreatitis. *Gastroenterol Clin North Am* 1990;19:849–861.

18. Marulendra S, Kirby D. Nutrition support in pancreatitis. *Nutr Clin Pract* 1995;10:45–52.
19. Balthazar EJ, Freeny PC, van Sonnenberg E. Imaging and intervention in acute pancreatitis. *Radiology* 1994;193:297–306.
20. Grant JP, James S, Grabowski V, Trexler KM. Total parenteral nutrition in pancreatic disease. *Ann Surg* 1984;200:627–631.
21. Van Gossum A, Memoyne M, Greig OD, Jeejeebhoy KN. Lipid associated total parenteral nutrition in patients with severe pancreatitis. *J Parent Ent Nutr* 1988;12:250–255.
22. Shaw JH, Wolfe RR. Glucose, fatty acid and urea kinetics in patients with severe pancreatitis: the response to substrate infusion and total parenteral nutrition. *Ann Surg* 1986;204:665–672.
23. Cerra FB. Hypermetabolism, organ failure and metabolic support. *Surgery* 1987;101:1–14.
24. Roth E, Funovies J, Muhkbacher F, et al. Metabolic disorders in severe abdominal sepsis. Glutamine deficiency in skeletal muscle. *Clin Nutr* 1982;1:25–41.
25. Bouffara YH, Delafosse BX, Annat GJ, et al. Energy Expenditure during severe acute pancreatitis. *J Parent Ent Nutr* 1989;13:26–29.
26. Mann S, Westenskow DR, Houtchens BA. Measured and predicted caloric expenditure in the acutely ill. *Crit Care Med* 1985;13:173–177.
27. Dickerson RN, Vehe KL, Mullen JL, Feurer ID. Resting energy expenditure in patients with pancreatitis. *Crit Care Med* 1991;19:484–490.
28. Roza AM, Shizgal HM. The Harris-Benedict equation reevaluated: resting energy requirements and the body cell mass. *Am J Clin Nutr* 1984;40:168–182.
29. Havala T, Shronts E, Cerra F. Nutritional support in acute pancreatitis. *Gastroenterol Clin North Am* 1989;18:525–542.
30. Mann S, Westerknow AR, Houtchens BA. Measured and predicted caloric expenditure in the acutely ill. *Crit Care Med* 1985;13:173–177.
31. Guzman S, Nervi F, Llanos O, et al. Impaired lipid clearance in patients with previous acute pancreatitis. *Gut* 1985;26:888–891.
32. Blackburn GL. In search of the "preferred fuel." *Nutr Clin Pract* 1989;4:3–5; also Long CL. Fuel preferences in the septic patient: glucose or lipid? *J Parent Ent Nutr* 1987;11:333–335.
33. Silberman M, Dixon NP, Gisenberg D. The safety and efficacy of a lipid-based system of parenteral nutrition in acute pancreatitis. *Am J Gastroenterol* 1982;22:494–497.
34. Kafarentzos FE, Daravias DA, Karatzas TM, et al. Total parenteral nutrition in severe acute pancreatitis. *J Am Coll Nutr* 1991;10:156–162.
35. Sitzmann JV, Steinborn PA, Zinner MJ, Cameron JL. Total parenteral nutrition and alternate energy substrates. *Surg Gynecol Obstet* 1989;168:311–317.
36. Steininger R, Karner J, Roth E, Langes K. Infusion of dipeptides as nutritional substrates for glutamine, tyrosine, and branched chain amino acids in patients with acute pancreatitis. *Metabolism* 1989;38:78–81.

37. Helton WS. Intravenous nutrition in patients with acute pancreatitis. In Rombeau JL, Caldwell MD, eds. *Clinical nutrition, parenteral nutrition.* Philadelphia: WB Saunders, 1993:442–461.
38. Stabile BE, Debas HT. Intravenous versus intraduodenal amino acids, fats, glucose as stimulants of pancreatic secretion. *Surg Forum* 1981;32:224–226.
39. Raasch RH, Hak LJ, Benaim V, et al. Effect of intravenous fat emulsion on experimental acute pancreatitis. *J Parent Ent Nutr* 1983;7:254–256.
40. Fried GM, Ogden WD, Rhea A, et al. Pancreatic protein secretion and gastrointestinal hormone release in response to parenteral amino acids and lipids in dogs. *Surgery* 1982;92:902–905.
41. Sax HC, Warner BW, Talamini MA, et al. Early total parenteral nutrition in acute pancreatitis: lack of beneficial effects. *Am J Surg* 1987;153:117–124.
42. Brennan MF, Pistero PWT, Posner M, et al. A prospective randomized trial of total parenteral nutrition after major pancreatic resection for malignancy. *Ann Surg* 1994;220:436–444.
43. Variyam ED, Fuller RK, Brown FM, et al. Effect of parenteral amino acids on human pancreatic exocrine secretin. *Digest Dis Sci* 1985;30:541–546.
44. Nosworthy J, Colodny AH, Eraklis AJ. Pancreatitis and intravenous fat: an association in patients with inflammatory bowel disease. *J Pediatr Surg* 1983;18:269–272.
45. Lashner BA, Kersner JB, Hanauer SB. Acute pancreatitis associated with high concentration of lipid emulsion during total parenteral nutrition therapy for Crohn's disease. *Gastroenterology* 1986;90:1039–1041.
46. Stabile BE, Borzetta M, Stubbs RS, et al. Intravenous mixed amino acids and fats do not stimulate exocrine pancreatic secretion. *J Physiol* 1984;246:G274–G280.
47. Bivens GP, Stein TA. Pancreatic enzyme secretion during intravenous fat emulsion. *J Parent Ent Nutr* 1987;11:60–62.
48. Seibowitz AB, O'Sullivan P, Iberti TJ. Intravenous fat emulsions and the pancreas: a review. *Mount Sinai J Med* 1992;59:38–42.
49. Bivins BA, Bell RM, Papp RP, et al. Pancreatic exocrine response to parenteral nutrition. *J Parent Ent Nutr* 1984;8:34–36.
50. Bush A, Bush J, Carlsen A, et al. Hyperlipidemia and pancreatis. *World J Surg* 1980;4:307–314.
51. Grundfest S, Steiger E, Selinkoff P, et al. The effect of intravenous fat emulsions in patients with pancreatic fistula. *J Parent Ent Nutr* 1980;4:27–31.
52. Kontwiek SJ, Tasler J, Cieszkowski M, et al. Intravenous amino acids and fat stimulate pancreatic secretion. *Am J Physiol* 1979;236:E678–E684.
53. Silberman H, Dixon NP, Eisenberg D. The safety of a lipid-based system of parenteral nutrition in acute pancreatitis. *Am J Gastroenterol* 1977;77:494–497.
54. Robin AP, Campbell R, Palani C, et al. Total parenteral nutrition during acute pancreatitis: Clinical experience with 156 patients. *World J Surg* 1990;14:572–579.
55. ASPEN Board of Directors. Practice guidelines pancreatitis: guidelines for the use of parenteral and enteral nutrition in adult and pediatric patients. *J Parent Ent Nutr* 1993;19(suppl):(16SA).

56. Kirby DF, Craig RM. The value of intense nutritional support in pancreatitis. *J Parent Ent Nutr* 1985;9:353–357.
57. Latifi R, McIntosh JK, Dudrick SJ. Nutritional management of acute and chronic pancreatitis. *Surg Clin North Am* 1991;71:579–595.
58. Pistero PWT, Ranson JHC. Nutritional support for acute pancreatitis. *Surg Gynecol Obstet* 1992;175:275–284.
59. Jackson MW, Schuman BM, Bowden TA, et al. The limited role of total parenteral nutrition in the management of pancreatic pseudocyst. *Am Surg* 1993;59:736–739.
60. Latifi R, Dudrick SJ. The effects of nutrient substrates in acute pancreatitis. In Latifi R, Dudrick SJ, eds. *Surgical nutrition: strategies in critically ill patients.* 1995:147–154.
61. Goodgame JT, Fischer JE. Parenteral nutrition in the treatment of acute pancreatitis: effect on complications and mortality. *Ann Surg* 1977;186:651–658.
62. Hennessey PJ, Black CT, Andrassy RJ. Nonenzymatic glycosylation of immunoglobulin G impairs complement fixation. *J Parent Ent Nutr* 1991;15:60–64.
63. McArdle AH, Echave W, Brown RA, et al. Effect of elemental diet on pancreatic secretion. *Am J Surg* 1974;128:690–692.
64. Cassim MM, Allardyce DB. Pancreatic secretion in response to jejunal feeding of elemental diet. *Ann Surg* 1974;2:228–231.
65. Wolfe BM, Keltner RM, Kaminski DL. The effect of an infraduodenal elemental diet on pancreatic secretion. *Surg Gynecol Obstet* 1975;140:241–245.
66. Kelly GA, Nahrwold DL. Pancreatic secretion in response to an elemental diet and intravenous hyperalimentation. *Surg Gynecol Obstet* 1976;143:87–91.
67. Ragins H, Levenson SM, Signer R, et al. Intrajejunal administration of an elemental diet at neutral pH avoids pancreatic stimulation. *Am J Surg* 1973;126:606–614.
68. Guan D, Ohta H, Green G. Rat pancreatic secretory response to infraduodenal infusion of elemental vs. polymeric defined formula diet. *J Parent Ent Nutr* 1994;18:335–339.
69. Voitle A, Brown RA, Echave V, et al. Use of an elemental diet in the treatment of complicated pancreatitis. *Am J Surg* 1973;125:223–227.
70. Feller JH, Brown RA, Toussaint GP, et al. Changing methods in the treatment of severe pancreatitis. *Am J Surg* 1974;127:196–201.
71. Bodoky G, Harsanyi L, Pap A, et al. Effect of enteral nutrition on exocrine pancreatic function. *Am J Surg* 1991;161:144–148.
72. Kudsk KA, Campbell SM, O'Brien T, et al. Postoperative jejunal feedings following complicated pancreatitis. *Nutr Clin Pract* 1990;5:14–17.
73. Parekh D, Lawson HH, Segal I. The role of total enteral nutrition in pancreatic disease. *So Afr J Surg* 1969;170:642–666.
74. Grant JP, Davey-McCrae J, Snyden PJ. Effect of enteral nutrition on human pancreatic secretions. *J Parent Ent Nutr* 1987;11:302–304.
75. McClave SA, Greene LM, Snider HL, et al. Should patients with acute pancreatitis receive early total enteral nutrition? *Gastroenterology* 1995;108:A739.
76. DeBeaux AC, Plester C, Fearon KCH. Flexible approach to nutritional support in severe acute pancreatitis. *Nutrition* 1994;10:246–249.

77. Caliari S, Benini L, Bonfante F, et al. Pancreatic extracts are necessary for the absorption of elemental and polymeric enteral diets in severe pancreatic insufficiency. *Scand J Gastroenterol* 1993;28:749–752.
78. Cerra FB. Nutrient modulation of inflammatory and immune function. *Am J Surg* 1991;161:230–234.

CHAPTER 22

Nutrition Support in the Patient with the Acquired Immunodeficiency Syndrome

J. Elizabeth Tuttle-Newhall, M.D. and Edward A. Mascioli, M.D.

INTRODUCTION

The acquired immunodeficiency syndrome (AIDS) is a disease characterized by progressive immunodysfunction, opportunistic infections, and eventual death. To date, it is estimated that 17.5 million people are infected with the human immunodeficiency virus (HIV) worldwide. It is predicted that by the year 2000, there will be 30 to 40 million people with AIDS.[1] Originally, AIDS was described as "slim disease" based on observations in Uganda of patients with severe wasting and HIV infection.[2] Malnutrition is one of the more common complications of AIDS. It occurs in the majority of patients with HIV infection if survival exceeds 5 years. The degree of malnutrition has been used as a sensitive predictor of survival in patients with AIDS. Similar to simple starvation, death usually occurs in patients with AIDS when lean tissue is 54% of normal or when body weight reaches 66% of predicted ideal.[3,4] A recent study has shown that nutritional status, as determined by body cell mass, is a more sensitive predictor of survival than CD4+ lymphocyte counts.[5]

Studies have verified that in certain patient populations infected with HIV and AIDS, the once thought inevitable weight loss can be halted and lean tissue stores repleted with aggressive nutritional intervention and support.[6] Although prognosis and length of survival are associated with the degree of malnutrition in the patient with HIV or AIDS, the evidence to support that nutritional repletion in the patient with AIDS changes the eventual outcome of the disease or decreases the associated morbidity is unclear. Nutritional specialists may be reluctant to use specialized regimes to support the patient with HIV until signs of wasting are apparent. At this point in the disease course, the potential benefits of that support may be limited.[7] Although nutritional support is not a curative therapy for the patient with HIV or AIDS, it can potentially reverse the immunodysfunction associated with malnutrition, restore lean tissue, increase tolerance and response to other therapies, and improve quality of life. As we review below, routine estimates of nutritional stability may be inadequate to accurately assess the AIDS patient's nutritional status. To formulate a logical plan of nutritional intervention for the patient with HIV or AIDS, the underlying metabolic disturbances, the effects of nutrition on immunocompetence, routes of nutritional support, and methods of augmenting weight gain all must be examined. Theoretically, if lean tissue can be restored, survival may be enhanced.

NUTRITIONAL ASSESSMENT AND WASTING

The prevalence of malnutrition has been documented to be as high as 50% in all hospitalized surgical or medical patients as determined by standard techniques of nutritional assessment.[8,9] Studies have documented malnutrition in the AIDS population admitted to hospitals to be as high as 90% as determined by an admission history of a 10% nonvolitional weight loss. Singer et al. have reported that in their in-hospital AIDS population with weight loss, only 30% received formal nutritional evaluation and intervention.[10] Although estimates of malnutrition in the in-hospital population may be exaggerated because of the severity of the underlying diseases requiring hospital admission, malnutrition is also high in the outpatient setting.[5] To accurately assess the nutritional status of the HIV-infected or the AIDS patient, we must review standard techniques of nutritional assessment and how they may be affected by the unique metabolic and body composition changes associated with the HIV infection.

Weight loss has long been used as a marker of malnutrition in non-HIV patient populations. A recent study has suggested that patterns of weight loss can also be used as a marker in the individual patient with AIDS as an indication of their clinical stability.[11] Most patients with AIDS have long periods of stable weight followed by periods of dramatic weight loss during acute illness. In patients with HIV, AIDS, and AIDS with active infection, resting energy expenditures (REE) have been found to be increased by 11%, 25%, and 29%, respectively. Patients with HIV or AIDS without active infection are able to sustain adequate calorie intake to maintain their weight despite the increase in REE; however, patients with AIDS and active infection reduce their caloric intake by 20% in the face of increased REE, leading to large and rapid weight losses.[12] Rapid weight loss in an otherwise stable patient could be used in this setting as a marker for undiagnosed infection. Large amounts of weight loss can occur over short periods of time. Macallan et al.[11] recently published their observations regarding rates of weight loss as related to the types of infection in the AIDS population studied. In their study population of both patients with HIV and AIDS, rapid weight loss, defined as greater than 4 kg lost in 4 months, occurred in patients with nongastrointestinal infections. A slower weight loss was noted in their patients with gastrointestinal pathogens as demonstrated by diarrhea and malabsorption. Weight gain was noted if and when the opportunistic infection was effectively treated. Progressive weight loss, however, continued if the infection was resistant to therapy.[11] It is common for patients with AIDS to have severe weight loss with the onset of a new catabolic infection. Once effective treatment is initiated, weight gain can occur. Over the time course of the disease, however, tissue repletion is not complete. The patient then enters a pathway of progressive debilitation and weight loss characterized by incomplete repletion of that weight loss and eventual disability (see Figure 22.1).[3,6]

Although weight trends are important in assessing the overall clinical status of the patient, they are not sensitive to underlying changes in body composition. The HIV infection is a metabolically active disease throughout the entire disease course. A brief viral-like syndrome associated with high levels of serum viral burden is followed by a clinically latent period when plasma viremia is difficult to detect. This clinically latent period is a misnomer in that, although detectable viremia disappears, viral activity persists in lymph tissue.[13] Even without clinical evidence of disease, subclinical viral activity causes metabolic derangements. Early in the course of the disease, metabolically active tissue, the body cell mass, is lost as documented by body impedance studies. Concurrently, there is an increase in total body water and a lesser increase in total body fat.[14] As the disease progresses, lean tissue, as measured by total body potassium content, is depleted out of proportion to weight loss in patients with AIDS. Extracellular water is increased while total intracellular water is decreased reflecting the loss of lean tissue. Body fat content is also decreased but to a much lesser degree in comparison to total body weight loss.[15] A linear relationship has been demonstrated between body weight and lean tissue loss and days until death as determined by total body potassium (Figure 22.2). Although weight loss as a whole has been demonstrated to correlate with length of survival, lean tissue mass is reflective of functioning cell mass. A loss of lean tissue is indicative of a loss of functioning cells.[3] In light of the underlying body composition changes, stable weight patterns and normal anthropometric measurements may be misleading in the assessment of the patient with AIDS.

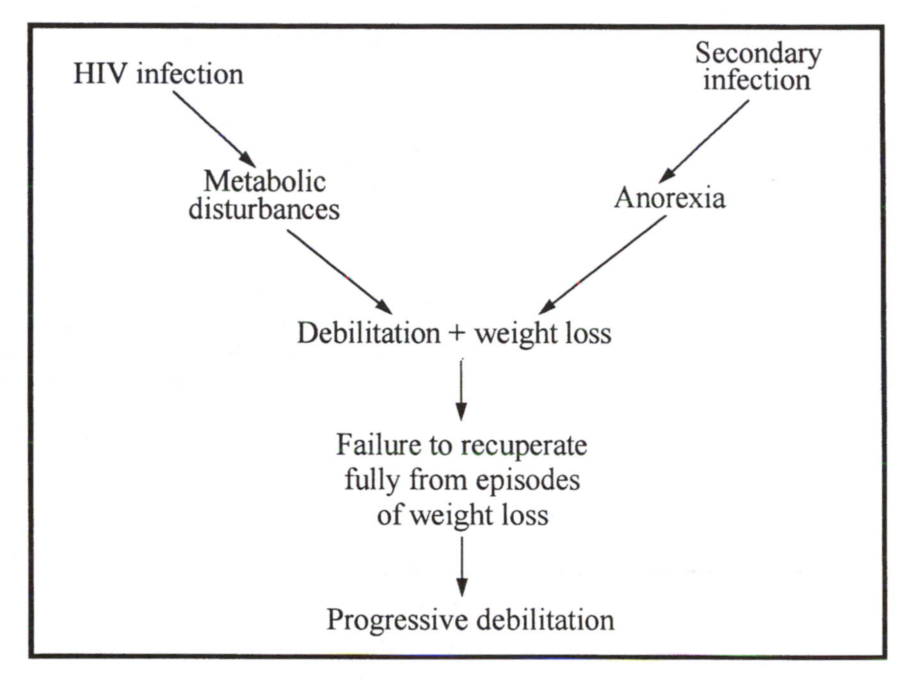

Figure 22.1. Cycle of debilitation. Reprinted with permission from Grunfeld and Feingold.[6]

ASSESSMENT OF LEAN TISSUE

A 24-hr urine collection for urine urea nitrogen (UUN) and creatinine will be more accurate in determining the patient's true lean tissue mass. As the degree of lean tissue mass reflects survival, accurate measurements are paramount to ascertain the success of the feeding regime. Protein is the major body substrate that contains nitrogen. Nitrogen waste and by-products are derived from the body's degradation of protein. A 24-hr urine collection can be used to determine baseline nitrogen excretion, nitrogen balance, and creatinine height index (CHI). Nitrogen balance is a measurement of nitrogen intake (protein intake divided by 6.25), minus nitrogen excreted, as measured in urine and estimated in stool.

$$\text{nitrogen balance} = \frac{\text{protein (g)}}{6.25} - (\text{UUN} + 4)$$

A positive balance is indicative of not only adequate protein and calorie intake, but anabolic metabolism and lean tissue accrual. Skeletal muscle contains creatinine in relatively constant amounts. Estimates have been made that it takes 18 to 20 kg

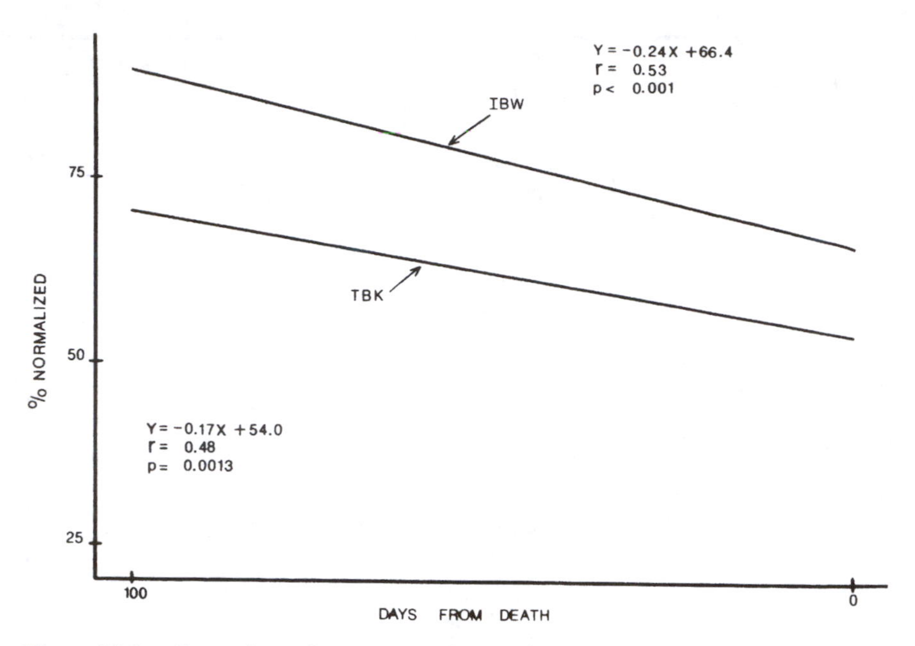

Figure 22.2. Comparison of percentage of normalized total body potassium counts and body weight as a percentage of ideal body weight and their relationship to the timing of death. Reprinted with permission from Kotler et al.[3]

of muscle to produce 1 g of creatinine daily.[16] Creatinine is excreted as a normal urinary metabolite but is reflective of lean tissue mass and up to 20% comes from dietary intake.[17] The CHI is a measurement of loss of lean tissue. It compares measured creatinine versus expected creatinine based on standards for height and sex. For men, expected creatinine excretion is ideal body weight multiplied by 23 mg/kg and women, 18 mg/kg. An index of 80% is reflective of moderate lean tissue loss and 60%, severe lean tissue loss.[18] As there are set standards for creatinine excretion, the amount of creatinine in the sample collection can also be used to determine whether or not the collection is complete and the information obtained is accurate and useful.

Not only is measurement of sequential nitrogen balances and CHIs more sensitive to determine baseline lean tissue stores, but it also represents a response to nutritional interventions. Nitrogen excretion can also be reflective of catabolic stress, and accordingly, indices have been created using 24-hr urine collection data to determine the degree of catabolic stress that the patient is experiencing.

$$\text{catabolic index} = \text{UUN} - [(0.5 \times \text{dietary protein}) \times 0.16) + 3\text{ g}]$$

This index reflects excessive lean tissue metabolism in times of catabolic stress and subtracts normal excretion from that measured. No stress results in a measurement of less than or equal to 0, moderate stress is less than 5, and severe stress is greater than 5.[19] A positive nitrogen balance is the goal of nutritional support and repletion. Unfortunately, a positive balance is dependent on the patient's clinical status and provision of adequate protein and calorie substrate. In times of catabolic stress, the counter regulatory mechanisms prevent lean tissue accrual. Nitrogen balance will not be maintained despite adequate substrate provision. Excessive protein supplementation, greater than 1.5 $g/kg \cdot day^{-1}$ will only contribute to ureagenesis. A 24-hr urine will reflect this in a larger nitrogen loss. When illness and stress resolve (or appropriate therapy is initiated), nitrogen balance is achieved and anabolism may begin. In the patient with AIDS, trends in nitrogen metabolism can guide nutritional intervention and perhaps, lead to earlier detection of new catabolic illness.

Serum albumin has long been used as a standard of nutritional assessment and has been used as a predictor of survival in the patients with AIDS. A serum albumin of less than 2.5 g/dL on initial nutritional assessment has been associated with a decreased length of survival.[10] Albumin, however, is not a sensitive nutritional measurement in any patient with catabolic stress. The metabolic response to injury increases vascular permeability and decreases hepatic production of albumin. Serum albumin levels fall significantly secondary to illness alone.[20] Its long serum half life and decline in the face of catabolic stress limits its ability to accurately reflect responses to nutritional intervention.

Although sequential weight measurements and anthropometry are important in the nutritional assessment of the patients with AIDS, underlying body composition changes may confound their meaning. Serum albumin is not a sensitive indicator of the nutritional status of the patient with AIDS or any patient with catabolic illness. It is, however, an excellent marker for morbidity and mortality. A 24-hr urine collection can give useful information regarding nitrogen balance, lean tissue mass, and the catabolic stress that a patient may be experiencing. With this knowledge, nutritional interventions can be assessed for their effectiveness.

MECHANISMS OF WASTING

The weight loss associated with AIDS has been given the term wasting. It is a complication often associated with AIDS. It is not, however, an inevitable consequence of the disease. Morbidity associated with wasting is significant to the individual patient. Wasting and associated malnutrition contribute to depression, weakness, inability to tolerate aggressive drug regimes such as chemotherapy, limitations of the activities of daily living, and the immunodysfunction associated with malnutrition. The mechanisms of wasting can be divided into those of inadequate energy and substrate availability and those causes related to abnormalities of metabolism. Inadequate substrate availability can result from anorexia or malabsorption.

Metabolic abnormalities are caused by alterations in metabolic rate and in protein and fat metabolism. The metabolic derangements that are present in the patient with HIV or AIDS are the opposite of what occurs in uncomplicated starvation. In simple starvation, there is a decrease in the metabolic rate, an increase in the utilization of fat as a fuel source, and a conservation of protein stores. Body fat is mobilized in times of starvation to provide energy substrate and visceral protein stores are maintained.[21,22] Changes noted early in the course of HIV infection include increased REE, decreased energy consumption, and increased protein wasting.[6,23] These changes in metabolism are similar to those found in the metabolic response to injury or starvation of the Kwashiorkor type. Energy requirements increase and protein stores become an increased source of fuel. Fat and glycogen are also mobilized, but to a lesser degree proportionally. Immediate immunologic concerns, such as the ability to launch an inflammatory response with acute phase reactants, become the metabolic priority.[21] Large amounts of nitrogen can be lost in the patient's urine indicating protein wasting. A 24-hr urine collection for UUN can reveal large losses of nitrogen per day indicating continued and severe catabolic stress.

Not only is there a disturbance in the metabolism of protein but fat as well. There is an elevation in the circulating level of serum triglycerides and a decrease in hepatic and adipose lipoprotein lipase. There also appears to be an increase in the synthesis of free fatty acids.[24] This inability to metabolize circulating triglycerides and the increase in free fatty acid production may be responsible for the modification of loss in adipose stores during times of catabolic stress. Although there is a net loss of adipose tissue in times of stress if the patient is not fed, in the patient with AIDS this loss is significantly less than expected. Hypertriglyceridemia is present throughout the course of the disease. Because of high serum levels, fat can be limited as an exogenous supplement. Although once suspected, there is no direct relationship between high serum levels of triglycerides and the degree of wasting. There also appears to be an increase in energy-consuming cycles of both hepatic glycogen and glucose.[24,25]

Inadequate substrate availability can be related to anorectic causes and nonanorectic causes. Anorexia in the patient with HIV or AIDS can be due to a multitude of factors, such as depression, side effects of antiviral medications, chemotherapy, systemic illnesses such as cancer and infections, and opportunistic infections of the gastrointestinal tract. The upper gastrointestinal tract is a common site for opportunistic infections that can seriously limit the patient's ability to consume adequate amounts of energy substrate. In light of altered metabolism, inadequate nutrient intake exacerbates the potential for substantial weight loss during times of illness. Malabsorption can also encourage anorexia. Unabsorbed nutrients in the distal small bowel and colon can provide negative feedback inhibiting the patient's desire to consume food.

The lower gastrointestinal tract in the patient with AIDS is also a common site for involvement with opportunistic pathogens. Malabsorption and diarrhea are common manifestations of AIDS. However, malabsorption of nutrients can occur without significant diarrhea. Objective documentation of malabsorption via D-xylose tests have confirmed malabsorption in those patients with and without diarrhea.[26]

MALNUTRITION AND IMMUNODYSFUNCTION

The causes of malnutrition in the patient with AIDS are multifactorial. From anorexia and malabsorption to specific metabolic abnormalities, these contribute to a state of depressed nutrition. Malnutrition has been known as the most common cause of immune dysfunction worldwide. Specific immune system abnormalities can be attributed to malnutrition alone. To appreciate the potential benefits of nutritionally repleting the patient with AIDS, these abnormalities must be reviewed. Starvation alone affects nonspecific immunity. Malnutrition can cause decreased chemotaxis of phagocytes but spares phagocytosis.[27,28] Specific immunity which is affected includes numbers of both T and B cells and their ability to produce cytokines and perform their effector functions. Serum antibody response is not normal and may have decreased affinity for known antigens and a decrease in the production of new antibodies. Secretory IgA concentrations are low in the mucosal barrier which may lead to an increased risk for pathogenic colonization of the gastrointestinal tract.[29] Isolated trace element deficiencies have also been associated with decreased immune function. Specifically, zinc and selenium have been reported to be decreased in patients with AIDS. Zinc deficiency leads to severe abnormalities in cell-mediated immunity. Severe zinc deficiencies can also lead to increased susceptibility to certain viral and bacterial pathogens. Adequate zinc replacement reverses these abnormalities in several weeks in those patients without AIDS.[30] Selenium deficiency has been documented to decrease T cell populations and their cytotoxic abilities. Selenium deficiency has also been associated with an increased risk for solid neoplastic disease and anemia with associated cardiomyopathy. Repletion of selenium in the patient without AIDS reverses these abnormalities.[31] The degree to which these starvation-associated immune dysfunctions contribute to the overall picture of immunoincompetence in AIDS is unclear at this time. Theoretically, if starvation can be reversed, specific immune abnormalities may be reversed. This in turn may lead to fewer secondary infections and hospitalizations.

NUTRITIONALLY SUPPORTING THE PATIENT WITH AIDS

As we have reviewed, lean tissue loss can be used to accurately predict the timing of death in the patient with AIDS.[3] The benefit of feeding the patient with AIDS must then depend on the ability to restore this lost lean tissue. Is this possible? Acute weight loss and increased nitrogen excretion have already been described in times of catabolic stress, such as secondary infections, leading to increased losses of lean tissue.[6,23] As the cycle of weight loss and secondary infections continue, the patient never fully regains the tissue lost and enters a cycle of progressive debilitation and eventual death.[6]

The potential effectiveness of nutritional repletion in the patient with AIDS is the ability to replete lost lean tissue or body cell mass. Body cell mass refers to that metabolically active tissue not including nonmetabolically active tissue such as bone mineral and extracellular fluids.[32] As 98% of total body potassium remains in active cells, studies can be done using total body potassium as an estimate of functioning cell mass. Results of several enteral and parenteral studies have shown that nutrition support can effectively maintain body composition, reverse lean tissue loss, and maintain weight in patients who suffer from cachexia alone.[6,33,34] Enteral provision of adequate protein and energy in those patients who are anorectic with preserved gastrointestinal absorptive capacity can lead to the restoration of lean tissue and body weight.[33] In patients with proven malabsorption, the efficacy of enteral support is limited. Parenteral nutrition can lead to lean tissue restoration if secondary infections are successfully treated or controlled.[34] However, in patients with untreated or untreatable secondary infections, lean tissue cannot be restored. The response to catabolic stress prevents lean tissue accrual. Nutritional intervention in the catabolically stressed patient with AIDS, however, may lessen lean tissue losses. Other potential benefits of nutrition support include reversing starvation-related immune dysfunction. The total number of lymphocytes have been found to be increased, including CD8+ cells in starved patient who are refed.[35] Unfortunately, CD4+ cells are not increased. These findings suggest that potentially reversing starvation in the patient with AIDS may improve immune function overall but will not affect the underlying effects of HIV.

Body weight restoration is also a benefit in patients with AIDS who receive nutrition support. Weight loss has been shown to adversely affect body image, self-esteem, mental performance and muscle function. In patients without secondary infections or in those patients with treatable infections, both parenteral and enteral nutrition support have been shown to be effective in restoring and maintaining weight.[33,34] In the catabolic patient with AIDS, nutrition support may reduce lean tissue loss, maintain that immune function related to nutritional stability, and modify weight loss that could occur without nutritional intervention.

Although there has been little investigation to date, patients with AIDS with opportunistic infections treatable or not, should receive nutrition support to preserve or restore lean tissue stores, augment weight gain, and potentially improve the quality of life. In the patient with untreatable secondary infection, weight gain can be augmented by a substantial increase in total body fat rather than lean tissue.[33,34]

ROUTES OF NUTRITIONAL SUPPORT

Selection criteria for nutritional support in the patient with AIDS is similar to that of patients without AIDS: presence or risk of significant weight loss (> 10% usual body weight), failure to maintain weight on a regular diet, failure of enteral intervention to maintain body weight, gastrointestinal malabsorption, and potential improvement in the quality and duration of life. With the underlying metabolic abnormalities and the potential for significant weight loss with illness, nutritional intervention should occur early. Once lean tissue and weight are lost, it is difficult to catch up. Two options are available to nutritionally support the patient with AIDS, enteral or parenteral. The benefits of enteral support are many: ease of support for both patient and caregiver, cost benefits, advantages of promoting gastrointestinal integrity and physiologic nutrient delivery. Contraindications to enteral support do exist (Table 22.1), but there are only a few absolute contraindications—bowel obstruction and failure of enteral support to maintain nutritional stability. Intractable vomiting and aspiration can be dealt with by using postpyloric feeding techniques and diarrhea is only a contraindication if a reversible cause cannot be found. Parenteral support has its specific uses and unique benefits. Not only can total parenteral nutrition (TPN) offer complete nutrition support, but there is an added benefit of metabolic support in those patients with large-volume diarrhea with complicating acid–base or electrolyte abnormalities.

ENTERAL SUPPORT

Enteral support can range from intermeal supplementation to complete support with nonvolitional tube feedings. There are many causes of decreased oral intake.

Table 22.1. Contraindications to Enteral Support

Absolute Contraindications	Relative Contraindications
Bowel obstruction	Intractable vomiting
Failure of enteral support	Aspiration
	Severe diarrhea

Chemotherapy and certain medications can induce nausea, opportunistic infections of the oropharynx and esophagus can limit oral intake due to pain or mechanical dysfunction, and depression and chronic illnesses can limit appetite. Dietary counseling and patient education early in the course of the disease can make the patient more aware of their disease process and the effects on their nutritional status. Education and awareness will ensure the patient's participation in increasing oral intake and in the issue of food safety. Supplementation of multivitamins and micronutrients is also warranted as deficiencies in vitamins and micronutrients have been reported.[31,32] Oral caloric supplementation, however, can be limited by gastrointestinal dysfunction. Malabsorption is common. Supplements with lactose may exacerbate the patient's diarrhea and cause uncomfortable gas bloating. This symptom can be relieved by either changing supplements to a nonlactose-containing formula or using lactase enzyme supplements. Lactase enzyme supplements may help alleviate some but not all of these symptoms. Likewise, fat is an excellent calorie source but often is the least well-absorbed of the major nutrients. Unfortunately, unabsorbed nutrients presented to the distal small bowel and colon can negatively inhibit appetite, thus exacerbating weight loss. If the patient has fat malabsorption, confirmed by a 24-hr fecal fat or a spot fecal smear, supplements that contain medium chain triglycerides (MCTs) as a fat source can be used. The MCTs are absorbed directly into the portal venous system bound to albumin. The absorption of MCTs is not lymphatic dependent, and they are also almost completely oxidized for energy and not stored as fat.[36] Unfortunately, MCTs are only found in supplements, thus committing the patient to expensive dietary sources. Oral diets can also be adjusted for severe diarrhea and malabsorption by limiting the patient to a solely elemental supplement orally. Elemental feedings are low-fat, but are limited by their palatability and subsequent patient compliance. At this time, there is no role for vegetarian or macrobiotic diets except to discourage their use. Limitations of protein sources may exacerbate protein calorie malnutrition and macrobiotic diets may lead to vitamin and trace element deficiencies.

Later in the disease course, when and if simple oral supplementation fails, nonvolitional support can be given via temporary nasogastric or nasojejunal feeding tubes. Tubes placed nasally, however, are not long-term solutions. Patient comfort and tube displacement can become issues as well as associated sinusitis from sinus occlusion. No longer than 6 weeks is recommended for nasally placed tubes. If enteral absorption of nutrients is confirmed and the need for support extends beyond 6 weeks, a longer term solution can be obtained. A percutaneous gastrostomy tube (PEG) or a percutaneous jejunostomy (PEJ) can assure long-term enteral access. It is important to note that PEGs can only be used once the issue of gastric tolerance and aspiration have been assessed. However, PEG and PEJ placements have a higher incidence of complications in the patient with AIDS than in non-HIV patients. Stomal infections have been reported to be as high as 21% in those patients with PEGs and HIV.[37] Percutaneous gastrostomy feedings enable patients to replete body cell mass, and PEG placement has been shown to facilitate earlier discharges to home in some populations of cancer patients.[37,38] Surgically placed gastrostomy or jejunostomy tubes are also an option in those patients who have had previous surgeries and are not candidates for percutaneous procedures. Jejunostomies should be reserved for those patients, who have failed attempts at gastric feedings.

Once the route of enteral access has been established, formula selection must be undertaken. Because of the increased incidence of malabsorption and diarrhea in the patient with AIDS, enteral support may occasionally exacerbate diarrhea leading to increased losses of volume and electrolytes as well as inadequate nutrient provision. Prior to concluding enteral nutrition failure, an exhaustive search for a reversible cause of diarrhea must be undertaken. Antidiarrheal agents are discouraged until infectious agents are ruled out to minimize the risk of toxic megacolon. As discussed with oral supplements, formulas low in lactose, high in MCTs, and low in fat are often necessary for the patient with diarrhea to provide adequate support. Because enteral formulas contain the recommended daily allowances for vitamins and minerals, supplementation of these nutrients is not indicated. There are many commercially available formulas, each with their own unique characteristics that may make them more attractive depending on the individual patients needs. Most formulas are 1 to 2 kcal/cc and vary in the amount and type of protein per liter. Fat sources are variable too. There are references available that describe formulary choices and their unique characteristics. Enteral support fails when weight loss continues despite adequate nutrient provision, exacerbation of diarrhea despite aggressive intervention, or the onset of an illness necessitates the initiation of parenteral support. Close monitoring of weight and sequential 24-hr urine collections can assess the success of each individual's nutritional regime.

PARENTERAL SUPPORT

For this discussion, peripheral parenteral nutrition will not be discussed except to discourage its use. Peripheral nutrition solutions are not optimal for the patient with AIDS secondary to their inability to provide adequate protein and calorie needs without large intravenous volumes. Electrolyte needs and acid-base requirements of the acutely ill AIDS patient or the chronically ill patient will quickly exhaust available peripheral access because of the hypertonicity of the required peripheral solution. Many AIDS patients have poor peripheral intravenous access from repeated bouts of illness and the administration of sclerosing medications, such that central access will most likely already be present for the administration of long-term antiviral or other chronic medications. Temporary central venous access can be obtained either by a percutaneously placed internal jugular or subclavian venous catheters. Most patients and caregivers prefer the subclavian site, as it has a higher degree of patient comfort and ease of dress. If access is required for more than 2 months, a long-term venous access device can be used.

There has been considerable concern regarding the use of long-term venous access devices in AIDS patients and the risk of increased infectious complications. Experience with such devices in other immunocompromised patients, such as oncology patients and bone marrow transplant patients, have shown that there is no increase in infectious complications.[39] Recently, studies have shown that long-term venous access is safe in patients with AIDS; however, there is a significantly increased risk of catheter-related infection. It appears that home total parenteral nutrition (HTPN) increases the risk of those infections as opposed to catheters being used for home antibiotic therapy.[40] Complication rates appear similar between Hickman®-type catheters and implantable ports, 0.47% per catheter days vs. 0.40 per 100 catheter days, respectively.[41] The decision regarding the type of long-term access device must depend on each individual's intravenous requirements and the type of care situation in which the patient lives.

Indications for parenteral support are the same for patients with AIDS as those without AIDS: failure of enteral support and/or the need for aggressive metabolic support. Also, TPN gives the physician and nutritional specialist the unique opportunity to improve the quality of life for a patient with an otherwise terminal disease. As previously discussed, weight loss and REE increase dramatically in patients with acute illness.[6,23] Acute catabolic illness with an inadequate functioning or nonfunctioning gastrointestinal tract or the need for concomitant nutritional and metabolic support should all be indications for total parenteral nutrition (TPN).

Once the route of support has been chosen, volume requirements, acid–base issues, electrolyte requirements, and protein and calorie needs should be calculated. Parenteral calculations should be customized for each patient. Standard formulas in the patient with AIDS are cumbersome and will not take into consideration the specific needs of the patient or consider the unique electrolyte, trace elements, or acid–base needs. The use of standard TPN formulas must be discouraged in the patient with AIDS.

Volume consideration is the first priority when calculating the TPN needs of the patient with AIDS. This should depend on the total daily requirements of fluid. These volume estimations must take into consideration maintenance needs based on the patient's weight and ongoing fluid losses. Correction of acid–base disorders can also be calculated in the TPN formulation. As diarrhea and malabsorption are common complications of HIV infection, so are acid–base disorders.[4,6,7,10] Diarrhea of small bowel origin can consist of large-volume losses as well as 30 to 50 mEq/L of HCO_3. These losses can cause a severe nonanion gap metabolic acidosis. The patient will increase the minute ventilation in an attempt to maintain a normal pH in the face of decreased buffering capacity.[42] The treatment of choice is to correct the underlying infectious etiology that causes the diarrhea. In the meantime, the lost buffer can be replaced on a continuous basis by adding the estimated base deficit and ongoing losses to the TPN solution as sodium or potassium acetate. Calculating base deficit is based on the formula

$$\text{base deficit} = (\text{body weight/kg} \times 0.5) \times (\text{expected serum } HCO_3 \text{ conc.} - \text{measured serum } HCO_3 \text{ conc.})$$

Once the existing deficit is calculated, half can be given over the first 24 hr and the remainder over the next 24 hr. Adjustments should be made for ongoing losses. Close monitoring with either venous or arterial blood gases is recommended. Metabolic alkalosis, on the other hand, is relatively uncommon in the patient with AIDS secondary to the predominance of small bowel pathogens; however, diarrhea of large bowel origin, steroids or diuretic therapy may cause an alkalosis due to chloride loss. Diarrhea of large bowel origin can also cause large-volume losses with the loss of chloride, 60 to 80 mEq/L. Metabolic alkalosis with a pH greater than 7.5 can be corrected by either potassium chloride replacement, volume expansion, or infusion of 0.1N hydrochloric acid in non-lipid-containing TPN. The total amount given over 24 hr should not exceed 100 mEq/day and the patient's pH should be followed closely. Large doses of histamine_2 receptor antagonists have been used in patients with short-gut syndrome to decrease diarrhea volumes.[43] Cimetidine or other H_2 blockers can be added to a daily TPN solution to accomplish the same task in the patient with AIDS. In decreasing the gastrointestinal output, acid–base disorders can be managed more easily.

CAUTION IN REFEEDING THE PATIENT WITH AIDS

There are potentially life-threatening consequences of refeeding a severely malnourished patient with AIDS. The refeeding syndrome is a phenomena of metabolic consequences in the starved individual due to physiologic intolerances of infusions of large volumes of glucose- and sodium-containing solutions, either enterally or parenterally. Patients at risk for the refeeding syndrome include any severely malnourished patient who has been deprived of adequate nutrition.[44] The danger in repleting the starved individual arises from the corresponding increases in insulin levels with the provision of glucose and calorie substrates. In the nonstressed starvation state, insulin levels are usually severely depressed. Insulin has powerful antinatriuretic properties such as salt and water retention. When the starved individual is presented with a large glucose load, insulin levels rise and sodium and water excretion decrease.[45] Extracellular volume can rapidly expand. This phenomena in the cachectic AIDS patient can lead to cardiac dysrhythmias and congestive heart failure. Hypophosphatemia, hypokalemia, and hypomagnesemia can all occur from intracellular anabolic shifts. This can exacerbate the cardiac dysfunction and lower the threshold for arrhythmias.[46] The optimal way to avoid this life-threatening complication of feeding is first to be aware of its potential existence. Second, nutritional repletion should initially limit volume, sodium, and calories provided. Glucose calories should be limited initially to the daily hepatic production of glucose, 100 to 150 g/day and increase slowly, taking care to monitor the patient's weight and clinical status.[21] Protein provision at 1.5 g/kg of body weight·day^{-1} poses no risk to the starved patient with normal renal and hepatic function. Serum electrolytes should be monitored closely during the first week of nutritional repletion and replaced aggressively as needed. Weight gain should be closely scrutinized. Acute increases in total body water will be reflected in an acute rise in body weight. Lean tissue cannot be replaced at a rate that exceeds 0.25 kg/week. Any weight gain greater than this must be suspected to be fat and/or fluid. Lipid as a calorie source may obviate the need for high levels of glucose especially in TPN formulas; however, with the already described abnormalities in lipid metabolism, triglyceride levels must be followed closely.

AUGMENTING WEIGHT GAIN AND LEAN TISSUE ACCRUAL

Nutritional repletion of the patient with AIDS is dependent upon treatment of symptoms causing anorexia, adequate provision of protein and calorie substrate and the treatment of reversible pathogens. Despite aggressive intervention, the potential exists for continued wasting. A variety of experimental agents has been investigated to augment weight gain and reverse the potential etiologies of wasting. These medications vary in their effects from stimulating appetite to lowering cytokine activity to reverse the wasting of catabolic stress (Table 22.2).

Table 22.2. Medications to Reverse Wasting

Medication	Proposed Mechanism	Body Composition Changes
Megesterol acetate	Appetite stimulant	↑Weight, ↑Fat, ?Lean tissue
Dronabinol	Appetite stimulant	↑Weight, (Studies pending)
Pentoxyfylline	Appetite stimulant Anti-TNF	+/– Weight gain, (Results of studies mixed)
Growth hormone	Enhances lean tissue accrual Anabolic protein hormone	↑Weight, ↑Lean Tissue (Effects reversed when therapy stopped)

The most studied of the appetite stimulants has been megesterol acetate, a synthetic progesterone. During treatment for breast cancer with this agent, patients noted an increase in weight.[47] Megesterol has not only been noted to be successful in promoting appetite in patients with AIDS, but also weight gain.[48] Its success, however, is dependent on a functioning gastrointestinal tract. It is not clear at this time, however, whether megesterol acetate can induce lean tissue accrual as well as fat in promoting weight gain. A major component of marijuana, dronabinol, has also been studied in patients with AIDS as an appetite stimulus. In normal volunteers and patients with cancer, dronabinol has been noted to increase appetite and weight gain.[49] Studies in patients with AIDS are still pending. Pentoxyfylline has also been studied as an appetite stimulant. Recent studies have shown that pentoxyfylline decreases tumor necrosis factor (TNF) messenger ribonucleic acid (mRNA) in the serum of treated AIDS patients. It has been shown that TNF decreases the efficacy of antiretroviral medication, enhances HIV activation, and potentially contributes to the wasting syndrome.[49] In clinical trials, however, results have been mixed. In some patients, weight gain was augmented, whereas others had an increased risk of bacterial infections.[50] At this time, pentoxyfylline is unproven in its efficacy to reduce wasting.

Growth hormone deficiency has been reported in AIDS patients.[51] Growth hormone is an endogenously occurring anabolic protein hormone that enhances lean tissue accrual and muscle strength.[52] Studies have shown that treatment with pharmacologic doses of growth hormone lead to weight gain with concomitant increases in lean body mass and total body water. Fat mass was noted to be decreased in the same study population. Nitrogen excretion, muscle power, and endurance were all reported to be increased; however, when therapy was stopped, over a six week period, all changes were reversed.[53] It is unclear what role growth hormone may play in the therapy for wasting at this time. Other therapies have been suggested, but again results are mixed. Glucocorticoids, insulinlike growth factor, testosterone supplementation and cyprohepatadine have been and are being evaluated for their effects on wasting. At this time, however, results are inconclusive.[54–56]

HOME TOTAL PARENTERAL NUTRITION THERAPY IN THE PATIENT WITH AIDS

The issue of HTPN therapy in the patient with AIDS is controversial. The decision to start this nutritional therapy must be based on a well-informed and frank discussion between the patient and his or her physician. The potential benefits, complications, and limitations must be discussed in an open environment of interchange. Objective data does not exist to support the potential benefits of restored immunodysfunction related to starvation, limitations of secondary infections, prolonged survival, or decreased rates of hospitalizations. There is data to support, however, the repletion of lean tissue in the AIDS patient with gastrointestinal dysfunction and the benefit of increased weight gain.[57] Weight gain has the benefits of increasing function, self-esteem, and reversing depression caused by an altered body image. Catheter infection rates are increased in the AIDS patient receiving HTPN and often can only be treated by catheter removal versus a trial of antibiotic therapy.[58] However, in the overall schema of therapy, the potential unique benefits of HTPN in certain patients outweigh the risks.[57,59] An unwavering policy refusing HTPN therapy to those patients with AIDS is unjustified in an environment of otherwise aggressive care. A double standard should not exist between unproven antiviral medications or experimental medications to treat opportunistic infections and nutrition support. Each decision to place the AIDS patient on HTPN must take into consideration the patient's wishes, the potential benefit, the patients stage of disease, and the home care situation. Caregivers must be involved in these decisions as HTPN is a labor-intensive therapy whose responsibility often falls to the home caregiver. Teams must follow HTPN patients on a much closer basis, adjusting formulae frequently as changing electrolytes and acid–base needs warrant. Likewise, the withdrawal of support must be the decision of the patient, caregiver, and physician dependent on the philosophy of aggressive care at that point in time. However, HTPN can be an effective therapy in select patients with AIDS.

CONCLUSION

With increasing numbers of patients with HIV and AIDS and the increased risk of malnutrition, the nutritional specialist will be asked more and more to become involved in the care of these complex patients. Weight loss is not an inevitable complication of this disease.[6] Lean tissue stores can be repleted in specific patient populations with aggressive nutritional intervention and follow-up. Although parameters of nutritional stability are more sensitive predictors of survival in the patient with AIDS, nutritional intervention has not been shown to change the eventual outcome of the disease process. Adequate nutrition is not a cure for this disease, however it can increase quality of life, potentially decrease the comorbidities of the disease process, and maintain weight status. It is clear that in the AIDS patient whose disease course is complicated by complex metabolic and acid–base disorders, parenteral nutrition can provide unique benefits. Nutritional support can and should play an integral role in the comprehensive care of the patient with AIDS.

REFERENCES

1. Merson MH. Slowing the spread of HIV: agenda for the 1990s. *Science* 1993;260(5112):1266–68.
2. Srwadda D, Sewankambo NK, Carswell JW, et al. Slim disease: a new disease in Uganda and its association with HTLV-III infection. *Lancet* 1985;2:849–852.
3. Kotler DP, Tierney AR, Wang J, et al. Magnitude of body-cell mass depletion and the timing of death from wasting in AIDS. *Am J Clin Nutr* 1989;50:444–447.
4. Chelbowski RT, Grovsner MB, Bernhard NH, et al. Nutritional status, gastrointestinal dysfunction and survival in patients with AIDS. *Am J Gastroenterol* 1989;4:1288–1293.
5. Suttmann U, Ockenga J, Selberg O, et al. Incidence and prognostic value of malnutrition and wasting in human immunodeficiency virus-infected outpatients. *J AIDS Human Retro* 1995;8:239–246.
6. Grunfeld C, Feingold KR. Metabolic disturbances and wasting in the acquired immunodeficiency syndrome. *N Engl J Med* 1992;327:329–337.
7. Bell SJ, Mascioli EA, Forse RA, Bistrian BR. Nutritional support and the human immunodeficiency virus (HIV). *Parasitology* 1993;107:S53–S67.
8. Bistrian BR, Blackburn GL, Hallowell E. et al. Protein status of general surgery patients. *JAMA* 1974;230:858–860.
9. Bistrian BR, Blackburn GL, Vitale J, et al. Prevalence of malnutrition in general medical patients. *JAMA* 1976;235:1567–1570.
10. Singer P, Katz DP, Dillon L, et al. Nutritional aspects of the acquired immunodeficiency syndrome. *Am J Gastroenterol* 1991;87:265–273.
11. Macallan DC, Noble C, Baldwin C, et al. Prospective analysis of patterns of weight change in stage IV human immunodeficiency virus infection. *Am J Clin Nutr* 1993;58:417–424.
12. Grunfeld C, Pang M, Shimizu L, et al. Resting energy expenditure, calorie intake, and short term weight change in human immunodeficiency virus infection and the acquired immunodeficiency syndrome. *Am J Clin Nutr* 1992;55:455–460.
13. Pantaleo G, Graziosi C, Fauci A. Mechanisms of disease: the immunopathogenesis of human immunodeficiency virus infection. *N Engl J Med* 1993;328(5):327–335.
14. Ott M. Lembecke B, Fischer H, et al. Early changes of body composition in human immunodeficiency virus infected patients tetrapolar body impedance analysis indicates significant malnutrition. *Am J Clin Nutr* 1993;57:15–29.
15. Kotler DP, Wang J, Pierson RN. Body composition studies in patients with the acquired immunodeficiency syndrome. *Am J Clin Nutr* 1985;42:1255–1265.
16. Forbes GB, Bruining GJ. Urinary creatinine excretion and lean body mass. *Am J Clin Nutr* 1976;29:1359–1366.
17. Bleiler RE, Schedl HP. Creatinine excretion: variability and relationships to diet and body size. *J Lab Clin Med* 1962;59:945–955.

18. Bistrian BR, Blackburn GL, Sherman M, et al. Therapeutic index of nutritional depletion in hospitalized patients. *Surg Gynecol Obstet* 1975;141:512–516.
19. Bistrian BR. A simple technique to estimate the severity of stress. *Surg Gynecol Obstet* 1979;148:675–78.
20. Klein S. The myth of serum albumin as a measure of nutritional status. *Gastroenterology* 1990;99:1845–1846.
21. Cahill GF. Starvation in man. *N Engl J Med* 1970;282:668–675.
22. Bistrian BR. Nutritional assessment of the hospitalized patient, a practical approach. In Wright RA, Heymsfield S, eds. *Nutritional assessment.* Boston: Blackwell Scientific, 1984:183–205.
23. Kotler DP, Tierney AR, Brenner SK, et al. Preservation of short term energy balance in clinically stable patients with AIDS. *Am J Clin Nutr* 1990;57:7–13.
24. Grunfeld C, Pang M, Doerrier W, et al. Lipids, lipoproteins, triglyceride clearance and cytokines in human immunodeficiency virus infection and the acquired immunodeficiency syndrome. *J Clin Endocrinol Metab* 1992;74:1045–1052.
25. Stein TP, Nutinsky C, Condoluci D, et al. Protein and energy substrate metabolism in AIDS patients. *Metabolism* 1990;39:876–881.
26. Dworkin B, Worsmer GP, Rosenthal WS, et al. Gastrointestinal manifestations of the acquired immunodeficiency syndrome. *Am J Gastroenterol* 1985;80:774–778.
27. Garre MA, Boles JM, Youninou PY. Current concepts in immune derangement due to under-nutrition. *J Parent Ent Nutr* 1987;11:309–311.
28. Douglas SD, Shopfer K. Phagocyte in protein calorie malnutrition. *Clin Exp Immun* 1976;17:121–28.
29. Chandra RK. Nutrition: the changing scene: nutrition, immunity and infection: present knowledge and future directions. *Lancet* 1983;1:688–691.
30. Chandra RK, Dayton DH. Trace element regulation of immunity and infection. *Nutr Res* 1982;2:721–733.
31. Dworkian BM, Rosenthal WS, Wormser GP, et al. Selenium deficiency in the acquired immunodeficiency syndrome. *J Parent Ent Nutr* 1986;10:405–407.
32. Heymsfield SB, Cuff PA, Kotler DP. AIDS enteral and parenteral nutrition support. In Kotler DP, ed. *Gastrointestinal and nutritional manifestations of AIDS.* New York: Raven Press, 1991:243–256.
33. Kotler DP, Tierney AR, Culpepper-Morgan JA, et al. Effect of total parenteral nutrition on body composition in patient with the acquired immunodeficiency syndrome. *J Parent Ent Nutr* 1990;14:454–458.
34. Kotler DP, Tierney AR, Wang J, et al. Enteral alimentation and repletion of body cell mass in malnourished patients with the acquired immunodeficiency syndrome. *Am J Clin Nutr* 1991;53:149–154.
35. Cunningham-Rundles S. Effects of nutritional status on immunologic dysfunction. *Am J Clin Nutr* 1982;32:1202–1210.
36. Isselbacher KJ. Mechanisms of absorption of long and medium chain triglycerides. In Senior JR, ed. *Medium chain triglycerides.* Philadelphia: University of Pennsylvania Press, 1986:21–33.

37. Cappell MS, Godil A. A multicenter case-controlled study of percutaneous endoscopic gastrostomy in HIV-seropositive patients. *Am J Gastroenterol* 1993;88(12):2059–2065.

38. Hunter JG, Laurenteno L, Shellito PC. Percutaneous endoscopic gastrostomy in head and neck cancer patients. *Ann Surg* 1989;210:42–46.

39. Harvey MP, Trent RJ, Joshua DE, et al. Complications associated with indwelling Hickman ® catheters in patients with hematological disorders. *Aust NZ J Med* 1986;16:211–221.

40. Mukau I, Talamini MA, Sitzmann JV, et al. Long-term central venous access vs. other home therapies: complications in patients with the acquired immunodeficiency syndrome. *J Parent End Nutr* 1992;16(5):455–459.

41. Pijl H, Frisseen PH. Experience with totally implantable venous access device (port-a-cath®) in patient with AIDS. *AIDS* 1992;6(7):709–713.

42. Schier RW, ed. *Renal and electrolyte disorders,* 3rd ed. Boston: Little, Brown, 1986:146–156.

43. Jacobsen O, Ladefoged K, Jarnum S. Effects of cimetidine on jejunostomy effluents in patients with severe short bowel syndrome. *Scand J Gastroenterol* 1980;21:824–828.

44. Soloman S, Kirby DF. The refeeding syndrome: a review. *J Parent Ent Nutr* 1990;14:90–97.

45. DeFronzo RA, Cooke CR, Andres, et al. The effect of insulin on renal handling of sodium, potassium, calcium and phosphate in man. *J Clin Invest* 1975;55:845–850.

46. Heymsfield SB, Bethel RA, Anslet JD, et al. Cardiac abnormalities in cachectic patients before and during nutritional repletion. *Am Heart J* 1975;95:584–589.

47. Tchekmedyian NS, Tait N, Moody M, et al. High dose megesterol acetate: a possible treatment for cachexia. *JAMA* 1987;257:1195–1198.

48. Saint-Marc T, Tourcine JC. 1993. Weight gain and appetite stimulation with medroxyprogesterone in AIDS patients with cachexia and anorexia. International Conference on AIDS, (abstract PO B362352).

49. Dezube BJ, Lederman J, Spritzler LG, et al. High dose pentoxyfylline in patients with AIDS: inhibition of tumor necrosis factor production. National Institute of Allergy and Infectious Diseases AIDS Clinical Trials Group. *J Inf Disease* 1995;171(6);1628–1632.

50. Landman D, Sarai A, Sathe SS. Use of pentoxyfylline therapy for patients with AIDS related wasting. *Clin Inf Disease* 1994;18(1):97–99.

51. Mg TT, O'Connell IP, Wilkins EG. Growth hormone deficiency coupled with hypogonadism in AIDS. *Clin Endocrinol* 1994;41(5):689–694.

52. Rudman P, Teller AG, Nagray HS, et al. Effects of human growth hormone in men over 60 years of age. *N Engl J Med* 1990;323(1):1–6.

53. Krentz AJ, Koster FT, Crist DM, et al. Anthropometric, metabolism and immunological effects of recombinant growth hormone in AIDS and AIDS-related complex. *J AIDS* 1993;6(3);245–251.

54. Nelson MR, Erskine D, Hawkins DA, et al. Treatment with corticosteroids: a risk factor for the development of clinical cytomegalovirus disease in AIDS. *AIDS* 1993;7:375–378.
55. Summerbell CD, Youle M, Mcdonald V, et al. Megesterol acetate vs. cyproheptadine in the treatment of weight loss associated with HIV infection. *Int J STD AIDS* 1992;3(4);278–280.
56. Coodley GO, Loveless MO, Merrill TM. The HIV wasting syndrome. *J AIDS* 1994;7:681–694.
57. Singer P, Rothkopf MM, Kvetan V, et al. Risks and benefits of home parenteral nutrition in the acquired immunodeficiency syndrome. *J Parent Ent Nutr* 1991;15: 75–79.
58. Mukau L, Talamani MA, Sitzmann JV, et al. Long-term central access Vs other home therapies: complications in patients with the acquired immunodeficiency syndrome. *J Parent Ent Nutr* 1992;16(5):455–459.
59. Tuttle-Newhall JE, Veerabagu MP, Mascioli EA, et al. Nutrition and metabolic management of AIDS during acute catabolic illness. *Nutrition* 1993;9(3):240–245.

Nutritional Oncology: A Proactive, Integrated Approach to the Cancer Patient

CHAPTER 23

Faith D. Ottery, M.D., Ph.D.

INTRODUCTION

Approximately 1.3 million new cancer cases will be diagnosed in 1996, with approximately 560,000 cancer deaths.[1] Loss of at least 5% of pre-illness weight is reported in one third of patients with malignancy at the time of initial presentation and is a nearly universal finding among patients with advanced cancer. Malnutrition frequently contributes to the cause of death in patients with cancer, with as many as 20% of cancer patients succumbing to progressive nutritional deterioration or inanition rather than to the malignancy per se.[2–5]

Failure to address the nutritional status of patients with cancer has several negative consequences that can adversely impact the outcomes, quality, and cost of care. The overall costs for cancer have been estimated at $104 billion per year with $35 billion for direct medical costs, $12 billion for morbidity costs (cost of lost productivity), and $57 billion for mortality costs.[6] It has been documented in a number of patient populations that malnourished patients have an average length of stay that is twice that of diagnosis-adjusted well-nourished patients.[7] Review of discharge diagnoses of oncology patients during 1993 and 1994 from the Fox Chase Cancer Center (FCCC) in Philadelphia were consistent with published data in noncancer patients.[8] Overall average length of stay for all FCCC patients was 5.8 days, for patients with discharge diagnoses of dehydration and/or malnutrition (in addition to their malignancy), 9.4 days, and for malnutrition alone, 13.4 days. Data suggest that once a patient becomes even borderline malnourished, the patient becomes a more costly one to care for.[9] The prognostic impact of weight loss per se is known in patients with cancer as well as in general hospitalized patients.

The major goals of supportive nutrition are adjunctive to the specific oncology treatment goal. Nutrition is supportive in the context of maintaining adequate nutritional status, body composition, performance status, immune function, and quality of life. When the patient is being treated with curative intent or with the goal of long-term survival, the aims of supportive nutrition include stabilizing or improving the patient's nutritional status as well as increasing the potential of a favorable response to therapy and enhancing recovery from any adverse effect of therapy. For those patients receiving palliative care, nutritional maintenance may help stabilize performance status, maintain activities of daily living, and improve the quality of life of the patient.

Weight loss and pronounced nutritional depletion are often considered to be inherent components of cancer. The most compelling reason for early supportive nutritional intervention in patients is to avoid irreversible nutritional and physiological deficits. It is generally recognized that functional deterioration based on muscle loss is poorly, if not minimally, reversible. The progressive decrease in physical performance that may occur can lead to increased bedrest and further debilitation. Malnutrition and weight loss affect the patient's physical status, the course of the disease, and response to treatment.[10]

Weight loss in the cancer patient can often be prevented (see below), but generally only if addressed proactively. This requires the adoption of a therapeutic strategy and routine clinical practice to maintain optimal nutrition in the presence of the cancer and its treatment. The use of simple, standardized assessment tools and algorithm-directed approaches facilitates such intervention. The Patient-Generated Subjective Global Assessment (PG-SGA) of Nutritional Status and the Algorithm of Optimal Nutritional Intervention are increasingly being used to direct the nutritional care of the oncology patient within the United States.[11–14]

USE OF NUTRITIONAL INTERVENTION IN CANCER PATIENTS

Questions concerning the effectiveness of nutritional care of the cancer patient have not been definitively answered. The basis for this is summarized below.

Inherent Part of Cancer

Malnutrition has generally been accepted as an inherent component of cancer. Several treatable impediments to adequate nutritional intake are seen in the majority of cancer patients which can be related to the tumor itself or the antineoplastic therapy. Appropriate pharmacologic, behavioral, or surgical treatment will alleviate many of these impediments. Just as one treats a cancer patient's diabetes or congestive heart failure as a separate disease from the cancer, so should one treat the malnutrition or symptoms impacting nutrient intake as separate from the cancer per se.

Inappropriate Study Design

Clinical trials of supportive nutrition in cancer have often times used inappropriate eligibility and ineligibility criteria—at times baseline malnourished patients were ineligible or nutritional intervention was not initiated until end-stage cancer and/or malnutrition. Supportive nutrition should not be put in the same category as phase I chemotherapy, that is, only used when all other treatment fails.

Quality of Intervention

Assessment of the quality of nutritional intervention regimens and the meeting of the individual patient's requirements have not generally been addressed in individual reports of nutrition support (i.e., parenteral and enteral nutrition), making it difficult to draw valid conclusions concerning effectiveness or lack thereof in many meta-analyses. The oncology literature is clear that ineffective doses or inappropriate choices of chemotherapy or inadequate radiation ports, fractionations, or total doses are ineffective in treating cancer. However, this consideration of the quality of the nutritional intervention is not consistent in a number of reports of the use of parenteral or enteral nutrition in treating malnutrition of the cancer patient.

Nutrition Support Is High-Technology Nutrition

The term nutrition support has generally come to mean high-technology nutrition support, usually parenteral nutrition, but it may also include enteral tube feedings. In the oncology patient, the concept of nutrition support is used primarily in the context of the severely malnourished, terminal, or end-stage patient rather than proactive, often oral intervention.

Cost

Nutritional intervention is generally considered to be a costly intervention and one that is to be avoided if possible. This concern, combined with poorly defined indicators for initiation of supportive nutrition, has led to delayed and/or inappropriate use of supportive nutritional intervention. Consideration of the use of nutritional counseling and aggressive symptom management (to encourage oral nutrition or lessened dependence on high technology nutritional regimens) is less often considered in the development of nutritional intervention protocols.

Poor Performance Status

Patients placed on nutritional intervention are frequently malnourished, with decreased performance status, marked decrease in muscle mass and function, and are generally considered to be poor candidates for any rehabilitation until nutritional status is improved. It has been known since the 1940s that bedrest for as little as 1 week is associated with significant loss of muscle volume and function.[15–17] Although function may improve with nutrition per se, mass loss is generally not reversible without a component of physical activity or exercise.

PROACTIVE NUTRITIONAL ASSESSMENT OF THE ONCOLOGY PATIENT

A number of tools, indices, functional assessments, and laboratory evaluations are available for use in nutritional assessment. Important components for the practicing oncologist or general physician caring for the oncology patient in using a nutritional assessment tool are (a) ease of use, cost-effectiveness, and reproducibility in several clinical settings; (b) ability to predict those patients who need nutritional intervention; and (c) little interobserver variability.

The Subjective Global Assessment (SGA) of nutritional status was developed by Jeejeebhoy, Baker, Detsky, and colleagues at the University of Toronto and published in a usable format in 1987.[18–20] An oncology modification was developed in 1993 by Ottery, with a later modification developed to be generated by the oncology patient while waiting for the oncology visit.[21] This latter format (the Patient-Generated Subjective Global Assessment or PG-SGA) is found in Figure 23.1. Although developed for the oncology patient, some clinicians have begun to use it for human immunodeficiency virus (HIV) and acquired immunodeficiency syndrome (AIDS) patients, geriatric patients, and general hospitalized patients on admission in the context of the screening for Joint Commission accreditation.[22]

The original SGA was initially used predominantly in surgical patients and has been used to prospectively determine the need for preoperative nutritional intervention.[23] The SGA was based on the hypothesis that risk associated with malnutrition can be minimized or reversed if intake to meet nutrient requirements is restored. The stage of nutritional risk or deficit is based on physiologic status rather than physical appearance alone.

Patient-Generated Subjective Global Assessment (PG-SGA) of Nutritional Status

*(**To the patient:** please check the box or fill in the space as indicated in the next four sections)*

A. History

1. Weight change
   Summary of my current and recent weight:
   I currently weigh about _____ pounds
   I am about _____ feet _____ inches tall
   A year ago I weighed about _____ pounds
   Six months ago I weighed about _____ pounds
   During the past 2 weeks, my weight has
   ❐ decreased ❐ not changed ❐ increased

2. Food intake:
   As compared to normal, I would rate my food intake during the past month as either:
   ❐ unchanged
   ❐ changed: ❐ more than usual
   ❐ less than usual
   I am now taking:
   ❐ little solid food ❐ only nutritional supplements
   ❐ only liquids ❐ very little, if anything

3. Symptoms: During the past 2 weeks, I have had the following problems that kept me from eating enough (check all that apply):
   ❐ no problems eating
   ❐ no appetite, just did not feel like eating
   ❐ nausea ❐ vomiting
   ❐ constipation ❐ diarrhea
   ❐ mouth sores ❐ dry mouth
   ❐ pain (where?) _____________
   ❐ things taste funny or have no taste
   ❐ smells bother me
   ❐ other_____________

4. Functional capacity:
   Over the past month, I would rate my activity as generally:
   ❐ normal, with no limitations
   ❐ not my normal self, but able to be up and about with fairly normal activities
   ❐ not feeling up to most things, but in bed less than half the day
   ❐ able to do little activity and spend most of the day in bed or chair
   ❐ pretty much bedridden, rarely out of bed

(The remainder of this form will be completed by your doctor, nurse, or therapist. Thank you.)

5. A. History (continued)
   Disease and its relationship to nutritional requirements
   Primary diagnosis (specify) _____________
   (Stage, if known)_____________
   Metabolic demand (stress) ❐ no stress ❐ low stress ❐ moderate stress ❐ high stress

   B. Physical (for each trait specify: 0 = normal, 1 = mild, 2 = moderate, or 3 = severe)
   ___ loss of subcutaneous fat (triceps, chest) ___ muscle wasting (quadriceps, deltoids)
   ___ ankle edema ___ sacral edema ___ ascites

   C. SGA rating (select one)
   ❐ A: well nourished
   ❐ B: moderately (or suspected) malnourished
   ❐ C: severely malnourished

Figure 23.1. The patent-generated subjective global assessment (PG-SGA) is a modification of the original SGA to be used in oncology patients with the initial four sections completed by the patient and the remaining sections completed by the oncology clinician (physician, nurse, or dietitian).

The oncology modification was developed to include earlier time point information and additional symptoms pertinent to the oncology patient. To be a useful tool in a busy oncology practice, it was decided that a patient-generated format would be most appropriate. The basis of this was fourfold.

1. Lack of time on the part of oncologists or oncology nurses to incorporate an additional assessment procedure or instrument.
2. Perception on the part of the patient and family that nutrition and weight loss are important in the overall oncology course. Patients and family are often frustrated by their perception that this is not being addressed by their physicians or health care team.
3. The nutrition screening/assessment would be an appropriate use of time while the patient is waiting for the oncologist and/or nurse. When in-serviced, the clinician component of the PG-SGA adds less than a minute to the overall clinic process and adds directly to the quality of nutritional and other components of supportive care.
4. In addition to outcome-based, cost-effective results, patient satisfaction is increasingly becoming an important component of physician and institution report cards in the managed care arena.

The PG-SGA can be used by any clinician (physician, nurse, dietitian) in an inpatient, outpatient, homecare, or hospice setting. The PG-SGA is filled out by the patient, who provides a current history of weight change, food intake, symptoms, and functional capacity. The form is then completed by the clinician (which adds approximately 1 min) with information and observations based on the physical examination. With the information provided, the patient is then categorized as one of the following: Stage A—well nourished; Stage B—moderate (or suspected of being) malnourished; Stage C—severely malnourished. The stages are based on those originally defined by use of the SGA. Although the ranking is subjective, Table 23.1 gives the general guidelines for each category. A scored

Table 23.1. Guidelines for Subjective Global Assessment Categories

Stage A	Stage B	Stage C
Well-nourished or Recent nonfluid weight gain and/or Improvement in components of history, e.g., improved symptoms, intake	> 5% weight loss within a few weeks No weight stabilization or weight gain Definite decrease in intake Mild subcutaneous tissue loss	Obvious signs of malnutrition (e.g., severe loss of subcutaneous tissue, possible edema) Clear and convincing evidence of weight loss

version is currently in development. The derived stage or category places the patient at a specific point in the algorithm in Figure 23-2.

ALGORITHM FOR OPTIMAL NUTRITIONAL INTERVENTION

The PG-SGA derived stage of nutritional status places the patient at a specific point in the algorithm. Although perhaps appearing somewhat complicated, the algorithm is quite simple once the clinician determines the SGA stage and determines the nutritional risk of the therapy. A single pathway point is determined by these two pieces of information. The actual treatment pathway is based on the determination of nutritional risk of the oncologic therapy that the patient is, or will be, receiving. The level of nutritional risk associated with any antineoplastic therapy is categorized as low risk or high risk depending on the frequency or intensity of nutritional impact symptoms that impede adequate nutritional intake. Nutritionally high-risk chemotherapy or biologic-response modifier therapy is that in which there is at least a 30% to 50% risk of grade 3 gastrointestinal toxicity or febrile reaction. Nutritionally high-risk radiation therapy is defined by the port site, fractionation protocol, and total dose, as well as whether the radiation therapy is part of a multimodality regimen.

Dotted lines from each point of deterioration determine the next level of intervention. The wide dashed lines from the Specialized Nutrition Intervention box refer to the pathways for those patients who need to be triaged directly to enteral tube feedings or parenteral nutrition rather than attempt the oral route. The thinner dashed line leading from the box for unsuccessful enteral support does not mean that parenteral nutrition is the only option but implies that if changes in the enteral regimen are not successful, parenteral nutrition is an option that should be considered in the patient who continues to undergo antineoplastic therapy.

Reassessment

It should be noted that reassessment throughout the oncology patient's course is important. Once management of the patient via the treatment algorithm is initiated, reassessment is indicated at regular intervals according to the patient's status and therapy. The importance of regular reassessment cannot be overstated. The responsible clinician should coordinate reassessment with each medical oncologist or radiation oncologist visit so that specific interventional decisions (e.g., pharmacologic treatment with nutritional impact symptoms) can be initiated as soon as problems are diagnosed. In other patient populations, e.g., surgery patients, a transition from SGA A to C is associated with a sevenfold increase

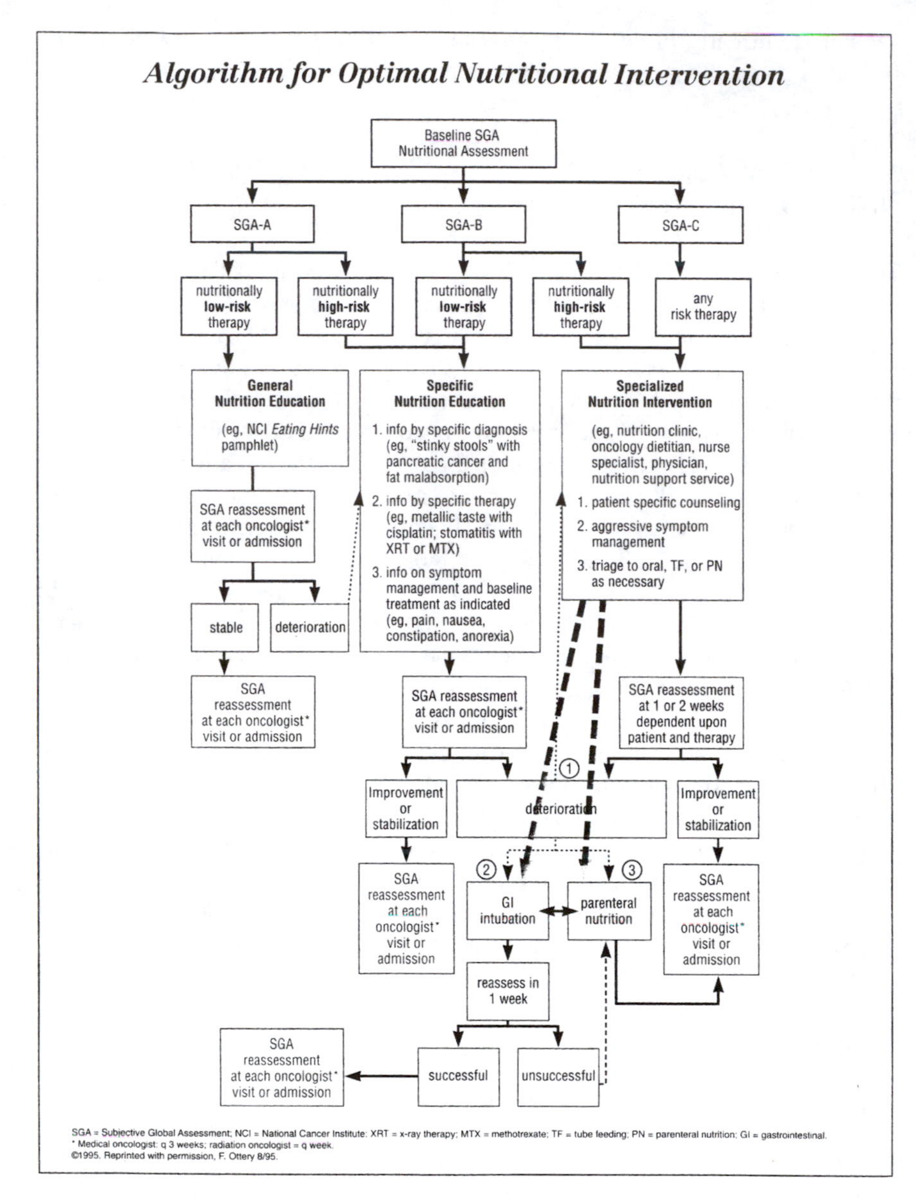

Figure 23.2. Algorithm of optimal nutritional intervention.

in both septic and nonseptic complications. Reassessment throughout the algorithm is coordinated with the medical or radiation oncologists visit (although the assessment can be done by any member of the clinical team) so that specific interventional decisions (e.g., pharmacologic treatment of nausea or anorexia) can be initiated as soon as problems are diagnosed and appropriate referrals made.

NUTRITIONAL INTERVENTION OPTIONS

Options of supportive nutrition are no different in the oncology patient than in any other population. Components of the decisionmaking process include the following:

1. Presence or absence of a functional gastrointestinal tract
2. Treatment plans: surgery, radiation, chemo/hormonal/biologic response modifier therapy
3. Degree of baseline nutritional deficit (e.g., as determined by the PG-SGA)
4. Issues of quality of life and prognosis
5. Issues of cost effectiveness and utility

COMPONENTS OF NUTRITIONAL INTERVENTION

As included in the algorithm, nutritional intervention routes include oral, enteral tube feedings, and parenteral nutrition. Education of the patient and family member or caregiver is extremely important in determining the success of any of the options. We would never consider sending a patient home on parenteral nutrition without adequate instructions; however, it is common to give a patient a pamphlet and a can of supplement and assume that this is adequate education for sending a patient home on oral nutrition. From this perspective, it may be easier to give generic teaching to a home enteral or parenteral patient, because these options may not be so dependent on individual patient variability. Components of successful oral intervention are included in Table 23.2.

Specific components of enteral and parenteral nutrition are not that dissimilar to other patient populations and these have been covered in a number of other chapters in this series. However, the following comments should be considered:

1. Nutrition must be considered in the overall scheme of oncology treatment. When a patient's abdomen is being closed at the end of a surgical procedure, the surgeon should consider whether a feeding tube placed at the time of closure will get the patient over the perioperative period and subsequent oncologic therapy more successfully. Additionally, any medical or radiation oncologist should feel comfort-

Table 23.2. Components of Successful Oral Intervention

AGGRESSIVE AND PROACTIVE SYMPTOM MANAGEMENT
GI symptoms: nausea and vomiting, constipation or diarrhea, mucositis/stomatitis, delayed gastric emptying/slowed GI transit time, food intolerances
Anorexia
Pain
Depression/anxiety/psychosocial considerations
INCLUSION OF THE FOLLOWING PRINCIPLES OF ORAL NUTRITION
Definition of calorie and protein goals
Removal of dietary restrictions
Management of sensory changes
Definition food intolerances with avoidance, treatment
Education of patient to think of food as medicine
Addressing patient issues of control and self-image
Timing of nutritional counseling and timing of trials of nutritional supplements to optimize compliance (i.e., timing separate from treatment session or during treatment related toxicity)
Addressing appropriate vitamin use in terms of timing and dose

able in suggesting the need for tube placement when talking with the surgeon preoperatively concerning long-term oncologic treatment plans.

2. If the patient has a baseline nutritional deficit, an attempt at oral nutrition as above with aggressive symptom management is feasible. Based on the algorithm, follow-up assessment, and triage to enteral or parenteral nutrition is necessary if the patient is unable to maintain weight with oral intervention.
3. In the oncology patient, start with 1.5 $g/kg \cdot day^{-1}$ of protein intake and use the Harris–Benedict (HB) equation for energy calculations. Consider all patients (beyond 3 to 5 days postoperatively) as metabolically medical and initially aim for approximately 1.2 to 1.3 times the requirements determined by the HB equation. All patients on enteral tube feeding and parenteral nutrition in the hospital should undergo a weekly 24-hr urine collection for urinary urea nitrogen (UUN) determination (e.g., every Sunday). All home parenteral nutrition patients similarly should collect a Sunday 24-hr UUN until they are on a stable regimen. Based on data from urine collections at Fox Chase Cancer Center,[24] we found that approximately 40% of the patients required a modification (in an upward direction) of the initial calculations. Although we do not know what a positive nitrogen balance means in a tumor-bearing host, we do know the meaning of a negative nitrogen balance. The UUN gives us a quantitative estimate of the increased needs of the patient.
4. Transferrin, for a number of reasons, appears to be a better laboratory parameter (in conjunction with the UUN) than is albumin for monitoring nutritional interven-

tion in the cancer patient. In addition to hydrational issues, there is evidence in animal models, that there is an increased albumin catabolism in tumor-bearing hosts that is independent of the nutritional status. Additionally, it is also superior to the shorter half-life transport proteins such as prealbumin, given the potential impact of short-lived neutropenic fevers. The latter may simply be too sensitive to address the global nutritional status in this patient population.

5. Transition of the cancer patient from parenteral nutrition to enteral tube feedings or oral nutrition is no different, in general, from the transition in other patient populations and should be the goal. Use of pancreatic enzymes (e.g., in pancreatic cancer, in short gut associated with ovarian cancer patients after several surgical resections, or in patients with malabsorption in chronic radiation enteritis) and antisecretalogues (e.g., H_2-blockers and octreotide) should be considered in transitional feeding attempts.
6. Discontinuation of enteral and parenteral support in the cancer patient is a topic that needs to be discussed in an open and supportive manner proactively, as with any other discussion of advanced directives. Patients and families who have a realistic understanding of the goals of nutritional care in the context of oncologic treatment are often accepting of potential discontinuation of aggressive nutritional support if it has been discussed previously. On occasion, complete discontinuation may come after a transitional period of hydration with parenteral vitamins. This discussion is imperative in the context of sending a patient home on enteral or parenteral nutrition.

HOME ENTERAL AND PARENTERAL NUTRITION SUPPORT

The number of active cancer patients on home parenteral nutrition (HPN) was shown to increase by 13%/year.[25] The annual survival rate was reported as 25% with active cancer, compared with 85% in patients with radiation enteritis, and 95% for patients with Crohn's disease. It should be noted, however, that this survival data is only for patients who remained on HPN, and may be somewhat misleading as demonstrated by data from Ottery et al. below. In the OASIS data, 50% of all cancer patients starting HPN were dead within 6 to 9 months, with the prognosis somewhat better in children. Also, in the OASIS data, 20% of active cancer patients did appear to do well, returning to full oral nutrition and experiencing complete rehabilitation. Additionally, it was noted that rehospitalization rates for HPN complications were the same (approximately 1/year) for cancer and noncancer patients.

Indications for HPN in cancer fall into three primary categories: (a) cancer that is cured but at the cost of permanent bowel dysfunction (e.g., short bowel, radiation enteritis), (b) patients undergoing aggressive therapy with severe gastrointestinal toxicity for potentially curable disease, and (c) incurable cancer with bowel obstruction. The latter is the most common indication. Whereas treatment in category a is the least controversial, indications in the others, especially category c are not so clear. The basis for prospective parenteral nutritional intervention has been well summarized by Maliakkal et al.[26]

The decision to place a patient on HPN requires a multidisciplinary assessment including nutritional indications, patient wishes, financial coverage, caregiver and support system availability, presence of enteral and/or parenteral access devices, physician philosophy, patient prognosis, patient and family/caregiver perceptions of the disease, and components of quality of life. Issues that need to be addressed prospectively concerning the use of HPN in cancer patients are included in Table 23.3.

The HPN experience at Fox Chase Cancer Center has been reviewed.[8] Sixty-five patients (26 female, 39 male) received HPN over a period of 6 years. At the time of review, 13 patients were still alive. Results in all patients revealed that 58% percent (38 patients) were able to taper off parenteral nutrition entirely or with transition to enteral tube feeding. Time on HPN ranged from 3 weeks to 5 years (68% ≥ 3 months, 8% ≥ 1 year). Of the patients able to successfully complete parenteral nutrition or transition to enteral tube feeding, 21 (55%) were able to be permanently off parenteral nutrition or have a long-term time off prior to death (i.e., greater than 1 year). Of deaths off parenteral nutrition, 10% were unrelated to the cancer diagnosis. Nutritional improvement (increased transferrin and removal from negative nitrogen balance) was seen in 100% and prevention of nutritional death and improvement in performance status and quality of life (including return to work or previous activity level) was seen in all appropriately chosen patients. Patients with disappointing results included the 8% who were too debilitated to undergo any additional therapy and for whom parenteral nutri-

Table 23.3. Issues for Consideration for Home Parenteral Nutrition

Are there alternatives to HPN, e.g., enteral nutrition, medical management, hydration?
Is the patient being treated for their cancer, with either curative or palliative intent?
In the patient with GI insufficiency, will the patient die of malnutrition rather than from a stable or indolent cancer?
What is the performance status and quality of life of the patient? Are these likely to be maintained or improved by HPN?
What are the co-morbid medical problems which can complicate nutritional therapy and potentially increase the risk of complication?
Is the patient a candidate for protocol or other antineoplastic therapy which is associated with prolonged GI toxicity which will not allow oral or enteral nutrition for prolonged periods?
Is the HPN being performed according to currently available standards and by a physician and/or nutrition team who knows what they are doing?

tion was started by the patient's oncologist in an attempt to improve their performance status enough to give further treatment.

The most important concept in addressing inappropriate use of parenteral nutrition in cancer patients is to address nutritional issues early in the patient's course, as discussed in the sections above. Often the need to address the severely cachectic patient stems from the oncologist's or other treating physician's failure to address nutritional risk and/or deficit when it is first a problem. Review of nutrition clinic results at Fox Chase over 4 years in 186 consecutive patients revealed that the average weight loss at time of nutrition clinic referral was 16.8%.[8] Despite this late referral pattern, maintenance or improvement of weight and visceral protein status was successful in 50% of patients during antineoplastic therapy through oral intake (food and supplements). When data from preterminal patients (those with a life expectancy of less than 6 weeks) were subtracted, the success rate in maintaining or improving visceral protein status and weight was nearly 80%. If this is possible in these significantly malnourished patients, proactive and early intervention should be successful in preventing the severe weight loss we have come to expect as part of the disease and/or therapy process.

If this proactive approach is impossible or has not occurred and one is now faced with a patient in the third category of HPN cancer patients (i.e., functional or mechanical bowel obstruction in a noncurable cancer patient), what are the options of therapy? In addition to hospitalization or HPN, the following are therapy options:

1. Decompressive gastrostomy and high jejunostomy with hydration.
2. Medical management including octreotide, anticholinergic and antiemetic medications, analgesics, etc.

Active participation in the appropriate decision making process of who should be placed on HPN requires good communication on the part of the physician with the patient, family, and other members of the medical team. This is necessary for the quality and ethics of care in an individual patient as well as in the context of limited health care resources.

SUMMARY

Integration of proactive nutritional assessment and intervention offers the opportunity to address issues of both quality of medical and oncologic care as well as cost-effectiveness issues. The Patient-Generated Subjective Global Assessment (PG-SGA) and an algorithm of nutritional intervention have been offered as well as a number of practical and philosophical aspects of nutritional aspects of oncology care.

REFERENCES

1. Parker SL, Tong Y, Bolden C, et al. 1995. *Cancer Statistics.* 46:5–27.
2. Ambrus JL, Ambrus CM, Mink IB, Picken JW. Causes of death in cancer patients. *J Med Clin Exp Theor* 1975;6:61–64.
3. Inagaki J, Rodrigues V, Body GP. Causes of death in cancer patients. *Cancer* 1974;33:658–573.
4. Klastersky J, Daneau D, Verhest A. Causes of death in patients with cancer. *Eur J Cancer* 1972;8:149–154.
5. Warren S. The immediate cause of death in cancer. *Am J Med Sci* 1932;184:610–615.
6. Cancer facts and figures 1996. Atlanta, Georgia. American Cancer Society, 1996.
7. Sproat KV, Russell CM, eds. *Malnutrition: a hidden cost in health care.* Columbus, OH: Products Division, Abbott Laboratories, 1994.
8. Ottery FD, Stofey J, Hagan M. Review of nutritional care in national cancer institute-designated comprehensive cancer center (NCI-CCC). Presented at 19th Clinical Congress, Miami, January 15–18, 1995.
9. Robinson G, Goldstein M, Levine GM. Impact of nutritional status on DRG length of stay. *J Parent Ent Nutr* 1987;11:49–51.
10. DeWys WD, Begg C, Lavin PT, et al. Prognostic effect of weight loss prior to chemotherapy in cancer patients. *Am J Med* 1980;69:491–497.
11. Ottery FD. Definition of standardized nutritional assessment and interventional pathways in oncology. *Nutrition* 1996;12(suppl 1):s15–S19.
12. Ottery FD. Supportive nutrition to prevent cachexia and improve quality of life. *Sem Oncol* 1995;22(suppl 13):98–111.
13. Ottery FD. Rethinking nutritional support of the Cancer patient: the new field of nutritional oncology. *Semin Oncol* 1994;21:770–778.
14. Ottery FD. Cancer cachexia: prevention, early diagnosis, and management. *Cancer Pract* 1994;2:123–131.
15. Deitrick JE, Whedon GO, Shorr E. Effects of immobilization upon various metabolic and physiologic functions of normal men. *Am J Med* 1948;4:3–36.
16. Stuart A, Shengraw RE, Prince MJ, et al. Bed-rest induced insulin resistance occurs primarily in muscle. *Metabolism* 1988;37:802–806.
17. Shangraw RE, Stuart A, Prince MJ, et al. Insulin responsiveness of protein metabolism in vivo following bedrest in humans. *Am J Physiol* 1988;255:E548–E558.
18. Baker JP, Detsky AS, Wesson DE, et al. Nutritional assessment: a comparison of clinical judgment and objective measurements. *N Engl J Med* 1982;306:969–972.

19. Detsky AS, Baker JP, O'Rourke K, et al. Predicting nutrition-associated complications for patients undergoing gastrointestinal surgery. *J Parent Ent Nutr* 1987;11:440–446.
20. Detsky AS, McLaughlin JR, Baker JP, et al. What is the subjective global assessment of nutritional status? *J Parent Ent* Nutr 1987;11:8–13.
21. Ottery FD. Modification of subjective global assessment (SGA) of nutritional status (NS) for oncology patients. Presented at the *19th Clinical Congress of the American Society for Parenteral and Enteral Nutrition,* Miami FL, January 15–18, 1995 (abstr 119).
22. Joint Commission on Accreditation of Healthcare Organizations: 1995. *Comprehensive Accreditation Manual for Hospitals.* Oakbrook Terrace, IL 1994.
23. Veterans Administration Cooperative Study Group. Perioperative total parenteral nutrition in surgical patients. *New Engl J Med* 1991;325:525–532.
24. Ottery FD. Unpublished data.
25. Howard L, Heaphey L, Fleming CR, et al. Four years of North American Registry home parenteral nutrition outcome data and their implications for patient management. *J Parent Ent Nutr* 1991;15:384–393.
26. Maliakkal RJ, Blackburn GL, Wilcutts, HD, et al. Optimal design of clinical outcome studies in nutrition and cancer: future directions. *J Parent Ent Nutr* 1992;16:112s–116s.

CHAPTER

24

Patients with Inflammatory Bowel Disease

Richard E. Karulf, M.D.

INTRODUCTION

Inflammatory bowel disease (IBD) is a broad category that encompasses two similar but distinct conditions: ulcerative colitis and Crohn's disease. The lack of a completely effective medical treatment for patients with inflammatory bowel disease has prompted innovative trials and research into new areas in search for a cure. Because there has been a postulated association of food antigens with the etiology of both ulcerative colitis and Crohn's disease and because malnutrition is a common problem in both groups, there have been many attempts to improve patient outcomes with food-group avoidance and nutritional therapy. The efficacy of nutrition therapy in treatment of active inflammatory bowel disease is, however, controversial. In general, remissions with enteral nutrition and parenteral nutrition are more common with active Crohn's disease than with ulcerative colitis.

PRESENTATION OF INFLAMMATORY BOWEL DISEASE

The presentation of patients with inflammatory bowel disease is most commonly associated with frequent episodes of bloody diarrhea, weight loss, and anorexia. Weight loss is a common feature in both Crohn's disease and ulcerative colitis, with approximately 70% to 80%[1] and 18% to 62%[2] of patients affected, respectively. The etiology of the weight loss is multifactorial. Decreased oral intake from anorexia, nausea, abdominal pain, and malaise, and drug-induced malabsorption from steroid and azulfidine therapy are common to both Crohn's disease and ulcerative colitis. In addition, there are increased nutritional requirements due to the presence of active inflammation, fever, infection, and increased gastrointestinal losses with bleeding and protein-losing enteropathy in both forms of inflammatory bowel disease.[3] Finally, patients with Crohn's disease are uniquely at risk for weight loss due to malabsorption from bacterial overgrowth, decreased bile salt concentrations, decreased absorptive surface area, lymphangiectasia, and mucosal cell disease.

NATURAL HISTORY OF INFLAMMATORY BOWEL DISEASE

The natural history of these two diseases with similar presenting symptoms is often vastly different. Ulcerative colitis patients display a spectrum of symptoms from mild proctitis or left-sided colitis, which is easily controlled with medical management, to toxic colitis with megacolon, which requires urgent surgical intervention. Crohn's disease is somewhat less likely to present with toxic colitis but is much more likely to develop fistulae from diseased segments of bowel into adjacent structures such as normal bowel, bladder, and vaginal cuff after hysterectomy. Patients with Crohn's disease are also prone to stricture formation and disease recurrence, which often result in multiple operations and intestinal resections. These factors have made it the most common cause of short-gut syndrome.[4]

For patients with chronic pancolonic involvement of ulcerative colitis, a significant risk of cancer occurs after 10 years of symptomatic disease. Crohn's disease has a similar increase in colon cancer risk but not until much later in the course of the disease.

Location

Although there are many characteristics that distinguish ulcerative colitis from Crohn's disease, one of the most obvious is pattern of disease. While Crohn's disease may occur at virtually any location in the digestive system between the mouth and anus, it is most frequently located in the ileocecal region (50%) followed by the ileum alone (30%) and the colon (20%). One unique feature of Crohn's disease is that it is noted to have areas of normal bowel (skip areas) in between areas of active disease.

Ulcerative colitis, on the other hand, begins as ulcerative proctitis and progresses without interruption into the more proximal colon. Although ulcerative colitis occurs exclusively in the colon and rectum, areas of small bowel inflammation can occur in the terminal ileum due to 'backwash' ileitis in 15% to 20% of patients.

Histologic Differences

Another important method of distinguishing the two forms of inflammatory bowel disease is by microscopic appearance. Crohn's disease is traditionally described as having full thickness penetration of the inflammatory process and is frequently associated with crypt abscesses. Ulcerative colitis has nontransmural inflammation and is not frequently associated with crypt abscesses. Even with these seemingly exclusive guidelines, up to 10% of patients with colitis cannot be characterized into either group and are referred to as having indeterminant colitis.

Etiology

The search for the underlying etiology of IBD has yielded many enticing possibilities, but none have proven durable in the face of meticulous inquiry. The interesting associations identified in these searches may yet provide insight into the actual etiology or treatment of IBD. There appears to be a genetic predisposition to IBD but also an environmental trigger or stimulus to prompt expression of the illness[5]. The exact nature of the stimulus or method of inheritance has not yet been defined.

Genetic factors. Many authors report that both ulcerative colitis and Crohn's disease patients tend to cluster in families and parts of the world. One article noted that 82% of a group of 188 patients with IBD were Ashkenazi Jews.[6] This genetic predisposition is more strongly associated with Crohn's disease with an observed increase in risk of close relatives that is 17 to 35 times higher than the general population.[7] Close relatives of patients with ulcerative colitis have a similar increased risk.

Suggestion of environmental influence in addition to genetic predisposition is provided by monozygotic twin studies in patients with IBD. It was shown that seven of eight twin pairs were concordant for Crohn's disease, whereas only five of eleven twin pairs were concordant for ulcerative colitis.[8,9] Exciting research into the human leukocyte antigen (HLA-B27) is ongoing as well as other major histocompatibility complex (MHC) markers looking for the genetic factors associated with IBD.

Dietary factors. Of the many factors that could trigger a person with a genetic predisposition for IBD, dietary factors have been widely studied. One study suggests that newly diagnosed patients have twice the normal intake of refined sugar as age-matched controls and that the relative risk of developing Crohn's disease with a high sugar diet is 4.6.[10] Another study suggests that there is a correlation of incidence of IBD with national margarine use.[11] There are, however, few reliable statements that can be made because of conflicting reports. As a result, no clear consensus has emerged regarding the role of breast feeding, fat, cereal (corn flakes), fiber, or sugar consumption with respect to the etiology of IBD.

Exogenous factors. Many other exogenous factors have been studied to determine if they have an association with IBD. Cigarette smoking is associated with an increased incidence of Crohn's disease by factors ranging from 1.8 to 4.8.[12] On the other hand, smoking appears to have a protective effect on patients with ulcerative colitis.[13] Other factors have been examined as possible exogenous factors such as toothpaste, medications, vitamins, and oral contraceptive pills, but no clear answer has resulted from these attempts.[12]

Infectious factors. The presence of granulomas in Crohn's disease and its similarity to animal forms of ileitis caused by *Mycobacterium paratuberculosis* has prompted a wide range of investigations into the infectious etiology of IBD. However, in spite of initially promising but later disappointing investigations into *Mycobacterium kanasasii,* bacteria, cell wall-deficient *Pseudomonas*-like organisms, viruses including cytomegalovirus and other small-RNA viruses, and cytotoxic proteins as possible etiologic agents, the etiology remains obscure.[14] Currently it is felt that IBDs are associated with both viral and bacterial flora but that the relationship of these flora to the etiology of the disease and the onset of symptoms has not been established.

PROBLEMS OF STUDYING INFLAMMATORY BOWEL DISEASE

There are problems unique to the study of the inflammatory bowel which impair conventional assessment. Spontaneous remissions are not unusual in patients with acute inflammatory bowel disease, especially in Crohn's disease, and make meaningful interpretation of response to therapy statistics difficult. In addition, multimodality therapy is common in patients that fail first-line medical management and terms such as drug resistance and failure of medical therapy are often poorly defined. When remissions occur, it is often difficult to isolate the cause of the clinical improvement. Many trials are retrospective or have no control group. Others use historic control groups in spite of changes in conservative medical management. Ethical constraints reasonably restrict trials from having placebo control groups, because two large prospective, randomized controlled studies have shown beneficial effect of medical management of Crohn's disease with steroids and sulfasalazine.[15,16]

The final problem in the study of IBD is the concept of bowel rest. Proponents claim that bowel rest allows healing of damaged bowel in patients with IBD by avoidance of antigenic intralumenal contents while providing adequate nutrition with total parenteral nutrition (TPN). Whether this allows normalization of immunological function and recovery of gut flora is not clear. Detractors from the use of bowel rest in patients with active IBD point out that villous atrophy is common in patients on TPN and that glutamine and fermentable fiber is not available to promote gut healing. With this viewpoint, bowel rest becomes bowel starvation, and a role for elemental diets is proposed to provide short-chain fatty acids and glutamine.[17]

USE OF NUTRITION THERAPY IN INFLAMMATORY BOWEL DISEASE

Articles related to the use of nutrition therapy in inflammatory bowel disease have been reviewed and a summary of the utility of each concept considered. Table 24.1 describes a scoring method for each topic, based on the availability and quality of research. A summary of the scoring of each topic based on these criteria is provided in table 24.2. Recent changes in the medical literature and unpublished articles may not be reflected in this review.

Table 24.1. Definition of Categories of Proof

Category	Definition
Level 1	Consistent results from multiple prospective, randomized, controlled studies with sufficient numbers to reach meaningful conclusions.
Level 2	Encouraging trend in multiple trials, results are either not statistically significant or appropriate trial can not be performed.
Level 3	Anecdotal evidence or small trials provide some support but falls short of providing conclusive proof or reliable trends. Reports may contradict each other.
Level 4	No meaningful conclusions can be drawn but future research is indicated. Often presented in concept or case report format.

Table 24.2. Level of Proof of Nutrition Therapy and IBD

Topic	Conclusion	Level of Proof	References
Primary prevention of IBD	none	4	18
Active Crohn's disease	—	—	—
Primary therapy with enteral nutrition	less benefit than steroid therapy	1	19,20
Primary therapy with parenteral nutrition	less benefit than steroid therapy	2	21
Adjuvant therapy	effective	2	22,23
Active ulcerative colitis	—	—	—
Primary therapy	minimal benefit	3	24,25,26,27
Adjuvant therapy	less benefit than steroid therapy	3	24,25,26,27
Perioperative care	important in select patients	2	22,28,29,30
Treatment of growth retardation	effective	1	31,32,33,34

PRIMARY PREVENTION OF INFLAMMATORY BOWEL DISEASE

Because many of the theories point to an intralumenal antigenic trigger to activate patients with a genetic predisposition to IBD, avoidance of food groups known to stimulate disease activity seems appropriate.[18] However, neither the genetic markers nor the antigen triggers have been specifically identified at this time. Although this may provide a very important means of disease prevention in the future, it can only be described as an interesting concept at this time.

CROHN'S DISEASE

Crohn's disease, in spite of the problems associated with studying IBD, is one of the most promising areas for the use of nutritional therapy. Multiple prospective, randomized, controlled trials have been performed comparing conventional medical management (usually defined as steroid therapy plus sulfasalazine) with medical management plus various forms of enteral and parenteral nutrition. Many articles have also examined parenteral nutrition as sole therapy in Crohn's disease. Finally, several authors have examined the use of nutritional therapy in helping to treat fistulous Crohn's disease.

Active Crohn's Disease—Primary Therapy with Enteral Nutrition

Nine recent prospective, randomized, controlled trials have been performed comparing total enteral nutrition with steroids in the treatment of active Crohn's disease.[19,20] With active Crohn's disease, only one of nine trials showed a significant advantage of steroid therapy over all types of enteral nutrition. All nine studies, however, suggested that steroids are better than enteral nutrition in inducing remissions in active Crohn's disease.[19] A review of these trials showed lower remission rates for elemental and whole-protein-based diets than steroid therapy in active Crohn's disease but results were not considered conclusive. Lowest remission rates were noted with the use of peptide-based diets when compared to steroid therapy. The quality of research available in this area is good and the overall conclusion is that steroid therapy is more effective than enteral nutrition in the management of active Crohn's disease.

Active Crohn's Disease—Primary Therapy with Parenteral Nutrition

With TPN as sole therapy in 25 patients with active Crohn's disease, there was a 36% surgical resection rate at 1 year. The cumulative relapse rate over 4 years was four times higher in patients receiving TPN as primary therapy compared with a retrospective study of patients undergoing surgical resection without previous TPN. They concluded that TPN is not an alternative to resection and should be reserved for patients with multifocal lesions when surgery is not advisable because of the risk of short-bowel syndrome.[21]

Active Crohn's Disease—Adjuvant Nutritional Therapy

Although there is debate about choice of appropriate primary therapy, most authors agree that there is a synergistic effect when using nutritional supplements and steroid therapy in active Crohn's disease. One report notes the prevention of protein loss and preservation of skeletal and respiratory muscle function with short courses of parenteral nutrition.[22] Similarly, modest weight gain is possible with total enteral nutrition in patients malnourished from Crohn's disease.[23]

ULCERATIVE COLITIS

Although there is room for optimistic use of nutritional therapy in patients with active Crohn's disease, there is much less enthusiasm for use in patients with ulcerative colitis. One study, contrasting patients with active IBD found 14 of

27 with ulcerative colitis but none of 16 patients with Crohn's disease to require urgent surgery while receiving medical management.[24] A similar study revealed 16 of 22 patients with ulcerative colitis compared with only 1 of 16 patients with Crohn's disease requiring surgery during treatment with parenteral nutrition and medical management of severe acute colitis.[25]

Active Ulcerative Colitis—Primary Nutritional Therapy

Few studies have specifically compared either enteral nutrition or parenteral nutrition with steroid therapy in patients with active ulcerative colitis. One author reviewed 10 studies in the literature and discovered that the pooled remission rate for medical management including use of parenteral nutrition was 81% (44 out of 54) for acute Crohn's disease and 44% (79 out of 178) for acute ulcerative colitis.[25] A single trial comparing enteral nutrition and parenteral nutrition in management of active ulcerative colitis showed no difference in remission rates but a significantly lower complication of therapy rate with parenteral nutrition.[26]

Elemental diet in steroid-dependent and steroid-refractory Crohn's disease showed 10 out of 16 patients with remission on 4 weeks of vivonex and taper of steroids. At one year only 3 out of 16 patients had sustained remissions after treatment.[27]

Active Ulcerative Colitis—Adjuvant Nutritional Therapy

The lack of response of ulcerative colitis to medical management limits the utility of enteral and parenteral nutrition as an adjunct to therapy. The majority of its usefulness is therefore in preparing a patient for surgery or as a supplement in the postoperative period.

PERIOPERATIVE USE OF ENTERAL AND PARENTERAL NUTRITION IN INFLAMMATORY BOWEL DISEASE

Use of enteral and parenteral nutrition in the perioperative care of patients with IBD has been the subject of minor controversy in the medical literature. There is evidence that preoperative nutrition therapy may prevent protein loss when preparing for surgery. One prospective study of 19 patients who were clinically malnourished and had acute exacerbations of inflammatory bowel disease were treated with intravenous nutrition (IVN). Measurements of various nutritional parameters were obtained before, during, and after 14 days of IVN. Prior to treatment, the patients with Crohn's disease had lost approximately 35% of their body protein stores with impairment of physiologic parameters by 20% to 40%

compared to a group of controls. After 4 days of IVN, there were improvements in all of the physiologic parameters but no significant change in total body protein. Short courses of IVN in these patients prevent protein loss and result in significant improvements in skeletal muscle and respiratory muscle function. Repletion of total body stores did not occur until later in convalescence during the anabolic phase of recovery.[22]

Reduction of the operative morbidity rate associated with surgery in malnourished IBD patients is an important concern. One study showed that patients with Crohn's disease and a low (less than 3.1 g/dL) serum albumin level had a 29% complication rate compared with 6% for those patients with a normal level.[28]

Although, proponents claim to be able to lower the significant operative morbidity associated with malnutrition, a minority of others question the need for delay and the use of nutritional supplementation when outcome figures are not altered. For example, bowel rest was not a major factor in achieving a remission and did not influence outcome during 1 year's follow-up in one study.[29] A second study of 46 consecutive patients with predominantly ileal Crohn's disease had only a 2.2% early and 6.5% late complication rate without the use of parenteral nutrition.[30]

GROWTH FAILURE IN CHILDREN WITH ACTIVE IBD

Approximately one third of children showed delayed linear growth and delayed onset of puberty with active Crohn's disease. Similar, but less dramatic effects on height and maturation, are noted in ulcerative colitis. Multiple authors have shown a resumption of growth with parenteral[31] and enteral[32] supplementation if initiated prior to epiphyseal closure.

There is some concern of the long-term effects of parenteral nutrition in children with IBD. One study compared children on long-term TPN with 22 age-matched controls with no chronic disease. Long-term TPN (but not short-term TPN) resulted in focal villus atrophy and a decrease in small intestine disaccharidase activity in 2 of 4 patients. Less thymidine was incorporated into biopsy specimens of patients on long-term TPN. This demonstrated a general agreement with animal studies suggesting a hypoplastic effect of long-term TPN.[33] As a result of this and the well-documented complications of prolonged intravenous access, most authors prefer enteral support for treatment of growth retardation in children with active IBD.

Because growth retardation often precedes the clinical onset of IBD, multiple endocrine abnormalities have been examined to explain the delayed growth. One value that is consistently low in children with growth retardation from IBD is somatomedin-C.[34] As nutritional status improves and growth rates return to normal, values of somatomedin-C also return to normal.

RECOMMENDATIONS AND CONCLUSIONS

The association of nutrition and IBD has many origins including assumptions about food antigens and the altered absorptive capacity of the gut. Patients with IBD are at substantial risk for the development of malnutrition from multiple factors. Medical management has produced acceptable remission rates for patients with Crohn's disease but has been somewhat less effective with ulcerative colitis. As a result, medical management with steroid therapy and sulfasalazine serves as the gold standard for nonoperative therapy. The use of food antigen avoidance as either primary prevention or a part of medical treatment has not been effective because of a lack of knowledge of the true etiology of IBD. Treatment of Crohn's disease with medical management results in higher remission rates than with enteral or parenteral nutrition as primary therapy. Use of nutrition therapy as adjunct therapy to conventional medical management has shown promise, especially when preparing a patient with IBD for surgery. Nutritional therapy has not provided a durable primary or adjunct therapy for patients with acute ulcerative colitis and the majority of these patients require surgery. One proven use of both enteral and parenteral nutrition therapy has been in young patients with growth retardation secondary to IBD. Uniform improvement in growth and development is noted after initiation of therapy. As our knowledge of IBD and malnutrition increases, we may see new and exciting methods of improving the quality of life and remission rate of patients with acute ulcerative colitis and Crohn's disease.

REFERENCES

1. van Pattern WN, Bargen JA, Dockerty MB, et al. *Regional enteritis. Gastroenterology* 1954;26:347–350.
2. Farmer RG, Hawk WA, Turnbull RG. Clinical features and natural history of Crohn's disease. *Gastroenterology* 1975;63:627–635.
3. Driscoll RH, Rosenberg IH. Total parenteral nutrition in inflammatory bowel disease. *Med Clin North Am* 1978;62(1):185–200.
4. Stokes MA, Hill GL. The short gut. *Curr Pract Surg* 1990;2:139–145.
5. Garland CF, Lilienfeld AM, Mendeloff AI, et al. Incidence rates of ulcerative colitis and Crohn's disease in fifteen areas of the United States. *Gastroenterology* 1981;81:1115–1124.
6. Gilat T, Rosen P. Epidemiology of Crohn's disease and ulcerative colitis: etiologic implications. *Isr J Med Sci* 1979;15:305–308.
7. Mayberry JF. Recent epidemiology of ulcerative colitis and Crohn's disease. *Int J Colorect Dis* 1989;4:59–66.

8. McConnell MB. Genetic aspects of idiopathic inflammatory bowel disease, in Kirsner JB, Shorter RG, ed. *Inflammatory Bowel Disease,* 3rd ed. Philadelphia: Lea and Febriger, 1988:87–95.
9. Weterman IT, Pena AS. Familial incidence of Crohn's disease in the Netherlands and a review of the literature. *Gastroenterology* 1984;86:449–452.
10. Stokes MA. Crohn's disease and nutrition. *Br J Surg* 1992;79:391–394.
11. Nordenvall B, Brostrom O, Hellers G. Entzündliche Darmkrankheiten und Nahrungsfette. *Dtsch Med Wocheschr* 1982;107:1900–1901.
12. Levine J. Exogenous factors in Crohn's disease—a critical review. *J Clin Gastroenterol* 1992;14(3):216–226.
13. Calkins BM. A meta-analysis of the role of smoking in inflammatory bowel disease. *Dig Dis Sci* 1989;34:1841–1854.
14. Gitnick G. Etiology of inflammatory bowel diseases: where have we been? where are we going? *Scand J Gastroenterol* 1990;25(suppl 175):93–96.
15. Summers RW, Switz DN, Sessions JT, et al. National cooperative Crohn's disease study: results of drug treatment. *Gastroenterology* 1979;77:847–869.
16. Malchow H, Ewe K, Brandes JW, et al. European cooperative Crohn's disease study (ECCDS): results of drug treatment. *Gastroenterology* 1989;86:249–266.
17. Culpepper-Morgan JA. Bowel rest or bowel starvation: defining the role for nutritional support in the treatment of inflammatory bowel diseases (editorial). *Am J Gastroenterol* 1991;86:269–271.
18. Bounous G. Elemental diets in the prophylaxis and therapy for intestinal lesions: an update. *Surgery* 1989;105:571–575.
19. Fernandez-Banares F, Cabre E, Esteve-Comas M, Gassull MA. How effective is enteral nutrition in inducing clinical remission in active Crohn's disease? A meta-analysis of randomized clinical trials. *J Parent Ent Nutr* 1995;19:356–364.
20. Griffiths AM, Ohlsson A, Sherman PM, Sutherland LR. Meta-analysis of enteral nutrition as a primary treatment of active Crohn's disease. *Gastroenterology* 1995;108:1056–1067.
21. Müller JM, Keller HW, Erasmi H, Pichlmaier H. Total parenteral nutrition as the sole therapy in Crohn's disease—a prospective study. *Br J Surg* 1983;70:40–43.
22. Christie PM, Hill GL. Effects of intravenous nutrition on nutrition and function in acute attacks of inflammatory bowel disease. *Gastroenterology* 1990;99:730–736.
23. Royall D, Greenberg GR, Allard JP, et al. Total enteral nutrition support improves body composition of patients with active Crohn's disease. *J Parent Ent Nutr* 1995;19:95–99.
24. McIntyre PB, Powell-Tuck J, Wood SR, et al. Controlled trial of bowel rest in the treatment of severe acute colitis. *Gut* 1986;27:481–485.
25. Sitzmann JV, Converse RL, Bayless TM. Favorable response to parenteral nutrition and medical therapy in Crohn's colitis: a report of 38 patients comparing severe Crohn's and ulcerative colitis. *Gastroenterology* 1990;99:1647–1652.

26. Gonzales-Huix F, Fernandez-Banares F, Esteve-Comas M, et al. Enteral versus parenteral nutrition as adjunct therapy in acute ulcerative colitis. *Am J Gastroenterol* 1993;88:227–232.
27. O'Brien CJ, Giaffer MH, Cann PA, Holdsworth CD. Elemental diet in steroid-dependent and steroid-refractory Crohn's disease. *Am J Gastroenterol* 1991; 86(11):1614–1618.
28. Lindor KD, Fleming CR, Ilstrup DM. Preoperative nutritional status and other factors that influence surgical outcome in patients with Crohn's Disease. *Mayo Clin Proc* 1985;60:393–396.
29. Greenberg GR, Fleming CR, Jeejeebhoy KN, et al. Controlled trial of bowel rest and nutritional support in the management of Crohn's disease. *Gut* 1988;29:1309–1315.
30. Steffes C, Fromm D. Is preoperative parenteral nutrition necessary for patients with predominantly ileal Crohn's disease? *Arch Surg* 1992;127:1210–1212.
31. Kelts DG, Grand RJ, Shen G, et al. Nutritional basis of growth failure in children and adolescents with Crohn's disease. *Gastroenterology* 1979;76:720–727.
32. Morin CL, Roulet M, Roy CC, et al. Continuous elemental enteral alimentation in children with Crohn's disease and growth failure. *Gastroenterology* 1980;79:1205–1210.
33. Rossi TM, Lee PC, Young C, Tjota A. Small intestine mucosal changes, including epithelial cell proliferative activity, of children receiving total parenteral nutrition (TPN). *Digest Dis Sci* 1993;38(9):1608–1613.
34. Kirschner BS, Sutton MM. Somatomedin-C levels in growth-impaired children and adolescents with chronic inflammatory bowel disease. *Gastroenterology* 1986; 91:830–836.

CHAPTER 25

Nutritional Assessment and Therapy in Abdominal Organ Transplantation

Jeffrey A. Lowell, M.D. and Mary Ellen Beindorff, R.D.

LIVER TRANSPLANTATION

Malnutrition is nearly ubiquitous in patients cared for by a liver transplant service. This is true of patients with both acute and chronic liver disease. This certainly is not surprising considering the fact that the liver is the major final common pathway for the utilization of most nutrients, playing a critical role in protein, carbohydrate, and lipid metabolism. With progressive liver failure, there is both a decrease in the number of functioning hepatocytes, as well as decreased delivery of nutrients to the remaining hepatocytes from extrahepatic shunting of portal blood flow. However, the importance of nutrition and most importantly the risks associated with severe malnutrition have only recently been recognized. Part of the recent improved success with liver transplantation is from improved patient selection and perioperative care. Malnutrition is one of the few risk factors in liver transplantation that is potentially reversible.

Causes of Malnutrition in Liver Transplant Candidates

Many factors contribute to malnutrition in patients who are candidates for liver transplantation. These patients have catabolic disease processes that impair their ability to absorb and synthesize proteins (Table 25.1). This is in part due to the hormonal imbalance associated with liver failure. Increased levels of insulin, glucagon, epinephrine, and cortisol all contribute to this catabolic state.[1] In addition, patients with end-stage liver disease (ESLD) will frequently have anorexia and/or dietary restrictions that compromise their ability to replete or even sustain their metabolic needs. The effect of this is proteolysis with inadequate protein resynthesis. Clinically, this leads to a significant decrease in albumin synthesis and other proteins, elevations in clotting times (e.g., prothrombin time) and skeletal muscle mass wasting. The branched-chain amino acids (leucine, isoleucine, and valine) are extensively catabolized for energy in skeletal muscle, being used for gluconeogenesis. This frequently leads to an imbalance in plasma amino acids, with relatively decreased levels of branched-chain amino acids (BCAAs), and increased levels of aromatic amino acids. This has been hypothesized to be a contributing factor in hepatic encephalopathy. Also, with liver failure, there is a diminished capacity to convert ammonia to urea.

Gluconeogenesis is maintained until the very end stages of liver failure. Prior to this, patients most commonly demonstrate glucose intolerance and hyperglycemia. This is in part due to peripheral insulin resistance. Hyperglucagonemia may significantly affect the liver's ability to store glycogen, which may in fact lead to an exaggerated hypoglycemic response to fasting. Because of decreased hepatic stores of glycogen, protein catabolism is accelerated to provide amino acids for gluconeogenesis. Therefore, fasting in a cirrhotic patient will lead to increased breakdown of fat and muscle.

Patients with ESLD will also demonstrate alterations in fat synthesis, transport, and absorption. The metabolism of long-chain fatty acids to ketones takes place primarily in the liver. With liver failure, there is an incomplete metabolism of these fatty acids, leading to an accumulation of short-chain fatty acids, the

Table 25.1. Metabolic Alterations in End-Stage Liver Disease

Increased plasma glucagon
Hyperinsulinemia
Increased plasma epinephrine and cortisol
Decreased liver and muscle carbohydrate stores
Accelerated gluconeogenesis
Hyperglycemia
Increased plasma aromatic amino acids
Decreased plasma branched-chain amino acids

intermediate metabolite. This increases the dependency on gluconeogenesis for substrate energy, and may limit the usefulness of supplemental fat calories in meeting the patients' metabolic needs. Steatorrhea is also quite common in patients with ESLD, perhaps in part due to decreased secretion of bile salts. Micronutrients and trace elements are also affected by liver dysfunction. These substances are commonly deficient in patients with ESLD, because of both impaired intake and metabolism.

Factors Contributing to Malnutrition

The etiology of malnutrition in patients with ESLD is clearly mutifactorial (Table 25.2). Nausea, vomiting, early satiety, and/or anorexia may be associated with medications or ascites causing delayed gastric emptying. In addition to poor intake, there may be maldigestion or malabsorption from decreased bile salts (secondary to cholestasis), bacterial overgrowth, or medication-induced losses (e.g., neomycin, lactulose, diuretics, cholestyramine). Bacterial overgrowth and coexistent intestinal diseases (such as inflammatory bowel disease or celiac sprue) can also contribute to malabsorption. Diets of these patients are commonly restricted in terms of volume, salt, and amounts of protein and may be unpalatable. Several investigators have demonstrated that patients with liver failure have increased catabolism from elevation in resting energy expenditure (REE).[2–7]

The high incidence of malnutrition is especially relevant in the context of

Table 25.2. Causes of Malnutrition in End-Stage Liver Disease

- Poor intake—aggravated by ascites, gastritis, GI bleeding, encephalopathy, or sepsis
  - Anorexia
  - Early satiety
  - Dietary restrictions—protein (encephalopathy), fat (malabsorption), calories (hyperglycemia), salt (ascites), fluid (ascites)
- Alterations in metabolism and storage
- Alterations in synthesis
- Malabsorption/maldigestion
  - Cholestasis, intraluminal bile deficiency
  - Pancreatic insufficiency
  - Mucosal disease
- Drug therapy effects
  - Neomycin (villous atrophy, diarrhea, zinc deficiency)
  - Lactulose (diarrhea, zinc deficiency)
  - Diuretics (potassium, magnesium, zinc deficiency)
  - Cholestyramine (diarrhea, fat-soluble vitamin deficiency)
- Iatrogenic protein losses—large volume paracenteses
  - 50–100 g of protein may be lost with each large-volume paracenteses, depending on volume and concentration

Table 25.3. Practical Nutritional Goals

Correct electrolyte abnormalities
Replace vitamins and minerals
Minimize sodium and water accumulation
Limit catabolism, and perhaps obtain positive nitrogen balance
Provide adequate calories to meet energy needs
Improve quantitative nutrition parameters
Avoid excesses of nutrients
Avoid nutrition associated complications

outcome analyses, which have demonstrated higher rates of morbidity and mortality in those patients with severe malnutrition. The goals of nutritional therapy are to limit catabolism, correct deficiencies, and maintain metabolic balance, without causing complications, such as worsening encephalopathy or catheter-related sepsis. The therapy that the patient with ESLD needs most is hepatic replacement with a functioning allograft. Transplantation should never be delayed to allow for nutritional repletion (Table 25.3).

Risks of Malnutrition in Liver Transplant Candidates

Malnutrition has long been known to complicate liver disease. The Child and Turcotte criteria to estimate mortality in cirrhotic patients who need portal decompressive procedures, includes a subjective assessment of nutrition as one of five equally weighted variables to estimate postoperative mortality. The risk of malnutrition in transplant patients was studied by Shaw et al.[8] A retrospective analysis was performed to identify predictors for survival 6 months posttransplant. One of the factors found to be a significant predictor was a subjective assessment of malnutrition (normal, mild-to-moderate, and severe malnutrition). The malnutrition score was one of six variables found to be highly correlated with patient survival ($p = 0.025$). Also, of the six variables, it was the only one that was potentially alterable, and not completely dependent on underlying hepatic function. The implication is that improved nutrition reduces posttransplant risk.

The finding that severe malnutrition complicates liver transplantation and leads to higher morbidity and mortality, has been reproduced in several series, in both adult and pediatric patients.[3,9] In a study by Pikul et al.[10] moderate-to-severe malnutrition was associated with prolonged ventilatory support, increased need for tracheostomy, and prolonged stay in intensive care, and prolonged hospital length of stay. Hasse et al.[11] reported similar results in adult patients. They found that the median initial hospitalization cost in severely malnourished patients was $13,827 more when compared to well-nourished patients ($p = 0.05$). Moukarzel et al.[9] found that growth retardation (presumably related to nutritional status) was one variable that significantly affected posttransplant mortality. Pikul et al.[10] reported a 79% incidence of malnutrition and found that the degree of malnutrition was correlated to the posttransplant mortality rate.

Prevalence of Malnutrition

The relationship between advanced liver disease and malnutrition is well known, especially in those patients with alcoholic liver disease and acute liver failure. One of the largest series is the VA Cooperative Study,[13] which found 100% prevalence of protein-calorie malnutrition in patients with alcoholic hepatitis. The degree of malnutrition correlated with the degree of liver dysfunction. DiCecco et al.[14] found evidence of malnutrition in all of 74 consecutive patients with ESLD presenting for liver transplant evaluation. Interestingly, the etiology for the liver disease correlated with a particular type of nutritional deficit. For example, patients with primary biliary cirrhosis had depleted fat and muscle stores with retained visceral protein stores. Patients with sclerosing cholangitis had the lowest muscle stores of all groups, but retained their body fat stores. The severest degree of malnutrition occurred in patients with acute hepatitis. In another series, more than 70% of 500 patients evaluated for liver transplantation demonstrated moderate or severe malnutrition.[11] A similarly high incidence of malnutrition among liver transplant candidates has been reported by several other authors.[1,5,7,15]

Nutritional Assessment

Nutritional assessment in patients with ESLD can be quite difficult. Most of the objective assessment parameters of nutritional status are affected by liver disease. The standard nutrition evaluation consists of a clinical evaluation (medical and dietary history and physical examination), and body composition analysis. However, the traditional tools used to evaluate patients' nutritional stores are fraught with elements attributable to their liver disease, which may confound their interpretation (Table 25.4).

A history of recent involuntary weight loss (5% over 1 month, or 10% over 6 months) is an important indicator of potential malnutrition. However, patients with ESLD who develop significant ascites or peripheral edema, may maintain their weight (or even gain weight) while continuing to lose muscle mass. Malnutrition is associated with a shift from the intravascular space to the extravascular space, with a concurrent decrease in lean body mass. Therefore, loss of lean body mass can occur without a significant change in weight. Other important questions in the medical history include determinations of whether the patient has experienced significant symptoms of anorexia, early satiety, diarrhea, or steatorrhea. A directed dietary history should determine the average daily calorie, protein, and fluid intake. Physical examination will commonly identify signs of protein–calorie malnutrition, such as muscle wasting (e.g., thenar and temporal

Table 25.4. Nutritional Assessment in Patients With End-Stage Liver Disease

Nutritional Parameter	Factors Affecting Interpretation
Weight	Edema
	Ascites
	Diuretics
Triceps Skinfold (TSF)	Variations in hydration
	Age
Creatinine-height index (CHI)	Age
	Renal failure
	Liver disease (creatine, precursors to creatinine synthesized in liver)
Nitrogen balance	Renal failure
Serum proteins (albumin, prealbumin transferrin, retinol-binding protein)	Variations in hydration
	Renal failure
	Malabsorption
	Iron stores
	Steroids
	Zinc deficiency
Immune function (total lymphocyte count, delayed-type hypersensitivity skin testing)	Infection
	Renal failure
	Immunosuppression (inflammatory bowel disease)

muscles) and loss of subcutaneous fat. There may also be more subtle signs of malnutrition, such as dry, brittle hair (zinc deficiency), flaky dermatitis (vitamin A deficiency), stomatitis (riboflavin deficiency), and neuromuscular irritability (magnesium deficiency) (Table 25.5).

Traditional measures of the somatic protein compartment include determination of midarm muscle circumference (designed to assess skeletal muscle mass),

Table 25.5. Physical Signs of Malnutrition in Liver Failure

	Clinical Sign or Symptom	Possible Deficiency
Hair	Dry, brittle	Zinc
Skin	Xerosis; flaky dermatitis;	Vitamin A, zinc, niacin
	increased bruising	Vitamin K
Eyes	Night blindness	Vitamin A
Mouth	Stomatitis; cheilosis; glossitis,	Riboflavin; vitamin B_{12}; folate, niacin
	altered taste	Magnesium, zinc
Extremities	Neuromuscular irritability; weakness	Magnesium; vitamin E, folate, patothenic acid, vitamins B_{12} and B_6, thiamine

and triceps skinfold thickness (designed to measure subcutaneous fat stores). As would be expected, fluid shifts and changes in hydration status will affect these measurements. However, the presence of excess body water because of edema accumulates to a less extent in the upper extremities. Comparisons with known standards should be avoided, because of the various confounding variables associated with liver failure. These measurements are best used for assessing changes in an individual patient over long periods of time. There are similar problems with the use of creatinine-height index (CHI) in patients with ESLD. Both malnutrition and aging are associated with a decline in lean body mass, and therefore a decrease in creatinine excretion. Hepatocytes are necessary to form creatinine from creatine, and therefore ESLD itself will alter the measures of CHI. Obviously, the development of hepatorenal syndrome may cause decreased creatinine excretion.

In general, visceral protein stores seem to correlate better with the degree of liver damage than with the degree of malnutrition, although the two certainly go hand in hand. For example, serum albumin levels are used both as markers of liver function as well as of nutritional status. Measurements of serum albumin are affected by hydration status, renal insufficiency (with proteinuria), and malabsorption. Albumin is a relatively insensitive marker for early malnutrition, because it has a large body pool (4 to 5 g/kg) and a long half-life (20 days).[1] Similarly, transferrin levels are affected by many of the same factors as albumin. Retinol-binding protein (RBP) levels may fall because of vitamin A deficiency, which may occur because of steatorrhea. Measurements of nitrogen balance may also be problematic in patients with ESLD. Urea and ammonia are retained, and urine output maybe compromised and/or urinary nitrogen excretion diminished.

Patients with ESLD also may have several reasons, in addition to malnutrition, for demonstrating decreased immunocompetence. Demonstration of anergy to delayed cutaneous hypersensitivity in these patients may be due to electrolyte imbalances, infection, renal insufficiency, or immunosuppressive medication (i.e., for inflammatory bowel disease), as well as malnutrition. For many of the same reasons, the total lymphocyte count may also be depressed. In one study, 60% of cirrhotic patients and 93% of patients with fulminant hepatic failure were anergic. Of these patients, 50% died in the hospital and 91% had documented bacterial sepsis in the week before death.[17] It is reasonable to assume that the patient with ESLD is an immunocompromised host.

There is a common pattern of vitamin, mineral, and electrolyte deficiencies in patients with ESLD. Hypokalemia is frequently present due to both the hyperaldosteronism associated with liver disease and diuretics. Loop diuretics also lead to magnesium and zinc deficiency, which are exacerbated by diarrhea and malabsorption. Levels of fat-soluble vitamins are commonly decreased in these patients.

Despite the numerous limitations of the standard battery of nutritional assessment tools, it is possible for the experienced clinician to make a very reasonable estimate of the nutritional reserves of patients with ESLD. The collective information from the history, physical exam, and laboratory data are assimilated, and a subjective global assessment (SGA) is formed.[18–24] This global assessment is actually highly reliable, reproducible, and predictive of nutrition-related morbidity.[26–31] Hasse et al.[31] evaluated the interrater reliability of SGA in liver transplant candidates and found an 80% agreement between the raters.

Another good predictor of the severity of malnutrition is the cause of the patient's ESLD, and the pace at which it progressed. For example, a patient with the rapid development of acute hepatic failure will have more nutritional reserve than a patient with long-standing sclerosing cholangitis and significant portal hypertension.[32]

Nutritional Requirements

Resting energy expenditure may be calculated using the Harris-Benedict equation. In general, most patients with ESLD will need 25 to 35 kcal/kg·day for total energy requirements, which is typically 1.2 to 1.4 × REE.[7–33] Calculations of weight should be based on either premorbid dry weight or ideal weight for height. The upper range applies to patients who may be particularly more catabolic, such as those with a recent infection. The REE is considered to be no different in patients with liver disease than in those without.[2,3,7,34] Other studies show that when REE is related to lean body mass, patients with advanced liver disease have increased REEs.[32–35] The excess catabolism associated with liver disease is frequently offset by the common overestimation of lean body mass from the body weight (because of excess total body water in patients with ESLD). Ascites, in and of itself, may be associated with an increased REE.[6] The most accurate method of determining caloric requirements is with the use of indirect calorimetry. This non–invasive test is especially useful to resist the common temptation to overfeed patients. Complications associated with overfeeding include the development of hyperglycemia, hepatic cholestasis, steatosis, and excess carbon dioxide production (and increased work of breathing to eliminate the extra carbon dioxide). The axiom, if a little (nutrition) is good, a lot is better, does not hold in this case. The proportion of nonprotein calories should be made up of 25% to 40% fat and 60% to 75% dextrose. The exact proportion is based on the individual patients tolerance. For example, a diabetic patient will receive a greater percentage of calories from fats, and a patient with hypertriglyceridemia will receive a greater percentage of calories from dextrose. Medium chain triglycerides may be useful in patients with malabsorption.

Considerations that are made in the determination of the amount and type of protein to be given include a history of protein intolerance, the presence or absence of encephalopathy, and the degree of malnutrition. The goal is to supply patients with as much protein as is tolerated without inducing encephalopathy.

All too often, patients with cirrhosis are automatically given a protein-restricted diet, without either a history of protein intolerance or encephalopathy. Many cirrhotic patients will tolerate normal amounts of dietary protein. It is optimal to give all catabolic patients 1.5 $g/kg{\cdot}day^{-1}$ of protein. In patients with severe encephalopathy, and malnutrition, the amount of protein given should be reduced, but not below a minimum of 1 $g/kg{\cdot}day^{-1}$. In some patients who have severe encephalopathy, but without severe malnutrition, it is reasonable to provide them with 0.8 mg/kg·day and then titrate additional protein based on tolerance. In patients who develop encephalopathy from nutritional support, occult sources of protein intake (i.e., gastrointestinal bleeding) should be searched for. The use of "hepatic" formulas that are enriched with BCAAs is controversial. This will be discussed further in the next section.

The severity of ascites and edema determine the level of sodium and fluid restriction. A level of 500 mg to 1000 mg of sodium and 1.0 to 1.5 L of fluid may be necessary for patients with severe edema and ascites. Unfortunately, this may lead to unpalatable diets and decrease nutrient intake. Virtually all patients with ESLD require supplementation with vitamins, minerals, and trace elements. The nutrients most frequently required are the B-complex vitamins, vitamin A, zinc, magnesium, and phosphorus. The doses necessary should be determined by the existing deficiencies and monitoring of blood levels. Potassium will commonly also need to be supplemented because of the use of diuretics (Table 25.6).

Nutritional Therapy

Enteral. Enteral nutrition supplementation is indicated in patients with ESLD who are unable to consume adequate nutrition to meet their caloric and protein requirements and who have a functioning gastrointestinal tract. The vast majority of patients with malnutrition associated with their liver failure are candidates for enteral dietary supplementation. As discussed previously, diarrhea is not uncommon in these patients, but should not be considered to be an absolute contraindication to tube feeding. Most diarrhea is medication related, and can be controlled with adjustments in the medical regimens. In patients who are unable

Table 25.6. Nutritional Goals for Patients With End-Stage Liver Disease

Total Calories: REE × 1.2–1.4/day. In general, 30 kcal/kg (ideal body weight)
Protein: 1.0–1.5 $g/kg{\cdot}day^{-1}$
30%–40% of nonprotein calories as fat
Liberal use of vitamins and minerals
Restrict iron and copper if overload exist
Adjust salt and water intake according to patient's volume and electrolyte status
Enteral nutrition preferred; oral or via small-bore silicone feeding tube (or both).
Parenteral nutrition only if enteral nutrition not possible

to sustain themselves with oral diets, with or without nutritional supplements (due to anorexia or early satiety), a nasoenteric feeding catheter is indicated. Because delayed gastric emptying is common in patients with ESLD (typically due to ascites), a nasojejunal tube is preferred to prevent complications, such as aspiration pneumonitis. This can usually be easily accomplished with fluoroscopic guidance. Small bore (8-gauge French) silicone feeding tubes are typically well tolerated, even in patients with esophageal varices.

In general, most patients do well with standard protein/amino dietary supplements. Desirable characteristics of enteral formulas include being lactose-free, casein-based, calorically dense (>1 kcal/mL), and low in sodium (<40 mEq/day). Formulas that come as powders allow their concentration to be increased beyond the standard dilution by adding less water. In patients who develop encephalopathy with the use of standard amino acid solutions, the use of formulas supplemented with BCAAs should be considered. Virtually all standard enteral nutritional supplements contain generous amounts of BCAAs. High-BCAA formulations should only be considered for those patients who have disabling encephalopathy while receiving the necessary amounts of protein.

In many patients, enteral tube feedings can be cycled overnight. This is particularly useful in patients who are at home, but require nutritional supplementation. It allows them to be free of the pump apparatus during the day and may also enhance their appetite during the day. This approach may be aided by the use of enteral products that have added caloric density, so that less volume needs to be infused over the 12- to 16-hr period.

In patients who are able to partially fulfill their nutritional goals with an oral diet, nutritional supplements need only be given to make up the difference between their intake and their calculated needs. Strategies to increase the patients oral intake include the use of small, frequent meals, and the use of liquid, calorically dense supplements.

Parenteral. Parenteral nutrition is indicated in those patients with ESLD who have malnutrition and are intolerant of feeding via the gastrointestinal tract. This would also include patients with active gastrointestinal bleeding. In determining the parenteral nutrition prescription, the patient's caloric and protein requirements are calculated, as is the total volume of parenteral fluid that is to be administered over the 24-hr period. The nonprotein calories of the nutrient admixture are made up of carbohydrates and fats. Excess administration of either of these may also be associated with potential complications. Dextrose infusion at a rate greater than 7 $mg/kg{\cdot}min^{-1}$ will lead to lipogenesis in the liver.[36–37] The rapid infusion of lipids is also problematic and may flood the reticuloendothelial system and fixed-tissue macrophages (e.g., Kupffer cells) and impair bacterial clearance. This is avoided by the continuous delivery of less than 1 $g/kg{\cdot}day^{-1}$ of lipids. During the initiation of parenteral nutrition, serum glucose and triglyceride levels should be closely monitored. Hyperglycemia (serum glucose >200 mg/dL) should be treated with supplemental insulin. This is initially given by subcutaneous sliding scale, and then can be added to the parenteral nutrition admixture. Triglyceride levels should be monitored to avoid levels of over 500 mg/dL.

The use of modified amino acid solutions is controversial. The vast majority of patients will tolerate standard parenteral amino acid solutions. Plasma concentrations of BCAAs may be decreased in patients with hepatic failure, although no unequivocal explanation for this has been established.[4] Solutions enriched with BCAA should be considered in patients who either have significant encephalopathy (Grade 2, or higher), or develop significant encephalopathy while receiving their protein needs with standard amino acid solutions. However, the evidence to support this strategy is not strong, and the cost can exceed by 10 to 14 times the cost of equivalent amounts of conventional amino acid solutions.[38–40] It is also important to note that it is possible to develop deficiencies of other amino acids, if solutions devoid of nonessential amino acids are given. Decreased BCAA levels in patients with cirrhosis without encephalopathy have been shown in several studies.[39,40] Other studies have demonstrated that decreased BCAA levels are significant in cirrhotic patients only when they are malnourished.[41] In dogs with portocaval shunts, encephalopathy does not occur without malnutrition.[42] Several experts feel that the decreased levels of BCAA in patients with advanced liver disease may be linked more closely to malnutrition than to hepatic encephalopathy.[25,43]

Several studies have attempted to study the use of BCAA in patients with advanced liver disease. Most of these studies have serious design flaws that seriously limit the ability to draw conclusions. The control group in many of these studies did not receive any amino acids or proteins during the study period. The majority of studies with reasonable control groups (standard amino acid solution), demonstrated significant improvement in nitrogen balance in both study arms. Most of these studies showed no significant differences in mortality or in the development of hepatic encephalopathy.[44] A meta-analysis of five randomized controlled trials using parenteral BCAA as therapy for hepatic encephalopathy concluded that treatment with BCAA was associated with a significant reduction in mortality. However, the role of improved nutrition was not considered in the outcome analysis, and may have played a significant role.[44] Despite the lack of overwhelming conclusive scientific proof, the fairly large positive clinical experience with the use of BCAAs in severely encephalopathic patients, makes its use in these patients quite reasonable.

The addition of vitamins and trace elements to the daily nutritional formula should not be overlooked. In addition to providing the recommended daily allowance of water- and fat-soluble vitamins, supplemental vitamin K is frequently given, especially in patients on antibiotics that may alter gut flora and thus endogenous reabsorption. Drugs may also be added to parenteral nutrition solutions. The continuous infusion of certain drugs has several pharmacokinetics, therapeutic and cost advantages over intermittent bolus administration. Medications commonly added to parenteral nutrition solutions include albumin, aminophylline, cimetidine or ranitidine, heparin, and insulin.[45]

Pretransplant Obesity

The presence of obesity in the patient with ESLD has recently been considered as a possible risk factor for liver transplantation. Two recent studies have concluded that although patients with severe obesity are at increased risk for perioperative morbidity, the authors concluded that liver transplantation was feasible in this population and that obesity should not be used as the only criteria to exclude patients from liver transplantation.[46,47] Perioperative complications that occur at an increased incidence in severely obese liver transplant recipients include wound infection, diabetes mellitus, hypertension, and respiratory infection and failure. It is important to remember that obese patients may also be malnourished with decreased protein stores.

Post-liver transplantation nutrition support

Nutrition support is equally essential after liver transplantation. Immediately postoperatively, the patient faces stress and catabolism from the extensive surgery as well as from high doses of steroids. During the first few weeks following transplant, considerations of graft function and other confounding circumstances such as infection and rejection or diminished renal and pulmonary function must be included in the nutritional assessment of the patient. In addition, biliary, intestinal, vascular, wound, and metabolic complications need to be considered.[18,48–50] Table 25.7 summarizes the nutrition implications of these variables.

Table 25.7. Short-term Complications Following Liver Transplant and their Nutritional Implications

Complication	Nutrition Implication
Rejection	Increased steroid doses may cause protein catabolism, diarrhea, vomiting, or decreased appetite
Infection	Antibiotics may cause nausea, vomiting, diarrhea, anorexia, taste changes Protein needs higher with sepsis
Renal insufficiency	May require adjustment in protein, potassium, sodium, and/or fluid
Intestinal complications	May require a change in method of providing nutrition
Biliary complications	May affect fat digestion
Vascular complications	If surgery required, nutrition support may be needed
Metabolic complications	May require changes in substrates or electrolytes

Energy needs. Catabolism occurs rapidly immediately posttransplant due to a number of factors, including preoperative malnutrition, surgical stresses, and steroid administration. The amount of catabolism may be increased even more in patients with sepsis or renal failure pre or posttransplant.[51] Caloric needs are usually estimated to be 30% to 75% greater than basal energy expenditure (BEE) using the Harris-Benedict equation.[52,54] However, some indirect calorimetric studies suggest that post-liver-transplant patients may not be hypermetabolic. These studies indicate that 75% of patients needed less than 20% above BEE.[32,55] Indirect calorimetry provides an inexpensive and noninvasive determination of a patient's caloric requirements.

Protein needs. Protein catabolism increases markedly immediately following liver transplantation. Large nitrogen losses also occur secondary to steroid therapy, surgical stresses, and muscle catabolism in a malnourished patient.[32,56] Protein goals are set at 1.3 to 2.0 g protein/kg dry body weight·day^{-1}.[52,54–56] Symptomatic uremia occurring in posttransplant patients with acute renal insufficiency, should be treated with hemodialysis.[56] It is not uncommon for physicians to fear worsening azotemia with full nutritional support and underfeed protein to patients with renal insufficiency. Dialysis should strongly be considered before significantly underfeeding these patients as their severe metabolic stress and catabolism will only increase without adequate nitrogen repletion.

Carbohydrate and fat requirements. Carbohydrates should provide 50% to 70% of nonprotein calories following liver transplantation. The remainder of the nonprotein calories are provided by lipids. However, hyperglycemia may necessitate the administration of a higher proportion of lipids. Supplemental insulin also may be necessary.

Electrolyte and micronutrient requirements. Careful attention must be given to electrolytes and micronutrients after liver transplantation. Sodium restriction is liberalized as increased losses occur, via abdominal drains, nasogastric tubes, biliary stents and urine. Micronutrients that tend to deplete rapidly after liver transplantation include magnesium and phosphorus. Diuretics and aggressive refeeding can also deplete potassium and magnesium levels. Cyclosporine can also increase magnesium losses and potassium retention. High cyclosporine levels may be particularly neurotoxic by lowering the seizure threshold, especially when associated with hypomagnesemia. Similarly, tacrolimus can also cause retention of potassium and precipitate renal dysfunction. Vitamin and mineral levels also need to be repleted especially in the malnourished patient.

Nutrition support. Frequently, nutrition support is required to provide nutrients to post-liver-transplant patients. Some well-nourished patients may be able to eat as early as 3 days postoperatively and will not need nutrition support.[50,54] Patients with a functional gastrointestinal tract who are unable to eat should

Table 25.8. Nutrition Support Regimes for the Short-term Liver Transplant Patient

Method	Nutrition Considerations
TPN	Requires a central venous catheter
	Appropriate when unable to use gastrointestinal tract for 5–7 days
	May need fluid restriction
	Best tolerated initially as a continuous infusion
	May need to increase to goal slowly and use a higher concentration of lipids if hyperglycemia a problem
Enteral	Feeds should be provided via small intestine
	Elemental or small peptide formulas may be tolerated best
	Intact protein formulas should be used for patients without digestive problems
	Cyclic feeds should be considered
Oral	Initiate diet when gastrointestinal activity returns
	Diet usually advanced to regular within 2 days
	Initiate sodium restriction for severe fluid retention and restrict sugar for hyperglycemia

Table 25.9. Parameters Monitored in the Short-term Liver Transplant Patient

Parameter	Observations
Weight	Weight can fluctuate with fluid changes
I & O	Determine fluid needs
BUN and ammonia	May increase with renal or hepatic dysfunction
Glucose	Can increase with stress and administration of steroids; may need to decrease glucose or increase insulin
Electrolytes	Can be altered by drugs and medical condition; alter and supplement as needed
Triglyceride	Can increase with stress and steroids
Metabolic rate	Indirect calorimetry accurately measures calorie needs
Nitrogen balance	Can be affected by altered renal or hepatic function
Calorie counts	Monitor intake and determine adequacy of nutrient intake

receive enteral tube feedings, usually through a soft, small-bore feeding tube placed beyond the pylorus.[50,54,57] Although gastric ileus and colonic ileus are common in the first few days posttransplant, immediate postoperative feedings into the small bowel have been successful. Total parenteral nutrition is reserved for patients whose gastrointestinal tracts cannot be used for nutrition. Suggested short-term nutrition support regimens are listed in Table 25.8. Table 25.9 lists parameters that should be monitored in the first few weeks following transplantation.

Long-term nutritional management of the liver transplant patient. Long-term nutrition goals for the liver-transplant patient differ greatly from the short-term goals. Macronutrient needs decrease after an initial anabolic stage. In patients with good liver allograft function, malnutrition is rare. Standard nutrition assessment techniques are appropriate for evaluating the long-term post-liver-transplant patient, as long as renal and hepatic functions are normal. Table 25.10 outlines nutrition recommendations for most post-liver-transplant patients. Some of the most common posttransplant nutrition-related problems include obesity, hyperlipidemia, hypertension, osteoporosis, and diabetes. Immunosuppressive medications may contribute to these complications and in addition can also lead to the development of hyperkalemia, hypomagnesemia, and hyperuricemia. Table 25.11 summarizes the nutritional side effects and treatment of the most commonly used immunosuppressive drugs.

Long-term nutritional problems. Excessive weight gain is one of the most prevalent long-term problems occurring after liver transplantation. Studies suggest that preillness obesity predicts post-liver-transplant obesity, although all patients may be at risk for excessive weight gain.[58,59] Excess weight is associated with increased risks of diabetes, cardiovascular disease, hypertension, hyperlipidemia, and abnormal liver function tests. Not uncommon is the histologic finding of micro- or macrovesicular steatosis in the obese patient after liver transplantation,

Table 25.10. Nutrition Recommendations for Liver Transplant Patients

Nutrient	Short-term Recommendations	Long-term Recommendations
Calories	120–130% of BEE or REE	Maintenance
Protein	1.3–2.0 g/kg·day^{-1}	1 g/kg·day^{-1}
Carbohydrate	50–80% of calories	50–70% of calories, restrict sugar
Fat	30% of calories, up to 50% of calories with hyperglycemia	<30% of calories <10% of calories as saturated fat
Calcium	800–1200 mg/day	1000–1500 mg/day
Sodium	2–4 g/day	3–4 g/day
Magnesium and phosphorus	Encourage intake of foods high in these nutrients; supplement as needed	Encourage intake of foods high in these nutrients; supplement as needed
Potassium	Supplement or restrict based on serum levels	Supplement or restrict based on serum levels
Other vitamins and minerals	Multivitamin/mineral supplement to RDA levels	Multivitamin/mineral supplement to RDA levels

Table 25.11. Common Nutritional Side Effects of Immunosuppressive Drugs and their Management

Drug	Side Effect	Management
Cyclosporine	Hypertension	Sodium restriction
	Hyperkalemia	Potassium restriction
	Hyperlipidemia	Limit fat and simple carbohydrate (CHO)
	Hyperglycemia	Decrease simple CHO intake
	Hyperuricemia	Increase fluid intake
	Neurotoxicity	Correct electrolyte abnormalities
	Hypomagnesemia	Increase magnesium intake
Prednisone	Hyperglycemia	Restrict simple CHO intake
	Increased appetite	Increase activity Monitor calorie intake
	Fluid and sodium retention	Limit sodium intake
	Calcium wasting	Supplementation 1–1.5 g/day
Azathioprine	Nausea, vomiting, sore throat and decreased taste	Adjust food/meals as needed; monitor intake for adequacy
	Macrocytic anemia	Folate supplements
	Mouth sores	May require bland soft foods
Tacrolimus	Hyperglycemia	Decrease simple CHO intake
	Hyperkalemia	Decrease potassium intake
	Nausea and vomiting	Adjust food/meals as needed Monitor intake for adequacy

who undergoes liver biopsy for the evaluation of abnormal liver function tests.[57–59] Excess weight also may precipitate or worsen arthritis and spontaneous osteopenic fractures. Prevention of excessive weight gain is encouraged by monitoring dietary intake, modifying dietary habits and behaviors, and promoting exercise after transplantation.

Cholesterol and triglycerides levels are commonly quite low at the time of liver transplantation in patients with noncholestatic liver diseases. Cyclosporine, steroids, and diuretics all may promote posttransplant hyperlipidemia.[58] In one study by Palmer et al., overweight patients had increased incidence of hyperlipidemia.[60] The combination of obesity, hyperlipidemia and hypertension is particularly concerning because of the risks of developing cardiovascular disease.

Liver transplant recipients require intensive nutritional counseling both prior to discharge from the hospital after their transplant and in follow-up as an outpatient. Most patients are encouraged to eat a high-calorie, high-protein diet right after transplant, and then told to minimize their calorie, fat, salt, and sugar intakes after they leave the hospital. Nutritional assessment will identify individual patient requirements and subsequent mode of therapy. Follow-up and monitoring each patient's progress can significantly contribute to the overall success of the transplant.

Summary

Pretransplant nutritional assessment in the patient with ESLD is problematic. The best system for nutritional assessment utilizes a global evaluation of the patient's nutritional reserves. Using such a technique, the vast majority of transplant candidates have been shown to have evidence of malnutrition. Several investigators have demonstrated the risk of significant malnutrition on posttransplant outcome. An aggressive approach to nutritional repletion is necessary before transplantation to improve patient's metabolic reserves, maintain their remaining hepatic function, and better their outcome after liver transplantation. After successful liver transplantation, continued attention is necessary to provide adequate macro and micro nutrients, and to avoid nutrition-associated complications.

RENAL TRANSPLANTATION

Many similarities to liver transplantation occur in the nutritional assessment and support given to patients before and after transplantations. There are, however, several important specific differences in nutritional support, and these will be discussed in the following sections.

Nutrition Support Immediately After Renal Transplantation

Acutely post-renal-transplant, nutritional support focuses on the provision of appropriate macronutrients and micronutrients. The acute posttransplant period refers to the first 4 to 6 weeks after surgery.

Energy needs. For the uncomplicated patient, caloric requirements should be 30 to 35 kcal/kg or 30% to 50% greater than basal energy expenditure (BEE). As with all patients, caloric needs are increased with infection, fever, or other metabolic stresses (e.g., rejection or urine leak).

Protein needs. Protein requirements are 1.3 to 1.5 g/kg body weight.[61,63] These levels are compatible with positive or neutral nitrogen balance, provided adequate calories are consumed.

Carbohydrate and fat requirements. Energy requirements are met by carbohydrates and fats in the renal transplant patient. Diets high in simple carbohydrates are undesirable, because glucose intolerance frequently occurs secondary to surgical stress and steroid therapy.[64] Also, many renal transplant recipients are diabetic. Diets or nutritional support consisting of complex carbohydrates and a relatively high amount of fat can often blunt the glucose intolerance associated with the acute postrenal transplant period. Steroid administration is associated with increased gluconeogenesis and can precipitate steroid-induced diabetes mellitus in the otherwise nondiabetic patient.[65] For patients not experiencing hyperglycemia, the diet generally prescribed in the acute transplant phase consists of 50% to 70% of total calories as carbohydrates, with 30% of calories as fat.

Fluids and electrolytes. Close monitoring of electrolytes is important following renal transplantation. Restriction of sodium, potassium, or both may be necessary in the acute phase, if fluid retention or hyperkalemia is present. Hyperkalemia can be caused by delayed renal allograft function (acute tubular necrosis, ATN) or by cyclosporine or tacrolimus even with high urine volume and good renal function. Attention should be given to drugs that may increase metabolic load of potassium such as beta blockers, angiotensin-converting-enzyme (ACE) inhibitors, and nonsteroidal antiinflammatory agents. Occasionally, potassium-wasting diuretics or a refeeding syndrome in a malnourished patient can precipitate hypokalemia.

Hypophosphatemia is a common finding posttransplant secondary to use of high-dose steroids, persistent hyperparathyroidism in the setting of normal renal function, and decreased renal tubular resorption of phosphate. Increased intake of high-phosphorous foods may not be adequate and oral replacement may be required. Hypomagnesemia is also a common finding secondary to cyclosporine-induced hypermagnesuria. Replacement of magnesium by diet alone is likely to be inadequate and enteral or parenteral supplementation usually is necessary. The efficacy of routine water-soluble vitamin supplementation has not been well studied; however, it is reasonable to continue a multivitamin during the acute posttransplant period.

An iron/total iron-binding capacity of less than 20%, or a plasma ferritin level of less than 100 ng/mL is indicative of iron deficiency. The studies should be obtained posttransplant because iron deficiency is commonly found in the dialysis population, particularly in conjunction with erythropoietin therapy.

Several days post-renal-transplant, after major fluid shifts have equilibrated, in patients with good renal allograft function, a reasonable minimum fluid intake is generally 2000 mL/day. Some patients may have persistently high obligate losses of fluids because of a prolonged period of high-output ATN. Because pretransplant patients are usually required to restrict their fluid intake, these patient often must be encouraged to drink enough fluid to match their urine output.

Nutrition Considerations in Acute Renal Allograft Rejection

In acute rejection, provision of optimal protein and calorie intake should be the primary nutritional concern. High-dose steroids produce an increase in the protein catabolic rate, and with rising serum creatinine and blood urea nitrogen (BUN), a common error is to restrict protein. Protein restriction at this stage may lead to severe catabolism. Protein intake of 1.3 to 1.5 g/kg·day^{-1} should be provided. A minimum of 30 to 35 kcal/kg should be provided during the acute cellular rejection therapy.

Long Term Nutritional Management Of The Renal Transplant Patient

Complications and nutritional needs in the long-term or chronic period after renal transplantation differ greatly from those in the acute phase. As with liver transplant recipients, the problems in the chronic phase relate more to overnutrition and include excessive weight gain, hyperlipidemia, hypertension, steroid-induced diabetes, and osteoporosis.

Obesity. Hyperphagia associated with steroid administration, along with a liberalized diet both contribute to a tendency for excessive weight gain in the posttransplant patient. Obesity may contribute to the development or acceleration of hypertension, hyperlipidemia, and diabetes in these patients.[66,67] In addition to limitations of caloric intake, preventive measures should include regular exercise and behavior modification programs. Frequent follow-up may optimize compliance to a weight-management program.

Hyperlipidemia. Atherosclerosis-induced vascular disease is second only to infection as the most common cause of death in renal transplant recipients.[67] Factors that contribute to posttransplant hyperlipidemia include increasing age, weight gain, steroid-induced diabetes, antihypertensive medicines (thiazide diuretics and beta blockers), and the nephrotic syndrome.[68,69] Steroids and cyclosporine also contribute to hyperlipidemia after a transplant procedure.[70] Several studies have shown that dietary manipulation can be effective in lipid reduction and should be attempted as a first-line therapy prior to the prescription of lipid-lowering medication.[71,72] A three-month trial of a carefully outlined diet, with regular follow-up should be considered before pharmacologic measures are begun.

Hypertension. Hypertension is common in renal transplant recipients, especially in the acute posttransplant period.[68] Steroid and cyclosporine use, renal artery stenosis, rejection, recurrence of primary renal disease, or the presence of the recipient's own diseased kidneys can contribute to hypertension.[68] Antihypertensive medications frequently are required after renal transplantation. Some controversy exists as to the efficacy of posttransplant sodium restriction, although a modest sodium restriction may be prudent in hypertensive renal transplant patients.

Steroid-induced diabetes mellitus. The incidence of steroid-induced diabetes varies depending on the immunosuppressive regimen. Factors implicated in the cause of diabetes include hereditary disposition, the use of steroids, cyclosporine, and tacrolimus.[61,68] Treatment includes reducing the dose of steroids as much as possible, reducing weight if necessary, eating an appropriate diet for diabetes, exercising, and using hypoglycemic agents or insulin.

Osteoporosis. Renal transplant patients are at risk for renal osteodystrophy before transplant. Osteoporosis has been associated with long-term steroid use as result of decreased intestinal absorption of calcium and low serum levels of active vitamin D. Steroid administration will also accelerate trabecular bone loss.[73] Other risk factors for osteoporosis include decreased estrogen levels in women, lack of exercise, and smoking. Careful monitoring of bone mineral losses is necessary after transplantation and treatment is indicated (calcium and vitamin D supplementation) for those patients with evidence of osteoporosis to decrease the incidence of a spontaneous fracture.

Summary

Nutritional therapy is important during all phases of kidney transplantation. Posttransplantation nutritional guidelines are summarized in Table 25.3. Adequate nutrition is required in the acute period to ensure proper healing and recovery; an appropriate diet can help prevent some of the long-term problems following renal transplantation.

PANCREAS TRANSPLANTATION

Diabetes mellitus affects approximately 5% of Americans and is the eighth leading cause of death.[74] The majority of patients with insulin-dependent diabetes mellitus (IDDM) will develop one or more complications (e.g., accelerated atherosclerosis, neuropathy, nephropathy, retinopathy) over their lifetime. Diabetes is the leading cause of renal failure and blindness in adults.[74] Evidence suggests that the microvascular complications of diabetes are related to prolonged hyperglycemia.[75,76] Methods to maintain euglycemia have included the use of autoregulating insulin pumps and pancreatic islet cell transplants.[77] Currently, neither technique has demonstrated significant success. However, successful vascularized pancreas transplantation can provide a consistent euglycemic state and active normalization of glycosylated hemoglobin levels.

Over the past several years, with refinements in surgical techniques and advances in immunosuppression, there has been a dramatic increase in the number of pancreas transplants. The vast majority of these transplants are performed as combined pancreas and kidney transplants in patients with IDDM who are on or soon approaching dialysis.

Complications that are unique to pancreas transplantation include metabolic acidosis and dehydration.[77] This is due to the obligate losses of sodium and bicarbonate in the urine from the transplanted duodenum, which is left attached to the pancreas to facilitate drainage of the exocrine secretions of the pancreas. Correction of the hyponatremia and metabolic acidosis may require supplementation with large amounts of sodium bicarbonate and fluids, given either enterally or parenterally, depending on the clinical situation.

Successful pancreas transplantation rapidly corrects glucose metabolism and result in a euglycemic state, freeing patients from the need to take exogenous insulin. Yet to be determined are the long-term effects on diabetic nephropathy, neuropathy, and retinopathy.

INTESTINAL TRANSPLANTATION

Intestinal failure is a process where the functioning gut mass has been reduced below the minimal amount necessary for the adequate digestion and absorption of food. This may be caused by either intestinal loss (short-bowel syndrome) or intestinal disease (e.g., chronic idiopathic intestinal pseudoobstruction, Crohn's disease, etc) (Table 25.12). The development of total parenteral nutrition (TPN) has led to the possibility of long-term survival of many infants and adults with intestinal failure. However, long-term TPN has several limitations, including the development of severe hepatobiliary dysfunction and irreversible chronic liver disease, especially in children. Also to be considered is the financial burden of long-term TPN use, which ranges between $100,000 and $250,000 per year. Other life-threatening complications may occur, such as catheter infections or central venous thrombosis. Metabolic disturbances are also common in these patients. Growth retardation has been associated with long-term TPN use in children.[78]

Lillehei described the first successful intestinal autotransplant in a canine model in 1959.[79] Unfortunately, early efforts at human intestinal transplantation were complicated by technical and immunologic failures. Within the past 5 years,

Table 25.12. Causes of Intestinal Failure

SMA thrombosis	Crohn's disease
SMA embolus	Trauma
Necrotizing enterocolitis	Radiation
Volvulus	Tumor (desmoid, polyposis)
Gastroschisis	Pseudoobstruction
Intestinal atresia	

efforts at human intestinal transplantation have again resumed, this time with significantly improved results. Technical aspects of the transplant procedure have now been refined. Grant et al. reported the first long-term survivor of an intestinal transplant.[80] This patient received a combined liver/intestine allograft. Since then, other centers have reported early successes with both combined liver/intestine transplants as well as isolated intestinal transplants. With increasing frequency, intestinal transplantation is being considered as an acceptable alternative therapy in some patients with intestinal failure. Several centers have reported patient and graft survival of 80% and 70% in patients who have received isolated intestinal transplants.[81]

Adults and children who have documented intestinal failure without potential for eventual long-term survival without TPN are candidates for intestinal transplantation. Intestinal failure is defined as any child following neonatal small bowel resection who is greater than 1 year old, who requires more than 50% of his/her caloric needs from TPN or, a child who is greater than 4 years old who requires more than 30% of his or her calories from TPN. Older children and adults receiving more than 50% of their nutritional requirements from TPN for more than 1 year should also be considered for intestinal transplantation. Other factors to be considered are venous access limitations (due to recurrent thrombosis), recurrent infectious complications related to central venous catheters, prolonged hospitalizations, growth retardation, and hepatobiliary dysfunction. Hepatic dysfunction associated with prolonged TPN use includes steatosis, cholestasis, fibrosis, and cirrhosis.

After successful intestinal transplantation, patients are begun almost immediately on elemental tube feedings delivered into the new small-bowel allograft via a jejunal catheter placed at the time of transplantation. With the recovery of gastrointestinal function, the volume of the feedings are increased and an oral diet is initiated. The tube feeding should be elemental, lactose-free, and contain glutamine.

Occasionally, after intestinal transplantation, children who have avoided food pretransplant (either from vomiting or diarrhea associated with eating), may need to learn (or relearn) how to eat. Other complications associated with intestinal transplantation include delayed gastric emptying, bacterial overgrowth, and dehydration associated with diarrhea. The majority of patients who undergo successful intestinal transplantation can be tapered off of parenteral nutrition within a few months.

CONCLUSION

Malnutrition is quite common in patients with either acute organ failure or those with decompensated end-stage disease. The recent success of abdominal organ transplantation can be attributed to many factors, including excellent perioperative care. Other than nutritional repletion and correction of metabolic derangements very few risk factors are able to be reversed prior to abdominal organ transplantation. Close attention to nutritional needs will undoubtedly contribute to improved success after abdominal organ transplantation.

ACKNOWLEDGMENT

The authors wish to thank Ms. Mary Gabriel for her expert assistance in the preparation of this manuscript.

REFERENCES

1. Shronts EP, Teasley KM, Thoele SL, et al. Nutrition support of the adult liver transplant candidate. *J Am Diet Assoc* 1987;87:441–51.
2. Shanbhogue RLK, Bistrian BR, Jenkins RL, et al. Resting energy expenditure in patients with end stage liver disease and in normal population. *J Parent Ent Nutr* 1987;11:305–308.
3. Hiyama DT, Fischer JE. Nutritional support in hepatic failure: current thought in practice. *Nutr Clin Pract* 1988;3:96–105.
4. Muller MJ, Lautz HU, Plogmann B, et al. Energy expenditure and substrate oxidation in patients with cirrhosis: the impact of cause, clinical staging and nutritional state. *Hepatology* 1992;15:782–794.
5. Latifi R, Killam RW, Dudrick SJ. Nutritional support in liver failure. *Surg Clin North Am* 1991;71:567–576.
6. Dolz C, Raurich JM, Ibanez J, et al. Ascites increases the resting energy expenditure in liver cirrhosis. *Gastroenterology* 1991;100:738–744.
7. Plevak DJ, DiCecco SR, Wiesner RH, et al. Nutritional support for liver transplantation: identifying caloric and protein requirements. *Mayo Clin Proc* 1994;69:225–230.
8. Shaw BW, Wood RP, Gordon RD, et al. Influence of selected patient variables and operative blood loss on six-month survival following liver transplantation. *Sem Liver Dis* 1985;5:385–393.
9. Moukarzel AA, Najm I, Vargas J, et al. Effect of nutritional status on outcome of orthotopic liver transplantation in pediatric patients. *Transplant Proc* 1990;22:1560–1563.
10. Pikul J, Sharpe MD, Lowndes R, et al. Degree of preoperative malnutrition is predictive of postoperative morbidity and mortality in liver transplant recipients. *Transplantation* 1994;57:469–472.

11. Hasse JM, Nutritional implications of liver transplantation. *Henry Ford Hosp Med J* 1990;38(4):235–240.
13. Mendenhall CL, Anderson S, Weesner RE. et al. Protein-calorie malnutrition associated with alcoholic hepatitis. *Am J Med* 1984;76:211–222.
14. DiCecco SR, Wieners EJ, Wiesner RH, et al. Assessment of nutritional status of patients with end-stage liver disease undergoing liver transplantation. *Mayo Clin Proc* 1989;64:95–102.
15. Hehir DJ, Jenkins RL, Bistrain BR, Blackburn GL. Nutrition in patients undergoing orthotopic liver transplant. *J Parent Ent Nutr* 1985;9:695–700.
16. Porayko MK, DiCecco S, O'Keefe SJD. Impact of malnutrition and its therapy on liver transplantation. *Sem Liver Dis* 1991;11:(4) 305–314.
17. O'Keefe SJD, Carraher ED, El-Zayadi AR, et al. Malnutrition and immunoincompetence in patients with liver disease. *Lancet* 1980;2:615–617.
18. Smith SL. Liver transplantation: implications for critical care nursing. *Heart Lung* 1985;14:617–627.
19. Donovan JP, Zetteman RK, Burnett DA, Sorrell MF. Preoperative evaluation, preparation, and timing of orthotopic liver transplantation in the adult. *Sem Liver Dis* 1989;9:168–175.
20. Sheets L. Liver transplantation. *Nurs Clin North Am* 1989;24:881–889.
21. Dickson ER, Evans LS, Koff RS, Sabesin SM. Who's a liver transplant candidate? *Patient Care* 1987;21:83–119.
22. Merli M, Romiti A, Riggio O, Capocaccia LO. Optimal nutritional indexes in chronic liver disease. *J Parent Ent Nutr* 1987;11:130S–134S.
23. Detsky AS, McLaughlin JR, Baker JP, et al. What is subjective global assessment of nutritional status? *J Parent Ent Nutr* 1987;11:8–13.
24. Munoz SJ. Nutritional therapies in liver disease. *Sem Liver Dis* 1991;11:(4) 278–290.
25. Baker JP, Detsky AS, Wesson DE, et al. Nutritional assessment: a comparison of clinical judgement and objective measurements. *N Engl J Med* 1982;306:969–973.
26. Detsky AS, Baker JP, Mendelson RA, et al. Evaluating the accuracy of nutritional assessment techniques applied to hospitalized patients: methodology and comparisons. *J Parent Ent Nutr* 1984;8:153–159.
27. Jeejeebhoy KN, Detsky AS, Baker JP. Assessment of nutritional status. *J Parent Ent Nutr* 1990;14:193S–196S.
28. Detsky AS, McLaughlin JR, Baker JP, et al. What is subjective global assessment of nutritional status? *J Parent Ent Nutr* 1987;11:8–13.
29. Baker JP, Detsky AS, Wesson DE, et al. Nutritional assessment: a comparison of clinical judgement and objective measurements. *N Engl J Med* 1982;306:969–972.
30. Hasse JM. Validity and reliability testing of a subjective nutritional assessment tool for adult liver transplant candidates (Thesis). Denton, TX: Texas Woman's University, 1989:1–78.
31. Hasse J, Strong S, Gorman MA, et al. Subjective global assessment—Alternative nutritional assessments technique for liver transplant candidates. *Nutrition* 1993; 9:339–343.

32. Shanbhogue RLK, Bistrian BR, Jenkins RL, et al. Increased protein catabolism without hypermetabolism after human orthotopic liver transplantation. *Surgery* 1987;101:146–149.
33. Mullen KD, Weber FL. Role of nutrition in hepatic encephalopathy. *Sem Liver Dis* 1991;11:292–304.
34. Vitale GC, Neill GD, Fenwick MK, et al. Body composition in the cirrhotic patient with ascites: assessment of total exchangeable sodium and potassium with simultaneous serum electrolyte determination. *Am Surg* 1985;51:675.
35. Shanbhogue RL, Bistrian BR, Jenkins RL., et al. Increased protein catabolism without hypermetabolism after human orthotopic liver transplantation. *Surgery* 1987; 101:146–149.
36. Shanbhogue RLK, Nompleggi D, Bell SJ, Blackburn GI. Nutritional support in surgery of the liver. In McDermott Jr, WV, ed. *Surgery of the Liver.* Boston: Blackwell Scientific, 1989;497–510.
37. Driscoll DF, Blackburn GR. Total parenteral nutrition 1990: a review of its current status in hospitalized patients and the need for patient-specific feeding. *Drugs* 1990;40:343–363.
38. Christie ML, Sack DM, Pomposelli J, Horst D. Enriched branched chain amino acid formula versus a casein based supplement in the treatment of cirrhosis. *J Parent Ent Nutr* 1986;9:671–678.
39. Morgan MY. Branched chain amino acids in the management of chronic liver disease: facts and fantasies. *J Hepatol* 1990;11:133–141.
40. Morgan MY, Marshall AW, Milsom JP, Sherlock S. Plasma amino acid patterns in liver disease. *Gut* 1982;23:362–370.
41. Merli M, Riggio O, Iapichino S, et al. Amino acid imbalance and malnutrition in liver cirrhosis. *Clin Nutr* 1985;4:249–253.
42. Thompson JS, Schafer EF, Haun J, Schafer GJ. Adequate diet prevents hepatic coma in dogs with Eck fistulas. *Surg Gynecol Obstet* 1986;162:126–130.
43. McCullough AJ, Mullen KD, Smanik EJ, et al. Nutritional therapy and liver disease. *Gastroenterol Clin North Am* 1989;18(3):619–643.
44. Naylor CD, O'Rourke K, Detsky AS, Baker JP. Parenteral nutrition with branched chain amino acids in hepatic encephalopathy. A meta-analysis. *Gastroenterology* 1989;97:1033–1042.
45. Lowell JA, Shaw BW. Selected topics in the critical care of liver transplant recipients. In *Transplantation of the Liver,* 2nd ed. Norwalk CT: Appleton and Lange, 1995;571–603.
46. Keeffe EB, Gettys C, Esquivel CO. Liver transplantation in patients with severe obesity. *Transplantation* 1994;57:309–311.
47. Blue LS, Hasse JM, Levy ML, et al. Accelerated improvement of alcoholic liver disease with enteral nutrition. *Gastroenterology* 1992;102:200–205.
48. Dindzans VJ, Schade RR, Van Thiel DH. Medical problems before and after transplantation. *Gastroenterol Clin North Am* 1988;17:19–31.

49. Koneru B, Tzakis A, Bowman J, et al. Postoperative surgical complications. *Gastroenterol Clin North Am.* 1988;17:71–91.
50. Grenvik A, Gordon R. Postoperative care and problems in liver transplantation. *Transplant Proc* 1987;10 (suppl):26–33.
51. Ohara M. National meeting of liver transplant dietitians; a summary. *Diet Nutr Support* 1989;11:13.
52. Dinga MA. Nutrition in liver transplantation. *Diet Nutr Support* 1987;8:5–13.
53. Fischer JE, Bower RH. Nutritional support in liver disease. *Surg Clin North Am* 1981;61:653–660.
54. Hasse J. Role of the dietitian in the nutrition management of adults after liver transplantation. *J Am Diet Assoc* 1991;91:473–476.
55. Delafosse B, Faure JL, Bouffard Y, et al. Liver transplantation: energy expenditure, nitrogen loss, and substrate utilization rate in the first two postoperative days. *Transplant Proc* 1989;21:2453–2454.
56. Foster PF, Williams JW, ed. Early postoperative care. In *Hepatic Transplantation.* Philadelphia PA: W.B. Saunders, 1990.
57. Calvey H, Davis J, Williams R. Nutritional management. In Calne RY, ed. *Liver Transplantation* New York: Grune & Stratton, 1983.
58. Munos SJ, Deems RO, Moritz MJ, et al. Hyperlipidemia and obesity in orthotopic liver transplantation. *Transplant Proc* 1991;23:1480–1483.
59. Hasse JM. Long-term nutritional problems in adult liver transplant recipients. *J Am Diet Assoc* 1990;90(suppl):A-36 (abstract).
60. Palmer M, Schaffner R, Thung SN. Excessive weight gain after liver transplantation. *Transplantation* 1991;51:797–800.
61. Rosenberg ME, Hostetter ED. Nutrition. In Toledo-Pereyra AH, ed. *Kidney Transplantation.* Philadelphia: FA Davis, 1988:169–186.
62. Hoy WE, Sargent JA, Freeman RB, et al. The influence of glucocorticoid dose on protein catabolism after renal transplantation. *Am J Med Sci* 1986;291(4):241–247.
63. Hoy WE, Sargent JA, Hall D, et al. Protein catabolism during the postoperative course after renal transplantation. *Am J Kidney Dis* 1985;5(3):186–190.
64. Whittier R, Evans E, Dutton S, et al. Nutrition in renal transplantation. *Am J Kidney Dis* 1985;6(6):405–411.
65. Twork A. Nutrition and kidney transplantation. *Diet Nutr Support* 1989;11(1):11–15.
66. Merion RM, Twork AM, Rosenberg L, et al. Obesity and renal transplantation. *Surg Gynecol Obstet* 1991;172:367–376.
67. Holley JL, Shapiro R, Lopatin WB, et al. Obesity as a risk factor following cadaveric renal transplantation. *Transplantation* 1990;49(2):387–389.
68. Rosenberg ME. Nutrition and transplantation. *Kidney* 1986;18(5):19–22.
69. Vathsala A, Weinberg RB, Schoenberg L, et al. Lipid abnormalities in cyclosporine-prednisone treated renal transplant recipients. *Transplantation* 1989;48(1):37–43.
70. Perez R. Managing nutrition problems in transplant patients. *Nutr Clin Pract* 1993; 8(1):28–32.

71. Shen SY, Lukens CW, Alongi SV, et al. Patient profile and effects of dietary therapy on post-transplant hyperlipidemia. *Kidney Int* 1983;24 (suppl 16):S147–S152.
72. Blue LS. Nutrition considerations in kidney transplantation. *Top Clin Nutr* 1992; 7(3):17–23.
73. Katz IA, Epstein S. Post-transplant bone disease. *J Bone Miner Res* 1992;7:123–126.
74. Harris M, Hadden WC, Knowles WC, Bennett PH. Prevalence of diabetes and impaired glucose tolerance and plasma glucose levels in the U.S. population aged 20–74 years. *Diabetes* 1987;36:523–534.
75. Nathan DM. Long-term complications of diabetes mellitus. *N Engl J Med* 1993; 328:1676–1685.
76. Chase HP, Jackson WE, Hoops SL, et al. Glucose control and the renal and retinal complications of insulin-dependent diabetes. *JAMA* 1989;261:1155–1160.
77. Sollinger HW, Sasaki TM, D'Alessandro AM, et al. Indications for enteric conversion after pancreas transplantation with bladder drainage. *Surgery* 1992;112:842–846.
78. Cooper A, Floyd TF, Ross III AJ, et al. Morbidity and mortality of short-bowel syndrome acquired in infancy: an update. *J Pediatr Surg* 1984;19:711–718.
79. Lillehei RC, Goot B, Miller FA. The physiologic response of the small bowel of the dog to ischemia including prolonged in vitro preservation of the bowel with successful replacement and survival. *Ann Surg* 1959;150:533–543.
80. Grant D, Wall W, Mimeault R, et al. Successful small bowel/liver transplantation. *Lancet* 1990;335:181–184.
81. Todo S, Tzakis AG, Abu-Elmagd K, et al. Cadaveric small bowel and small bowel–liver transplantation in humans. *Transplantation* 1992;53:369–376.

The Severely Obese Patient: An Approach to Management

CHAPTER 26

Donald R. Duerksen, M.D. and
R. Armour Forse, M.D., Ph.D.

INTRODUCTION

The recent National Health and Nutrition Examination Survey (NHANES III) conducted from 1988 to 1991 has emphasized an increase in the prevalence of overweight adults in the United States.[1] In adults over 20 years of age, 33.4% were estimated to be overweight, an 8% increase from the 1976 to 1980 survey. Mean body weight increased 3.6 kg over this time period. Significant comorbidities and an increased mortality rate are associated with obesity and in particular, for morbidly obese individuals. For both generalist and specialist physicians, nutritional management of the obese patient has become an increasingly important clinical issue. This chapter focuses on an approach to the management of obesity and specifically the subset of severely obese individuals.

ASSESSMENT

Degree and Type of Obesity

The medical assessment of the obese patient requires an initial measure of the degree of excess weight. In adults, body fat increases with age and in males ranges from 12% to 25% of body weight, whereas in females it ranges from 20% to 35%. Obesity can be defined as an excess of body fat—in males >25% and in females >30%. Body composition studies may be performed to evaluate body fat, but are not necessarily available or practical. While different methods of grading and assessing obesity have been used in the literature, a clinically useful measure is the body mass index (BMI), defined as the weight/height^2 (kg/m^2). Table 26.1 lists six grades of obesity according to this system.[2] An increased BMI is associated with both an increased mortality and an increased morbidity. In this chapter, we want to focus particularly on the approach to and management of the severely obese patient, those with a BMI >35. Although this subset of individuals makes up a smaller percentage of the overall population of overweight individuals, they have a high associated comorbidity and mortality and thus require aggressive therapeutic intervention.

Patients may also be subclassified according to regional distribution of body fat.[3] Android or central obesity results from abdominal deposition of body fat and is more common in men. This distribution is also associated with visceral deposition of adipose tissue. In contrast, gynoid or femoral-gluteal obesity results from deposition of excess fat in the thighs, hips, and buttocks, and is more frequent in women. Android obesity is associated with an increased risk of developing cardiovascular disease, breast cancer, and stroke. Insulin resistance and diabetes mellitus are also increased. Women with android obesity demonstrate impaired glucose tolerance and increased insulin secretion following an oral glucose load when compared with women of similar total body fat but a gynoid distribution.

Imaging studies with computed tomography may be a reliable way of quantifying the visceral deposition of fat, determining comorbid risks (such as cardiovascular and cerebrovascular disease), and monitoring response to weight reduction therapy. Clinically, the degree of visceral obesity may be estimated by the waist

Table 26.1. Classification of Obesity

Classification	BMI[a] (kg/m^2)	Grade of Obesity	% IBW[b] (males)	%IBW[b] (females)
Supermorbid	>50	6	225	245
Super Obesity	45	5	200	220
Morbid Obesity	40	4	180	195
Medically Significant	35	3	160	170
Obesity	30	2	135	145
Overweight	25	1	110	120
Ideal Body Weight	22	0	100	100

[a]BMI, body mass index.

[b]IBW, ideal body weight.

Table 26.2. Risk Factors in the Obese Patient

Body mass index (BMI)
Pattern of regional fat distribution
Early age of onset of obesity
Family history of obesity
Race (greater risk in Native American, Hispanic, African-American)
Personal history of obesity-related problems

to hip ratio (WHR). This is measured directly using a tape measure around the narrowest point above the umbilicus and the widest point below the hips. A WHR of more than 1.0 in men and 0.8 in women has been associated with an increased risk of ischemic heart disease, stroke, and death.

Table 26.2 summarizes other factors that may modify the associated morbidity and mortality of the individual obese patient. Progressive childhood obesity is a significant risk factor for the development of obesity in adults and for subsequent increased comorbidity risks and increased mortality. Approximately 50% of overweight children remain obese as adults. This form of obesity is not the most common type of obesity, as less than one third of obese adults were obese as children.

Studies on monozygotic and dizygotic twins and of adopted children have enabled estimations of the relative contributions of genetic and environmental factors to the development of obesity. The best estimates from these studies suggest that these two factors are of approximate equal importance in obesity development.

Associated Risks and Comorbid Illnesses

One of the prime reasons for encouraging weight reduction, through medical and/or surgical intervention, is to lower the obesity-associated increased morbidities and increased overall mortality.[4] Table 26.3 lists the associated morbidities by organ system.

Most studies including the Framingham study have demonstrated that obese individuals are at increased risk for coronary artery disease (CAD). This is not surprising as obesity is associated with an increase in cardiac risk factors including hypertension, non-insulin-dependent diabetes mellitus (NIDDM), and altered lipid status. A syndrome of obese cardiomyopathy with abnormalities in atrial and ventricular filling and left ventricular function has been described. These patients are also at increased risk for sudden death,

The risks of developing hypertension are 2.0 to 8.0 times greater in obese people and may be particularly high in younger individuals under age 45. Changes in cardiovascular and neuroendocrine function associated with obesity likely account for this association. These include increased cardiac output, increased sodium retention as a result of hyperinsulinemia, and altered noradrenergic activ-

Table 26.3. Conditions Associated with Obesity

Cardiovascular	Gastrointestinal	Endocrine
coronary artery disease	cholelithiasis	increased very low density lipoproteins
cardiomyopathy	Gastroesophageal reflux disease	decreased high density lipoproteins
hypertension	hepatic steatosis	hyperinsulinemia and insulin resistance
congestive heart failure	Renal	diabetes mellitus
cor pulmonale	proteinuria	Postsurgical Complications
Cerebrovascular	renal vein thrombosis	increased risk for postoperative infections
Transient Ischemic attack, stroke	Musculoskeletal	Gynecologic/obstetric
Peripheral vascular	osteoarthritis	menstrual disorders
varicose veins	gout	increased obstetric risk
deep vein thromboses	Neoplasm	Psychosocial
venous stasis	Females: breast, cervix, ovarian, endometrium, gallbladder	poor self-image
Dermatologic	Males: colon/rectal, prostate	discrimination
acanthosis Nigricans	Pulmonary	social isolation
hirsuitism	obstructive sleep apnea	susceptibility to psychoneuroses
intertrigo	primary alveolar hypoventilation	increased suicide risk
Endocrine	restrictive lung disease	Increased Mortality
hypercholesterolemia		cancer
hypertriglyceridemia		cardiovascular disease
increased Low density lipoproteins		cerebrovascular disease
		sudden death

ity and opiate suppression. Hypertension is associated with central obesity and correlates with changes in weight. Studies have demonstrated that weight loss reduces blood pressure while subsequent regain of weight results in blood pressure increases.

The chance of developing diabetes doubles for every 20% increase of body weight.[5] Insulin resistance in muscle and liver is common in obese individuals. Peripheral glucose uptake is decreased, whereas hepatic glucose output is increased. Obese individuals have a decrease in the number of insulin receptors, but there also appears to be a postreceptor defect that accounts for a major part of the insulin resistance. Although basal insulin levels are increased in obese subjects, the meal response (percentage increase) is decreased compared with lean controls, further compromising blood sugar control. BMI and WHR are strongly correlated with the development of NIDDM. The risk of type II diabetes also increases with age, family history of diabetes, and central obesity.

Increased triglycerides and in particular very low density lipoproteins (VLDL) can be seen in obese subjects. Weight reduction frequently results in a decrease in plasma triglycerides. Although hypercholesterolemia also occurs, its association with obesity is less common. High density lipoprotein (HDL) cholesterol is usually low in these individuals and is an independent risk factor for coronary artery disease.

A significant respiratory complication of obesity is the obesity-hypoventilation or Pickwickian syndrome. Massive obesity reduces the compliance of the chest wall and decreases the functional residual capacity. Many patients also develop obstructive sleep apnea (OSA), although sleep-induced hypoventilation may occur in the absence of OSA. Complications resulting from these abnormalities include daytime somnolence, chronic hypoxemia and hypercapnea, polycythemia, pulmonary hypertension, and right heart failure.

Life insurance studies were the first to suggest a relationship between obesity and mortality. Since this time, numerous studies have evaluated this question. In general, long-term follow-up studies have found an increased mortality correlating with a BMI above 25. The relative mortality risk for both men and women increases as BMI rises. Therefore, for mildly obese individuals with a BMI of 25 to 30, the mortality ratio is 1.1 to 1.3 whereas for the morbidly obese (BMI 35 to 40), this ratio increases to 1.7 to 2.5 for age-matched controls. The average relative risk of mortality for middle-aged men with a BMI >26 after adjusting for smoking, age, and physical activity is 1.67. The increase in mortality is due to an increased incidence of diabetes mellitus, cerebrovascular disease, cardiovascular disease, malignancy, and an increase in sudden death. The malignancies that have been associated with obesity include gallbladder, biliary tract, endometrium, ovary, breast, and cervix in women, and colon and prostate in men. The strongest association appears to be for endometrial carcinoma. Obese postmenopausal women are at increased risk for the development of breast cancer. The mortality ratio for cancer in persons more than 40% overweight is 1.33 for men and 1.55 for women.

Secondary Causes of Obesity

Secondary causes of obesity are uncommon but should always be considered in the initial assessment, particularly in childhood onset obesity. Endocrine causes include hypothyroidism, Cushings disease, and insulinomas. It is unusual for these disorders to present with obesity as the only manifestation. Even rarer causes such as Froelich's syndrome, Laurence-Moon-Biedl syndrome, and Prader-Willi syndrome are usually diagnosed in childhood and are associated with hypogonadism. Medications such as phenothiazines, tricyclic antidepressants, antiepileptics, and steroids may result in iatrogenic obesity.

Diet and Exercise History

A detailed history of dietary intake is important. An estimation of total daily caloric intake and fat intake should be performed by an experienced dietitian. It must be remembered, however, that there may be a significant discrepancy between reported and actual intake in obese subjects.[6] The average daily fat intake of most North Americans is 35% to 40% of caloric intake. There is evidence that obese individuals consume a diet higher in fats and lower in carbohydrates as compared to nonobese individuals. General dietary recommendations are to reduce fat intake to below 30% of total caloric intake. Specific patterns of intake are also important to identify possible interventions predicated on altering behavioral norms. For example, binge eating is common in a subgroup of obese individuals.

In general, overweight individuals have difficulty in developing an effective physical activity program. Exercise does not generally result in dramatic weight loss when utilized in isolation, but can be extremely important in promoting mild weight loss and its maintenance, reducing risks of disease, and contributing to a sense of well-being. A baseline of the individual's activity and previous attempts to improve this level should be noted.

Psychological and Behavioral Effects

A widespread belief is that obesity results from psychological dysfunction. When compared with normal weight individuals, an increased incidence of psychopathology has not been identified in morbidly obese individuals. There is also no particular personality type associated with marked obesity. However, overweight individuals are more depressed, self-conscious, and less self-assertive in matters relating to eating and weight. A common personality trait that may influence eating patterns and behavior is the passive-aggressive or passive-dependent personality type. This could impact upon therapeutic efforts, which may be met with procrastination, withdrawal, or other indirect obstructionist measures. Many morbidly obese patients have experienced childhood trauma including illness, parental loss, physical abuse, or sexual abuse. Identification and support for those patients with such premorbid antecedents are critical for the development of a successful weight-loss program.

Finally, severe obesity also impacts significantly on many social aspects. Social isolation, discrimination, and low self-image may all be present in obese individuals. Ultimately, the suicide risk is increased in these patients. It is important for the health care worker to identify the effect of obesity on intimate relationships, job satisfaction, and current life goals and to determine the methods of coping of the individual.

In summary, assessment of the individual patient requires an integrated evaluation of the degree of obesity and associated risk factors and comorbidities, along with an assessment of the patient's own perceptions and lifestyle limitations prior to the development of a treatment plan. We suggest the following approach be taken in assessing the obese individual.

1. Measure height and weight; calculate BMI
2. Assess waist/hip ratio
3. Risk factor assessment—diabetes, lipid status, hypertension
4. Assess for other risk factors, comorbid conditions
5. Consider secondary causes of obesity
6. Determine psychological and behavioral effects of obesity on the individual
7. Dietary and exercise history

MANAGEMENT

From the preceding discussion, it is evident that the medically significant obese individual is at a much greater risk, and therapeutic intervention is necessary. Recently, Leibel et al. demonstrated that maintenance of a reduced or elevated body weight is associated with a compensatory change in energy expenditure, opposing the body weight which is different from the usual weight.[7] Thus the body has a natural mechanism to oppose weight reduction. Weight reduction therapies are therefore inherently challenging for both the patient and clinician. The different modalities available to treat morbid obesity have recently been extensively reviewed.[8] A summary of these is given below.

Diet Therapy

Caloric intake is critically important in the development of obesity. As a general rule, an obese individual requires approximately 10 kcal/lb for weight maintenance. Predicting energy requirements for the individual patient may be difficult however, as there is significant variability in energy expenditure. The rate of weight loss depends upon the difference between energy intake and energy expenditure.

Very low calorie diets. Approximately 25% of men and 39% of women are trying to lose weight by reducing caloric intake in the United States. In 1959, a study following 100 consecutive patients treated by conventional diets reported that only 12% lost more than 20 lb and only 6% maintained this reduced weight at 1 year. In an effort to improve upon the frequent failure of conventional diet therapy in reducing weight, very low calorie diets (VLCD) were introduced in the late 1960s and early 1970s.[9] There is no universally agreed upon definition of a VLCD. Some investigators have proposed that any diet providing less than

Table 26.4. Complications of Very-Low-Calorie Diets (VLCD)

Binge Eating
Depression
Anemia
Cholelithiasis
Hyperuricemia

800 kcal/day be considered in this category. As stature and severity of obesity are important variables, more clinically appropriate definitions include 10 kcal/kg ideal body weight per day or a diet that provides 50% or less of the patient's daily resting energy expenditure.

The goal of the VLCD is to induce weight loss yet prevent a significant loss of lean tissue. Approximately 25% of excess weight in obese individuals is lean tissue, and one fifth of this is protein. Therefore significant weight loss will result in some degree of obligatory protein loss. Dietary protein is critical for minimizing the loss of lean tissue and 75 g/ day for men and 55 g/ day for women has been shown to induce a positive nitrogen balance within 2 to 3 weeks. This diet should include a multivitamin, 3 to 5 g of sodium chloride, 2 to 3 g of potassium. A full liquid diet may be more successful in preventing violations of the VLCD than a diet of conventional food. The weight loss induced by a VLCD adhered to for 12 to 16 weeks is approximately 20 kg. The percentage of lean body mass loss is approximately 15% to 25% and is unlikely to contribute to any major complications.

When used in the appropriate clinical setting under medical supervision, VLCDs are considered relatively safe. Complications of VLCDs are listed in Table 26.4. In most cases, after the first few days, hunger is not a major symptom of these patients. Some investigators have been concerned about possible adverse effects of cycles of weight loss and weight gain induced by these diets. Although some evidence suggests this may be detrimental with increased morbidity and mortality, this is in dispute. More definitive studies are needed to address this important clinical issue.

In general, the use of VLCDs should be restricted to adults less than 65 years of age who are more than 30% overweight. Patients with major organ dysfunction should be excluded from such therapy (Table 26.5).

While the short-term benefit of VLCDs has been proven, the long-term results have not been well studied. In general, however, there is a substantial likelihood of weight gain in the first 2 years following treatment. The long-term success with diet therapy alone may be as low as 5% in morbidly obese individuals. This has led to trials combining behavioral therapy and drug therapy, which may improve long-term benefits.

Table 26.5. Contraindications to Very-Low-Calorie Diets

Recent myocardial infarction
Cardiac conduction disorder
A history of cerebrovascular, renal, or hepatic disease
Malignancy
Type 1 diabetes
Pregnancy
Significant depression
Substance abuse disorders

Behavioral Therapy

The major goal of behavioral therapy is to help obese individuals identify and modify inappropriate diet and eating habits that have contributed to their obesity. The most appropriate patients to consider this therapy in are those in whom cognitive and emotional factors exacerbate difficulties with weight control. Commonly, group therapy provided by a behavioral psychologist and combined with a dietary approach such as a VLCD are employed. Long-term weight loss is more likely to be achieved when VLCDs are combined with behavioral techniques. In one study, 32% of patients maintained weight loss at 1 year with a program of diet and behavioral therapy compared with 5% receiving diet therapy only. Thus, behavioral therapy does not work alone but is a valuable adjunct both to inducing and maintaining long-term weight loss when combined with other modalities. Long-term results are once again lacking, with maintenance of weight loss a major concern.

Pharmacotherapy

The most widely used drugs for the treatment of obesity have been the appetite suppressants.[10] Amphetamine was the first drug marketed for this purpose, and since then numerous other medications have been developed and are listed in Table 26.6. The main classes of medications used in clinical practice are the centrally acting catecholaminergic and serotoninergic compounds. These anorectic agents have been studied in short-term, usually less than 12 weeks, clinical trials. On average, weight loss has been approximately half a pound a week when compared with placebo.[11] Longer term studies with the serotoninergic medications have demonstrated small increases in weight loss compared with placebo. Because of the unimpressive results of single drug therapy, recent trials are studying the effect of combination therapy. Preliminary results demonstrate maximum weight loss at six months with persistence of weight loss for as long as medication compliance is maintained. These combinations are still considered experimental

Table 26.6. Drugs used in the Management of Obesity

Appetite Suppressants	Drugs influencing absorption
Centrally acting catecholaminergic	Enzyme inhibitors
Phenylpropanolamine, ephedrine	Acarbose
Diethylpropion, phentermine, mazindol	Orlistat
Benzphetamine, chlorphenterimine	Undigestible fat
Amphetamine, metamphetamine	Olestra
Centrally acting serotoninergic	
Fenfluramine	
Fluoxetine	
Combined adrenergic and serotoninergic	
Sibutramine	

but offer some promise for future use, particularly in association with other modalities such as behavior modification, very low calorie diets, and surgery. Of concern is the weight gain that occurs when medications are stopped.

The long-term side effects of these drugs have not been adequately defined because of the short-term nature of most clinical trials. Side effects that have been seen include depression and short-term memory loss, dry mouth, tachycardia, and increased blood pressure.

Several experimental medications have been developed that decrease intestinal absorption of important nutrients. Acarbose is an intestinal disaccharidase inhibitor while orlistat is an inhibitor of lipase. Olestra is an indigestible fat produced by esterifying sucrose with fatty acids to form a sucrose polyester. It has recently been approved by the FDA. Clinical trials with these agents alone or in combination with other modalities are awaited.

Exercise Therapy

In comparison to severe dietary restriction, the impact of exercise alone on weight loss is fairly minimal. However, exercise therapy may be useful in the management of obesity for several reasons. First of all, exercise enhances energy expenditure; secondly, it reduces energy intake; and finally, it may modify the health risks associated with obesity. Exercise has a beneficial effect on body fat distribution and reduces the waist-to-hip ratio. It also helps to maintain lean body mass. Overweight men and women who are active and fit have lower morbidity and mortality rates than sedentary overweight persons. Exercise may be particularly beneficial in the long-term maintenance of weight loss. A negotiated treatment plan between physician and patient with a psychological and behavioral commitment to exercise is most likely to be effective. Low to moderate exercise may be safely administered in conjunction with a VLCD. The emphasis of exercise programs should be on adherence rather than maximizing weight loss as this is more likely to lead to a sustained effort.

Surgery

The most effective method of reducing weight in the severely obese is through surgical treatment.[12,13] The indications for surgical management are serious medical complications from obesity such as hypertension, diabetes, and the obesity/hypoventilation syndrome and/or total body weight 100 lb above insurance weight standards for height. Patients considered for surgery should be refractory to the more conservative measures described above prior to consideration of surgery. Psychological stability and a social support system are important factors to consider in selecting patients for this procedure. To optimize surgical outcome, a psychiatric evaluation is advisable, identifying psychiatric illnesses that might preclude surgery and behavioral or personality patterns that might influence an adverse outcome. Proper patient selection and preparation are important factors in optimizing the major goal of surgery—significant weight loss and improved quality of life. Program compliance and sustaining altered diet and eating patterns will enhance the surgical success.

Historically, several different types of operations have been used in weight reduction. In the past, gastrointestinal bypass operations were performed but, because of a large number of complications including malabsorption, chronic liver disease, and arthropathy, have now been abandoned. Recently, the most popular operations include the vertical banded gastroplasty, and the vertical gastric bypass with roux-en-y gastrojejunostomy. Randomized trials have demonstrated that the gastric bypass procedure is more effective in inducing weight loss than gastroplasty alone.

On average, patients lose 40% to 70% of their excess weight, most of this in the first year. Five-year follow up studies have demonstrated a sustained loss of 60% of excess weight with gastric bypass operations. Equally important, risk factors such as diabetes, hypertension, hyperlipidemia, arthritis, and sleep apnea

Table 26.7. Improvement of Comorbidities Following Bariatric Surgery

Comorbidity	Medication Free (%)	Reduced Medication (%)
Diabetes	90–95	100
Hypertension	60–65	90
Osteoarthritis	?	>90
Congestive heart failure	60	90
Sleep apnea	100	100
Dyslipidemia	70	85

have improved following this weight loss. Ninety percent of patients are able to become independent of antihypertensive medications and insulin. Symptoms of degenerative arthritis have resolved or improved in over 90% of patients and symptoms of severe sleep apnea have been reported to improve in 100% of patients with significant weight loss. Quality of life also improves considerably with many patients reporting an improved self-image.

The perioperative mortality associated with this surgery is less than 1% in most centers. Immediate postoperative complications occurring in 1% to 10% of patients include anastomotic leaks, wound infections, stomal stenosis, and incisional hernias. There is a 10% incidence of developing cholecystitis necessitating cholecystectomy. Long-term complications include vitamin and mineral deficiencies, particularly iron and vitamin B_{12}, nausea and vomiting, diarrhea, and the "dumping syndrome" (a symptom complex associated with eating food containing processed sugar). These complications and side effects occur in up to 20% of patients.

NUTRITION SUPPORT OF THE HOSPITALIZED MORBIDLY OBESE PATIENT

Metabolic Response to Injury

The metabolic response to injury in the nonobese person is marked by activation of the counter-regulatory hormones including epinephrine, glucagon, cortisol, and growth hormone, as well as numerous cytokines including interleukin 1 and 6 and tumor necrosis factor. The end result is an increase in basal metabolic rate and body temperature. With respect to substrate utilization, nitrogen catabolism is increased, hyperglycemia and insulin resistance are common, and there is an accelerated hydrolysis of triglycerides, resulting in an increased turnover of free fatty acids.

In severe obesity there may be several significant differences in the metabolic response to injury, although few studies have addressed this important question. Jeevanandam et al.[14] have demonstrated an increased mobilization of protein and a decreased catabolism of fat in stressed obese individuals as compared with the stressed nonobese individuals. The clinical significance of these findings need to be further investigated.

Energy and Protein Requirements

Traditional methods, such as the Harris-Benedict equation, predict energy expenditure based on weight. Numerous studies have demonstrated a gross overestimate of energy expenditure if these are applied to severely obese patients. An estimate based on ideal body weight (IBW) tends to underestimate the actual resting energy expenditure. Several alternative estimates have been developed. As a portion of the excess weight in severely obese individuals is metabolically active, a correction of 25% of the difference between the ideal and actual weight has been added to the ideal body weight to obtain a corrected body weight. Several equations have also been developed, but these are more cumbersome to use in routine clinical practice. The recommended protein requirements range from 1.5 to 2.0 g/kg IBW.

Hypocaloric Feeding

The use of hypocaloric feeding has been studied in the nonstressed obese population. In the stressed hospitalized patient, however, studies are more limited. There are several potential benefits of hypocaloric feeding in the severely obese.[15] These include better control of blood glucose, which may lower the risk for infectious complications, facilitate weaning in the ventilator-dependent patient, and improve fluid balance in the fluid overloaded individual. Because obese stressed individuals appear to have a greater protein catabolism and lesser ability to mobilize fat, the effect of hypocaloric feeding on nitrogen balance is an important clinical question.

Two clinical studies have addressed the question of hypocaloric feeding in the obese. Dickerson et al.[16] fed thirteen patients an average of 1434 total calories per day which was 72% of measured caloric intake. The total protein intake averaged 2.1 g/kg IBW. These patients attained a positive nitrogen balance and had a good clinical outcome with respect to wound healing and closure of fistulae. In a more recent study by Burge et al.[17] 16 obese patients (>130% IBW) were randomized to receive hypocaloric (50% nonprotein calories) or caloric (100% resting energy expenditure) TPN. Each group received approximately 1.25 g/kg protein. Mean nitrogen balance was similar between these two groups, demonstrating the feasibility of hypocaloric feeding.

What practical recommendations can be made regarding nutritional support of the stressed obese individual? Based on the above data, and until more studies are available, we would suggest the following:

1. Estimate the energy expenditure with one of the following formulas:
   i. Estimated energy expenditure = ABWT × 25 kcal / kg, where adjusted body weight (ABWT) = ideal body weight (IBW) + (current body weight (CBWT) – IBW)/4
   ii. Estimated energy expenditure = 8 to 10 kcal/lb
2. If possible confirm energy expenditure with the use of indirect calorimetry.
3. Provide calories to approximately 300 to 500 kcal below the numbers obtained by items 1 or 2 to provide hypocaloric nutritional support.

4. Provide protein at 2.0 g/kg of IBW.
5. Determine nitrogen balance by urinary nitrogen assessment and adjust protein supplementation accordingly.
6. Lipids should not routinely be infused, and approximately 200 to 400 g of dextrose should be supplied to approximate glucose production after an overnight fast.

SUMMARY

Severe obesity is a significant health problem contributing to a significantly morbidity and mortality. A multidisciplinary approach is necessary to properly assess and effectively treat severely obese individuals. Currently, surgical therapy is the most effective therapy in inducing and maintaining weight loss in the severely obese patient. Multimodality therapy including VLCDs, combination pharmacotherapy, behavior and exercise therapy in addition to surgery hold promise for the future. For the hospitalized obese patient requiring nutritional support, mildly hypocaloric feeding with adequate protein supplementation and close nutritional follow up is recommended.

REFERENCES

1. Kuczmarski RJ, Flegal KM, Campbell SM, et al. Increasing prevalence of overweight among US adults. *JAMA* 1994;272:205–211.
2. Kanders BS, Forse RA, Blackburn GL. Obesity. In Rakel RE, ed. *Conn's Current Therapy*. Philadelphia: 1991:Saunders 524–531.
3. Wadden TA, VanItallie TB, eds. *Treatment of the seriously obese patient*. New York: Guilford Press, 1992.
4. Grundy SM, Baarnett JP. Metabolic and health complications of obesity. *Disease-a-Month* 1990;12:645–696.
5. Colditz GA, Willett WC, Rotnitzky A, Manson JE. Weight gain as a risk factor for clinical diabetes mellitus in women. *Ann Intern Med* 1995;122:481–486.
6. Lichtman SW, Pisarska K, Berman ER, et al. Discrepancy between self-reported and actual caloric intake and exercise in obese subjects. *N Engl J Med* 1992;327:1893–1898.
7. Leibel RL, Rosenbaum M, Hirsch J. Changes in energy expenditure from altered body weight. *N Engl J Med* 1995;332:621–628.
8. Danford D, Fletcher SW. Methods for voluntary weight loss and control. *Ann Intern Med* 1993;119:641–764.
9. National Task Force on the Prevention of Obesity. Very low-calorie diets. *JAMA* 1993;270:967–974.

10. Silverstone T. Appetite suppressants: a review. *Drugs* 1992;43:820–836.
11. Goldstein DJ, Rampey AH, Dornseif BE, et al. Fluoxetine: a randomized clinical trial in the maintenance of weight loss. *Obesity Res* 1993;1:92–98.
12. Consensus Development Panel. Gastrointestinal surgery for severe obesity. *Ann Intern Med* 1991;115:956–961.
13. Sugarman HJ, Kellum JM, Engle KM, et al. Gastric bypass for treating severe obesity. *Am J Clin Nutr* 1992;55:560S–566S.
14. Jeevanandam M, Young DH, Schiller WR. Obesity and the metabolic response to severe multiple trauma in man. *J Clin Invest* 1991;87:262–269.
15. Baxter JK, Bistrian BR. Moderate hypocaloric parenteral nutrition in the critically ill obese patient. Nutrition in Clinical Practice 1989;4:133–135.
16. Dickerson RN, Rosato EF, Mullen JL. Net protein anabolism with hypocaloric parenteral nutrition in obese stressed patients. *Am J Clin Nutr* 1986;44:747–755.
17. Burge JC, Goon A, Choban PS, Flanebaum L. Efficacy of hypocaloric total parenteral nutrition in hospitalized obese patients: a prospective, double blind randomized trial. *J Parent Ent Nutr* 1994;18:203–207.

CHAPTER

27 Nutritional Support for the Critically Ill

Scott A. Shikora, M.D.

INTRODUCTION

The critically ill represent a subset of patients that are often the most difficult to support nutritionally. In general, they manifest significant physiologic alterations from normal metabolic homeostasis as a consequence of injury or severe illness. These stress responses, as they are called, are generally characterized by hypermetabolism, hypercatabolism, persistent lean body carcass wasting, and increased nutrient requirements. For these patients, the development of protein-calorie malnutrition is almost inevitable. Increased nutrient demand is often seen in the setting of inadequate intake. The ability to deliver sufficient nutrients is commonly frustrated by such problems as volume limitations, access difficulties, preexisting medical conditions, gastrointestinal dysfunction, invasive procedures, and acute organ failure. On the other hand, despite even the most sophisticated attempts at nutrient support, the hallmark of critical illness is persistent wasting of the lean body mass and malnutrition. The introduction of newer technologies into the critical care arena has only magnified this dilemma. Survival after acute injury is prolonged in ever older, more frail individuals who often have compromised nutritional reserves prior to their acute illness. It's truly ironic that the ability to prolong survival in the intensive care unit (ICU) is also responsible for exacerbating the nutritional debilitation. In addition, it should come as no surprise that the degree of malnutrition closely coincides with increased morbidity and mortality.[1]

In the last several years, nutrition support has made great strides in improving the delivery of nutrients. The benefits of early feeding, and specifically, early enteral nutrition recently has become increasingly clear. In addition, the ability to modulate the metabolic pathways with patient-specific and disease-specific nutritional therapies offers a great potential to affect outcome. It seems obvious that we are entering an era in critical care medicine where nutritional and metabolic support for critically ill or injured patients will be utilized for an even greater role than just the appropriate provision of nutrients.

This review describes the current strategies for providing the timely and appropriate nutritional support for this difficult group of patients. To properly discuss this topic, it is necessary to detail the metabolic response to injury, the limitations of the nutritional assessment, and review strategies for utilizing total parenteral and enteral nutrition to provide the best nutritional and metabolic support for the severely traumatized or critically ill patient.

METABOLIC RESPONSE TO INJURY

The metabolic response to injury or severe illness employs a characteristic network of physiologic responses designed to maximize the chances for recovery and survival (Table 27.1). After any injurious event, whether it is surgery, trauma, sepsis, or an inflammatory process, there is a cytokine- and hormonally mediated mobilization of endogenous substrates. These nutrients are utilized as oxidative fuels and for the synthetic building blocks essential for the stabilization of organ function, maintenance of immunocompetency, repair of injured tissue, and ultimately recovery.[2–3] It is known that the stress response is also responsible for many of the metabolic consequences of critical illness, namely the fluid overload, hypercatabolism, hypermetabolism, and the glucose intolerance.

More than 50 years ago, Cuthbertson and Tilson[4] described this phenomenon as a two-phase process. The Ebb phase occurs immediately after injury and lasts approximately 24 to 48 hr. During this short-lived period, there is an increase in sympathetic activity and a stimulation of the hypothalamic–pituitary axis.[5] This period is characterized by marked hypometabolism and decreased oxygen consumption.[5] It is now recognized that these findings are primarily the result of hypovolemia leading to decreased cardiac output and inadequate oxygen transport to the tissues.[6] In contrast to the ebb phase, the second phase called the acute or flow phase is one of hypermetabolism, catabolism, and increased oxygen consumption.[7] These mechanisms are thought to be mediated by cytokine release

Table 27.1. The Metabolic Alterations of the Stress Response[a]

Metabolic Parameters	Rate Compared to Normal
Resting energy expenditure	I
Oxygen consumption	I
Carbohydrate metabolism	
Blood sugar concentration	I
Gluconeogenesis	I
Glycogenolysis	I
Tissue glucose uptake/oxidation	I
Fat metabolism	
Ketogenesis	N/D
Lipolysis	I
Tissue uptake/oxidation	I
Protein metabolism	
Net synthesis	D
Net breakdown	I
Hepatic synthesis	I
Muscle synthesis	D
Ureagenesis	I

[a]These changes are essential to guarantee an adequate availability of substrates for healing. I, increased; D, decreased; and N, equal to normal.

and afferent nerve signals from the injured tissues.[8] During this period, there is an active liberation of endogenous substrates such as glycogen-derived glucose, skeletal muscle-derived and labile amino acids, and adipose tissue fatty acids.[2,3,9,10] Increased sympathetic and hypothalamic–pituitary activity stimulate the release of the catecholamines and the counterregulatory hormones, namely, the glucocorticoids, glucagon, growth hormone, prolactin, and aldosterone. These hormones act synergistically to promote lipolysis, muscle protein breakdown, insulin resistance and glycogenolysis.[3,5,11,12] Because the glycogen stores are limited, this source of glucose is quickly exhausted.[2] The need for readily available glucose will then depend on enhanced muscle protein breakdown to provide amino acids for hepatic gluconeogenesis.[2,3,5] In response to the rising plasma glucose levels, insulin secretion is increased. However, the hormone's actions on glucose metabolism are greatly restricted by the profound insulin resistance.[2,5,11]

The net effect of these metabolic pathways is the liberation of peripherally stored substrates to meet the energy requirements of the major organ systems. Each substrate plays an important physiologic role in the stress response. Glucose is an important fuel for the central nervous system, the wound, and the immune system, all of which are metabolically active during stress.[2,3,7] The fatty acids function to provide energy for cardiac and skeletal muscle, the liver, and many other tissues.[2,3,7] Although some liberated amino acids are utilized for gluconeogenesis, the majority are required for the synthesis of the acute phase proteins, for thermogenesis, and as precursors for tissue repair.[2,3,7,10] In states of semistarvation, the ongoing demand for amino acids can lead to marked wasting of lean body tissue. The healthy human body stores approximately 100 g of labile protein nitrogen and is designed to withstand about 1 week of stress without feeding.[13] In the critically ill, where catabolism is more intense and nutrient stores are likely to be inadequate, storage fuels may be more rapidly exhausted. Further protein wasting without supplementation can lead to decreased protein synthesis, organ dysfunction, immunodeficiency, sepsis, and ultimately, death.[3,5,9]

During the period of net protein breakdown, there is a significant loss of protein nitrogen, which is converted into urea and excreted in the urine.[5,10] While total body protein synthesis is decreased (muscle amino acid incorporation is suppressed), hepatic and other local synthesis is actually increased.[3,5,9] The persistently elevated urine nitrogen and negative nitrogen balance reflect the degree of stress and the ongoing endogenous amino acid utilization. Even with sufficient supplementation, a positive nitrogen balance can usually not be achieved.[5,10,14] This has been demonstrated by Shaw and Wolfe in a study involving trauma and septic patients.[15] In that investigation, the administration of parenteral nutrition only succeeded in cutting net protein loss in half.

Simply stated, protein-calorie malnutrition is associated with increased morbidity and mortality.[1,16,17] The adverse effects of nutrient deprivation on cellular function involves more than just the lack of available nutrient substrate. It also causes alterations in ion gradients, membrane potentials, production of high-energy compounds, and antioxidant defense.[18] Using sophisticated neutron activation analysis to measure body composition, Hill has been able to accurately identify the degree of patient malnutrition.[19] He demonstrated that most physiologic functions become impaired when 20% of body protein is depleted. Malnutrition and wasting have been shown to affect most organ systems including respiratory,[20] cardiac,[21,22] and the immune system.[23] Many of the components of the immune system from the macrophage to the mature lymphocytes are compromised.[24–27] This can greatly increase the susceptibility to septic events, the major cause of death in this population. Respiratory and cardiac dysfunction can manifest as prolonged ventilator dependence, recurrent pneumonia, myocardial infarction, congestive heart failure, and recurrent arrthymias. These untoward events dramatically increase mortality, morbidity, and length of stay with its inherent complications.

Like the macronutrients, micronutrients are redistributed during stress. Zinc, a cofactor for several enzymatic processes, is taken up by the liver.[2] Iron is sequestered by both the liver and the bone marrow, thereby decreasing the amount of iron in the blood available to the iron-dependent pathogenic microorganisms.[2]

Calcium, magnesium, phosphorus, chromium, copper, selenium, and manganese are also important for many enzymatic processes and are needed in greater amounts.

Recovery, should it occur, leads to a decrease in metabolic rate, a restoration of appetite, a replenishing of the body energy stores, and a rebuilding of the lost lean body mass. The cytokine, catecholamine, and counter-regulatory hormone influences subside, and insulin again becomes the primary nutrient regulatory hormone. There is also a significant uptake of amino acids in the muscle for synthesis and glucose for the production of glycogen and triglycerides. Typically, urinary nitrogen falls and the nitrogen balance (provided that protein is supplied) becomes positive.[2] With uncomplicated stress, recovery usually occurs between 5 and 10 days after injury. However, the stress response can persist in patients with recurrent or persistent sepsis or organ dysfunction. In these circumstances, nutrient reserves can be exhausted leading to organ failure and death.

NUTRITIONAL ASSESSMENT

Performing a comprehensive nutritional assessment (discussed in detail elsewhere in this book) on patients entering the ICU to evaluate nutritional status is not terribly important. Irrespective of the nutritional status, nutrition support should commence early in all critically ill or traumatized patients. Studies have been able to demonstrate an improvement in outcome when feeding is begun within 48 to 72 hr of ICU admission.[28] In addition, many of the parameters evaluated for the nutritional assessment, such as serum albumin, patient weight, etc. will be affected by their illness. Fluid overload, peripheral edema, and lean body wasting hamper the ability to use body weight as a measure of body mass. Metabolic derangements associated with the stress response and fluid status alter serum protein concentrations.[2,29–31] For these patients, the components of the nutritional assessment offer more information as measures of the degree of stress imposed by the injury than as markers of nutrient reserves. For example, most critically ill patients become hypoalbuminemic as a result of their illness rather than from acute changes in nutritional status. Nevertheless, studies have confirmed that hypoalbuminemia correlates with increased morbidity and mortality.[29–33]

CALORIC REQUIREMENTS

Currently, it is generally accepted that calorie provision should not be provided in excess; instead, it should be titrated to match actual metabolic requirements.[34] The provision of calories in excess of measured needs has not been shown to improve outcome or promote lean body tissue growth in acutely ill patients.

Therefore, reliable methods for estimating or measuring caloric requirements are necessary. Many formulas have been proposed to estimate caloric goals. One of the most widely used, the Harris–Benedict equation[35] predicts resting energy expenditure based on patient age, sex, height, and weight (Figure 27.1). Several stress factors have been subsequently incorporated into the equation to account for the increased metabolic demand of stress, sepsis, or the thermic effect of feeding. These factors increase the predicted caloric requirements from 20% for patients recovering from minor surgery, up to 100% for patients after major thermal burns. Although the addition of such factors has been shown in some studies to improve the accuracy for predicting caloric goals, other reports dispute this claim.[36,37] The magnitude of the increase in the metabolic rate seen in stressed patients is extremely variable and rarely is it to the degree assumed by the stress factors. Surprisingly, only approximately 60% of stressed patients actually increase their energy expenditure above baseline and approximately 15% are found to be hypometabolic.[38] Patients with lower-than-expected energy expenditures tend to be those who are mechanically ventilated, heavily sedated, and/or paralyzed.

Another weakness of the Harris–Benedict equation and related formulas is the reliance on patient weight and/or body surface area to approximate metabolically active tissue. Weight can vary according to body fluid status, particularly in critically ill patients who commonly retain significant amounts of fluid. Catabolic illnesses have been shown to increase extracellular fluid while decreasing both fat and lean body mass.[18,39] An increase in body water will increase both body weight and surface area without increasing metabolically active tissue. The use of such formulas in this setting can lead to the overestimation of metabolic

$$\text{Males:} \quad REE = 66.5 + 13.8W + 5.0H - 6.8A$$
$$\text{Females:} \quad REE = 655.1 + 9.6W + 1.8H - 4.7A$$

Stress factors	Adjustments for REE (%)
Mild starvation	0.85–1.00
Postoperative (no complications)	1.00–1.05
Peritonitis	1.05–1.25
Cancer	1.10–1.45
Severe head injury	1.30–1.50
Multiple trauma	1.25–1.75
Severe infection	1.50–1.75
Thermal injury	1.25–2.00

Figure 27.1. The Harris-Benedict equation for estimating resting energy expenditure (REE). W = weight in kg, H = height in cm, and A = age in years. The stress factors have been subsequently added to compensate for the increased energy expenditure of illness.

rate.[40] As an alternative, Shanbhogue et al.[41] recommend using the 24-hr urinary creatinine excretion to estimate the caloric needs of the critically ill malnourished.

The best method for determining the caloric needs for the critically ill is the measurement of energy expenditure by indirect calorimetry.[9,34] Using a portable gas analyzer or "metabolic cart," resting energy expenditure can be easily obtained at the bedside. In addition, the simultaneous measurement of the respiratory quotient (RQ) by the metabolic cart is often helpful in determining the mixed-fuel requirements. Using actual measurements, it has been determined that the average-sized patient lying comfortably in bed requires approximately 23 kcal/kg body weight each day.[2] This increases only slightly with most elective surgical procedures.[34,42] The energy requirements of the most hypermetabolic, those patients with major sepsis or significant thermal burn, increases only to approximately 40 to 50 kcal/kg $\cdot$ day^{-1}.[2,43,44] Today, in most modern burn units, with aggressive wound care and sophisticated antibiotic regimens, even these high levels are rare. Therefore, most critically ill patients can be given 25 to 30 kcal/kg $\cdot$ day^{-1}.[43] The estimated calorie requirement always can be verified or adjusted according to patient condition by actual measurements.

The need to provide calories to meet requirements cannot be overemphasized. Significantly underfeeding leads to continued weight loss, poor healing, sepsis, protein wasting, organ failure, and ultimately death. Overfeeding can be even more troubling and is associated with numerous complications (Table 27.2). Whether nutrients are administered parenterally or enterally, the best nutritional goal remains the provision of the appropriate number of calories required to meet energy demands and protein to reduce nitrogen losses.

MACRONUTRIENT REQUIREMENTS

Protein

After the initial fluid and electrolyte replacement, the provision of protein is central to successful nutritional intervention. Dietary amino acids are necessary as building blocks for wound healing, for the maintenance of organ integrity and immunocompetence, and as an energy fuel.[2,3] Like caloric requirements, the estimation of protein needs must be accurate. Insufficient administration can lead to excessive nitrogen wasting,[9,45] the result of which, as previously stated, is organ failure, sepsis, and subsequent death. The overprovision of protein does not enhance uptake and may lead to increased ureagenesis which can cause renal injury in some patients.[9,45,46] Recent studies have supported the administration of approximately 1.5 to 2.0 g protein/kg body weight $\cdot$ day^{-1}.[2,8,43,45,47] Only rarely do patients require more. The ability to measure 24-hr urinary nitrogen excretion enables the clinician to estimate needs on the basis of nitrogen losses (Figure

Table 27.2. The Potential Hazards of Overfeeding[a]

Nutrient	Metabolic Derangement
Glucose	Hyperglycemia
	Hyperosmolar state
	Osmotic diuresis
	Dehydration
	Immunosuppression
	Hepatic steatosis
	Ventilatory alterations
	Increased oxygen consumption
	Increased carbon dioxide consumption
	Increased minute ventilation
	Increased response to hypercapnea
	Increased response to hypoxia
	Increased resting energy expenditure
Lipid	Immunosuppression
	Increased prostaglandin production
	Hypercholesterolemia
	Hyperlipidemia
	Impaired liver function
	Ventilatory alterations
	Decreased arterial oxygen saturation
	Decreased pulmonary diffusion capacity
	Increased alveolar-arterial PO_2 gradient
Amino acids	Ureagenesis
	Hyperchloremic acidosis
	Ventilatory alterations
	Increased oxygen consumption
	Increased minute ventilation
	Increased ventilatory drive
	Increased response to hypercapnea
	Increased response to hypoxia
	Increased resting energy expenditure

[a]The excessive administration of the three macronutrients can have profound effects on metabolic rate and overall outcome.

27.2) Total nitrogen excretion is the sum of the urine urea nitrogen, fecal, skin, and respiratory tract nitrogen losses (a factor of 2 to 4), and any abnormal nitrogen loses such as fistula output, wound exudates, dialysates, diarrhea, or gastric tube losses.[2] Unfortunately, the estimation of total body nitrogen losses based upon urinary nitrogen is a gross estimate and dependent upon stable renal function.

The optimal treatment strategy is to match losses during the catabolic phase of illness and to replete lost lean body mass during the anabolic phase. Serial

$$\mathbf{Nbal(g/d)} = \frac{\mathbf{Protein\ Intake(g/d)}}{\mathbf{6.25}} - \mathbf{[UUN\ (g/d) + ANL\ (g/d) + 4]}$$

Figure 27.2. Formula for calculating nitrogen balance (Nbal). UUN, urine urea nitogen. Abnormal nitrogen losses (ANL) include fistula output, wound exudates, dialysates, diarrhea, or gastric losses.

nitrogen balance measurements can be performed, for example, weekly, and protein provision adjusted as indicated. This is particularly important for critically injured patients where dietary nitrogen requirements may vary as their condition changes. Unfortunately, the loss of lean body mass will persist despite the provision of adequate protein until the stress response subsides.[16,39]

In certain subgroups of patients, such as those with renal or hepatic disease, protein administration might need to be kept intentionally below usual estimations because of impaired nitrogen metabolism. The provision of protein in levels exceeding 0.8 to 1.0 $g/kg \cdot day^{-1}$ can lead to worsening renal function or encephalopathy in patients with renal or hepatic disease, respectively. However, the usual protein requirements for the critically ill or injured is still 1.5 to 2.0 $g/kg \cdot day^{-1}$. Therefore, prolonged protein restriction can lead to protein malnutrition with all its complications. For these patients it is far better to initiate dialysis or accept encephalopathy than risk protein depletion.

Carbohydrate

As previously described, glucose is the preferred fuel for many metabolically active organ systems including the central nervous system, the immune system, and the healing wound.[2,3,7] In times of stress or injury, the importance of a readily available supply of glucose is magnified. When availability is deficient, gluconeogenesis is enhanced to produce the needed glucose. This results in increased protein catabolism and further nitrogen wasting with its inherent complications.

Although carbohydrates usually provide the majority of the administered calories, excessive provision is not without risks. The overconsumption of parenteral glucose has been shown to enhance lipogenesis, cause hepatic steatosis and subsequent hepatic dysfunction, exacerbate hyperglycemia in glucose-intolerant patients, and increase the production of carbon dioxide, which potentially can compromise respiration in patients with pulmonary disease.[9,20,34,43,48,49] Hyperglycemia, an all too common sequela of stress in both diabetics and nondiabetic patients, has been shown to adversely effect immunologic function.[50–54] These derangements can improve with better glycemic control.

The optimal amount of carbohydrate utilization has been determined. Wolfe et al.[55] have shown that patients do not oxidize more than 5 to 7 mg of glucose/kg of body weight $\cdot$ min^{-1} given intravenously. Higher rates of administration can lead to storage. Initially, the excess is used to replete glycogen stores in the liver. Because this is a finite glucose pool, it is quickly replenished, and the body then converts all excess glucose into triglyceride. Therefore, in most circumstances, the rate of infusion should not exceed 4 to 5 $mg/kg \cdot min^{-1}$. In the average-sized patient, daily provision should be no greater than 300 to 400 g/day.[43]

Fat

With the addition of fat to both enteral and parenteral formulas, caloric goals can be achieved with less potential for carbohydrate overfeeding. Of the total calories, 30% to 40% are usually supplied by lipids. However, as with the other nutritional substrates, it must be used in a judicious manner. The provision of lipids, traditionally as long-chain triglycerides, has been associated with immunologic abnormalities including reticuloendothelial system dysfunction, impaired phagocytosis, and increased gram-negative bacterial survival.[20,56,57] Long-chain triglycerides, the usual lipid source in parenteral nutrition and enteral formulas, is cleared by the lymphatics and then engulfed by the kupffer cells in the liver. When administered discontinuously over a short time period, it hampers the ability of the kupffer cells to clear pathogens.[44] Jensen et al. have shown that these same lipids given intravenously over 24 hr continuously and at concentrations less than 50% of the nonprotein calories is safe.[58]

The type of fatty acid utilized also seems to be important. The omega-6 polyunsaturated fatty acids (PUFAs), traditionally the fat source in both parenteral and enteral formulations, have been implicated to cause both structural alterations and chemical mediation of the immune system. Structurally, the fatty acid composition of the cell membrane phospholipids are altered. These changes are thought to impair cell division and hormonal signal transduction. Chemical mediation is secondary to the overproduction of the eicosanoids, physiologically active arachidonic acid metabolites, which are classified as prostaglandins (PGs) and leukotrienes (LT). Excess dietary omega-6 PUFA increases the level of cell membrane arachidonic acid, which in turn is utilized to synthesize the 2-series PG and the 4-series LT. They play a significant role in immunosuppression, tumorigenesis, and enhancing inflammation.[56]

MICRONUTRIENT and VITAMIN REQUIREMENTS

Micronutrients

The micronutrients including magnesium, zinc, calcium, phosphorus, copper, and selenium are crucial for maintaining nutritional competence. Many play important roles in enzymatic processes. With an adequate dietary intake and normal metabolism, deficiencies of the micronutrients are unusual. With critical illness, oral nutrient intake is interrupted, metabolism is altered, requirements are greater, and abnormal losses of the micronutrients are not uncommon. Although both parenteral and enteral formulations provide adequate amounts of these substances for most patients, daily intake may be insufficient for the critically ill (Table 27.3). When requirements and losses are greater than intake, deficiencies develop. Therefore, in this setting, serum levels should be monitored closely and supplementation performed accordingly. However, little data exists concerning the appropriate level of supplementation for the critically injured.

Magnesium is an extremely important trace element. Normal serum concentrations seem to be necessary for the sodium-potassium ATPase pump. Deficiencies may lead to cardiac, vascular, neurologic, and electrolyte abnormalities (hypocalcemia, hypokalemia), and an increase in mortality for the critically ill.[59] Magnesium deficiencies have been noted to occur in almost 60% of the critically ill.[60] Because only 1% of total body magnesium is extracellular, deficiencies can occur even in the setting of normal serum magnesium concentrations.[61] Zinc is important for many enzymatic reactions. Deficiencies can impair platelet aggregation, cell chemotaxis, and wound healing.[2] In addition to being crucial for normal metabolism, copper, zinc, and selenium may have important roles as antioxidants.[62]

Vitamins and Antioxidants

Vitamins and antioxidants are also critical for normal metabolic processes and may become deficient as a consequence of critical illness. Most vitamins are added to both parenteral and enteral formulations to meet standard daily requirements (Table 27.4). Vitamin K is an exception and must be given by intramuscular injection. Recently, there has been increased interest in the provision of vitamins E and C as antioxidants.[62] Further research is necessary to prove that the supplementation of the diet with antioxidants will improve outcome.

As with the trace elements, little data exist to guide appropriate supplementa-

Table 27.3. Suggested Daily Mineral Supplementation for the Critically Ill

Mineral	RDA	Recommendation for the Critically Ill Patient
Iron	10 mg	—
Zinc	12–15 mg	50 mg
Copper	2–3 mg	—
Chromium	0.05–0.2 mg	—
Selenium	55–70 ug	100 ug
Manganese	0.15–0.8 mg	25–50 mg
Molybdenum	0.15–0.5 mg	0.2–0.5 mg

Source: Data compiled from multiple sources.

Table 27.4. Suggested Daily Vitamin Supplementation for the Critically Ill

Vitamin	RDA	Recommendation for the Critically Ill Patient
	FAT-SOLUBLE VITAMINS	
A	1000 IU	2500–10,000 IU
D	400 IU	400 IU
E	12–15 IU	400 IU
K	1 μg/kg	1 mg
	WATER-SOLUBLE VITAMINS	
B_1 (thiamine)	1–1.5 mg	10 mg
B_2 (riboflavin)	1.1–1.8 mg	10 mg
B_6 (pyridoxine)	1.6–2.2 mg	20 mg
B_{12}	2–3 μg	20 mg
Niacin	14–20 mg	200 mg
Pantothenic acid	4–10 mg	100 mg
Biotin	80–100 μg	5 mg
Folic acid	180–200 μg	2 mg
C (ascorbic acid)	45–60 mg	1000 mg

Source: Data compiled from multiple sources.

tion in the critically ill. Often, recommendations are presented that represent safe levels of vitamin administration.

It is important also to be mindful of the potential for medications to alter vitamin and mineral requirements. These interactions are reviewed elsewhere in this book.

ROUTE OF ADMINISTRATION

Since the late 1960s, total parenteral nutrition (TPN) has been the gold standard for nutritional support. It can be administered safely and has become an integral part of the care of malnourished or critically injured or septic patients. Although present day parenteral nutrition can substitute nutritionally for the interruption of oral nutrition, gut mucosal atrophy and immune dysfunction leading to translocation is not prevented.[63–65] A stream of luminal nutrients in the gut seems to be essential.

To prevent the invasion of microbes into the systemic circulation, the small intestine relies on both a physical barrier and an immunologic shield. The physical barrier is created by the tight junctions between intact epithelial cells, gastric acid, digestive enzymes, mucus production, intestinal motility and a normal bacterial flora. The intestinal immune system, often referred to as the gut-associated lymphoid tissue (GALT), consists of cells such as lymphocytes and macrophages situated throughout the intestinal wall and the production of the antibody immunoglobulin A (IgA), which is secreted into the lumen and prevents the adherence of microbes to the mucosa, a step necessary for invasion.[66]

The spread of pathogens or their products through the intestinal wall and into the systemic circulation is referred to as *translocation*.[67] Factors thought to promote translocation include luminal bacterial overgrowth, impaired host defense mechanisms, protein-calorie malnutrition, trauma, critical illness, or any process that leads to mucosal atrophy such as interruption of the luminal nutrient stream.[68] Translocation is believed to be associated with and maybe even the underlying cause of the nosocomial sepsis and multisystem organ dysfunction seen with critical illness.[67,69]

Enteral nutrition has recently gained new-found popularity. A large body of literature describes benefits in terms of cost, lower complication rate, and favorable effects on metabolic and immune function. The infusion of nutrients intraluminally has been shown to be important for maintaining gut mucosal integrity, a key factor for organism homeostasis and immunologic competence.[70,71] The provision of nutrients intraluminally has also been shown to attenuate the stress response. The acute-phase reactive protein secretion is blunted and higher levels of synthetic proteins are seen.[72] Lowry demonstrated lower counterregulatory hormone and C-reactive protein levels in subjects fed enterally after exposure to endotoxin versus those fed parenterally.[70] The net effect is a lower rate of catabolism and energy expenditure.

Improvements in outcome have also been reported. Studies including those by Moore and coworkers have demonstrated in trauma patients improved nitrogen balance, lower infection rates, and improved wound healing with early enteral nutrition.[73,74] In similar work, Kudsk et al. demonstrated a significantly lower infection rate in critically ill trauma patients who were provided nutrition enterally.[75] A recent meta-analysis of eight published studies by Moore's group also concluded that high-risk surgical patients fed enterally had reduced septic complications compared with patients given parenteral nutrition.[76]

Other benefits seen with enterally administered nutrition include improved gallbladder contraction leading to a reduction in the likelihood of gallstone formation and acalculous cholecystitis, increased pancreatic stimulation with a reduction in sluggish secretion and functional insufficiency, and improved gut healing after surgical anastomosis.[77]

Most intensivists support the early institution of enteral nutrition for critically ill patients. In general, it is believed that the early initiation of intestinal feeding would attenuate the stress response, decrease the loss of lean body tissue from catabolism, and improve patient outcome. Unfortunately, these theories have never been conclusively proven to be true. Although a number of investigational studies have demonstrated improved metabolic parameters and patient outcome when comparing early enteral nutrition to early parenteral feeding, there are only a few published reports that evaluate the differences between early and delayed enteral nutrition.

Grahm et al. compared early jejunal feeding (within 36 hr of injury) to late gastric feeding in head-injured patients and demonstrated a reduced incidence of bacterial infections and hospital stay in the early fed group.[28] Chiarelli and coworkers demonstrated that early feeding (within hours of injury) led to reduced urinary catecholamines and plasma glucagon levels and improved nitrogen balance.[78] In contrast, Eyer and coworkers evaluated the effects of early enteral feeding (within 24 hr of injury) versus initiating feedings after 72 hr. They found that early feeding did not blunt the stress response or alter patient outcome.[79]

Despite the lack of strong clinical evidence to support early gut feeding, most intensivists would support initiating enteral nutrition with the first 48 hr of injury or surgery in a stable patient, because it is usually well tolerated and may improve outcome.[80] Unfortunately, enterally administered nutrition can be difficult in the critically ill. Volume restriction, gastric ileus, lack of jejunal access, and severe metabolic abnormalities may impair the ability to adequately utilize the gastrointestinal tract. In these circumstances, it is still advantageous to drip even small quantities of nutrients enterally while providing the bulk of the nutritional and metabolic support parenterally (combined approach). With time and improvement in patient status, nutrient provision can be slowly converted from parenteral to entirely enteral.

Currently there are a wide variety of enteral formulas on the market (Table 27.5). These products differ in protein source and form, fat content and source, caloric density, and osmolarity. Carbohydrate source is less important because most commercial formulas are lactose free and many other types of carbohydrate are generally well tolerated. Presently, there is no consensus in the literature regarding the optimal choice of enteral product for the critically ill. As patients and illnesses vary widely, it could be said that the formula of choice is that which the patient best tolerates.

Formulas are usually categorized based on their contents and applications. These include polymeric (encompasses standard, high calorie/high nitrogen, fiber enriched, and disease specific), peptide, special nutrient, and modular. Even in the critically ill, a standard isotonic formula can usually be well-tolerated. The high calorie/high nitrogen formulas require significant digestion and have a higher osmolality but may be useful for the fluid-restricted patient. In most cases, postpyloric administration is necessary, because many critically ill patients manifest gastric dysmotility.[81] The high fat content and hyperosmolality of some formulas may also contribute to gastric retention. Fiber-containing products may

Table 27.5. Selected Enteral Formulas for Critically Ill Patients

Formula	Cal/cc	Prot (g/L)	Fat (%)	mOsm	Vol to meet US RDA
		PEPTIDE FORMULAS			
Vital HN[1]	1.0	42	10	500	1500
Reabilan HN[2]	1.3	58	35	490	1875
Peptamen VHP[3]	1.0	62.4	35	430	1000
		STANDARD FORMULAS			
Osmolyte HN[1]	1.06	44.4	30	300	1321
Promote[1]	1.0	62.5	21	300	1250
Sustacal[4]	1. 01	61	21	650	1060
		HIGH NITOGEN/HIGH CALORIE FORMULAS			
Ensure Plus HN[1]	1.5	63	30	650	947
Two Cal HN[1]	2.0	84	41	690	947
Isocal HCN[4]	2.0	75	46	690	1000
		FIBER ENRICHED FORMULAS			
Jevity[1]	1.06	44	30	310	1321
Compleat[5]	1.07	43	31	300	1500
Ultracal[4]	1.06	44	37	310	1250
		SPECIAL NUTRIENT FORMULAS			
Alitraq[1]	1.0	53	14	575	1500
Impact[5]	1.0	56	25	375	1500
Perative[1]	1.3	67	25	385	1155

This list represents only a sampling of the many commercial products on the market. A number of equally useful formulas not on the list are also available. 1 = Ross, 2 = Elan Pharma, 3 = Clintec, 4 = Mead Johnson, and 5 = Sandoz.

also be introduced in the ICU setting; however, tolerance needs to be established gradually to avoid the associated side effects, including gas production and abdominal bloating.

Disease-specific products are also popular. These are predominantly polymeric formulas in which the macronutrients, free water, and electrolytes have been modified to accommodate the specific requirements of certain disease states such as liver failure, renal insufficiency, intestinal malabsorption, diabetes, and respiratory failure. Renal-failure formulas are useful, because they do not significantly increase the need for hemodialysis and are relatively concentrated. Hepatic-failure formulas contain di- and tripeptides as well as higher concentrations of branch chain amino acids and lower concentrations of aromatic amino acids. Both of these types of formulas are relatively low protein. For the critically ill where higher nitrogen requirements exist, standard, nondisease specific products are usually necessary and very adequate.

Modular enteral products can be used to augment specific macronutrients requirements at a sensible cost. Specific products exist to provide additional protein, carbohydrate, and fat to enteral formulas. The use of modular products greatly expands the flexibility of standard feeds to better support a wide variety of patients.

NUTRICEUTICALS

It has recently become apparent that certain nutrients are indispensable for maintaining the intestinal barrier to microbes and for modulating metabolic processes. The term *nutriceuticals* has been coined to describe the use of nutrients as pharmacologic agents rather than as energy substrates. These nutrients and a review of the literature describing their effects are discussed in detail elsewhere in this book. The use of specially formulated enteral products enriched with one or more of these beneficial nutrients is becoming increasingly utilized for the critically ill. In human studies, enteral formulas enriched with arginine, omega-3 fatty acids, and nucleotides have been shown to improve immunologic parameters, decrease the incidence of infection, and shorten the length of hospitalization in postoperative patients with gastrointestinal malignancies[82,83] and in the critically ill or traumatized.[84,85] The results of these studies and others like them suggests a potential benefit from the use of these specialized formulations to enhance immune response and ultimately improve patient outcome. As this field of nutritional support grows, it may become apparent that the provision of these specialized nutrients may reflect more of a pharmaceutical use than a nutritional.

HORMONAL MANIPULATION

As previously described, adequate nutritional intervention in the critically injured is often inadequate. Recently, there has been great interest in attempting to manipulate the stress response to attenuate the metabolic and catabolic rates so that nitrogen and calorie shortfalls can be minimized. One such mediator, growth hormone, has been shown to increase protein synthesis, spare lean body tissue, and promote fat mobilization even when hypocaloric feeding is administered.[86]

Growth hormone has also been shown to promote wound healing in perioperative, fasted, and tumor-bearing rats and to enhance postoperative immune function.[87] It is hoped that growth hormone can be used to optimize the benefits of adjuvant nutritional support in patients who have the highest requirements yet the least potential to meet those requirements.

CONCLUSION

Critically ill or traumatized patients present a unique array of problems for the clinician. Due to the persistent hypermetabolism, and wasting of catabolic illness, they are at significant risk for developing severe protein-calorie malnutrition, which can further complicate their clinical course. Despite increased nutritional requirements, these patients are often unable to be adequately fed secondary to gastrointestinal dysfunction, fluid restrictions, hyperglycemia, and other limitations such as organ failure. Even when nutritional supplementation is successfully provided, loss of lean body tissue continues. In this population, there is little tolerance for error. Gross overfeeding is no longer acceptable and may be more harmful than the starvation it is meant to remedy. Nutrient supplementation should be early and precise. The gastrointestinal tract should always be considered for nourishment solely or in conjunction with parenteral nutrition. The use of special nutrients may be considered to augment, attenuate, or modify the metabolic and immune responses. The proper use of nutritional support can be central to the ever-improving survival rates in the ICU despite the appearance of ever-more complicated and elderly patients.

REFERENCES

1. Seltzer MH, Slocum BA, Cataldi-Betcher EL, et al. Instant assessment: absolute weight loss and surgical mortality. *J Parent Ent Nutr* 1982;6:218–221.
2. Blackburn GI. Nutrition in surgical patients. In Hardy JD, Kukora JS, Pass HI, eds. *Hardy's Textbook Of Surgery*, 2nd ed. Philadelphia: J.B. Lippincott, 1988:86–104.
3. Hensle TW, Askanazi J. Metabolism and nutrition in the perioperative period. *J Urol* 1988;139:229–239.
4. Cuthbertson DP, Tilstone WJ. Metabolism during the post-injury period. *Adv Clin Chem* 1969;12:1–55.
5. Douglas RG, Shaw JHF. Metabolic response to sepsis and trauma. *Br J Surg* 1989;76:115–122.
6. Bessey PR, Downey RS, Monafo WW. Metabolic response to injury and critical illness. In Civetta JM, Taylor RW, Kirby RR, eds. *Critical Care*, 2nd ed. Philadelphia: J.B. Lippincott, 1992:527–540.

7. Wilmore DW, Aulick LH. Systemic responses to injury and the healing wound. *J Parent Ent Nutr* 1980;4:147–151.
8. Cerra FB, Upson D, Angelico R, et al. Branched chains support postoperative protein synthesis. *Surgery* 1982;92:192–198.
9. Cerra FB. Hypermetabolism, organ failure, and metabolic support. *Surgery* 1987;92:1–14.
10. Rennie MJ, Harrison R. Effects of injury, disease, and malnutrition on protein metabolism in man: unanswered questions. *Lancet* 1984;5:323–325.
11. Alberti KGMM, Batstone GF, Foster KJ, et al. Relative role of various hormones in mediating the metabolic response to injury. *J Parent Ent Nutr* 1980;4:141–146.
12. Kenney PR. Neuroendocrine response to volume loss. *Trauma Q* 1986;2:18–27.
13. Shikora SA, Blackburn GL. Nutritional consequences of major gastrointestinal surgery. Patient outcome and starvation. *Surg Clin North Am* 1991;71:509–521.
14. Shaw JHF. Influence of stress, depletion, and/or malignant disease on the responsiveness of surgical patients to total parenteral nutrition. *Am J Clin Nutr* 1988;48:144–147.
15. Shaw JHF, Wolfe RR. An integrated analysis of glucose, fat, and protein metabolism in severely traumatized patients. Studies in the basal state and the response to total parenteral nutrition. *Ann Surg* 1989;209:63–72.
16. Studley HO. Percentage of weight loss: a basic indicator of surgical risk in patients with chronic peptic ulcer. *JAMA* 1936;166:458–460.
17. Dempsey DT, Mullen JL, Buzby GP. The link between nutritional status and clinical outcome: can nutritional intervention modify it? *Am J Clin Nutr* 1988;47:352–356.
18. Zaloga GP. Frontiers in critical care. *New Horizons* 1994;2:121–122.
19. Hill GL. Body composition research. Implications for the practice of clinical nutrition. *J Parent Ent Nutr* 1992;16:197–218.
20. McMahon MM, Benotti PN, Bistrian BR. A clinical application of exercise physiology and nutritional support for the mechanically ventilated patient. *J Parent Ent Nutr* 1990;14:538–542.
21. Abel RM, Grimes JB, Alonso D, et al. Adverse hemodynamic and ultrastructural changes in dog hearts subjected to protein-calorie malnutrition. *Am Heart J* 1979;97:733–744.
22. Heymsfield SB, Bethel RA, Ansley JD, et al. Cardiac abnormalities in cachetic patients before and during repletion. *Am Heart J.* 1978:95:584–594.
23. Haw MP, Bell SJ, Blackburn GL. Potential of parenteral and enteral nutrition in inflammation and immune dysfunction: a new challenge for dietitians. *J Am Diet Assoc* 1991;91:701–706.
24. Rose AH, Holt PG, Turner KJ. The effect of a low protein diet on the immunogenic activity of murine peritoneal macrophages. *Int Arch Allergy Appl Immunol* 1982;67:356–361.
25. Redmond HP, Leon P, Lieberman MD, et al. Impaired macrophage function in severe protein-energy malnutrition *Arch Surg* 1991;126:192–196.

26. Chandra RK. Interactions of nutrition, infection, and immune response. Immunocompetence in nutritional deficiency, methodological considerations and intervention strategies. *Acta Paediatr Scand* 1979;68:137–144.
27. Charlomagno MA, Alito AE, Rife SU, et al. B-cell immune response during total protein deprivation. *Acta Physiol Latinoam* 1980;30:187–192.
28. Grahm TW, Zadrozny DB, Harrington T. The benefits of early jejunal hyperalimentation in the head-injury patient. *Neurosurgery* 1989;25:729–735.
29. Boosalis MG, Ott L, Levine AS, et al. Relationship of visceral proteins to nutritional status in chronic and acute stress. *Crit Care Med* 1989;17:741–747.
30. Meguid MM, Campos AC, Hammond WG. Nutritional support in surgical practice: part I. *Am Surg* 1990;159:345–358.
31. Agarwal N, Acevedo F, Leighton LS, et al. Predictive ability of various nutritional variables for mortality in elderly people. *Am J Clin Nutr* 1988;48:1173–1178.
32. Seltzer MH, Bastidas JA, Cooper DM, et al. Instant nutritional assessment. *J Parent Ent Nutr* 1979;3:157–159.
33. Reinhardt GF, Mycofsky JW, Wilkens DB, et al. Incidence and mortality of hypoalbuminemic patients in hospitalized veterans. *J Parent Ent Nutr* 1980;4:357–359.
34. Hill GL, Church J. Energy and protein requirements of general surgical patients requiring intravenous nutrition. *Br J Surg* 1984;71:1–19.
35. Harris JA, Benedict FG. *Standard basal metabolism constants for physiologists and clinicians, a biometric study of basal metabolism in man.* Philadelphia: JB Lippincott. 1919:223–250.
36. Quebbeman EJ, Ausman RK, Schneider TC. A re-evaluation of energy expenditure during parenteral nutrition. *Ann Surg* 1982;195:282–285.
37. Weissman C, Kemper M, Askanazi J, et al. Resting metabolic rate of the critically ill patient: measured versus predicted. *Anesthesiology* 1986;64:670–679.
38. Makk LJK, McClave SA, Creech PW, et al. Clinical application of the metabolic cart to the delivery of total parenteral nutrition. *Crit Care Med* 1990;18:1320–1327.
39. Wilmore DW. Catabolic illness. Strategies for enhancing recovery. *N Engl J Med* 1991;325:695–702.
40. Damask MC, Schwarz Y, Weissman C. Energy measurements and requirements of critically ill patients. *Crit Care Clin* 1987;3:71–73.
41. Shanbhogue LKR, Bistrian BR, Jones C, et al. Resting energy expenditure in patients with end-stage liver disease. *J Parent Ent Nutr* 1987;11:305–308.
42. Daly JM, Dudrick SJ, Copeland EM. Intravenous hyperalimentation. Effect on delayed cutaneous hypersensitivity in cancer patients. *Ann Surg* 1980;192:587–592.
43. Shanbhogue LKR, Chwals WJ, Weintraub M. Parenteral nutrition in the surgical patient. *Br J Surg* 1987;74:172–180.
44. Elwyn DH. Nutritional requirements of adult surgical patients. *Crit Care Med* 1980;8:9–20.
45. Wolfe RR, Goodenough RD, Burke JF, et al. Response of protein and urea kinetics to different levels of protein intake. *Ann Surg* 1983;197:163–171.

46. Shaw JHF, Wolfe RR. Whole-body protein kinetics in patients with early and advanced gastrointestinal cancer: the response to glucose infusion and total parenteral nutrition. *Surgery* 1988;103:148–155.
47. Shizgal HM. Nutritional complications in the surgical patient. In Hardy JD, ed. *Complications in surgery and their Management*, 4th ed. Philadelphia: WB Saunders, 1981:245–279.
48. Askanazi J, Rosenbaum SH, Hyman AI, et al. Respiratory changes induced by the large glucose loads of total parenteral nutrition. *JAMA* 1980;243:1444–1447.
49. Baker AL, Rosenberg IH. Hepatic complications of total parenteral nutrition. *Am J Med* 1987;82:489–497.
50. McMahon M, Manji N, Driscoll DF, et al. Parenteral nutrition in patients with diabetes mellitus: theoretical and practical considerations. *J Parent Ent Nutr* 1989;13:545–553.
51. McMurray JF. Wound healing with diabetes mellitus. Better glucose control for better wound healing in diabetes. *Surg Clin North Am* 1984;64:769–778.
52. Bagdade JD, Stewart M, Walters E. Impaired granulocyte adherence. A reversible defect in host defense in patients with poorly controlled diabetes. *Diabetes* 1978;27:677–681.
53. Jones RL, Peterson CM. Hematologic alterations in diabetes mellitus. *Am J Med* 1981; 70:339–352.
54. Hostetter MK. Handicaps to host defense. Effects of hyperglycemia on C3 and *Candida albicans*. *Diabetes* 1990;39:271–275.
55. Wolfe RR, O'Donnell TF, Stone MD, et al. Investigation of factors determining the optimal glucose infusion rate in total parenteral nutrition. *Metabolism* 1980;29:892–900.
56. Wan JM-F, Teo TC, Babayan VK, et al. Invited comment: lipids and the development of immune dysfunction and infection. *J Parent Ent. Nutr* 1988;12(suppl):43s–52s.
57. Alexander JW, Saito H, Trocki O, et al. The importance of lipid type in the diet after burn injury. *Ann Surg* 1986;204:1–8.
58. Jensen GL, Mascioli EA, Seidner DL. Parenteral infusion of long- and medium-chain triglycerides and reticuloendothelial system function in man. *J Parent Ent Nutr* 1990;14:467–471.
59. Rubeiz GJ, Thill-Baharozian M, Hardie D, et al. Association of hypomagnesemia and mortality in acutely ill medical patients. *Crit Care Med* 1993;21:203–209.
60. Chernow B, Bamberger S, Stoiko M, et al. Hypomagnesemia in post-operative intensive care patients. *Chest* 1989;95:391–397.
61. Olerich MA, Rude RK. Should we supplement magnesium in critically ill patients? *New Horizons* 1994;2:186–192.
62. Grimble RF. Nutritional antioxidants and the modulation of inflammation: theory and practice. *New Horizons* 1994;2:175–185.
63. O'Dwyer ST, Smith RJ, Hwang TL, et al. Maintenance of small bowel mucosa with glutamine-enriched parenteral nutrition. *J Parent Ent Nutr* 1989;13:579–585.

64. Grant JP, Snyder BA. Use of L-glutamine in total parenteral nutrition. *J Surg Res* 1988;44:506–513.
65. Wilmore DW, Smith RJ, O'Dwyer ST, et al. The gut: a central organ after surgical stress. *Surgery* 1988;104:917–923.
66. Alverdy JC. Effects of glutamine-supplemented diets on immunology of the gut. *J Parent Ent Nutr* 1990;14(suppl):109S–113S.
67. Deitch EA. Multiple organ failure. Pathophysiology and potential future therapy. *Ann. Surg* 1992;216:117–134.
68. Johnson LR, Copeland EM, Dudrick JT, et al. Structural and hormonal alterations in the gastrointestinal tract of parenterally fed rats. *Gastroenterology* 1975;68:1177–1183.
69. Carrico J, Meakin JL. Multiple organ failure syndrome. *Arch Surg* 1986;121:196–208.
70. Lowry SF. The route of feeding influences injury responses. *J Trauma* 1990;30:S10–S15.
71. Alverdy JC. The GI tract as an immunologic organ. *Contemp Surg* 1989;35(suppl):14–19.
72. Petersen SR, Sheldon GF, Carpenter G. Failure of hyperalimentation to enhance survival in malnourished rats with *E. coli*-hemoglobin adjuvant peritonitis. *Surg Forum* 1979;30:60–61.
73. Moore FA, Moore EE, Jones TN, et al. TEN versus TPN following major abdominal trauma-reduced septic morbidity. *J Trauma* 1989;29:916–923.
74. Moore EE, Jones TN. Benefits of immediate jejunostomy feeding after major abdominal trauma-A prospective, randomized study. *J Trauma* 1986;26:874–881.
75. Kudsk KA, Croce MA, Fabian TC, et al. Enteral versus parenteral feeding. Effects on septic morbidity after blunt and penetrating abdominal trauma. *Ann Surg* 1992;215:503–513.
76. Moore FA, Feliciano DV, Andrassy RJ, et al. Early enteral feeding, compared with parenteral, reduces postoperative septic complications. The results of a meta-analysis. *Ann Surg* 1992;216:172–183.
77. McClave SA, Lowen CC, Snider HL. Immunonutrition and enteral hyperalimentation of critically ill patients. *Digest Dis Sci* 1992;37:1153–1161.
78. Chiarelli A, Enzi G, Casadei A, et al. Very early nutrition supplementation in burned patients. *Am J Clin Nutr* 1990;51:1035–1039.
79. Eyer SD, Micon LT, Konstantinides FN, et al. Early enteral feeding does not attenuate metabolic response after blunt injury. *J Trauma* 1993;34:639–644.
80. Minard G, Kudsk KA. Is early feeding beneficial? How early is early? *New Horizons* 1994;2:156–163.
81. Nompleggi D, Teo TC, Blackburn GL, et al. Human recombinant interleukin-1 decreases gastric emptying in the rat. *Gastroenterology* 1988;94:A326.
82. Daly JM, Lieberman MD, Goldfine J, et al. Enteral nutrition with supplemental argnine, RNA, and omega-3 fatty acids in patients after operation: Immunologic, metabolic, and clinical outcome. *Surgery* 1992;112;56–67.

83. Kemen M, Senkal M, Homann HH, et al. Early postoperative enteral nutrition with arginine-omega-3 fatty acids and ribonucleic acid-supplemented diet versus placebo in cancer patients: an immunologic evaluation of Impact®. *Crit Care Med* 1995;23;652–659.

84. Moore FA, Moore EE, Kudsk KA, et al. Clinical benefits of an immune-enhancing diet for early postinjury enteral feeding. *J Trauma* 1994;37;607–615.

85. Bower RH, Cerra FB, Bershadsky B, et al. Early enteral administration of a formula (Impact®) supplemented with arginine, nucleotides, and fish oil in intensive care unit patients: results of a multicenter, prospective, randomized, clinical trial. *Crit Care Med* 1995;23;436–449.

86. Jiang ZM, He GZ, Zhang SY, et al. Low-dose growth hormone and hypocaloric nutrition attenuate the protein-catabolic response after major operation. *Ann Surg* 1989;210:513–524.

87. Chwals WJ, Bistrian BR. Role of exogenous growth hormone and insulin-like growth factor I in malnutrition and acute metabolic stress: a hypothesis. *J Parent Ent Nutr* 1991;19:1317–1327.

CHAPTER

28

Nutrition in Trauma and Burns

David W. Voigt, M.D., John Fitzpatrick, M.D., and Basil A. Pruitt, Jr., M.D.

INTRODUCTION

Injury evokes an increase in metabolic rate that is proportional to the severity of the injury. The greatest increases in metabolic rate occur in patients with extensive burns (i.e., burns involving 50% or more of the body surface). The hypermetabolism caused by injury induces a catabolic state characterized by erosion of lean body mass (proteolysis), negative nitrogen balance, altered glucose metabolism, and alteration of substrate utilization as manifested by a respiratory quotient (RQ) near 0.7.[1] Because the hypermetabolism is wound directed, it is essential to provide the nutrients that will satisfy both energy and substrate needs to prevent autocannibalism, maintain muscle strength, preserve organ function, reduce the risk of infection, hasten wound healing, accelerate convalescence, and optimize survival.

The metabolic rate of injured patients changes in a biphasic manner.[2] The "ebb" phase occurs prior to and during the early part of the resuscitation phase. During this phase, oxygen consumption is depressed to a variable degree. After resuscitation, the metabolic rate of the patient steadily rises, peaking in burn patients between 6 and 10 days postinjury. This is the hypermetabolic, catabolic, or flow phase of injury. The metabolic rate slowly returns toward basal levels, but does not reach preinjury levels until well into convalescence. In the burn patient, this hypermetabolism is not simply a stress response, because growth hormone is not elevated despite increased plasma levels of cortisol and catecholamines. In fact, there appears to be a blunted response of growth hormone to insulin, hypoglycemia, arginine infusion, and nocturnal sleep. Resting metabolic rate after injury appears to be independent of thyroid hormone levels and to be under the influence of catecholamines, cortisol, and glucagon.[3] Catecholamines and cortisol are associated with elevated gluconeogenesis as well as protein and fat breakdown. This accounts for the increased glucose production and utilization documented in trauma patients. A trend toward hyperglycemia accompanies the stress of trauma. This is accounted for by a resistance to the action of insulin in the peripheral tissues.[4]

The metabolic response to trauma is not homogeneous throughout the body. In fact it is not even homogeneous within an injured limb. A severe wound causes an increase in the total body oxygen consumption and cardiac output to support enhanced delivery of blood and nutrients, particularly glucose, to the wound. The "total body response" to injury is wound directed such that in a burned limb the skin or granulation tissue, not the underlying muscle receives the increased blood flow.[5–7] This provides the glucose that is obligatorily utilized by the healing tissues. One might conclude that by closing the burn wound the metabolic rate would be decreased. In a sheep model of burn injury, if the entire burn was excised to fascia and closed with full-thickness skin from other sheep the metabolic rate did return to normal.[8] In humans, however, excision and grafting of burn wounds does not appear to decrease the metabolic rate.[9–10]

When patients with mechanical trauma, and/or burn trauma become septic, their metabolic rate increases to an even greater extent.[11] It is interesting that recovering trauma patients who develop sepsis have reversion of hormonal status to the early postinjury pattern. To prevent this most severe hypermetabolism, prevention as well as prompt diagnosis and treatment of infection are important components of the nutritional care of these patients. This includes placing patients with open wounds involving more than 20% total body surface area (TBSA) burn in a single patient room where all entering the room wear surgical caps, shoe covers, masks, sterile gloves, and clean gowns, changing central venous catheters every 3 days, and monitoring of scheduled surveillance cultures.

The hypermetabolism in the injured patient leads to wasting of the body's protein stores. If exogenous calories are not supplied to the injured patient, protein break down provides 15% to 20% of the caloric expenditure.[12] This accounts for the obligatory protein catabolism seen in trauma patents who receive inadequate nutrition. The goal of nutrition support should include minimizing protein losses and supporting the increased metabolic demands of healing wounds. Burn patients have protein losses in addition to those associated with catabolic proteolysis.

Burn patients leach protein across the burn wound with this source of loss being greatest during the first 3 days postburn. Waxman et al.[13] found that daily protein loss from the burn can be estimated by:

$$\text{WPL} = 0.3 \times \text{BSA} \times \% \text{ burn (for postburn days 1–3)}$$
$$\text{WPL} = 0.1 \times \text{BSA} \times \% \text{ burn (for postburn days 4–16)}$$

where WPL is wound protein loss in grams nitrogen per day and BSA is body surface area in meters squared.

Nutritional support must provide the amino acids and the energy needed for protein synthesis. When properly administered, this results in a positive nitrogen balance without the complications of overfeeding. Nutrition support, however, does not stop the catabolic process induced by the neurohumorally mediated metabolic stress response accompanying burns, surgery, or trauma.[14] Maintenance of lean body mass facilitates rehabilitation and may help prevent certain complications, such as superior mesenteric artery (SMA) syndrome, which may occur in very malnourished patients.

Several studies have been performed to examine the effect of nutrition support on morbidity and mortality.[15–16] These studies demonstrated that nutrition support significantly reduced morbidity and mortality only in patients who were severely malnourished prior to the performance of major surgical procedures. Maintenance of lean body mass by appropriate nutrition support has been shown to be effective in reducing hospital costs by decreasing the length of stay.[17]

Attempts to decrease or reverse the catabolic response to trauma by using drug or hormone therapies have been reported. These efforts have included growth hormone, insulin-like growth factor-1 (IGF-1), insulin, beta-blockade, beta-agonist, and narcotics. None of these therapies have shown enough success to be brought into general use for all trauma patients. For example, studies using human growth hormone have shown mixed results. Growth hormone has been shown to stimulate fat mobilization and increase the proportion of fat used as fuel.[18] It has also been demonstrated that, with adequate nutrient loading, human growth hormone helps to preserve protoplasmic mass and improve nitrogen retention in the catabolic phase. This was true as long as an augmented insulin production was present.[19] Growth hormone's increased lipolysis, insulin antagonism, and failure in some patients to reverse catabolism make it less than the ideal agent with which to counter postinjury catabolism. Growth hormone's failure to reverse catabolism in some patients may be explained by a reduced effectiveness of its ability to stimulate the release of IGF-1 in those patients.[20] The short-term anabolic effects of recombinant IGF-1 have been studied in eight burn patients. Given as an infusion of 20 μm/kg$\cdot$hr^{-1} for three days, it was found to inhibit lysine oxidation, decrease protein breakdown despite the inhibition of insulin secretion, and cause no increase in glucose disposal. However, IGF-1 did not reverse the catabolic state as indicated by the fact that all patients remained in negative nitrogen balance.[21] Therefore, it has been suggested that IGF-1 be administered simultaneously with growth hormone to take advantage of the combined anabolic effects.[22] This awaits prospective trials to determine if this would be effective.

Insulin infusion in combination with infusion of sufficient glucose to maintain euglycemia has also been used to modulate the catabolic response to injury.[23] The protein-sparing effect of such treatment appears to be due, at least in part, to facilitated uptake of branched-chain amino acids.[24] The catabolic response may be ameliorated by providing the trauma patient with branched-chain amino acids in conjunction with insulin and glucose.

Both beta-blockers and beta-agonists have been touted as exerting beneficial effects in the hypermetabolic patient. Infusion of a nonselective beta-blocker (propranolol) is associated with a decrease in lipid oxidation but not in carbohydrate or protein oxidation.[25] The decrease in lipid oxidation is a result of a relative stimulation of the beta-2-receptors since a selective beta-1-blocker (metoprolol) did not reduce lipolysis. Despite a decrease in lipolysis, propranolol does not significantly affect total body protein catabolism.[26] Conversely, beta-agonists decrease protein catabolism in animal models.[27–28] Unfortunately, in humans treated with such agents the protein sparing is not certain, and in fact there may be an increase in muscle breakdown.[29] In the animal model, the protein is primarily spared in slow-twitch muscle fibers and the decreased lipolysis is primarily involving brown fat. Humans have no clear distinction between fast-twitch and slow-twitch fibers and we have little if any brown fat. This could account for the discrepancy between the animal and the human studies. Therefore, the clinical usefulness of beta-agonists remains undefined.

Oxygen consumption was shown to be decreased in critically ill patients under controlled ventilation who received 0.5 mg/kg of morphine. The decreased oxygen consumption appeared to be due to decreased oxygen demand resulting from the sedation and analgesia provided by this dose of morphine.[30] However, in patients with severe head trauma, the use of narcotics has not significantly affected their metabolic rate.[31] Although experimentally there appears to be a decrease in oxygen demand with high doses of morphine, this has not been shown to equate to sparing of lean body mass. Moreover, the required dosage level of narcotic would necessitate controlled ventilation.

ASSESSMENT OF NUTRITIONAL STATUS

When initiating any form of nutrition support, one must have a basic understanding of the patient's baseline nutritional status. One must remember that trauma patients and burn victims may have preexisting nutritional deficits. The elderly, very young, handicapped, economically disadvantaged, chemically or alcohol dependent, and mentally disturbed are overrepresented in the population of injured patients, particularly among burn patients. It is important to identify preexisting nutritional deficiencies such as the thiamine deficiency often present in alcoholics. That possibility prompts us to provide patients suspected of alcohol addiction with multivitamins including thiamine and folate before giving intravenous glucose-containing solutions such as D_5W used during the second 24-hr period of a burn patient's resuscitation. Providing thiamine before giving intravenous glucose avoids precipitating Wernicke–Korsakoff disease. Trace mineral deficiencies, when present, can usually be corrected by the trace minerals provided in the nutritional formulations commercially available. If mineral or vitamin deficiencies are so severe as to cause skin changes, glucose intolerance, anemia, or poor wound healing, additional supplementation may be needed.

Laboratory measurements of nutritional status are not useful in the trauma patient because of injury-induced changes. Also because of edema and tissue damage, anthropometric measurements may not be helpful. A detailed nutritional history, if obtainable, is probably the best way to assess the patients nutritional status. A history of recent weight change is especially important.[32] The admission height and weight may aid in determining premorbid nutritional status. They are also important for calculating the patient's nutritional needs, as discussed later.

ROUTES OF NUTRITION SUPPORT

Three practical routes exist for providing nutrition support, each with certain advantages and disadvantages. Enteral nutrition is the least expensive and most physiologic way to supply nutrients to the trauma patient. Although diarrhea sometimes complicates enteral feeding, glucose intolerance and septic complications common to parenteral nutrition are almost entirely avoided. Intact protein and peptides stimulate hormonal responses favorable to nitrogen utilization, wound repair, and the support of liver function.[33–34] Complex carbohydrates are the principal source of short-chain fatty acids (the preferential fuel of the colon) and the source of fiber, which is important for stimulating gut motility. Complex carbohydrates are only available for enteral administration. Parenteral nutrition is limited to providing amino acids and simple sugars as protein and carbohydrate sources. Central (total) parenteral nutrition is capable of providing all nutritional requirements and is an acceptable alternative when injury temporarily prevents the utilization of the gastrointestinal tract for nutritional support. Peripheral parenteral nutrition is typically incapable of providing sufficient nutritional support to the hypermetabolic trauma patient because of the limited concentrations of carbohydrate and amino acids that can be tolerated by peripheral veins without causing sclerosis.

Numerous modes of access are available for each type of nutrition support and complications of access are site-specific. Enteral nutrition can be provided per os in the alert patient who possesses an intact gag reflex and is able to cooperate, by nasojejunal or nasogastric tubes or by jejunostomy or gastrostomy tubes placed at the time of laparotomy or percutaneously. Jejunostomy and gastrostomy tubes placed by laparotomy carry the risks of a surgical procedure (anesthesia, hemorrhage, infection, improper placement of the tube, and site leakage), whereas those that are placed percutaneously carry the additional risks of separation of the gastric wall from the anterior abdominal wall, abdominal wall necrosis, and inadvertent organ perforation. Although nutrients can be delivered into the stomach, feedings must be stopped for operative procedures and are dependent on gastric emptying for success. The delivery of nutrients into the distal duodenum, or preferably the proximal jejunum, is sufficiently distal that aspiration is less likely than with gastric feedings. Consequently feeding into the small bowel, which is unaffected by gastric emptying, can be continued perioperatively. Total parenteral nutrition may be provided by percutaneous placement of an intravenous catheter into the central venous circulation via the internal jugular, external jugular, supraclavicular subclavian, infraclavicular subclavian, or femoral vein routes. Soft indwelling catheters may also be placed into the deep facial vein or any of the above veins operatively. Percutaneous indwelling central catheter (PICC) lines also may be threaded into the central circulation from a peripheral venous site such as the basilic vein, but such access is seldom used in a trauma patient. Risks associated with central venous catheters include catheter sepsis, thrombosis, pneumothorax, inadvertent vessel perforation, improper catheter placement, hemothorax, and arrhythmia. To minimize the catheter sepsis rate in burn patients, the catheters should not be changed over a wire, and the catheter site should be changed every 3 days. Both leaving catheters in place longer than 3 days and changing catheters over a wire may be better tolerated by the nonburn trauma patient, but we continue to change catheters to a new site every 3 days in our nonburn trauma patients.

CALCULATION OF NEEDS FOR NONBURN TRAUMA

Numerous formulas exist to estimate the nutritional needs of various types of trauma patients. Due to the nonhomogeneous nature of mechanical trauma and the variability of the evoked response, the formulas that are usually used to calculate either the basal or resting metabolic rate of an uninjured person of that size and age include the addition of, or multiplication by, a stress factor to determine the total predicted needs of the patient. At the United States Army Institute of Surgical Research (USAISR), the patient's actual weight (premorbid weight if known or admission weight if knowledge of premorbid weight is unavailable) is used in formulas to calculate their nutritional needs, provided the patient weighs less than 125% of desired body weight (DBW).

$$\text{DBW for males} = 106 \text{ lb} + [6 \text{ lb} \times (\text{number of inches of height above 5 ft})]$$
$$\text{DBW for females} = 100 \text{ lb} + [5 \text{ lb} \times (\text{number of inches of height above 5 ft})]$$

If a patient is identified as exceeding 125% of their desired body weight, their adjusted desired body weight (ADBW) is used in place of actual body weight (ABW) in the formulas to calculate their nutritional needs.

$$\text{ADBW} = [(\text{ABW} - \text{DBW}) \times 0.25] + \text{DBW}$$

It is tempting to minimize the calories provided to obese trauma patients. Obese patients; however, respond to trauma by mobilizing relatively more protein and less fat compared with normal subjects. To spare protein, this is best managed by providing enough carbohydrate calories based on the above weight adjustments and the following calculations.[35] Basal energy expenditure (BEE) are then usually calculated using the Harris–Benedict equations.[36]

$$\text{BEE female} = 655.1 + 9.6\text{W} + 1.9\text{H} - 4.7\text{A}$$
$$\text{BEE male} = 66.5 + 13.8\text{W} + 5\text{H} - 6.8\text{A}$$
$$\text{BEE infant} = 22.1 + 31.1\text{W} + 1.2\text{H}$$

where W is weight (kg), H is height (cm), and A is age (years). The estimate of resting energy expenditure (REE) is based on oxygen consumption ($\dot{V}O_2$)

$$\dot{V}O_2 \text{ (mL/min)} = \text{CO (L/min)} \times [CaO_2 \text{ (mL/L)} - CvO_2 \text{ (mL/L)}]$$
$$\text{MR (kcal/hr)} = \dot{V}O_2 \times 60 \text{ (min/hr)} \times 0.00483 \text{ kcal/mL}$$

where CO is cardiac output, CaO_2 is arterial oxygen content, CvO_2 is mixed venous oxygen content, and MR is metabolic rate. Final estimates of need are made by multiplying the basal rate by Long's stress and activity factors.[37]

$$\text{Total energy utilization} = \text{BEE} \times \text{activity factor} \times \text{stress factor.}$$

where the activity factors are

$$\text{Bed rest} = 1.2$$
$$\text{Out of bed} = 1.3$$

and the stress factors are

Elective surgery = 1.2
Skeletal trauma = 1.3
Sepsis = 1.6
Burn > 45% total body surface area = 2.1

These estimates are often inaccurate in the severely injured and critically ill, and a better estimate of need can be made by measuring oxygen consumption and calculating the resting energy expenditure by indirect calorimetry using a metabolic cart or a thermodilution pulmonary artery catheter. Indirect calorimetry has the advantage of measuring the CO_2 production which permits calculation of the respiratory quotient (RQ) and provides an estimate of the utilization of the calories provided. RQ's over 1.0 indicate that lipogenesis is occurring and that excessive carbohydrate calories are being administered.

CALCULATION OF NEEDS FOR BURNS

Burns are a unique injury in that they can be easily graded in severity in terms of the TBSA. This characteristic lends itself to accurate estimation of nutritional needs and several formulas exist for the calculation of those needs (Table 28.1). All of these formulas are estimates, not exact calculations, and nutritional support

Table 28.1. Common Nutritional Formulas[a]

Formula	Age	kcal/day
ISR[b]	any	$\{BMR\times[0.89142+(0.01335\times TBSA)]\}\times M^2\times 24\times AF$
		ISR Male BMR = $54.337821-1.19961\times A+0.02548\times A^2-0.00018\times A^3$
		ISR Female BMR = $54.74942-1.54884\times A+0.03586\times A^2-0.00026\times A^3$
Curreri	0–1	$BMR+15\times\%TBSA$
	1–3	$BMR+25\times TBSA$
	4–15	$BMR+40\times TBSA$
	16–59	$25\times W+40\times TBSA$
	60+	$20\times W+65\times TBSA$
Galveston I	1–13	$1800/m^2+1300/m^2$ burned
Galveston II	14–20	$1800/m^2+1500/m^2$ burned
Galveston III	<1, >25% TBSA	$2100/m^2+1000/m^2$ burned

[a]BMR is basal metabolic rate, TBSA is percent total body surface area burned, W is weight (kg), A is age (years), m^2 is total body surface area (square meter), AF is activity factor (we use 1.25)

[b]The current ISR formula shows a linear relationship to burn size where the previous formula had a curvilinear relationship. The older formula would overestimate energy requirements of patients treated by today's method.[38]

should be monitored by indirect calorimetry and nitrogen balance studies to tailor nutritional support to the patient's individual needs. The formulas for adult patients are designed to maintain body weight and thus are inappropriate for children; the children's formulas designed for all age groups include sufficient calories to allow for growth as well.

FORMULAS AND SPECIFIC NUTRIENTS

After the energy requirements have been determined, the requirement for specific nutrients must be determined. These requirements include protein, carbohydrate, and fat. The optimal protein requirement for the catabolic patient has not been determined. Recommendations for the trauma patient are for 1.5 to 4 g of protein/kg body weight administered with nonprotein calorie–nitrogen ratios of from 80:1 to 150:1. Burn patients should be provided 1.5 to 3 g of protein/kg of body weight daily with a calorie–nitrogen ratios of from 100:1 to 150:1. Protein supplements can be added to the commercially available formulas to increase the protein content if necessary.

Glucose is important in trauma patients, both to avoid excessive protein catabolism and to provide the preferred wound metabolite. Despite glucose's importance for wound healing, injured patients cannot utilize more than 5 to 7 $mg/kg \cdot min^{-1}$ of carbohydrate. Carbohydrate administered in excess of this rate should be avoided because it increases carbon dioxide production and results in net fat synthesis.

Fat is an essential nutrient used to provide nonprotein calories, prevent essential fatty acid deficiency, and deliver fat-soluble vitamins. Prevention of essential fatty acid deficiency can be accomplished by providing 4% of the daily energy requirements as linoleic acid.[39] The optimal type of fat to provide nutritional supplementation is still under investigation. Because of ease of absorption from the gastrointestinal tract, medium-chain triglycerides are utilized in many of the commercially available enteral formulas. Long-chain triglycerides are the lipid source used in parenteral formulations in the United States. The amount of fat to provide to parenterally nourished patients is still a matter of debate. Recommendations range from 10% to 30%.[40–42] Fat is less efficient in sparing nitrogen than carbohydrate and may have deleterious effects.[41–43] Because fat is less efficient and many have deleterious effects and glucose is the fuel required by the wound, we provide the nonprotein calories as dextrose. We recognize the need to provide essential fatty acids to parenterally nourished patients and give them 500 mL of a 20% fat emulsion intravenously twice a week.

The requirements for carbohydrate, protein, and fat can usually be satisfied with the commercially available enteral formulas and the additional enteral protein supplements can be administered along with the commercially available enteral formulas if necessary. If parenteral nutrition is required, the nonprotein calories are provided as dextrose in solutions that contain up to 35% dextrose and keep osmolarity to less than 2000 mOsmol/L. Parenterally, we provide protein needs with a commercially available amino acid formula that delivers all the essential amino acids as well as alanine, arginine, L-aspartic acid, L-glutamic acid, histidine, proline, serine, *N*-acetyl-L-tryosine, and glycine. As discussed above we give intravenous fat twice a week if the patient requires parenteral nutrition.

VITAMINS AND MICRONUTRIENTS

Appropriate supplementation of vitamins and micronutrients is essential to permit efficient utilization of protein and energy provided by enteral or parenteral nutrition. Factors that have been shown to be important for wound healing, energy metabolism, and protein synthesis are vitamin A, vitamin C, and zinc.

Vitamin A, which is an important cofactor in wound healing, promotes epithelial integrity and glycoprotein synthesis. Vitamin C is also important in wound healing (collagen synthesis) and is a cofactor in adenosine triphosphate (ATP) production. Zinc is a cofactor in many different enzyme systems and is particularly important for protein synthesis, collagen formation, and cell division.

The clinical importance of other trace metals such as chromium, copper, and selenium remains undefined. Some of their uses have been identified. Chromium promotes glucose tolerance, especially in diabetics. Copper is an important cofactor for cytochrome-c oxidase (ATP production) and collagen cross-linking. Selenium is essential for the function of glutathione peroxidase, which reduces intracellular hydroperoxidases involved in free-radical scavenging.

Current recommendations for vitamin supplementation in patients with extensive burns are listed below.[44]

0–3 YEARS	4–ADULT
5000 IU vitamin A qd	10,000 IU vitamin A qd
250 mg vitamin C bid	500 mg vitamin C bid
1 children's multivitamin qd	1 multivitamin qd

SPECIAL FORMULATIONS

Special formulations have been proposed for patients experiencing severe physiologic stress. Most of these preparations are fortified with branched-chain amino acids (BCAA). Brennan et al. reviewed the cumulative data on the use of BCAA. They concluded that despite positive effects on nitrogen-metabolism the use of

BCAA-enriched solutions did not affect outcome.[45] Other special formulas fortified with specific amino acids such as arginine and glutamine have also been evaluated. It has been reported that nutritional regimens containing supplemental arginine improved nitrogen balance and decreased the incidence of infection in trauma patients.[46]

Glutamine represents 60% of the pool of free intracellular amino acids, but this is reduced by 50% during catabolic stress. This depletion of glutamine appears resistant to replacement by conventional nutritional support. Glutamine-supplemented diets may have an impact in specific clinical settings.[47] Studies using glutamine-enhanced parenteral formulas show a more positive nitrogen balance in the noninjured catabolic patient. However, glutamine-enhanced parenteral formulas are not able to influence the intracellular concentrations of glutamine following severe injury.[48] Patients requiring parenteral nutrition who suffer from sepsis or multisystem organ failure states may benefit from glutamine and arginine supplementation.[49] Enterally nourished patients have not derived a significant benefit from supplementation with arginine and glutamine. The final recommendations on the use of special formulations to modulate the metabolic demands of physiologic stress await further clinical trials.

Occasionally, acute renal failure accompanies severe trauma or illness. Acute renal failure increases the requirement for energy and protein.[50] Fulfilling these increased requirements appears to be more important then the substrate used for providing the nutritional support.[51] Patients requiring dialysis can be provided commercially available standard formulations based upon their metabolic demands. Consideration should be given to using one of the commercially available essential amino acid formulations for patients in acute renal failure who do not require dialysis. This may decrease the total protein load and minimize uremia in this subgroup of renal-failure patients.

The traumatized patient can manifest hepatic dysfunction due to preexisting cirrhosis, complications of their injury, or sepsis. Hepatic dysfunction is associated with an increase in catabolism. It can be challenging to provide the volume of nutrients required to meet metabolic needs, because these patients may develop total body water and sodium overload. Special commercially available formulations fortified with BCAA may be beneficial in decreasing blood urea nitrogen production, hepatic encephalopathy, and possibly mortality in these patients.[52–53]

INITIATION OF NUTRITION SUPPORT

Nutrition support other than oral intake is indicated in patients with burns greater than 30% TBSA, undergoing multiple operations over a relatively short time, having a continued need for ventilatory support, having a mental status that is incompatible with oral nutrition, or who are unable to take oral nutrition because of facial or abdominal injury.[54] The ileus associated with burns or trauma has a minimal effect on the small intestine; therefore, by delivering nutrition directly into the small intestine, full nutritional support usually can be provided without resorting to TPN.

There is a consensus that it is desirable to use the intestine as the route of nutrition. Studies indicate that immediate enteral nutrition provided through a nasogastric tube is safe and effective in burn patients.[55–57] At the USAISR, enteral nutrition is begun within the first 72 hr of hospitalization. This is done with a nasoenteric tube positioned such that the tip is beyond the ligament of Treitz. This is usually accomplished with the aid of fluoroscopy. The patients are taken to the fluoroscopy suite for this procedure; therefore, they must be stable enough to be out of the burn intensive care unit. Depending on the TBSA burn and the patients comorbid illnesses, this can safely be done between 24 and 72 hr postburn. If placement under fluoroscopy is unsuccessful then an endoscope is used to aid in placing the tip of the nasoenteric tube beyond the ligament of Treitz. By placing the tip of the tube beyond the ligament of Treitz, the patient can be safely provided nutrition continuously, even while undergoing operative interventions.[58–59] A nasogastric tube is also placed and is used to decompress the stomach, infuse antacids for stress ulcer prophylaxis, and monitor for reflux of enteral formula from the small intestine.

ADVANCING THE ENTERAL DIET

Patients should be started on half strength formula at a rate of 25 to 50 mL/hr. The rate of delivery of the formula is increased by 25 mL every 8 hr until the target rate determined by one of the formulas discussed earlier is obtained. Eight hours after achieving the target rate, the concentration of the formula is increased to three-quarters strength, and if that is tolerated to full strength 8 hr later. If the patient becomes intolerant of the increases, as demonstrated by presence of glucose in the nasogastric aspirate, abdominal distention, or diarrhea, the infusion rate should be reduced to a tolerated level and thereafter advanced more slowly to allow for the intestine to adapt to the formula being provided. It is rarely necessary to exceed an infusion rate of 150 mL/hr, and patients usually do not tolerate higher rates.

COMPLICATIONS OF ENTERAL NUTRITION

The most common complication of enteral nutrition is diarrhea. This can be minimized by using lactose-free formula, isotonic solutions, and avoiding bolus feeding by delivering the formula continuously over 24 hr. In 20% of the cases, the cause of diarrhea in the enterally nourished patient is the enteral nutrition.[60] When this occurs during initiation of enteral nutrition, it is attributable to advancing the feeding at too rapid a rate. This can be easily treated by slowing the rate of advancement. Medications are another likely cause of diarrhea in the enterally nourished patient. Some of the most common of these are magnesium-containing antacids that produce an osmotic diarrhea, sorbitol containing elixirs that produce an osmotic diarrhea, and antibiotics that can change the microflora of the colon. There are many other drugs that can cause diarrhea, and the medication list of any patient having diarrhea should be reviewed and any offending agents removed. Other causes of diarrhea are less common but must be considered in the differential diagnosis. One of these less common causes is fecal impaction, which can be easily diagnosed by a simple rectal exam and treated by disimpaction.

Infectious colitis must also be kept in the differential diagnosis of diarrhea in the critically ill trauma patient. This is diagnosed by culture of the diarrhea stool or in the case of Clostridium difficle by detecting *C. difficle* toxin in the stool or by visualizing psuedomembranes on sigmoidoscopy. Infectious colitis is treated by antibiotics directed at the offending organism. Hypoalbuminemia is another rare cause of diarrhea that can occur when the serum albumin level is less than 2.6 g/dL.[61] This can be treated by utilizing peptide-based enteral formulas, TPN, or infusion of salt poor albumin. If no cause can be identified, changing to an enteral formula containing fiber polysaccharide may resolve the diarrhea. If the addition of fiber is unsuccessful in controlling diarrhea, consideration can be given to using an antidiarrheal agent if iatrogenic and infectious causes have been ruled out.

The following approach to the critically ill patient who develops diarrhea may be helpful. If the diarrhea occurs during the advancement of the enteral feedings, the rate of advancement is slowed to the last tolerated rate and then advanced more slowly. If the diarrhea occurs after the patient has already been tolerating enteral nutrition then the patient's medications should be reviewed for possible offending agents and a stool specimen should be sent for culture and *C. difficle* toxin. If the diarrhea persists after removing medications that could possibly be causing the diarrhea, the stool studies fail to show an infectious etiology, and the sigmoidoscopy fails to show pseudomembranes, then the patient could be started on a formula that will provide more fiber polysaccharides. If that is unsuccessful and several stool cultures are negative for infection, then an antidiarrheal agent should be administered.

Other complications of enteral nutrition include tube dislodgment, tube obstruction, aspiration pneumonia, hyperkalemia, azotemia, hypokalemia, nasopharyngeal irritation, sinusitis, mucosal erosions, bleeding, cramping, distention, vomiting, hyperglycemia, tube erosion through the wall of a hollow viscus, microbiological contamination of enteral diets, and inadvertant placement of the feeding tube in the bronchial tree (See Table 28.2).

Table 28.2. Complications of Enteral Nutrition

Complication	Prevention/Treatment
Diarrhea	See text
Tube dislodgement/malplacement	Verify placement of tubes by physical exam on insertion and daily thereafter.
	Verify placement of tubes by Roentgenogram on placement, twice a week thereafter, and after all major transports.
Tube obstruction	Avoid placing medications through tubes.
	Flush tubes frequently.
	Neutralize acidic gastric contents with H_2 blockers and antacids.
	Obstructed tubes can often be cleared by flushing ginger ale or cola through the tube with a 3-cc syringe or failing that a J-wire can be used to unclog the tube.
Aspiration pneumonia	Protect the airway with a cuffed endotracheal tube.
	Elevate the head of bed.
	Place tube tip past the ligament of Treitz.
	Decompress the stomach with a nasogastric tube.
Nasopharyngeal irritation	Apply topical anesthetics.
Sinusitis	Apply nasal decongestants and saline nose drops.
	If suppurative, treat by drainage and antibiotics directed by the culture results of the fluid drained.
Bacterial contamination of formula	Attention to detail in preparation and handling of formulas.
	Use prepackaged products.
	Do not allow formula to "hang" for extended periods especially in warm rooms (change bags at least every 8 hr)
Mucosal erosions	Prevention of gastrointestinal hemorrhage with H_2 blocking agents and antacids.
Distention/fullness/vomiting	Use nasogastric tube decompression.
	Advance rates of infusion only as tolerated.
	If the patient cannot tolerate enteral nutrition for a period of more than 72 hr, then consider utilizing TPN until the patient can tolerate enteral nutrition.
Dehydration	Monitor urine output and electrolytes.
	Adjust with electrolyte free water either enterally or intravenously.

Source: From Cataldi-Betcher et al.[62]

The superior mesenteric artery (SMA) syndrome is a complication of insufficient nutrient support. The cause of this uncommon complication is weight loss and specifically, loss of the retroperitoneal fat pad. It should be suspected if the volume of the nasogastric aspirate increases without any other signs of intestinal dysfunction such as abdominal distention, diarrhea, constipation, obstipation, abnormal bowel sounds, or abdominal tenderness. It should also be suspected in patients tolerating enteral nutrition who have emesis when converted to oral (PO) nutrition. The diagnosis is confirmed by an upper gastrointestinal contrast study. This will show a delay in the transit of contrast at the third portion of the duodenum. The delay will resolve or at least markedly improve with the patient placed in the left lateral decubitus position. The SMA syndrome is usually easily treated by positioning a feeding tube beyond the point of obstruction and providing gastric drainage. Although this can be accomplished by a nasojejunal feeding tube and a nasogastric tube, a PEG/PEJ is sometimes more beneficial for the patient who is actively participating in rehabilitation. Only rarely is TPN required to meet the patient's nutritional needs until the duodenal obstruction resolves.[63] Resolution is confirmed by a decrease in the volume of nasogastric drainage and a normal upper gastrointestinal contrast study.

ADVANCING PARENTERAL NUTRITION

Patients who cannot be fed enterally can be provided their estimated caloric need through a central venous catheter. Utilization rates typically limit glucose infusion to 5 $mg/kg \cdot min^{-1}$ or less. The use of hypertonic solutions to administer the needed calories can result in endovascular injury and thrombophlebitis at the site of infusion. The increased evaporative water loss of burn patients commonly necessitates addition of electrolyte free water, which reduces the hypertonicity of the infused solutions and decreases endothelial damage. The TPN is initiated at a concentration that provides half the estimated glucose need and a rate that meets the patient's 24-hr fluid needs. If the patient tolerates the TPN at half strength without significant elevations in serum glucose, the formula is advanced to the full strength after the first 24 hr and continued at the same rate. Serum glucose should be monitored by a bedside glucometer at least every 2 hr during the initiation of TPN. Glucose intolerance is treated with intravenous insulin. Because the critically ill trauma patient can have rapid changes in their metabolic status or hormonal milieu, it can be wasteful to put insulin in the TPN solution, because the entire bag of TPN containing insulin would need to be discarded if the patient becomes hypoglycemic. An insulin drip separate from the TPN can be easily adjusted in response to changes in the patient's metabolic demands and hormonal milieu.

COMPLICATIONS OF PARENTERAL NUTRITION

Parenteral nutrition carries all the complications of nutritional support that enteral nutrition carries with the additional complications of the central venous access required to deliver the TPN solutions. Total parenteral nutrition requires careful monitoring of serum triglycerides. Serum triglycerides should be measured prior to starting TPN, for the first three days of TPN, and then if stable, at least twice a week thereafter. Abnormally high triglyceride levels require a decrease in the level of nutritional support provided by TPN. Intravenous glucose often causes more hyperglycemia than enterally administered glucose. Serum glucose must be carefully monitored especially when starting or discontinuing TPN. Meeting the patient's nutritional needs with TPN is more difficult than with enteral regimens, especially if fat is not used as an additional nonprotein calorie source. For the reasons discussed earlier in the section on formulas and specific nutrients, we resist using intravenous fat except to prevent or correct essential fatty acid deficiencies.

EVALUATION OF PROGRESS

Both enteral and parenteral nutrition can contribute to derangements in serum glucose, serum triglycerides, serum electrolytes, and hydration. There is also the potential with both routes of administration provide too much or too little nutrient support. Continually evaluating the patient's nutritional status is important. Daily weights are one of the tools used to evaluate the success of nutritional therapy and to monitor hydration status. However, until the trauma or burn patient reaches the convalescent state, daily weight is more indicative of volume status than nutritional status. Changes in weight of more than 500 g/day are associated with a change in body water and not in lean body mass.[64] Calorie counts help determine if the patient is receiving all the calories needed. Daily weights and calorie counts are augmented by several laboratory tests. Nitrogen balance, derived by subtracting the amount of nitrogen loss from the amount of nitrogen intake, should be measured once a week. Nitrogen intake can be determined on the basis of the nitrogen content of the nutrient formulation provided and the amount administered.

$$\text{nitrogen loss (g/24 hr)} = 1.25 \times \text{UUN} + 2.0 + \text{WPL}$$

where UUN is urinary urea nitrogen from a 24-hr urine collection, WPL is wound protein loss (calculations described in the introduction of this chapter), and 2.0 is the correction factor for losses in stool.

The goal of nutrition support is to minimize loss of lean body mass and if possible attain positive nitrogen balance. A negative nitrogen balance necessitates reassessment of the caloric requirement by indirect calorimetry. The caloric and protein supplementation is then adjusted accordingly. Monitoring nitrogen balance is more cost effective than measuring serum visceral protein concentrations to assess the protein response to nutritional support.[65–66]

Serum glucose should be monitored by a bedside glucometer. Blood glucose levels can be indirectly monitored by testing the urine for the presence of glucose with a reagent strip at the bedside. Critically ill patients should have serum glucose measured at least daily to confirm either of the two bedside methods. Hyperglycemia is treated with insulin delivered intravenously, because subcutaneous injections are not a reliable method to deliver insulin to patients with peripheral edema and decreased peripheral perfusion. The sudden onset of hyperglycemia in a patient who has tolerated a given carbohydrate load should prompt a search for occult infection.

To avoid the complications of hypertriglyceridemia, serum triglycerides should be monitored. Elevated levels of serum triglycerides indicate that the amount of calories provided as fat should be decreased. However, if too little fat is provided the patient may manifest signs of fatty acid deficiency. The clinical manifestations of essential fatty acid deficiency are: a diffuse dermatitis involving dryness, desquamation, and thickening of the skin, fatty liver, impaired water balance, diminished would healing, hemolytic anemia, and enhanced platelet aggregation. The diagnosis is confirmed by a serum triene–tetraene ratio greater than 0.4.

A full chemistry profile to include serum calcium, magnesium, potassium, phosphorus, blood urea nitrogen, creatinine, amylase, lipase, and liver enzymes should be obtained at least twice a week. Periodically obtaining a full chemistry profile will alert the physician to derangements in electrolytes before they become clinically evident and identify the early stage of pancreatitis or cholecystitis before the patient demonstrates a systemic inflammatory response. When alterations occur in liver enzymes, the possibility of overfeeding or cholecystitis, especially acalculus cholecystitis, should be considered.[67] The possibility of overfeeding is evaluated easily with indirect calorimetry. An REE of less than 80% of the calories provided or an RQ > 0.95 are indicative of overfeeding. Cholecystitis is often difficult to diagnose in the critically ill trauma or burn patient. Ultrasound can be done at the bedside and may help in determining if the elevation of liver enzymes is a result of cholecystitis or biliary obstruction. A full discussion of diagnosis of cholecystitis in the critically ill trauma patient is beyond the scope of this chapter.

A potential complication of overfeeding is uremia. If the blood urea nitrogen increases, the amount of protein provided in the diet may need to be decreased. A reassessment of the patient's nutritional requirements should be performed.

As an aid in monitoring the adequacy of the nutritional support, indirect calorimetry is used to measure the resting energy expenditure (REE) and respiratory quotient (RQ) every other week or when dictated by changes in nitrogen balance or the occurrence of uremia. The calories delivered should provide 125% of the measured REE. The 25% in excess of the REE accounts for the additional metabolic demands of dressing changes and physical activity. A RQ of <0.8 measured by indirect calorimetry suggests underfeeding and a value of 0.95 to 1.0 suggests overfeeding. If the RQ is <0.8 and the calories being delivered to the patient are less than 125% of the measured REE, then the patients caloric needs are not being met, and the nutrition support should be increased. If the RQ is >0.95 and the calories being delivered are in excess of the 125% of the measured REE, then the nutrition support should be decreased. The REE measured by indirect calorimetry is more reliable than the RQ. If the REE and RQ disagree as to the nutritional state of the patient, the REE should weigh heavier in decisions regarding changing the patients nutritional regimen. Certainly the physician should consider all information including REE, RQ, nitrogen balance, and daily weights before making any changes in nutritional support.

CONCLUSIONS

Proper nutrition is important to the outcome of major trauma and burn patients. Nutrition should be provided as early in the patient's course as possible. Enteral nutrition should be implemented if at all possible and can be supplemented or replaced by parenteral nutrition if necessary. A careful nutritional assessment and estimation of the patient's nutritional requirements should be accomplished on admission and repeatedly, as needed, throughout the treatment course. Reassessment and monitoring permit modifications in the nutrition support program as necessitated by the patient's condition. This will ensure that the complications of postinjury catabolism as well as the complications of the metabolic support are minimized and that the patient's outcome is optimized.

REFERENCES

1. Cuthbertson DP. Observations on disturbance of metabolism produced by injury to limbs. *Q J Med* 1931;2:233–246.
2. Tredget EE, Yu YM. The metabolic effect of the thermal injury. *World J Surg* 1992;16:68–79.
3. Vaughan GM, Becker RA, Unger RH, et al. Nonthyroidal control of metabolism after burn injury: possible role of glucagon. *Metabolism* 1985;34:637–541.
4. Thomas R, Aikawa N, Burke JF. Insulin resistance in peripheral tissues after a burn injury. *Surgery* 1979;86:742–747.

5. Wilmore DW, Aulick LH, Mason Jr. AD, Pruitt Jr. BA. Influence of the burn wound on local and systemic responses to injury. *Ann Surg* 1977;186:444–458.
6. Aulick LH, Wilmore DW, Mason Jr. AD, Pruitt Jr. BA. Muscle blood flow following thermal injury. *Ann Surg* 1978;188:778–782.
7. Downey RS, Monafo WW, Karl IE, et al. Protein dynamics in skeletal muscle after trauma: local and systemic effects. *Surgery* 1986;99:265–273.
8. Lalonde C, Demling RH. The effect of complete burn wound excision and closure on postburn oxygen consumption. *Surgery* 1977;102:862–868.
9. Rutan TC, Herndon, DN, Van Osten T, Abston S. Metabolic rate alterations in early excision and grafting versus conservative treatment. *J Trauma* 1986;26:140–142.
10. Ireton-Jones CS, Turner WW, Baxter CR. The effect of burn wound excision on measured energy expenditure and urinary nitrogen excretion. *J Trauma* 1987;27:217–220.
11. Frankenfield DC, Wiles CE, Bagley S, Siegel JH. Relationships between resting and total energy expenditure in injured and septic patients. *Crit Care Med* 1994;22:1796–1804.
12. Wolfe RR, Durket MJ, Allsop JR, Burke JI. Glucose metabolism in severely burned patients. *Metab Clin Exp* 1979;28:1031–1039.
13. Waxman K, Rebello T, Pinderski L, et al. Protein loss across burn wounds. *J Trauma* 1987;27:136–140.
14. Konstanides FN. Nitrogen balance studies in clinical nutrition. *Nutr Clin Pract* 1992;7:231–238.
15. Buzby GP, Blovin G, Colling CL. Perioperative total parenteral nutrition in surgical patients. *N Engl J Med* 1991;325:525–532.
16. Mughal MM, Meguid MM. The effect of nutritional status on morbidity after elective surgery for benign gastrointestinal disease. *J Parent Ent Nutr* 1987;11:140–143.
17. Askanazi J, Hensle TW, Starker PM, et al. Effect of immediate postoperative nutritional support on length of hospitalization. *Ann Surg* 1986;203:236–239.
18. Roe CF, Kinney JM. The influence of human growth hormone on energy sources in convalescence. *Surg Forum* 1962;13:369–371.
19. Wilmore DW, Moylan Jr. JA, Bristow BF, et al. Anabolic effects of human growth hormone and high caloric feedings following thermal injury. *Surg Gynecol Obstet* 1974;138:875–884.
20. Kimbrough TD, Shernan S, Ziegler TR, et al. Insulin-like growth factor response is comparable following intravenous and subcutaneous administration of growth hormone. *J Surg Res* 1991;51:472–476.
21. Cioffi WG, Gore DC, Rue III LW, et al. Insulin-like growth factor-1 lowers protein oxidation in patients with thermal injury. *Ann Surg* 1994;220:3, 310–316, discussion 316–319.
22. Chawls WJ, Bistrain BR. Role of exogenous growth hormone and insulin-like growth factor 1 in malnutrition and acute metabolic stress: a hypothesis. *Crit Care Med* 19:1317–1322.

23. Hinton P, Allison SP, Littlejohn S, Lloyd J. Insulin and glucose to reduce catabolic response to injury in burned patients. *Lancet* 1971;17:767–769.
24. Brooks DC, Bessey PQ, Black PR, et al. Insulin stimulates branched chain amino acid uptake and diminishes nitrogen flux form skeletal muscle of injured patients. *J Surg Res* 1986;40:395–405.
25. Bretenstein E, Chiloero RL, Jequier E, et al. Effects of beta-blockade on energy metabolism following burns. *Burns* 1990;16:259–264.
26. Herndon DN, Nguyen TT, Wolfe RR, et al. Lipolysis in burned patients is stimulated by the beta 2-receptor for catecholamines. *Arch Surg* 1994;129:1301–1304.
27. Nelson JL, Chalk CL, Warden GD. Anabolic impact of cimaterol in conjunction with enteral nutrition following burn trauma. *J Trauma* 1995;38:237–241.
28. Chance WT, Von Allmen D, Benson D, et al. Clenbuterol decreases catabolism and increases hypermetabolism in burned rats. *J Trauma* 1991;31:365–370.
29. Shaw JH, Wolfe RR. Metabolic intervention in surgical patients: an assessment of the effect of somatostatin, ranitidine, naloxone, diclophenac, dipyridamole, or salutamol infusion on energy and protein kinetics in surgical patients using stable and radioisotopes. *Ann Surg* 1988;207:274–282.
30. Rouby JJ, Eurin B, Glaser P. Hemodynamic and metabolic effects of morphine in the critically ill. *Circulation* 1981;64:53–59.
31. Raurich JM, Ibanex J. Metabolic rate in severe head trauma. *J Parent Ent Nutr* 1994;18:521–524.
32. Blackburn GL, Bistrian BR, Maini BS, et al. Nutritional and metabolic assessment of the hospitalized patient. *J Parent Ent Nutr* 1977;1:11–22.
33. Moore FA, Moore EE, Jones TN, et al. TEN versus TPN following major abdominal trauma—reduced septic morbidity. *J Trauma* 1989;29:916–923.
34. Peterson VM, Moore EE, Jones TN, et al. Total enteral nutrition versus total parenteral nutrition after major torso injury: attenuation of hepatic protein reprioritization. *Surgery* 1988;104:199–207.
35. Jeevanandam M, Young DH, Schiller WR. Obesity and the metabolic response to severe multiple trauma in man. *J Clin Invest* 1991;87:262–269.
36. Harris JA, Benedict FG. *Biometric studies of basal metabolism in man.* Pub. No. 279. Washington, DC: Carnegie Institute, 1979.
37. Long CL, Schaffel N, Geiger JW, et al. Metabolic response to injury and illness: Estimation of energy and protein needs from indirect calorimetry and nitrogen balance. *J Parent Ent Nutr* 1979;3:452–456.
38. Carlson DE, Cioffi Jr. WG, Mason AD, et al. Resting energy expenditure in patients with thermal injuries. *Surg Gynecol Obstet* 1992;174:270–276.
39. Apelgren KN, Wilmore DW. Nutritional care of the critically ill patient. *Surg Clin North Am* 1983;63:497–507.
40. Serog P, Baigts F, Apfelbaum M, et al. Energy and nitrogen balances in 24 severely burned patients receiving 4 isocaloric diets of about 10 MJ/m^2/day (2392 kcal/m^2/day). *Burns Therm Inj* 1983;9:422–427.

41. Long JM, Wilmore AD, Mason Jr. AD, et al. Effect of carbohydrate and fat intake on nitrogen excretion during total intravenous feeding. *Ann Surg* 1977;185:417–422.
42. Gottschlich MM, Alexander JW. Fat kinetics and recommended dietary intake in burns. *J Parent Ent Nutr* 1987;11:80–85.
43. Alexander JW. Nutrition and infection (new perspective for an old problem). *Arch Surg* 1987;121:966–972.
44. Gottschlich MM, Warden GD. Vitamin supplementation in the patient with burns. *J Burn Care Rehab* 1990;11:275–279.
45. Brennan MF, Cerra F, Daly JM, et al. Report of a research workshop: branched chain amino acid in stress and injury. *J Parent Ent Nutr* 1986;10:446–452.
46. Brown RO, Hunt H, Mowatt-Larssen CA, et al. Comparison of specialized and standard enteral formulas in trauma patients. *Pharmacotherapy* 1994;14:314–321.
47. Souba WW, Herskowitz K, Austgen TR, et al. Glutamine nutrition: theoretical considerations and therapeutic impact. *J Parent Ent Nutr* 1990;14(5, suppl):237S–243S.
48. Furst P, Albers S, Stehle P. Glutamine-containing dipeptides in parenteral nutrition. *J Parent Ent Nutr* 1990;14(4, suppl):118S–124S.
49. Baur AE. Nutrition and metabolism in sepsis and multisystem organ failure. *Surg Clin North Am* 1991;71:549–565.
50. Asbach HW, Stoeckel H, Schuler HW, et al. The treatment of hypercatabolic acute renal failure by adequate nutrition and hemodialysis. *ACTA Anaesthesiol Scand* 1974;18:255–263.
51. Bartlett RH, Dechert RE, Mault JR. Measurement of metabolism in multiple organ failure. *Surgery* 1982;92:771–779.
52. Cerra FB, Cheung NK, Fischer JE, et al. Disease-specific amino acid infusion (F080) in hepatic encephalopathy. *J Parent Ent Nutr* 1985;9:288–295.
53. Strauss E, Santos WR, Davsilva EC, et al. A randomized controlled clinical trial for the evaluation of the efficacy of an enriched-branched-chain amino acid solution compared to neomycin in hepatic encephalopathy. *Hepatology* 1983;3:862–864.
54. Carlson DE, Jordan BS. Implementing nutritional therapy in the thermally injured patient. *Crit Care Nurs Clin North Am* 1991;3:221–235.
55. Moore FA, Felician DV, Andrassy RJ, et al. Early enteral feeding, compared with parenteral reduces postoperative septic complications—the results of a meta-analysis. *Ann Surg* 1992;216:172–183.
56. McDonald WS, Sharpp CW, Deitch EA. Immediate enteral feeding in burn patients is safe and effective. *Ann Surg* 1991;213:2177–2183.
57. Hansbrough WB, Hasnsbrough JF. Success of immediate intragastric feeding of patients with burns. *J Burn Care Rehab* 1993;14:512–516.
58. Buescher T, Cioffi WG, Becker W, et al. Perioperative enteral feedings. *Proc Am Burn Assoc* 1990;22:162.
59. Jenkins M, Gottschlich M, Warden GD. Enteral feeding during operative procedures in thermal injuries. *J Burn Care Rehab* 1994;15:199–205.

60. Edes TE, Walk BE, Austin JL. Diarrhea in tube-fed patients: feeding formula not necessarily the cause. *Am J Med* 1990;88:91–93.
61. Brinson RR, Kolts BE. Hypoalbuniemia as an indicator of diarrheal incidence in critically ill patients. *Crit Care Med* 1987;15:506–509.
62. Cataldi-Betcher EL, Seltzer MH, Slocum BA, Jones KW. Complications occurring during enteral nutrition support: a prospective study. *J Parent Ent Nutr* 1983;7:546–552.
63. Milner EA, Cioffi WG, McManus WF, Pruitt Jr. BA. Superior mesenteric artery syndrome in a burn patient. *Nutr Clin Prac* 1993;8:264–266.
64. Kinney JM, Long CL, Gump FE, Duke Jr. JH. Tissue composition of weight loss in surgical patients I. Elective operation. *Ann Surg* 1968;168:459–474.
65. Milner EA, Cioffi WG, Mason Jr. AD, et al. Accuracy of urinary urea nitrogen for predicting total urinary nitrogen in thermally injured patients. *J Parent Ent Nutr* 1993;17:414–416.
66. Carlson DE, Cioffi Jr. WG, Mason Jr. AD, et al. Evaluation of serum visceral protein levels as indicators of nitrogen balance in thermally injured patients. *J Parent Ent Nutr* 1991;15:440–444.
67. Munster AM, Goodwin MN, Pruitt Jr. BA. Acalculous cholecystitis in burned patients. *Am J Surg* 1971;122:591–593.

CHAPTER

29

Nutrition Support and Pregnancy

Roger C. Andersen, MD., M.P.H., Michael C. Moore, M.D., Laurie Mello Udine, R.D., C.N.S.D., Harrison D. Willcutts, M.D., and H. David Willcutts, Jr.

INTRODUCTION

Nutrition support during pregnancy can cover a broad range of topics. Our discussion focuses on three main areas: maternal nutritional requirements, approaches to nutritional intervention, and experiential discussions from the community setting. Knowledge on the use of both enteral and parenteral nutrition has improved greatly over the past 20 years. Improvements in catheters, formulations, and care guidelines have allowed once problematic therapies to be prescribed with confidence.

In an ideal setting, nutrition support can be maintained with proper nutritional counseling, however, many circumstances can arise during the course of pregnancy that require more involved intervention. While considering nutritional intervention, the goal is always to minimize risk to both the mother and fetus, recognizing that poor nutritional status itself can represent a significant risk. In the following pages, we describe how nutritional intervention has been initiated, monitored, and completed safely and effectively. We emphasize the total nutritional needs in pregnancy taken from our experience of approximately 100 hyperemesis patients treated with TPN with safety and benefit. Starvation benefits no disease state, but is particularly risky in the pregnant patient.

NUTRITIONAL REQUIREMENTS IN PREGNANCY

Over the past 20 years, there has been a renewed interest in the area of maternal nutrition. The relationship of maternal nutrition to fetal well being is an area that can be positively influenced by behavioral and lifestyle changes. Preconception or during the initial prenatal visit, a nutritional risk assessment should be performed. This will allow identification of women who are nutritionally at risk during pregnancy. Appropriate referral to a registered dietitian or nutritionist can then be made.[1,2] Examples of women who may be at an increased risk include those who are over- or underweight, vegetarian, adolescent, diabetic, or have substance abuse problems. By increasing access to prenatal care along with increased screening and intervention for behavioral risk factors, nutritional adequacy during pregnancy and lactation can be improved.[3]

Extensive knowledge of nutrition in pregnancy is still lacking in part because it is difficult to design ethical studies. Retrospective data is available from war years, when strict rationing of food to the population was imposed. Individual case reports have contributed additional information. The primary national supplemental food program is the WIC (Women, Infants and Children) Program, and is administered by state health agencies. The National WIC Evaluation found substantial benefits to the program, including: increased birth weight, decreased pre-term delivery, reduced late fetal death, earlier and more compliant prenatal care, and improved maternal nutrition.[4,5]

Weight Gain

Recommendations for appropriate weight gain have varied among different groups. The American College of Obstetricians and Gynecologists recommended that pregnant women of normal weight gain approximately 10 to 12 kg (22 to 27 lb) during pregnancy. The National Academy of Sciences put forth more specific recommendations for gestational weight gain. These recommendations have focused on desirable ranges of weight rather than a single target for total weight gain. They have identified five different categories of women and suggested ranges for weight gain for each category. These values range from 15 to 45 lbs. depending on the category (underweight, normal, overweight, twins).[1]

Energy

The Food and Nutrition Board has recommended a daily calorie increase of 300 kcal throughout pregnancy. However, it is important to recognize that energy needs differ from one woman to another. It is therefore not appropriate to make a single recommendation for all pregnant women. The best method of assessing individual energy status is by evaluating the rate of weight gain. In the first trimester, weight gain is expected to be between 0.9 and 2.2 kg. In the second and third trimesters, the average weight gain is 0.34 to 0.5 kg/week. During pregnancy, the body adapts to available energy stores, sparing energy for fetal growth when supplies are limited, adjusting basal metabolism and fat deposition if they are abundant.[6]

Protein

Approximately 925 g of protein are deposited in a normal weight fetus and in the maternal accessory tissues. Thus, the 1989 RDA recommendations are for 10 g of protein a day over the nonpregnant state.[1]

Iron

During normal pregnancy, the average woman needs to absorb approximately 3 mg/day of elemental iron. Because only 10% to 30% of the dietary intake of iron gets absorbed, the level of ingested iron must be substantially increased. The average American diet provides 12 to 14 mg of iron per day. To ensure that there is sufficient iron absorbed to satisfy the demands of pregnancy, a total intake of 30 mg/day of iron is required in most cases.[7] Most prenatal vitamin and mineral supplements provide this recommended level of iron in the form of ferrous salts. The serum ferritin level and total iron is regarded as a useful index of the status of iron stores during pregnancy. Periodic estimates during pregnancy can help to determine the need for additional supplementation, particularly with a low hemoglobin or low hematocrit.

Calcium

Current recommended calcium intake in pregnancy and lactation is 1200 mg/day.[8] Hormonal factors in pregnancy cause the promotion of progressive calcium retention. Fetal use of calcium will peak during the third trimester with the formation of teeth and skeletal growth. Some studies have suggested that calcium supplementation in pregnancy has been associated with a reduction in pregnancy-induced hypertension,[9] as well as a reduced risk of preterm delivery in populations at risk for low calcium intake.[10] The role of maternal calcium intake on breast-milk calcium and infant bone growth are not understood, and more research is needed before general advice is given to women to increase their calcium intake during pregnancy and lactation.[11]

Sodium

It was common practice in the past to restrict sodium intake in the pregnant woman suffering from edema. It is now known that moderate edema is normal during pregnancy. Total pregnancy demands for sodium are approximately 950 mEq. This is usually met by the typical diet. Increased sodium intake has not been found to cause an increased incidence of toxemia nor has sodium restriction helped to prevent or reverse it. In general, salting food to taste will provide adequate dietary sodium for the pregnant woman.

Other Nutrients

Periconceptual folic acid of 0.4 mg/day has been recommended by the U.S. Public Health Service to lower the incidence of neural tube defects (NTDs). Several studies have found this to be successful in lowering the incidence of NTDs.[12] Unfortunately, the average diet contains less than 0.2 mg/day of folate. With the exception of folic acid and iron, routine supplementation with other vitamin and mineral preparations is not necessary. The best source of adequate nutrients is a balanced diet. Correcting inadequate dietary habits by vitamin and mineral preparations should be discouraged. If, however, there are those who do not have an adequate diet or are in a high-risk category they should start supplements in the first trimester.

EFFECTS OF POTENTIALLY HARMFUL SUBSTANCES

Although not directly related to nutritional status, some common dietary habits can present the fetus with potentially harmful substances. Concerns over other substances with no proven negative impact, can cause stress and anxiety for the pregnant women through the course of pregnancy. Any nutritional counseling should be prepared to discuss and answer questions relating to these issues.

Alcohol adversely affects fetal development and is now recognized as causing the fetal alcohol syndrome (FAS). This occurs in one or two infants per 1000 live births in the United States. Moderate drinkers may have offspring with only some features of FAS. It has been reported that fewer than two drinks weekly had no detectable effects of fetal outcome.

Caffeine is a stimulant drug that crosses the placenta. Studies of caffeine use in pregnancy have had mixed results. Current work indicates increased risk associated with caffeine use. It would make sense though to advise women to use caffeine in moderation if they choose to use it at all.

Aspartame is a frequently used artificial sweetener used in carbonated beverages. Concern has been raised over the added phenylalanine. In people who do not have phenylketonuria, adequate phenylalanine hydroxylase activity occurs in the liver to prevent any substantial rise in serum phenylalanine levels. There is no data at this time to suggest that aspartame-containing products are associated with adverse pregnancy outcomes.

ENTERAL NUTRITION IN PREGNANCY

There have been numerous case reports using enteral feedings to support patients during pregnancy.[13,14] In general, the indications for enteral nutrition in pregnancy are similar to that of all patients and include, but are not limited to neurological disease, hypermetabolism due to trauma, and psychiatric disease. As long as the patient's gastrointestinal tract is functioning and her intake is inadequate to meet maternal and fetal needs, enteral nutrition should be considered. Because the complications associated with enteral nutrition are less than those with parenteral nutrition, whenever possible enteral nutrition should be elected.

The American Society for Parenteral and Enteral Nutrition (ASPEN) has developed practice guidelines for the use of parenteral and enteral nutrition during pregnancy that are based on fair-to-good research-based evidence and expert opinion.[15] The guidelines for pregnancy are as follows:

1. Enteral tube feeding or TPN may be useful for preventing nutritional deficits and promoting adequate weight gain in pregnant women with impaired intake or absorption resulting from complications of pregnancy (hyperemesis gravidarum) or from illnesses unrelated to pregnancy. In particular, gain of less than 1 kg/month in underweight or normal weight women or 0.5 kg/month in obese women during the second and third trimesters frequently indicates inadequate nutritional status.
2. The benefits of enteral nutrition for patients with hyperemesis should be balanced against the risk of tube displacement and pulmonary aspiration.
3. Maternal euglycemia should be maintained during enteral and parenteral nutrition support.
4. Progressive maternal weight gain (using established guidelines for reference data) and fetal growth should be monitored during nutrition support.[16]

Nutrition Support

The goals of nutrition support during pregnancy include achieving a normal rate of weight gain appropriate to pregravid weight, maintaining appropriate vitamin and mineral intakes, achieving positive nitrogen balance, and avoiding metabolic complications.[17] Although recommendations exist for weight gain throughout pregnancy, there are no specific recommendations for daily kcal/kg or per Body Mass Index.[18] The 1989 recommended dietary allowance (RDA) during pregnancy in a normally nourished woman is an additional 300 kcal/day during the second and third trimesters.

The RDA for protein was decreased from 30 additional g/day to 10 additional g/day.[19] Adding 10 g to assessed needs should provide adequate protein unless there are increased losses. The fluid and micronutrient needs should be based on individual assessment considering the increased needs for vitamins and minerals during pregnancy.

Nutritional goals can be met through enteral nutrition when attention is paid to the special needs of the pregnant women.

Enteral Nutrition and Hyperemesis Gravidarum

Hyperemesis gravidarum (HG) occurs in up to 2% of pregnancies and is characterized by severe nausea and vomiting. The condition places patients in a high-risk condition due to the nutritional consequences resulting from decreased nutrient intake and increased nutrient loss. As a result, dehydration as well as electrolyte and metabolic imbalances may ensue depending on the severity of the condition.[20] Poor maternal weight gain increases the risk of delivery of a low-birth-weight infant, which is associated with an increase in poor neonatal outcome and neonatal death.[21] In addition to HG, other maternal medical conditions may require the use of nutrition support during pregnancy. Both enteral tube feeding and TPN have been used safely and effectively during pregnancy.[22–26]

Most treatment of HG has focused on intravenous fluid therapy and/or parenteral nutrition, avoidance of oral intake or use of a low-fat solid diet, psychological therapy, and antiemetic drug therapy. However, there are reports of successfully treating patients with HG using a continuous infusion of an isoosmolar enteral feeding.[27–29]

Patients who were fed intragastrically versus postpylorically appeared to experience more symptom relief. The intragastric feeding itself reduced nausea in the majority of patients within 1 to 2 hr.[28] In some cases, antiemetic therapy may be discontinued; however, not all patients appear to respond to this treatment.[29] Nasogastric tube placement may be preferred because of the potential buffering effect of feeding into the stomach and because transpyloric positioning requiring fluoroscopic placement may increase environmental exposure at a time of rapid fetal development.

An isoosmolar formula infused at 20 to 50 mL/hr can be initiated while the patient remains NPO (nothing per os). If, after 12 hrs, tolerance has been established, the formula can be gradually increased until the estimated nutritional needs are met. Oral intake should be advanced slowly, with careful attention paid to tolerance when the patient is being weaned off the tube feeding.

In the Gulley et al. study,[28] aspiration was not determined to be a complication of the therapy; however, nasopharyngeal irritation and tube displacement were noted. In general, complications frequently associated with enteral nutrition may appear in this population. Specific complications related to enteral feeding in the pregnant patient include: gastric retention, diarrhea, and hyperglycemia.[30]

Although there is limited clinical experience, the use of enteral nutrition support in HG has been shown to be safe and effective and should be considered when diet, psychological counseling, and drug therapy has failed.

HYPEREMESIS GRAVIDARUM IN THE COMMUNITY HOSPITAL AND HOME SETTING

Background

Interest in nutrition support for the hyperemesis gravidarum patient by my obstetrical colleagues became acute in 1989. A local malpractice case involving a patient with hyperemesis gravidarum was settled in favor of the plaintiff who became so malnourished during her pregnancy that she required admission to a local hospital with bizarre neurological symptoms. In time, she was diagnosed as having Werneicke's encephalopathy and eventually was left with permanent defects. Even the expert witness for the defense felt that the delay in treatment of the plaintiff was "indefensible." Since this time, we have monitored and intervened nutritionally with over 60 patients. Below is a description of our experience and results from actual community practice.

Referral Recommendations

Most women with hyperemesis gravidarum are able to ingest several hundred calories (mostly carbohydrate calories) per day. However, they are unable to consume enough calories and vitamins to ensure proper fetal growth and development. My obstetrical colleagues usually refer their patients who are not gaining weight, or who have lost weight due to hyperemesis. We recommend referral for nutrition support before a 10-lb weight loss occurs.

Initiation of nutrition support is typically performed in the office setting, requiring no hospital admission. As always, full discussion of the risks and benefits should be made to the patient and family prior to initiation of treatment. Over the past five years, approximately 15 patients each year have been identified as needing nutritional support. This is 0.75% of the total deliveries at our hospital. When nonvolitional intervention was required, all opted for intravenous rather than enteral support.

The Nutrition Support Team

Because therapy begins as an outpatient, communication and confidence between the physicians, homecare pharmacy and visiting nurses is essential. Periodic meetings are recommended so that there is a satisfactory comfort level among all health care providers.

After initiation of therapy, we monitor patients through direct nursing interaction and blood chemistry results. Our weekly laboratory work typically includes a CBC and a standard blood chemistry profile (SMA-20, etc.). PT (prothrombin time), PTT (partial thromboplastin time), copper, zinc, magnesium, and ferritin levels are done monthly. Each week the patient formula is reviewed against the most current laboratory results, adjusting patient-specific micronutrients as needed. Much of our macronutrient additives and nutritional guidelines have been adapted from the work done by MacBurney et al.[30]

All our patients received intravenous nutrition support via a peripherally inserted central catheter (PICC) inserted into the subclavicular area by the home-care nurse. Because of the relatively short treatment duration, PICC lines have proven the simplest route for both the patient and our care team. Our formulations reflect lower osmolarity because of the subclavicular placement.

Usually 5% final dextrose concentrations are used at first and slowly increased to 7.5%, with a maximum of 10%. Higher dextrose concentrations are not needed because the patients can consume some oral calories which mostly are carbohydrates. Glucometer readings are measured 1 hr into the infusion and 1 hr after the infusion. The entire volume is given 10 to 12 hrs/day to allow the patient to continue daily activity as normally as possible. A sample of a TPN formula in an HG patient is shown in Table 29.1. Our initial experience of 16 patients was presented at the ASPEN Conference in 1993. Since then, an additional 47 patients have completed therapy.

Of 63 patients, 59 (94%) had live births with a mean weight of 7.19 ± 1.98 lb. There were three sets of twins and four miscarriages, including two of three

Table 29.1. Sample of a Successful Initial Hyperemesis Gravidarum TPN Formula

Large Volume	% Solution	Solution (mL)	Final solution Concentration	Recommended Changes
Amino acids	10	800		
Dextrose	70	300		
Lipids	20	500		
Sterile water		500		
Total volume: 2214 mL				Infuse at a rate of 185 mL/hr over 12 hours.

Additives[a]	Per Liter/Bag	Changes	Additives[a]	Per Liter/Bag	Changes
Sodium Acetate	80 mEq		Zinc	mg	
Sodium Chloride	100 mEq		Copper	mg	
Potassium Acetate	mEq		Selenium	100 µm	
Potassium Chloride	mEq		Trace Metals	5 mL	
Pot. Phos. (PO_4)	40 mm		MVC 9+3	10 mL	qd.
Calcium Gluconate	18 mEq		Heparin	U	
Magnesium Sulfate	20 mEq		Iron Dextran	mg	
			Vitamin K	mg	
			Insulin	U	

[a]Additives ordered per liter are based on amino acid and dextrose volumes only.

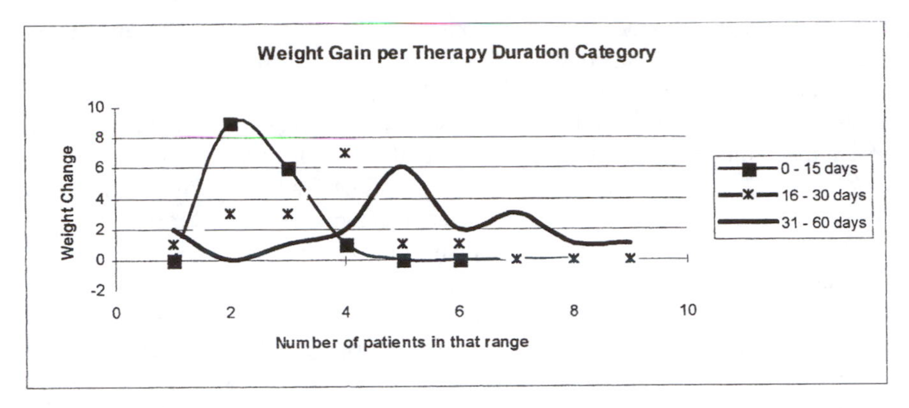

Figure 29.1. Therapy duration versus weight gain.

patients with clinically significant gallbladder disease. One patient underwent successful laproscopic cholecystectomy during her second trimester and had a successful birth outcome.

There were no episodes of PICC line sepsis. The average length of therapy was 41 days. Albumin levels averaged 3.33 ± 0.38 g at the conclusion of therapy. Apgar scores averaged 8.27/8.84. The average weight gain of the mothers was 11.3 ± 8.6 lb.

In our group of patients, home parenteral nutrition support was associated with reduced hospitalization, an increase in the weight of the mother, and a positive birth outcome with no physical abnormalities in the live births. As one might expect, longer therapy duration was associated with more significant weight gain while on therapy (see Figure 29.1).

Implementation

The concept of parenteral nutrition is not emphasized in most obstetrics–gynecology residency programs despite the fact that a relatively significant number of pregnant women develop hyperemesis gravidarum (up to 2% of pregnancies). Prior to our involvement, the obstetrician/gynecologist often admitted and occasionally readmitted this type of patient to the hospital several times during her pregnancy. This causes stress and quality-of-life disruptions for the pregnant woman as well as presenting potential risks, as discussed above.

The evolution of the homecare industry has helped to allow this type of care to be provided effectively in the home environment. The sophistication in the delivery of a quality service by homecare companies has become an expected norm. To ensure you receive this level of quality, use of Joint Commission on Accreditation of Healthcare Organizations (JCAHO) accredited companies is strongly recommended.

The homecare company chosen should be able to offer a wide range of services for the complete patient care. Some of these services typically include: third-party-payor approval, patient training, PICC line insertion, parenteral feeding formula recommendations, as well as day-to-day clinical care and monitoring of the patient. As we have developed relationships with homecare companies, the process of starting and monitoring patients has become a smooth, efficient, and effective process for both our staff and our patients.

The trust bond between physician and homecare company is essential for successful patient management. It is much easier to care for these patients at home than in hospital, in large part because of the expertise of the nursing, pharmacy and other clinical support personnel. The concept of homecare in the 1990s has advanced tremendously since the time when the physician alone made house calls a generation ago. Homecare has become a high-tech option for the physician in which a variety of clinical expertise is regularly available to assist in patient management of varying case complexity.

REIMBURSEMENT CONSIDERATIONS

In my experience, third-party payors have been universally favorable in their decision to reimburse the costs for nutrition support in the hyperemesis gravidarum patient. Proper nutrition support early in pregnancy reduces the risk of congenital anomalies and neural tube defects; support later in pregnancy reduces the risk of low birth weight (LBW) infants, small for gestational age (SGA) infants, and infants born with an underdeveloped central nervous system, including brain cell size and number. The total cost to the third-party payor is less and the benefit to the fetus is greater, if nutrition support is begun early. Two-thirds of all neonatal deaths occur among LBW (less than 2500 g) infants.

SUMMARY

Nutrition has been clearly shown as a critical area of concern during pregnancy. When dietary counseling alone is insufficient, there are viable alternatives for nonvolitional nutritional intervention. Our community-based experience of approximately 100 patients has seen the successful use of both PICC lines and central lines for TPN administration, allowing both high and low osmolarity formulations.

While weighing the risk versus the benefit with the patient and care team, our experience recognized a positive impact for both patient and family, with little to no complications. The use of enteral and parenteral support as outlined in this chapter, has been a documented successful treatment in pregnancy malnutrition, particularly in the hyperemesis gravidarum patient population.

REFERENCES

1. American Academy of Pediatrics and American College of Obstetricians and Gynecologists. *Guidelines for Prenatal Care,* Elk Grove Village, IL: AAP, 1988.
2. American College of Obstetricians and Gynecologists. *Standards for Obstetric-Gynecologic Services,* Washington, DC: ACOG, 1989.
3. Olson, CM. Promoting positive nutritional practices during pregnancy and lactation. *Am J Clin Nutr* 1994;59(2, suppl):525S–531S.
4. Rush D. *The National WIC Evaluation: An Evaluation of the Special Supplemental Food Program for Women, Infants and Children.* Vols. 1 and 2. Research Triangle Institute and New York State Research Foundation for Mental Hygiene, 1986.
5. Rush D, et al. The national WIC evaluation: evaluation of the special supplemental food program for women, infants and children. *Am J Clin Nutr* 1988;48(suppl):389–519.
6. King JC, Butte NF, Bronstein MN. Energy metabolism during pregnancy: influence of maternal energy status. *Am J Clin Nutr* 1994 Feb.;59(2, suppl):439S–445S.
7. Subcommittee on Nutritional Status and Weight Gain During Pregnancy. Nutrition during pregnancy, *Nutr Today* 1990;25:13–22.
8. National Research Council, National Academy of Sciences, Committee on Dietary Allowances, Food and Nutrition Board. *Recommended Dietary Allowances,* 10th ed., Washington, DC: National Academy Press, 1989:60–62, 201.
9. Belizan JM, Villar J, Repke J. The relationship between calcium intake and pregnancy-induced hypertension: Up-to-date evidence. *Am J Obstet Gynecol* 1988;158:898–902.
10. Martaugh MA, Weingart S. Individual nutrient effects on length of gestation and pregnancy outcome. *Sem Perinatal* 1995;19:197–210.
11. Prentice A. Maternal calcium requirements during pregnancy and lactation. *Am J Clin Nutr* 1994;59(2, suppl):477S–483S.
12. Rush D. Periconceptional folate and neural tube defect. *Am J Clin Nutr* 1994;59(2, suppl):511S–516S.
13. Brown RO, Vehe KL, Kaufman PA, et al. Long-term enteral nutrition support in a pregnant patient following head trauma. *Nutr Clin Pract* 1989;4:101–104.
14. Nelson JH, McLean WT. Management of Landry–Guillain–Barre syndrome in pregnancy. *Obstet Gynecol* 1985;65:25S–29S.
15. American Society for Parenteral and Enteral Nutrition. Guidelines for the use of parenteral and enteral nutrition in adult and pediatric patients. *J Parent Ent Nutr* 1993;17:1SA.
16. American Society for Parenteral and Enteral Nutrition. Guidelines for the use of parenteral and enteral nutrition in adult and pediatric patients. *J Parent Ent Nutr* 1993;17:22SA–23SA.
17. MacBurney M, Wilmore DW. Parenteral nutrition in pregnancy. In Rombeau J, Caldwell M, eds. *Parenteral nutrition,* 2nd ed. Philadelphia: W.B. Saunders, 1992.

18. Institute of Medicine, Subcommittee on Nutritional Status and Weight Gain During Pregnancy. *Nutrition in pregnancy.* Washington, DC: National Academy Press, 1990.
19. Food and Nutrition Board. *Recommended Dietary Allowances,* 10th ed. Washington, DC: National Academy of Science, 1989.
20. Newman V, Fullerton JT, Anderson PO. Clinical advances in the management of severe nausea and vomiting during pregnancy. *J Obstet Gynecol Neonatal Nurs* 1993;22:483–490.
21. Abrams BF, Laros RK. Pre-pregnancy weight, weight gain, and birth weight. *Am J Obstet Gynecol* 1986;154:503–509.
22. Levine MG, Esser D. Total parentaral nutrition for the treatment of severe hyperemesis gravidarum: maternal nutritional effects and fetal outcome. *Obstet Gynecol* 1988;72:102–107.
23. Landye ST. Successful enteral support of a pregnant comatose patient: a case study. *J Am Diet Assoc* 1988;88:718–720.
24. Nugent FW, Rajala M, O'Shea R, et al. Total parental nutrition in pregnancy: conception to delivery. *J Parent Ent Nutr* 1987;11:535–427.
25. Nuutinen LS, Alahuhta SM, Heikkinen JE. Nutrition during ten-week life support with successful fetal outcome in a case with fatal maternal brain damage. *J Parent Ent Nutr* 1989;13:432–435.
26. Wolk RA, Rayburn WF. Parenteral nutrition in obstetrical patients. *Nutr Clin Pract* 1990;5:139–152.
27. Boyce RA. Enteral nutrition in hyperemesis gravidarum: a new development. *J Am Diet Assoc* 1992;6:733–736.
28. Gulley RM, Vander Pleog N, Gulley JM. Treatment of hyperemesis gravidarum with nasogastric feeding. *Nutr Clin Pract* 1993;8:33–35.
29. Barclay BA. Experience with enteral nutrition in the treatment of hyperemesis gravidarum. *Nutr Clin Pract* 1990;5:153–155.
30. MacBurney M. Nutrition support in pregnancy. In Gottschlich M, Matarese L, Shronts E, eds. *Nutrition Support Dietetics Core Cirriculum,* 2nd ed. Silver Spring, MD: ASPEN, 1993.

Pediatric Nutrition Support

Susan S. Baker, M.D., Ph.D.,
Robert D. Baker, Jr., M.D., Ph.D., and
Anne Davis, M.S., R.D., C.N.S.D.

INTRODUCTION

The simple statement that children are not little adults encompasses a wide assortment of factors, such as a gastrointestinal tract that may not be completely developed, unlearned feeding behaviors, and the additional nutrients required to support growth. Children are never alone as patients. They are part of a family and often parental and sibling needs must be included in plans for nutrition support. And so, the entire approach to nutrition support is different for children than for adults and this approach varies with the age of the child.

Nutrition support can be considered to include supplements, tube feedings, and total parenteral support.

GASTROINTESTINAL DEVELOPMENT

Mechanical Function

Sucking is a complex activity involving the cranial nerves and the appreciation of tactile stimulation of facial skin. It is partially conscious, partially unconscious, and requires the coordination of sucking, breathing, and swallowing. Swallowing develops before sucking at approximately 11 weeks of gestational age.[1] Mouthing movements, early attempts at sucking, begin by approximately 18 weeks. After 34 to 35 weeks sucking and swallowing rapidly develop.[2] Thus, infants born before 33 or 34 weeks of gestation do not have coordinated sucking activities and cannot rely on sucking for nutrient intake. Rather, they must be fed by tube. Nonnutritive sucking is sucking activity not associated with the intake of nutrients. Infants offered the opportunity for nonnutritive sucking during tube feeding demonstrate increased oxygenation measured transcutaneously.[3] Premature infants allowed nonnutritive sucking gain weight, transition to oral feeds, and leave the hospital earlier than infants not offered nonnutritive sucking.[4–6] The physiologic basis for this observation is unclear. Nonnutritive sucking may decrease crying, which in turn results in decreased activity and energy expenditure.

Gastric emptying is defined as the time it takes for one half of the gastric contents to leave the stomach. This time depends on the volume and composition of the test meal. There is no standardized method to assess gastric emptying time, thus, it is difficult to compare studies.

It is thought that gastric emptying is delayed in the neonate, especially infants born prematurely.[7] However, no studies have compared adults and neonates using identical liquid meals in volumes normalized for weight.

Motility depends on the integration of neural, smooth muscle, and neural-humoral mechanisms. Few developmental studies have been conducted on humans because of ethical considerations. However, before 30 weeks gestation, there is little movement of contrast beyond the stomach.[8] By 34 weeks, contrast moves into the colon. Duodenal contraction rate increases as a function of gestational age and is associated with an increase in contractions per burst.[9] Both gastric emptying and intestinal motility can be delayed in ill infants and children. Conversely, small enteral feedings result in improved feeding tolerance and earlier progression to full enteral intake when compared to infants receiving delayed enteral feedings.[10,11] Thus, the early delivery of some enteral feedings improves gastrointestinal motility and tolerance to enteral nutrition.

Digestion

The absorption of fat depends on lipases, bile acids, and enterocyte uptake. In premature infants, only 65% to 70% of ingested fat is absorbed, whereas 85% of ingested fat is absorbed in term infants less than 6 months of age. It is not until approximately 6 months of age that fat absorption approaches that of normal adults (more than 90% of ingested fat).[12] In a series of experiments, Fredrikzon et al.[13] demonstrated that duodenal lipase activity is lower in infants than adults and does not increase after a test meal containing a standard amount of fat per kilogram of body weight. Hamosh et al.[14] found lipase present in gastric aspirates at 26 weeks of gestational age. They proposed that the lingual glands were the source of the lipase and the lipase was important for fat digestion in the neonate. Lipase is present in biopsies of the gastric fundus at 11 weeks gestational age and reaches adult levels by 3 months of age.[15] Breast fed infants have the additional advantage of human milk lipase, which aids in the digestion of triglycerides.

Bile acids act to solubilize lipids. However, in normal infants the bile salt pool and average rate of cholate synthesis is less than in normal adults.[12] Infants born prematurely have a more marked reduction in bile acid pool size when compared to term infants.[16]

The digestion of carbohydrates by infants depends on enzymes that hydrolyze starches and disaccharides. The duodenal concentration of amylase is low during infancy.[16] However, infants can digest starch because mammary and salivary amylase and brush-border glucoamylase may be involved in the digestion of starch. Further digestion of starches by colonic hydrolysis into short-chain fatty acids may offer some energy from undigested starches that reach the large bowel.[17] Sucrase, isomaltase, and maltase activities are present at 10 weeks and reach adult levels by 30 weeks of gestational age.[18] By 28 to 34 weeks of gestation, lactase activity is present at approximately one third of the activity level found in term infants. Lactase activity peaks at birth and by adulthood is lower than in newborn infants, nevertheless, infants retain 96% of lactose intake and show no clinical sign of intolerance.[19] Monosaccharides such as glucose and galactose are absorbed primarily by a sodium-dependent active transport process, which is present as early as the 10th gestational week. In the human newborn, the capacity for glucose absorption per unit weight of intestinal tissue is only 50% to 60% of the adult level.[20] Fructose, in contrast, is absorbed by sodium-independent facilitated diffusion.

The digestion of protein begins with proteolysis in the stomach. The infant has the capacity to produce gastric acid, intrinsic factor, and gastrin from approximately 18 weeks of gestational age.[21] However, little protein hydrolysis occurs in the stomach of infants.[22] The pancreas of neonates secretes trypsin at approximately the same level as adults.[23] Chymotrypsin secretion is approximately 50% to 60% of the adult level. Nevertheless, neonates can digest and absorb adequate protein.[24] Premature infants require higher levels of protein intake, but they have a more limited capacity to digest proteins. Care is essential when supplying excess dietary protein to premature infants, because it may increase their metabolic load and result in uremia, metabolic acidosis, and neurologic disturbances.[25] In addition, undigested proteins may present a source of antigens to the neonate to which allergy can develop.[26]

Barrier Function

The gastrointestinal tract functions as an important barrier against the passage of viruses, bacteria, and antigenic proteins as well as an absorptive surface.

However, the barrier is not absolute,[27] permitting the passage of some molecules. Under specific conditions[28,29] the barrier may become more permeable. Infants born prematurely demonstrate a higher intestinal permeability than do infants born at term.[30,31] The higher permeability of the gastrointestinal tract can lead to uptake of potentially harmful infectious agents and antigens.

Behavior

Tuchman[32] recently reviewed the oropharyngeal and esophageal complications of tube feeding from a developmental point of view. His comments are appropriate for children supported by parenteral nutrition as well. The development of feeding itself is a complex process dependent on anatomic structures and mechanical function. For infants or children who receive long-term nutrition support that bypasses the mouth, initiating or reinstituting oral feeding may be difficult. Frequently these children have experienced a critical illness which transmits fear and perhaps pain or discomfort around feeding. Parents may focus on feeding, as the only act in which they can participate. Illingworth and Lister[33] describe a critical period during which feeding behaviors develop. Once this critical time has passed, it may be difficult to learn the skill or behavior of eating. Many difficult problems can be avoided by an awareness of their importance and institution of a program of oral stimulation at the time nutrition support commences.

GROWTH

Growth is extremely rapid in infancy; infants double their birth weight by 6 months of age and triple it by 12 months of age. Thereafter, growth proceeds at a steady rate of approximately 5 g/day, until the adolescent growth spurt. In addition to this rapid growth, changes occur in body composition.[34] Fuel reserves may be marginal, particularly in prematurely born or sick infants and the presence of disease may require additional nutrients. During the initial stages of nutrition support in very sick or injured children growth itself may not be a reasonable goal. However, with time after the acute process, growth becomes important if children are to achieve their full developmental and intellectual potential. Later intellectual achievement may depend on the provision of adequate nutrients in infancy. For some nutrients, it is unclear whether the later correction of deficits can completely compensate for earlier lack. An example of this is iron deficiency in infancy and intellectual achievement at age 5 years.[35–37]

NUTRITION SUPPORT

Developmental immaturity and growth are two major factors that must be considered in nutrition support for children. Careful attention to the provision of nutrients to meet these additional needs of children will determine the success of nutrition support. Figure 30.1 is an algorithm that is useful in deciding the route of nutrition support. Often both enteral and parenteral support are necessary for a time until all nutrition can be delivered via the gastrointestinal tract.

Before providing nutrition support, several steps are necessary (Table 30.1). First an assessment of nutritional needs must be performed, goals must be set,

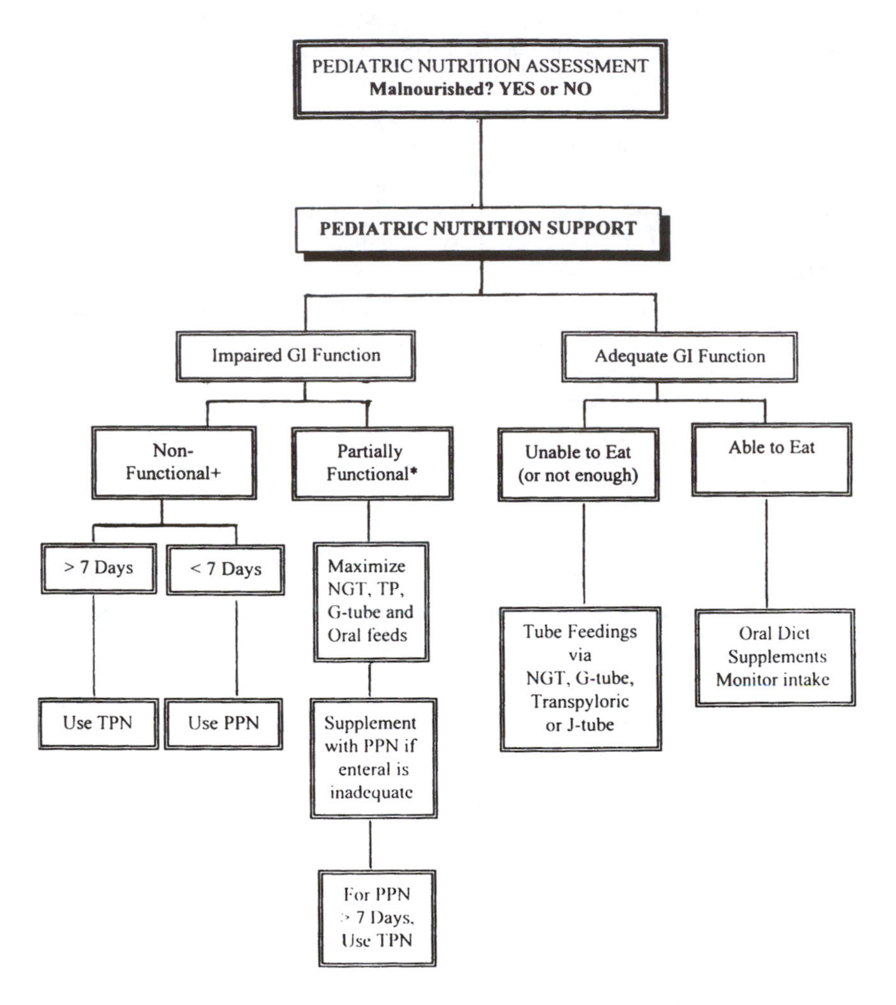

Figure 30.1 Decision making for pediatric nutrition support.

Table 30.1 Initiation of Nutrition Support

- Assess patient
  - Anthropomorphics
  - Biochemistry
  - Determine if malnourished
  - Determine level of stress
  - Determine expected level of gastrointestinal function
  - Consider predicted changes in above (planned surgery, chemotherapy, etc.)
- Set goals
  - Deliver at least enough nutrients to supply needs for basal metabolism
  - Rehabilitate malnutrition
  - Correct mineral/electrolyte imbalance
  - Support growth
  - Initiate catch-up growth
  - Support through metabolic stress
  - Account for nutrient needs of disease processes (cystic fibrosis, inflammatory bowel disease, etc.)
  - Minimal enteral feeds to support normal gastrointestinal mucosa
  - Enteral feeds to potentiate gastrointestinal development (?)
- Design written therapeutic plan
  - What nutrient solution: parenteral, enteral, or both
  - Solution composition for parenteral
  - Concentration of solution (add modular nutrients?) for enteral
  - Planned rate of advancement
  - Rate of delivery of solution
  - How administered:
    - For parenteral, continuous, or cycled
    - For enteral, use bolus, continuous administration, cycled administration, or a combination
  - Site of delivery:
    - For parenteral, central, or peripheral
    - For enteral, stomach, small bowel
  - Determine tools used to monitor each patient
- Monitor
  - Daily review flow sheets
  - Record amount of solution administered at rate administered
  - Record fluid composition
  - Calculate percent of nutrients from therapeutic plan actually delivered
  - Record adverse reactions
  - Record all medications administered enterally
- Reassess
  - Repeat initial assessment
  - Reevaluate goals in light of nutritional therapy and changes in metabolic/disease state
  - Reaffirm goals and therapeutic plan
  - Modify plan

Table 30.2. Nutritional Assessment

Historic	Biochemical
Recall dietary history	Electrolytes
Three day food diary	Acid base status
Calorie count in hospital	Minerals
Anthropomorphics	Visceral proteins
Weight	Renal function
Height	Liver function
Weight for height	Trace elements
Body mass index	Vitamins
Head circumference	Serum triglycerides
Midarm muscle circumference	
Triceps skinfold thickness	
Z-score for each measurement	

and a therapeutic plan must be developed. Once nutrition support begins, careful monitoring and reassessment on a frequent basis must occur.

For infants and children, a nutritional assessment consists of measurements of weight, height, head circumference, triceps skinfold thickness, and midarm muscle circumference (Table 30.2). In addition to the usual anthropometric measures, this table includes Z-scores, which denote the units of standard deviation from the median. The use of Z-scores permits the comparison of an observation to the normal curve. It detects movement toward or away from the median and is more sensitive than percentile changes.[38] In addition, the measurement of hemoglobin, hematocrit, and serum proteins and liver and renal function tests are helpful.

The nutritional status of a child is important, because children who are malnourished are at greater risk for infections and metabolic problems. In addition, malnourished children may require more of all nutrients, especially phosphorus and potassium.[39] Table 30.3 lists some criteria for malnutrition. Table 30-4 lists some useful criteria for deciding when to intervene with nutrition support.

Table 30.3. Criteria for Malnourished State

Weight for height < 70% of standard
Height age < 85% of standard
Weight age < 64% of standard
Plus
Weight loss of ≥ 10% of body weight within 1 to 4 weeks
Decrease in two growth channels within 1 to 4 weeks
No weight gain for 1 to 3 months in a prepubescent child
Triceps skinfolds < 5th percentile for age and/or obvious low fat stores
Any of the above plus hypoalbuminemia (serum albumin < 2.0 mg/dL, pre albumin < 11.0 mg/dL)

Table 30.4 When to Begin Nutrition Support

Condition	Begin Nutrition Support
Malnourished child	
Low birth weight, premature	Immediately
<1 year of age	by 24 hours
1 to 5 years of age	by 48 hours
5 to 18 years	by 72 hours
Well-nourished child	
<1 year of age	by 3 to 5 days
1 to 5 years of age	by 5 to 7 days
5 to 18 years	by 5 to 7 days
Optimize available intravenous fluid with 10% dextrose initially	

Nutritional goals vary with the degree of malnutrition, level of metabolic stress, and specific nutrient deficiencies that may exist. In general, at least the basal metabolic requirements need to be supplied on a daily basis. Additional needs can be met by increasing the concentration of nutrients, the rate at which they are delivered, or the length of time over which they are administered. Specific deficiencies, such as iron, calcium, phosphorus, zinc, and so on can be corrected by providing extra nutrients into the parenteral solution, or as supplements with enteral feeds.

The therapeutic plan consists of a written set of orders that clearly establishes the type of nutrition support and its implementation. If enteral or parenteral nutrition support is to be supplied, the concentration of the enteral or parenteral fluid, how that concentration is achieved, the volume of the nutrient solution to be delivered, the rate at which it must be administered, the length of time over which it must be administered and any additives, such as vitamins, minerals, or trace elements that must be included are clearly written.

Reassessment is vital to the successful delivery of nutrition support. It consists of an ongoing evaluation of the patient with respect to anthropometries, biochemistries, delivery of nutrients (did the patient actually receive the volume and concentration of nutrition solutions ordered?), and changes in disease state.

Supplements

Children with normally functioning gastrointestinal tracts who do not eat enough or have not been offered appropriate foods may develop failure to thrive. The addition of high calorie foods to the diet may be all that is necessary to improve nutrition and achieve catch up growth. If poor growth occurs in infancy because of inadequate intake or increased losses (for example, reflux), the first step in nutrition support can be concentration of the formula (Table 30.5) or addition of breast milk fortifiers to expressed mother's milk. Children who require only additional foods or concentration of their formula must receive the same assessment as those who are candidates for parenteral nutrition, since even single nutrient deficiencies, such as zinc, may alter growth.[40–43] If, with the addition of high-calorie foods, catch up growth does not occur, oral supplements may be considered.

Enteral Feeds

The enteral route is the preferred route for nutrition support unless a specific contraindication exists. The reasons the enteral route is preferred are numerous and have been reviewed recently.[44] Essentially, feeds infused into the gastrointestinal tract prevent atrophy, may induce maturation, may prevent breakdown of the

Table 30.5 Example for Increasing the Caloric Density of Infant Formula[a]

Step	Caloric Density
Start	20 kcal/oz (0.67 kcal/mL) [standard infant formula]
	↕
Step 1:	Add 4 kcal/oz (0.13 kcal/mL) by formula concentration *[One 13 oz can of liquid infant formula concentrate plus 8 oz of water]* 24 kcal/oz (0.8 Kcal/mL)
	↕
Step 2:	Add 4 kcal/oz (0.13 kcal/mL) of carbohydrate *[3 tablespoons of powdered polycose per 500 mL (16.6 oz)]* 28 kcal/oz (0.93 kcal/mL)
	↕
Step 3:	Add 4 kcal/oz (0.13 kcal/mL) of fat *[15 mL (1 tablespoon) Microlipid per 500 mL (16.6 oz)]* 32 kcal/oz (1.06 kcal/mL)
	↕
Step 4:	Add 4 kcal/oz (0.13 kcal/mL) by formula concentration *[One 13 oz can of liquid infant formula concentrate plus 5.5 oz of water]*
Finish[b]	36 kcal/oz (1.2 kcal/mL)

[a]Nutrient Guidelines: carbohydrate, 35 to 65% of total kcal; protein, 05 to 16% of total kcal; fat, 30 to 55% of total kcal.

[b]Final formula:	36 calories per ounce	(1.2 calories/mL)
	Base formula, 28 kcal/oz	(7% as protein kcal)
	Carbohydrate, 4 kcal/oz	(45% as carbohydrate kcal)
	Fat 4 kcal/oz	(48% as fat kcal)
	Estimated Renal Solute Load	135 mOsm/L
	Estimated Osmolality	429 mOsm/kg water
	Estimated Free Water	80.5%

gastrointestinal barrier, and may directly supply nutrients to enterocytes. The enteral route permits the provision of some nutrients that may be important to the gastrointestinal tract and are not available in a parenteral form. The gastrointestinal tract and its vascular system modulates nutrients before they reach the systemic circulation by passage through the portal circulation. This modulation does not occur if nutrients are infused directly into a systemic vein.

In certain situations, it may not be possible to supply all necessary nutrients enterally. Nevertheless, as much as possible of the total nutritional needs should be supplied enterally. Gastric and colonic hypomotility may occur in severely injured children, whereas the small intestine continues to maintain good motility and absorption. Bowel sounds, produced by the movement of air through the intestines, may not always be a good indicator of small bowel function, especially if a nasogastric tube is on constant suction. Instillation of air into the small bowel via a duodenal tube may restore bowel sounds.[45] Sick and injured children who are vomiting or children with malnutrition and/or vomiting and reflux may be fed enterally into the small bowel if the stomach is decompressed. Even very small amounts of hypotonic nutrition solutions may be beneficial to the gastrointestinal tract.[46,47]

Table 30.6 lists criteria for initiating enteral feedings in children with chronic problems and Table 30.7 lists some indications for using enteral feedings. Once enteral nutrition support is identified as the nutrition support of choice, the nutritional formula (Table 30.8), concentration, route of delivery (Table 30.9), and rate of advancement (Table 30.10) must be established.

Enteral Formulas

The choice for an enteral formula must be individualized since factors such as nutritional requirements, fluid requirements, age, medical condition, gastroin-

Table 30.6 Indications for Enteral Nutrition Support

Impaired energy consumption	From 50% to 60% of recommended daily amount despite high-calorie supplements plus total feeding time more than 4 to 6 hr/day
plus	
Severe and deteriorating wasting	Weight for height > 2 standard deviations below the mean plus skinfold thickness below the 5th percentile
and/or	
Depressed linear growth	Fall in height velocity > 0.3 standard deviations per year or height velocity < 5.0 cm/year or decrease in height velocity of at least 2 cm from the preceding year during early to midpuberty

Source: Reproduced with permission from Davis.[48]

Table 30.7. Some Indications for Enteral Feedings

Limited ability to eat
Neurologic disorders
Acquired immunodeficiency syndrome
Facial trauma
Tumors of face, mouth, or esophagus
Injury of face, mouth, esophagus
Congenital abnormalities of face, mouth, esophagus
Prematurity, less than 34 weeks gestational age
Inability to meet full nutrient needs orally
Increased metabolic needs
Anorexia, especially from chronic disease
Psychological disorders
Altered absorption or metabolism
Chronic diarrhea
Short small bowel
Inflammatory bowel disease
Glycogen storage disease types I and III
Gastroesophageal reflux
Pseudoobstruction
Pancreatitis
Amino or organic acidopathies

Source: Adapted from Davis.[48]

testinal function, food intolerance, and so on must be considered. In general, pediatricians group enteral formulas according to the ages of children for whom they will be used: Premature infants, term infants, children aged 1 to 10 years, and children older than 10 years.

Premature infants. Formulas designed for premature infants have higher calorie, protein, vitamin, and mineral concentrations and lower lactose than term-infant formulas. These formulas also contain more medium chain triglycerides (MCTs). Because of concerns over gastrointestinal immaturity, the formulas designed for premature infants are recommended for infants with weights of approximately 2.0 to 2.5 kg with normal alkaline phosphatase and serum albumin levels. These specially designed formulas supply adequate nutrients for premature infants when used as the sole nutrient source. The formulas are 24 kcal/oz (0.8 kcal/mL). Breast milk is not adequate as a sole nutrient source for premature infants. Breast milk must be supplemented with calories, protein, vitamins, and minerals, including zinc, calcium, and phosphorus. Human milk fortifiers can be added to human milk to achieve adequate enteral nutrition. Nevertheless, premature infants are at risk for rickets and trace elements deficiencies.[49,50]

Table 30.8. Pediatric Enteral Formulas

Clinical Condition	Formula Description
Premature Infant	Premature formula—12% protein, contains MCT oil, carbohydrate, lactose/glucose polymers, calcium and phosphorus
Term Infant	
Primary or secondary lactose intolerance	Lactose-free cow's milk formula
Primary or secondary lactose intolerance or cow's milk sensitivity	Lactose-free, soy or hydrolyzed protein formula (sucrose- and corn-free)
Renal or cardiac disease	Low electrolyte/renal solute load formula
Steatorrhea associated with bile acid deficiency, ileal resection, or lymphatic anomalies	MCT oil containing formula
Cow's milk protein and soy protein sensitivity; abnormal nutrient absorption, digestion and transport intractable diarrhea or protein-calorie malnutrition	Hypoallergenic, hydrolyzed casein, lactose- and sucrose-free
1–10 years	
Tube feeding	Complete nutrition in 1100 mL, Intact protein, 1.0 kcal/mL, glutein-free, lactose-free, isotonic, appropriate minerals for children
Oral supplement	
Over 10 years	
Normal GI function, varying medical conditions	Hypercaloric formula
Fluid restriction	
Abnormal bowel movement	Added fiber
Pulmonary problems/diabetes	High-fat formula
High-stress trauma, sepsis, burns	Hypercaloric, high-protein formula
Lactose intolerance	Lactose-free formula
Compromised GI/pancreatic function	Chemically defined formula
Protein allergy	Elemental formula
Impaired renal function	Low protein, branched-chain amino acid enriched

Term infants. Term infants with normal gastrointestinal function can be fed human milk or infant formula for the first year of life. If human milk is used as the sole nutrient source, supplementation may be necessary at approximately 6 months of age when the infant's need for iron outstrips the quantity in human milk. If human milk is used for continuous tube feedings, care must be taken to

Table 30.9. Routes for Delivery of Enteral Nutrition

Site	Advantages	Disadvantages	Complications
Nasogastric	Short-term No surgery needed	Vomiting, reflux	Tube dislodgement Possible tactile sensitivity
Passed by mother or child nightly	No visible tube during the day	Difficult for mother/child, may irritate nares	May interfere with development of oral motor skills
Indwelling	Infrequently passed	Visible during the day	
Gastrostomy	More stable tube placement, greater patient mobility, doesn't interfere with development of oral motor skills	Requires invasive procedure, may require local skin care	Stomal infections, abdominal leak, tube misplacement
Percutaneous Gastrostomy (PEG)	Easy placement, rapid institution of feedings	Vomiting, reflux not prevented	Tethered colon
Surgery	General anesthesia, longer time to institution of feedings	Vomiting, reflux not prevented	Complications of surgery
Jejunal	Bypasses stomach and pylorus, decreased risk of aspiration, decompress stomach while feeding	Continuous feeds	Tube dislodgement, bacterial overgrowth, malabsorption, dumping
Nasojejunal	No surgery needed	May interfere with development of oral motor skills	
Surgically placed	More stable, no visible tube on face	General anesthesia, complications associated with surgery	

Table 30.10 Guidelines for the Initiation and Advancement of Enteral Feeds

Age	Initial Infusion	Advances	Final Goal
		CONTINUOUS FEEDINGS	
Preterm	1–2 mL/kg·hr^{-1}	10–20 mL/kg·day^{-1}	120–175 mL/kg·day^{-1}
0–12 mo	1–2 mL/kg·hr^{-1}	1–2 mL/kg·q2–8 hr	6 mL/kg·hr^{-1}
1–6 yr	1 mL/kg·hr^{-1}	1 mL/kg·q2–8 hr	4–6 mL/kg·hr^{-1}
> 7 yr	25 mL/hr	25 mL/q2–8 hr	100–150 mL/hr
		INTERMITTENT FEEDINGS	
Preterm (>1200 g)	2–4 mL/kg·feed^{-1}	2–4 mL/feed	120–175 mL/kg·day^{-1}
0–12 mo	10–15 mL/kg·q2–3 hr (30–60 mL)	10–30 mL/feed	20–30 mL/kg·q4–5 hr
1–6 yr	5–10 mL/kg·q2–3 hr (60–90 mL)	30–45 mL/feed	15–20 mL/kg·q4–5 hr
> 7 yr	90–120 mL·q3–4 hr	60–90 mL/feed	300–480 mL/kg·q4–5 hr

Source: Reproduced with permission from Davis.[71]

adjust for nutrient losses caused by separation of fat and adherence of the fat to the tubing.[51–53] Approximately 20% of fat from human milk adheres to the tubing, and those calories that must be considered when calculating enteral feedings. Simply flushing the tubing with water may cause a fat bolus. Adherence of fat to tubing is not a problem with short, intermittent feedings.[54]

Standard infant formulas are available in powder, liquid concentrate, and ready-to-feed forms. In general the powdered forms are the least expensive and the ready-to-feed the most expensive. A proliferation of infant formulas has made it difficult to keep up with each new product. There are, however, major groupings with respect to protein, carbohydrate, fat, and mineral content.

PROTEIN. The most commonly used infant formulas are based on cow's milk protein and can be predominately whey (the liquid portion of milk) or casein (the curd or solid portion). Human milk has proportionately more whey than casein. Soy-based formulas use soy bean protein with the addition of methionine and heat treatment to reduce the activity of trypsin inhibitors and hemagglutinin normally found in soy protein isolates. Because soy-based formulas are also lactose free they are indicated for infants with cow's milk protein intolerance and lactase deficiency. Hydrolyzed protein formulas are made from cow's milk casein or cow's milk whey protein. The protein is hydrolyzed, or partially hydrolyzed, to polypeptides. Hydrolyzed protein formulas can be used for infants with protein intolerance, pancreatic insufficiency, and after gut injury. If food allergy is the basis for the use of a hydrolyzed formula, caution must be exercised to use only formulas that have been shown clinically to be nonallergenic.[55] Partially hydrolyzed protein formulas may contain polypeptides of sufficient length to cause allergy.[56,57]

CARBOHYDRATE. Lactose is the main carbohydrate constituent of human milk as well as most formulas based on cow's milk. For lactose intolerance, whether congenital or acquired, a cow's milk-based lactose-free formula is available. Soy-based formulas are lactose free, as are most of the hydrolyzed protein formulas.

FAT. The fat content of infant formulas can vary with the product and the manufacture. Some fat combinations may be more desirable than others, especially for the long-term growth and development of normal infants. However, for sick or injured infants, the nutritional issues often revolve around the content of MCT. Medium-chain triglyceride (MCT) oil is important in the management of such problems as steatorrhea, chylous ascites, chylothorax, intestinal lymphangiectasia, and ileal resections. Infants on formulas containing 86% MCT oil must be carefully monitored, because they may develop essential fatty acid deficiency with long-term use.[58,59]

IRON. Infant formulas can be purchased with or without iron supplementation. All infant formulas have adequate iron except for cow's milk-based formulas that are labeled low-iron formula. The indications for the use of a low-iron formula in infancy are almost nonexistent. Low-iron formulas should not be prescribed. Interestingly, the soy formulas and the hydrolyzed formulas have adequate iron, as do all the infant formulas manufactured in England and Europe.

CALORIC DENSITY. The caloric density of infant formulas is 20 kcal/oz (0.67 kcal/mL), the same as human milk. The formulas can be concentrated, either by decreasing the amount of water in which they are diluted or by using modular nutrients. Infant formulas can be safely concentrated to 1.0 kcal/mL by adding less water to concentrated liquid or powdered preparations.[60] Because concentrated infant formulas contain less water than standard formulas, careful attention to free water requirements is necessary. If water intake is inadequate, the capacity of the kidneys to concentrate and excrete renal solute may be exceeded and the infant may become dehydrated. The renal solute load (RSL) consists of nonmetabolizable nutrients, electrolytes, and metabolic products such as urea. The RSL can be estimated as follows:

$$\text{RSL (mOsm)} = [\text{protein (g)} \times 4] + [\text{Na (mEq)} + \text{K (mEq)} + \text{Cl (mEq)}]$$

To estimate the potential renal solute load (PRSL) formula fed to low-birth weight infants, the following formula is used:[61]

$$\text{PRSL (mOsm)} = \frac{\text{protein (g)}}{0.175} + [\text{Na (mEq)} + \text{K (mEq)} + \text{Cl (mEq)} + \frac{\text{P (mg)}}{31}]$$

For infants with increased fluid losses (fever, diarrhea, sweating), with impaired renal concentrating ability, or who are unable to express thirst, dehydration is a risk. Additional free water can be given as a bolus as needed. Caloric supplements, such as fats and carbohydrates do not increase the RSL. The osmolality of standard infant formulas (90.67 kcal/ml) ranges from 150 to 380 mOsm/kg. The American Academy of Pediatrics recommends that the osmolality of infant formulas be less than 460 mOsm/kg.[62]

Infant formulas can also be concentrated by adding modular nutrients. Modular nutrients can be used to increase the caloric density of the formula by adding carbohydrate or fat modules, and to increase the protein content of the diet. Modular nutrients can also be used to alter the nutrient composition of diets. Whenever nutrient modules are used, the final composition of the diet must be calculated, because a specific nutrient may be diluted or otherwise inadequate in the final diet. For example, the most commonly added nutrient module[63] is a calorie source, such as carbohydrate or fat. If an additional 10 kcal/oz (0.33 kcal/ml) is added to a limited volume of formula, the protein, vitamins, and minerals may be inadequate in the final diet. In contrast, the addition of a noncarbohydrate, nonfat nutrient module may increase the RSL of the final diet.

CHILDREN AGED 1 TO 10 YEARS. Enteral nutrition formulas for children are more calorically dense than infant formulas, but contain less protein, sodium, potassium, chloride, and magnesium than adult formulas. However, they contain more iron, zinc, calcium, phosphorus, and vitamin D than adult formulas.[48] Thus, adult preparations should not be used for children unless there is a specific indication. If such an indication exists, the children must be carefully monitored and supplements of zinc, iron, calcium, phosphorus, and vitamins given. Blenderized diets, either commercially available, or prepared in the home can be used for enteral nutrition support. These feedings generally have a higher osmolality and viscosity than prepared formulas. Home prepared blenderized diets are cheaper than prepared formulas and may have psychosocial importance for the family. However, home blenderized diets may have higher bacterial counts, require daily labor, may cause hypernatremic dehydration, and can result in specific nutrient deficiencies if not carefully monitored by a nutritionist.[64,65]

CHILDREN OVER 10 YEARS. Children over 10 years of age can be fed an adult formula.

ENTERAL NUTRITION

Enteral feedings can be initiated in children who are on parenteral nutrition support with no or little enteral intake for prolonged periods of time or in children who are on a full diet (Table 30.11). In general, a slow continuous infusion of nutrients is better tolerated as an initial prescription for enteral feeds than bolus feeds. Over time, slow continuous feeds, such as nocturnal feeds are preferable, no matter what the indication for enteral nutrition support, since continuous feeds are associated with better nutrient absorption and less aspiration than bolus feeds.[66–70]

Table 30.11 Example of Transition from Parenteral to Enteral Nutrition (4 kg infant)[a]

Day	Transition[b]	mL /hr	kcal/kg	g protein/kg	Enteral Goal (%)
	Parenteral nutrition (PN): 2% amino acid, 2.5% fat, 20% dextrose, 0.93 NP kcal/mL	20	111	2.4	0
1	FS formula = 0.67 kcal/mL	4	16	0.4	16
	PN = 0.93 NP kcal/mL	20	111	2.4	
2	FS formula	8	32	0.7	32
	PN	16	89	1.9	
3	FS formula	12	48	1.1	48
	PN	12	67	1.4	
4	FS formula	16	64	1.4	64
	PN	8	44	1.0	
5	FS formula	20	80	1.8	80
	PN	4	96	0.5	
6	FS formula	24	96	2.2	100
	Discontinue PN				
7	FS formula—concentrate to 24 kcal/oz (0.7 kcal/mL)	24	115	2.6	100

Source: Reproduced with permission from Davis.[71]

[a]Estimated daily needs:

Calories 100–120 kcal/kg·day^{-1}

Protein 2–3 g/kg·day^{-1}

Fluid 100–175 mL/kg·day^{-1} (17–30 mL/hr)

Selected enteral formula

Hydrolyzed protein, 60% MCT, lactose free, elemental

Formula concentration is full strength (FS) (20 kcal/oz, 0.67 kcal/mL)

Abbreviations: MCT, medium-chain triglycerides; NP, nonprotein.

[b]Formula volume and concentration should not be increased simultaneously.

Transitions

Parenteral to enteral nutrition. There are few controlled studies comparing possible feeding schedules. In general, an isotonic, full-strength, lactose-free formula is well tolerated as a first enteral feeding. If the gastrointestinal tract has not been used for 5 to 7 days, if gastrointestinal injury has occurred, or if malabsorption or food allergy are present, then a hydrolyzed protein-based (elemental) formula should be chosen. Controversy exists over whether a diluted formula should be used. No well-controlled studies exist that demonstrate an advantage to the use of a dilute formula. Greater caloric intake can be achieved when full-strength formula is used and our experience suggests a full-strength formula is generally well tolerated. We initiate enteral nutrition with low-volume, full-strength formula. The transition from parenteral to enteral nutrition is gradual.[71] Approximately 1 week is required for an uncomplicated transition from parenteral to enteral nutrition. In some patients with short small bowel or intractable diarrhea, the transition time to full enteral feeds may be very long.[72]

Once formula tolerance is established at 15% to 50% of the desired volume, the formula concentration may be gradually increased. Generally, enteral feedings begin at 1 to 2 $mL/kg \cdot hr^{-1}$ and are advanced (Tables 30.11 and 30.12). As enteral feedings reach 35% to 50% of calculated enteral requirements, parenteral nutrition may be tapered. Tapering parenteral nutrition while advancing enteral feeds prevents rebound hypoglycemia and fluid overload. When enteral feedings are providing 75% to 100% of enteral requirements, parenteral nutrition may be discontinued. When parenteral fluids are finally stopped, the enteral infusion must meet the child's full fluid requirement.

Initiation of enteral feedings. The initiation of enteral feeds in infants or children with an intact gastrointestinal tract who have not been on parenteral nutrition is similar to the transition from parenteral to enteral nutrition. Generally a polymeric, isotonic, undiluted formula is well tolerated. The feedings can begin as a nocturnal supplement (for example, in children with failure to thrive) at 1 to 2 $mL/kg \cdot hr^{-1}$ and after 2 to 4 hr increasing the rate by 1 to 2 $mL/kg \cdot hr^{-1}$ until the desired volume is achieved. This process may take more than one night. Alternatively, continuous, 24-hr feedings can begin and the feeds cycled after the full volume is reached.

Continuous to intermittent feedings. Intermittent feedings are administered several times a day, each feeding lasts for 15 to 45 min. Continuous enteral feeds are indicated for transpyloric feeding, for patients with no enteral intake for more than 5 to 7 days, for patients with limited absorption due to bowel resection, gastrointestinal injury, or malabsorption, and for patients who are at risk for aspiration or who do not tolerate intermittent feedings. Intermittent feedings

Table 30.12 Transition from Parenteral to Enteral Nutrition (40-kg adolescent)[a]

Day	Transition[b]	mL/hr	kcal/kg	g protein/kg	Enteral Goal (%)
	Parenteral nutrition (PN): 3% amino acid, 4% fat, 17.5% dextrose, 1.0 NP kcal/ml	90	54	1.6	0
1	FS formula (30 kcal/oz, 1.0 kcal/ml)	25	15	0.5	25
	PN (1.0 NP kcal/mL)	90	54	1.6	
2	FS formula	50	30	0.9	50
	PN	65	39	1.2	75
3	FS formula	75	45	1.4	75
	PN	40	20	0.7	59
4	FS formula	100	60	1.9	100
	Discontinue PN				

Source: Reproduced with permission from Davis.[71]

[a]Estimated daily needs:

Calories 50–60 kcal/kg·day^{-1}

Protein 1–2 g/kg·day^{-1}

Fluid 47–75 mL/kg·day^{-1} (80–125 mL/hr)

Selected enteral formula

Semielemental (350 mOsm/kg), peptides, 35% fat with MCT, lactose free

Formula is full strength (FS)

allow patients to be mobile; they simulate oral feeds and are associated with a more natural hunger–satiety cycle than continuous feeds.[73] However, compared to continuous feeds, diarrhea, cramping, dumping syndrome, delayed gastric emptying, and aspiration are more likely with bolus feeds. Also, better nutrient absorption occurs with continuous feeds.[69] Overlapping continuous with intermittent feeds is economically and nutritionally more efficient than sudden discontinuation of continuous feeds. Optimum infusion rates for intermittent feeds are not greater than 30 mL/min.[74]

Continuous to cycled feedings. Cycled feeding refers to the administration of enteral tube feedings by continuous drip infusion for a period of less than 24 hr. Usually uninterrupted infusions are 8 to 18 hr in duration, depending on individual patient requirements and tolerance (Tables 30.13 and 30.14). Cycled feedings are delivered with a pump and are usually administered overnight to enable the infant or child to feed orally during the day. Cycled feedings are associated with less abdominal discomfort than intermittent feedings and are better absorbed.[69] Infants normally tolerate up to 6 mL/kg·hr^{-1} over a 24 hr period, but they also can tolerate approximately 10 to 12 mL/kg·hr^{-1} cycled over an 8- to 12-hr period. Children receiving 3 to 5 mL/kg·hr^{-1} over 24 hr can progress to

Table 30.13 Transition/Combination from Continuous Enteral Feeds to Intermittent and Cycled Enteral Feeds in a 4-kg Infant[a]

Day	Formula	mL/hr	Hours Infused	kcal/kg	g protein/kg
		CONTINUOUS TO INTERMITTENT			
	0.7 kcal/mL formula	24	24	115	2.6
8	0.7 kcal/mL formula	40	q2 h	96	2.1
9	0.7 kcal/mL formula	60	q3 h	96	2.1
10	0.7 kcal/mL formula	80	q4 h	96	2.1
		CONTINUOUS TO CYCLED			
	0.7 kcal/mL formula	24	24	115	2.6
8	0.7 kcal/mL formula	28	20	112	2.5
9	0.7 kcal/mL formula	32	18	115	2.6
10	0.7 kcal/mL formula	36	16	115	2.6
11	0.7 kcal/mL formula	40	14	112	2.5
		CONTINUOUS TO CYCLED AND INTERMITTENT (I)			
	0.7 kcal/mL formula	24	24	115	2.6
8	C-formula[b]	28	17	95	2.1
	I-formula	28	3 feeds	17	0.4
9	C-formula	32	14	90	2.0
	I-formula	38	3 feeds	23	0.5
10	C-formula	36	12	86	1.9
	I-formula	48	3 feeds	29	0.6
11	C-formula	40	10	80	1.8
	I-formula	60	3 feeds	36	0.8

Source: Reproduced with permission from Davis.[71]

[a]Estimated daily needs:

Calories 100–120 kcal/kg·day^{-1}

Protein 2–3 g/kg·day^{-1}

Fluid 100–175 mL/kg·day^{-1}

[b]Formula concentration is 0.7 kcal/mL.

a cycled infusion of 120 to 150 mL/hr. When maximal volume is reached, as assessed by patient tolerance, the caloric density of the formula can be increased (1.0 to 2.0 kcal/mL). This method is often used in patients with cystic fibrosis, tracheoesophageal fistula, bronchopulmonary dysplasia, and inflammatory bowel disease. If children receive nocturnal cycled feeds, the head of the bed should be elevated and, if reflux is present bethanecol, metoclopramide, or cisapride may be useful.

Enteral to oral feedings. The goal of all nutrition in pediatrics is to supply nutrients for optimal growth and development while supporting developmentally appropriate eating behavior. Often the development of the normal eating process is interrupted and delayed.[75,76] Development of mouth sensitivity with distinct oral defensive behaviors and lack of the hunger–satiety cycle due to around-the-clock feedings can make the transition from tube to oral feeds long and difficult. If children are deprived of appropriate oral stimulation during critical development phases, feeding difficulties will arise.[33] Initiating oral feedings in children who have been fed by tube can evoke a resistant or fearful response, such as gagging, choking, or vomiting.[75] Thus, feeding programs for infants must include a speech therapist or occupational therapist to assure a vital oral component is in place at the time tube feedings are started. This will help both the infant and family with oral stimulation of feeding behaviors. See Tuchman[32] for an in depth review of this topic with specific treatment recommendations.

The presence of medical complications is a critical factor influencing the length of time to full transition to oral feeding; the process may vary from days to years. Table 30.15 lists important evaluations that must be in place before initiating a transition to oral feeding.

To begin oral feeds, normalize feedings to occur approximately at times for meals and snacks. Connect the tube feeding with the process of eating. Introduce the sensation of hunger followed by feeding to alleviate that sensation. This is a gradual process during which behavioral problems may be prominent. The abrupt discontinuation of tube feeds to stimulate oral intake is rarely successful.

Table 30.14 Transition of Continuous Enteral Feeds to Cycled/Intermittent Enteral Feeds in a 40-kg Adolescent[a]

Day	Transition to Cycled Feeds	mL/hr	Hours Infused	kcal/kg	g protein/kg
	1.0 kcal/mL semielemental formula	100	24	60	1.9
8	1.5 kcal/mL polymeric[b] formula	100	16	60	2.2
9	1.5 kcal/mL polymeric formula	125	12	56	2.0
10	1.5 kcal/mL polymeric[c] formula	150	10	56	2.0

Source: Reproduced with permission from Davis.[71]

[a]Estimated daily needs:

Calories 40–60 kcal/kg·day^{-1}

Protein 1–2 g/kg·day^{-1}

Fluid 47–75 mL/kg·day^{-1}

Enteral formula selection—from semielemental (1.0 kcal/mL) to polymeric, lactose-free formula (1.5 kcal/mL) with 77% free water, higher calorie for volume tolerance.

[b]Formula volume and concentration not increased simultaneously.

[c]Day 10 of cycled tube feeding provides 1155 mL of free water; if tube feeds are the only intake source, an additional 725 mL of water/day will be needed.

Table 30.15 Evaluation before Transition to Oral Feedings

Factor	Status
Original indication for enteral nutrition	Not improved—do not attempt to transition to oral feedings. Try to resolve initial problem first Improved or resolved—begin transition
Quality of oral motor skills	Determine level of skills: can child suck, chew? can child make a bolus? can child initiate coordinated swallowing? can child swallow without aspirating? If skills are not adequate, do not begin transition. Resolve skill problem, if possible.
Parent readiness	Parents agree with transition Parent taught skills Support for parents—process can be long and frustrating and will fail if parents do not have adequate support Support for behavioral problems, which are common and include food aversion, negative or disruptive behaviors during feeding, refusal to self-feed, and power struggles

Careful monitoring of fluids, anthropomorphics, vitamin, and mineral intake as well as motor skill is vital for the transition to full oral feeds.[77]

Monitoring of enteral feeds

To properly manage infants and children on enteral feeds, food tolerance, metabolic, mechanical, gastrointestinal, nutritional, and growth parameters must be carefully and frequently assessed. Table 30.16 outlines parameters useful to monitor children on enteral nutrition support. Tolerance to enteral feeds is followed by noting the presence or absence of vomiting, diarrhea, and abdominal distention. Gastric residuals are not routinely checked unless the child is at risk for aspiration. A single high gastric residual (more than 1.5 to 2 times the hourly rate of formula administration) alone is not enough reason to discontinue enteral feeds. Medications instilled through the feeding tube that decrease gastric motility, such as paralytic agents and morphine or use of a high osmolar formula may alter gastrointestinal motility. Decreasing the infusion rate by half for a few hours may be sufficient to resolve the problem with high gastric residuals. If not, the use of transpyloric feeds or metoclopramide or cisapride can be considered.

For children on long-term enteral feedings, careful monitoring of growth is essential, because the goal for enteral feeding therapy is to promote normal growth and development. Mechanical complications, such as occluded tubes and malposition of tubes generally require replacement of the apparatus with a new one.

Table 30.16 Monitoring Pediatric Nutrition Support

Parameter	Initially	Later/Home
Growth		
Weight	daily	daily to weekly
Height	weekly	weekly
Head circumference	weekly	weekly
Body composition		monthly
Laboratory[a]		
Electrolytes	daily to weekly	weekly to monthly
BUN/Cr	weekly	monthly
Ca, PO_4, Mg	twice weekly	weekly to monthly
Acid-base status	until stable	weekly to monthly
Alb/pre-Alb/transferrin	weekly	weekly to monthly
Glucose	daily	weekly to monthly
Triglycerides	while increasing lipid	weekly to monthly
CBC with diff	weekly	weekly to monthly
Fe/TIBC/retic count	as indicated	as indicated
Trace elements	monthly	every 2 to 4 months
Platelet count	weekly	as indicated
Liver function tests	weekly	weekly to monthly
Folate/vitamin B_{12}	as indicated	as indicated
Fat soluble vitamins	as indicated	every 6 to 12 months
Carnitine	as indicated	as indicated
Urinalysis		
Glucose	2 to 6 times/day	daily to weekly
Additional Parameters for Enteral Feeds		
Gastrointestinal		
Gastric residuals	2 hourly	prn
Vomiting	daily	prn
Stools:		
Frequency	daily	prn
Reducing substances	initially	prn
Ova/parasites	prn	prn

[a]For metabolically unstable patients: labs need to be checked more frequently.

Sources: Levy, Winters, and Heird[107] and Kerner.[108]

Complications

Table 30.17 lists potential complications of tube feedings and management suggestions. These complications are similar to those that occur in adults and are reviewed in depth by Davis.[48]

Table 30.17 Possible Complications of Tube Feedings

Problem	Management Suggestion
	MECHANICAL
Tubes	
Improper size	Change tube to appropriate size
Improper placement	Change placement
Aspiration	Elevate head of bed 30°–45°, confirm tube placement, use continuous feed rather than bolus, consider antireflux medication
Occlusion	Flush before and after intermittent feedings and every 8 hr with continuous feedings, Thoroughly mix powder additives, use only liquid preparations of medications, crush medications well if must use a pill
Medications	Use liquid whenever possible, assess physical compatibility of drug/formula, avoid mixing formulas with liquid medications with pH < 5.0.
	METABOLIC
Overhydration, malnutrition, refeeding	Monitor input and output, include all other fluid intake (oral, intravenous)
Dehydration	Evaluate osmolality of formula, provide more fluids
Diarrhea	Culture stools, for *C. difficile* toxin titer, evaluate osmolality of all medications, could medications be acting as GI stimulants, evaluate osmolality of formula, test stools for pH and reducing substances, give additional fluids
Hyperglycemia	Slow or stop feeding, monitor blood sugar, reduce carbohydrate, give insulin if diabetic
Hyperkalemia	Change formula, give potassium binders, give insulin and glucose, correct acidosis
Hyperphosphatemia	Change formula, use phosphate binder, give calcium supplement
Hypokalemia	Give potassium, monitor electrolytes, evaluate adequacy of formula
Hypophosphatemia	Give phosphorus, evaluate formula
Hyponatremia	Evaluate fluid balance; if overhydrated restrict fluids, evaluate adequacy of formula

continued on next page

Table 30.17 *Continued*

Problem	Management Suggestion
Fatty acid deficiency	Change formula, add modular fat, add 5 mL safflower oil
Abnormal liver function tests	Determine etiology, evaluate formula in light of liver status
Rapid or excessive weight gain	Evaluate electrolytes, evaluate fluid balance, decrease amount or concentration of formula
Azotemia	Decrease protein
Congestive heart failure	Decrease sodium, slow rate, give diuretics
Inappropriate weight gain	Evaluate macro- and micronutrient intake, monitor daily input and output, correct deficiencies
GASTROINTESTINAL	
Diarrhea	
Mucosal atrophy	Use isotonic or dilute hypertonic formula, start at a slow rate and increase gradually
Medications (changes motility, flora, increases osmolality when given with feedings)	Change time medication is given, type, or prescribe an antidiarrheal agent, check sorbitol content of medication
Hyper osmolar solution	Dilute to isotonicity and slowly increase concentration
Rapid delivery	Slow rate and gradually increase
Bacterial contamination	Use aseptic preparation techniques and limit time the formula hangs in bag
Intolerance of formula component	Use formula without intolerant component (e.g., if lactose intolerant, use lactose-free formula)
Malabsorption	Use elemental or semi-elemental formula, MCT oil
Gastric residuals	
High osmolality	Dilute to isotonicity and slowly increase concentration
High fat content	Consider changing formula to one having < 30% to 40% total calories from fat
Intermittent feedings	Consider continuous feeding
Medications that slow peristalsis	If unable to stop medication, consider addition of medications that stimulate peristalsis
Gastroparesis	Small bowel feedings
Gastric residuals on initiating feedings	Advance slowly

continued on next page

Table 30.17 *Continued*

Problem	Management Suggestion
Nausea and vomiting	
Rate too fast	Slow rate
High osmolality	Dilute formula to isotonicity, then gradually increase
Mechanical problems	Check tube placement
Delayed gastric emptying	Consider metoclopramide, patient position, transpyloric feeds, continuous infusion, isotonic formula
Medications given with feeding	Consider changing time of medication, check contents of medications
Obstruction	Stop feeding
Patient positioning	Elevate head of bed
Delayed gastric emptying	Consider medication, reposition patient, transpyloric feeds, continuous infusion, isotonic formula
Constipation	
Inadequate fluids	Monitor fluid balance, increase fluids by increasing the rate and decreasing the formula
Inadequate fiber	Consider formula with fiber or fiber supplement
Inactivity	Encourage activity
Obstruction	Stop feeds
Fecal impaction	Disimpact, consider stool softeners
	DEVELOPMENTAL
Prevent delayed feeding skills development	Nonnutritive sucking, offer small amounts of food from spoon, fluids from cup, develop an association between oral activity and satiety
Food refusal	Consult an occupational or speech therapist

Source: Adapted with permission from Davis.[48]

Infections. Careful attention to cleanliness of the gastrostomy tube site promotes healthy tissue. Dirty occlusive dressings with formula leakage promote bacterial growth and increase the risk of infections. By using commercially prepared products or confining the time home blenderized diets and breast milk infusions are in the delivery system to 4 hr, bacterial growth can be limited.[78,79] Delivery systems should be changed every 24 hr and not reused.

Mechanical. Mechanical complications are similar to those that occur in adult patients on enteral feedings. Nasogastric or enteric tubes can cause erosion of the nares or esophagus. Gastrointestinal perforation can occur with the use of stiff tubes or improper placement of those with a removable stylet. These problems can be lessened by careful attention to tube position, choosing a small, soft tube and using trained personnel to introduce tubes with stylets. In addition, nasoduodenal or jejunal tubes can be placed under fluoroscopic guidance. The incidence of tube occlusion can be lessened by infusing only liquid preparations of all nutrients and medications through feeding tubes and thorough flushing of the tubes. Often tubes can simply lie within a gastric fold and a gentle water flush with repositioning of the patient will resolve the occlusion. Pancreatic enzymes, cranberry juice, water, carbonated beverages, and papain have been used to declog tubes.[80–85]

Metabolic. Over- and underhydration can occur with enteral feedings. Overhydration can occur as the transition is made from parenteral to enteral feedings, or if excessive free water flushes are used, especially in small children or infants. Children with renal, cardiac, hepatic, or other diseases may require fluid restriction and are at greater risk for fluid overload than children who do not require fluid restriction.

Dehydration can occur if inadequate fluids are given, if the child has increased losses, such as occur with diarrhea or if hyperosmolar feedings are given. Dehydration can be prevented by calculating the fluid in the final feeding and assuring adequate free water is supplied.

Other metabolic complications, such as electrolyte imbalance and hypoglycemia can occur if feedings are more than 6 to 8 hr apart. Children on long-term enteral nutrition support can develop calcium, phosphorus, vitamin D, zinc and other deficiencies with chronic enteral feedings.[86–88]

Gastrointestinal. Gastrointestinal complications include nausea, vomiting, diarrhea, and constipation. Sometimes the smell of the formula may make children nauseous. Flavorings and mixing the formula at a distance so children cannot smell it may be helpful. Vomiting can result because of improper tube placement, slow gastrointestinal motility, rapid infusion rate, hyperosmolality, and medications. Reflux in infants may occur. It can be treated by positioning the infant so the head of the bed is at a 30° to 45° angle, using a slow infusion rate and possibly using bethanecol, metoclopramide, or cisapride.

Constipation can be a problem for children on enteral feeds and can be helped by choosing a formula with fiber, giving adequate fluids, and encouraging as much physical activity as possible. At times stool softeners are useful.

Diarrhea can be a problem for children on enteral feedings. Before the enteral feeding itself is assumed to be the cause of the diarrhea, infections, as pseudomenbraneous colitis, or medications that increase gastrointestinal motility or have a high osmolality should be considered.[89] The use of a formula with fiber can be helpful. For children with a short small bowel, agents that slow gastrointestinal motility or bind bile salts may be considered.

PARENTERAL NUTRITION

With an understanding of the importance of luminal contents to gastrointestinal function, indications for parenteral nutrition support dwindled. If specific contraindications for enteral nutrition exist, or if the total nutrition requirement cannot be delivered enterally, the parenteral route can be used. Figure 30.1 is an algorithm that can be useful in deciding whether or not to use parenteral nutrition support. If adequate gastrointestinal function is present then either supplements or tube feeds should be used. If the function of the gastrointestinal tract is impaired, parenteral nutrition support can be considered. Table 30.18 lists some situations where the gastrointestinal tract may be partially functional. If the gastrointestinal tract is partially functional, it should still be used to deliver some of the nutrient requirements, because there is considerable evidence that even small amounts of enteral nutrition may be beneficial.[46,47]

Table 30.18 lists some situations where the gastrointestinal tract is not functional. If the enteral route cannot be quickly established, parenteral nutrition should begin. If parenteral nutrition support will be required for less than 7 days and the child is not malnourished, peripheral support can be used. If the child is malnourished or if it is anticipated that parenteral nutrition will be required longer than 7 days, the parenteral nutrition should be administered into a central vessel; otherwise adequate nutrients to rehabilitate the malnourished state or

Table 30.18 Gastrointestinal Function

Partially Functional
Cannot meet nutrient requirements after maximize enteral support
Burns
Multiorgan failure
Malabsorption—Short small bowel, intractable diarrhea, villous atrophy
Risk of aspiration when small bowel feedings are not possible
Nonfunctional
Paralytic ileus
Chronic intractable vomiting and cannot access small bowel
Small bowel ischemia
Necrotizing enterocolitis
Severe acute pancreatitis
Gastrointestinal surgery—gastroschisis, omphalocele, multiple intestinal atresias, etc., until the enteral route is accessible

support growth cannot be assured. Unless it is anticipated that parenteral will be needed for more than 3 to 5 days, little benefit derives from its institution.[90]

Parenteral Formulas

Parenteral nutrition, whether delivered peripherally or via a central line consists of water, protein, carbohydrate, lipids, electrolytes, minerals, trace elements, vitamins, and other additives. The solutions can be formulated so all the nutrients are in a single bag or the fat can be delivered by a separate system (piggy-backed) onto the solution already containing protein, carbohydrates, and so on. The advantages and disadvantages of each approach have been reviewed.[91,92] The production of two-chamber bags, which keep the fat and other solution separate until the wall between the chambers is disrupted permit easy administration of combined solutions in the home with a longer shelf life.

Several steps are involved in the initiation, advancement, and final cycling of parenteral nutrition. They involve calculation of fluid requirements (Table 30.19), calories (Table 30.20) and protein (Table 30.21). Electrolytes can be estimated using Table 30.21, although if increased losses occur, such as with an ileostomy, diarrhea, or fistula, or increased requirements exist, such as with malnutrition, these must be included in the estimation. Table 30.22 is a case study of the step-by-step initiation and advancement of parenteral nutrition.

Premature infants. Infants born prematurely do not have the benefits of the full third trimester of intrauterine life, and thus have high nutrient requirements. It is difficult at times, to assure intrauterine rates of growth and accretion of minerals in prematurely born infants who may have other medical problems. It is estimated that infants who receive no amino acid intake lost from 130 to 180 $mg/kg{\cdot}day^{-1}$ of nitrogen, meaning these infants loose from 0.8 to 1.1 $g/kg{\cdot}day^{-1}$, equivalent to a daily loss of at least 1% of endogenous protein.[93] Because this loss is considered to be significant and cumulative early nutrition support intervention is advisable.[94] Prematurely born infants are a diverse group, consisting of infants who are small for gestational age as well as appropriate for gestation age but immature. Thus, it may be difficult to set general recommendations for prematurely born infants as a group. Stable, growing, low-birth-weight infants

Table 30.19 Estimation of Fluid Requirements for Parenteral Nutrition

Body Weight (kg)	Fluid Requirement/Day[a]
1 to 10	100 mL/kg
11 to 20	1000 mL plus 50 mL/kg for each kg > 10 kg
> 20	1500 mL plus 20 mL/kg for each kg > 20 kg

[a]Increase final estimation by 12% for each degree Celsius by which body temperature is greater than 37.5°C.

Table 30.20 Estimation of Caloric Requirements for Parenteral Nutrition

1. Estimate caloric needs based on 1.0 kcal/1.0 ml of estimated fluid requirement

OR

2. Estimate caloric needs based on basal energy metabolism[104]:

Basal Energy Metabolism							
Age 1 week to 10 months		Age 11 to 36 months			Age 3 to 16 years		
Metabolic Rate			Metabolic Rate kcal/hr			Metabolic Rate kcal/hr	
Weight (kg)	kcal/hr Male or Female	Weight (kg)	Male	Female	Weight (kg)	Male	Female
3.5	8.4	9.0	22.0	21.2	15	35.8	33.3
4.0	9.5	9.5	22.8	22.0	20	39.7	37.4
4.5	10.5	10.0	23.6	22.8	25	43.6	41.5
5.0	11.6	10.5	24.4	23.6	30	47.5	45.5
5.5	12.7	11.0	25.2	24.4	35	51.3	49.6
6.0	13.8	11.5	26.0	25.2	40	55.2	53.7
6.5	14.9	12.0	26.8	26.0	45	59.1	57.8
7.0	16.0	12.5	27.6	26.9	50	63.0	61.9
7.5	17.1	13.0	28.4	27.7	55	66.9	66.0
8.0	18.2	13.5	29.2	28.5	60	70.8	70.0
8.5	19.3	14.0	30.0	29.3	65	74.7	74.0
9.0	20.4	14.5	30.8	30.1	70	78.6	78.1
9.5	21.4	15.0	31.6	30.9	75	82.5	82.2
10.0	22.5	15.5	32.4	31.7			
10.5	23.6	16.0	33.2	32.6			
11.0	24.7	16.5	34.0	33.4			

Activity Factors[a]		Stress Factors[a]			
Paralyzed	1.0	Surgery	1.1–1.2	Burn	1.5–2.5
Confined to bed	1.1	Infection	1.2–1.6	Starvation	0.7
Ambulatory	1.2–1.3	Trauma	1.1–1.8	Growth Failure	1.5–2.0

Calculate estimated energy requirement (kcal/day) = Basal energy metabolism × activity factor × stress factor

[a]Activity and stress factors from Page et al.[105]

Table 30.21 Suggested Daily Amounts of Nutrients for Maintenance Pediatric Parenteral Nutrition

Nutrient	Infants and Toddlers	Children	Adolescents
Calories	80–120	60–90	30–75
Dextrose (mg/kg·min^{-1})	7–21	5.5–19	3.5–14
Protein (g/kg)	2.5–3.0	1.5–2.4	1.5–2.5
Fat (g/kg)	0.5–4.0	1.0–3.0	1.0–3.0
Sodium (mEq/kg)	2–5	2–5	60–150 mEq/day
Potassium (mEq/kg)	2–5	2–5	70–180 mEq/day
Chloride (mEq/kg)	2–5	2–5	60–150 mEq/day
Magnesium (mEq/kg) (125 mg/mEq)	0.25–1.0	0.25–1.0	8–32 mEq/day
Calcium (mEq/kg) (20 mg/mEq)	0.5–4.0	0.5–3.0	10–40 mEq/day
Phosphorus (mmol/kg) (31 mg/mmol)	0.5–2.0	0.5–2.0	9–30 mmol/day
Zinc (µg/kg)[a]	50–250	50	5 mg/day
Copper (µg/kg)[a]	20	20	300 µg/day
Chromium (µg/kg)[a]	0.20	0.20	5.0 µg/day
Manganese (µg/kg)[a]	1.0	1.0	50 µg/day
Selenium (µg/kg)[a]	2.0	2.0	30 µg/day
Molybdenum (µg/kg)[a]	0.25	0.25	5.0 µg/day
Iodide (µg/kg)[a]	1.0	1.0	1.0 µg/day
Pediatric multivitamin[b]	1–3 kg: 3.3 mL/day	3–39 kg: 5 mL/day	> 40 kg: use adult MVI

Source: Adapted from Forlaw et al.[109]

[a]Data from Green et al.[110]

[b]Pediatric multivitamin contains: vitamins A, D, E, B_1, B_2, B_6, B_{12}, C, K, folic acid, biotin, dexapanthenol in amounts based upon the current FDA recommendations for pediatric intravenous vitamins from the AMA Nutrition Advisory Group.

Table 30.22 Case Study of Initiation and Advancement of Parenteral Nutrition

CASE:	KS is a 24-month-old female admitted for trauma suffered in a motor vehicle accident when she was not in a infant car seat. She had a history of small bowel ischemia, secondary to a volvulus, resulting in small bowel resection and placement of an ileostomy approximately 2 months prior to the trauma. Admission weight is 10.2 kg and height is 85 cm. Two months ago she weighed 11.2 kg. Daily ileostomy output is approximately 550 ml. All laboratory is normal except for a low serum prealbumin and albumin.
ASSESSMENT:	KS will require parenteral nutrition for a prolonged time, she has a 9% decrease in weight (currently 86% of ideal body weight) and ongoing losses through the ileostomy.
STEP 1:	Set goals: Provide nutrients for anabolism (recent history of weight loss and hypoalbuminemia) Calories: 100–120 kcal/kg Protein: 2.0–3.0 $g/kg \cdot day^{-1}$ Adequate fluid, electrolytes, minerals, trace elements and vitamins to support anabolism
STEP 2:	Estimate fluid needs: 100 mL/kg × kg = 1000 mL 50 mL/kg × 0.2 kg = 10 mL Ileostomy losses = 550 mL Total fluids over 24 hr at 65 mL/hr = 1560 mL
STEP 3:	Estimate caloric requirements: One kcal/mL = 1560 kcal/day Or Estimated energy requirement = basal energy metabolism × activity factor × stress factor: Estimated energy requirement = (24.2)(24 hr) × 1.1 × 1.8 = 1150 kcal/day
STEP 4:	Decide on nutrition solution content Aim for a micronutrient content that mirrors dietary intake: 7–12% of total calories from protein 4 kcal/g 35–60% of total calories from carbohydrate 3.4 kcal/g 30–60% of total calories from fat 2 kcal/mL
STEP 5:	Estimate initial fluid: Calculate a starting solution: Dextrose: Start with 10% dextrose ____ mL TPN × ____ % = ____ g carbohydrate/day KS: 1560 mL × 0.10 = 156 g carbohydrate/day 156 g – 10.2 kg – 1440 min = 10.6 $mg\ carbohydrate/kg \cdot min^{-1}$ 156 g × 3.4 kcal/g = 530 kcal (52 kcal/kg)

continued on next page

Table 30.22 *Continued*

STEP 5:
(continued)

Fat: ____ gm/kg·day^{-1} × ____ kg × mL lipid/g = mL/day
KS: 1.0 g/kg·day^{-1} × 10.2 kg × 5 mL × 51 mL/day
51 mL × 2 kcal/mL = 102 kcal (10 kcal/kg)

For piggyback infusion rate = ____ mL/hr × ____ hr/day
KS: 2.1 mL/hr × 24 hr/day

Protein: ____ g/kg·day^{-1} × ____ kg = ____ g protein/day
____ g protein/day = ____ % amino acids
____ mL TPN volume

KS: (75%) 2.2 g/kg·day^{-1} × 10.2 kg = 22.4 g protein/day
22.4 g protein/day = 1.4% amino acids
1560 mL TPN
22.4 × 4 kcal/g = 90 kcal (9 kcal/kg)
22.4/6.25 = 3.58 g nitrogen

Total: 1560 mL at 65 mL/hr with 1.4% amino acids, 10% dextrose, and 51 mL 20% lipid to provide 722 kcal (71 kcal/kg) or 158 mL/kg

STEP 6:

Calculate the Electrolytes

	Per Liter
Potassium 20 mEq (1.96 mEq/kg)	13 mEq
Sodium 77 mEq (7.6 mEq/kg)	49 mEq
Calcium gluconate 10 mEq (1.0 mEq/kg)	6.5 mEq
$MgSO_4$ 5 mEq (0.5 mEq/kg)	3.2 mEq
Phosphorus 20 mmol (2.0 mmol/kg)	13 mmol
Cl:Acetate Ratio	1:1

Other Additives Per Day:
2 mg zinc   20 mg ranitidine
20 μg selenium   2.5 mcg molybdenum

STEP 7:

Advance the Solution—Day 2:
1560 mL at 65 mL/hr with
2% Amino acids (3.0 g protein/kg) = 30.6 gm = 122 kcal
12.5% Dextrose = 195 gm carbohydrate
(13 mg carbohydrate/kg·min^{-1}) = 663 kcal
102 ml lipid at 4.3 mL/hr
(2.0 gm fat/kg) = 204 kcal

Total   989 kcal
(97 kcal/kg)

continued on next page

Table 30.22 *Continued*

STEP 7: *(continued)*	Advance the Solution—Day 3: 1560 mL @ 65 mL/hr with 2% Amino acids = 122 kcal 15% Dextrose = 234 gm CHO (16 mg CH/kg·min^{-1}) = 795 kcal 153 mL lipid @ 6.4 mL/hr (3.0 gm fat/kg) = 306 kcal Total 1223 kcal (120 kcal/kg) 10% kcal as protein 65% kcal as carbohydrate 25% kcal as fat		
STEP 8:	Can cycle solutions if patient is metabolically stable and demonstrates consistent weight gain on continuous parenteral nutrition		
	<u>18 hr ON, 6 hr OFF</u> 55 mL/hr × 1 hr 90 mL/hr × 15 hr 55 mL/hr × 1 hr 25 mL/hr	Blood Sugar finger sticks: (1) 1 hr after maximum rate is started (2) Midpoint time of maximum infusion rate (3) 1 hr after TPN is discontinued (4) Midpoint time pt is off TPN	
	<u>14 hr ON, 10 hr OFF</u> 75 mL/hr × 1 hr 125 mL/hr × 11 hr 75 mL/hr × 1 hr 35 mL/hr × 1 hr		
SUMMARY:	Initiation	Dextrose	10% dextrose (7–8 mg carbohydrate/kg·min^{-1})
		Amino acids	50–100% of goal
		Lipid (20%)	0.5–1.0 g/kg·day^{-1} (2.5–5 mL/kg·day^{-1})
	Advancement	Dextrose	5% dextrose/day (2–4 mg carbohydrate/kg·min^{-1})
		Lipid (20%)	0.5–1.0 g/kg·day^{-1} (2.5–5 mL/kg·day^{-1})
	Usual Upper Limits	Dextrose	17 mg/kg·min^{-1}
		Peripheral	12.5% dextrose
		Central	25–35% dextrose
		Amino acids	3.0 g/kg·day^{-1}
		Lipid	3.0 g/kg·day^{-1}
STEP 9:	Monitor		
STEP 10:	Reassess—modify above as indicated by recovery from trauma, repletion of malnourished state, development of sepsis, etc.		

require parenteral amino acid intake of approximately 3 g/kg·day^{-1} to assure accretion of nitrogen at the intrauterine rate. The addition of an energy intake of 80 kcal/kg·day^{-1} results in a rate of weight gain approximating the intrauterine rate.[95] In general, these findings are applicable to smaller, sicker infants.[96–98]

Intravenous lipid can be safely used to provide essential fatty acids and as a calorie source. If the lipid is infused at a rate less than or equal to the rate of lipid hydrolysis, triglyceride will not accumulate in the serum. Heird and Gomez[94] recommend initiating lipid at a rate of 0.5 g/kg·day^{-1} and gradually increasing to 2 to 3 g/kg·day^{-1}.

In addition to the amount of protein and calories, the composition of nutrients are important to prematurely born infants. For example, Helms et al.[99] demonstrated that a parenteral amino acid preparation designed to normalize serum amino acid levels resulted in greater weight gain and more positive nitrogen balance in preterm infants when compared to a similar group of infants receiving an isocaloric, isonitrogenous parenteral solution that contained a standard amino acid preparation. Taurine, cysteine, tyrosine, arginine, glutamine, choline, inositol, nucleotides, and carnitine may be conditionally essential, especially in prematurely born infants. These nutrients and their importance to infants have been reviewed.[100]

Unless there is a specific contraindication to the use of enteral feedings, enteral feeds should begin as soon as possible since even hypocaloric, low-volume enteral substrate has a positive effect on subsequent feeding and length of hospital stay.[48,47,101]

Term infants. Term infants have the benefit of the third trimester to more fully develop than infants born prematurely. Term infants, however, are developmentally immature compared to the older child and adult, have less nutrient stores, and may have congenital infections or abnormalities that preclude the use of enteral feeds. The issues important to initiating and maintaining parenteral nutrition are essentially the same for term neonates as for premature infants. The schedule to monitor term infants is also the same as for premature infants. Although not specifically studied in term infants, it is likely that even small enteral feeds are beneficial, as they are for prematurely born infants.

Children. Parenteral nutrition is used in older children because of a congenital problem, trauma, or autoimmune disease of the gastrointestinal tract. If parenteral nutrition is necessary to sustain life indefinitely and no chance of gastrointestinal recovery exists, small bowel transplantation can be considered.

Monitoring of Parenteral Nutrition

To properly manage infants and children on parenteral nutrition, careful attention to metabolic status, growth parameters, and a vigilance regarding infections must be kept. Parenteral nutrition support is expensive and the complications, such as malposition of the line, requiring operating room time, anesthesiologist, and surgeon can be expensive, as well as dangerous to correct. Infections can be difficult to clear in children on long-term parenteral nutrition who might have seeded a thrombus or are immune compromised and require long-term treatment with expensive antibiotics. Table 30.16 outlines a useful schedule for monitoring children on parenteral nutrition support.

Complications

Metabolic. To minimize metabolic complications in pediatric patients receiving parenteral nutrition, two safeguards should be in place: The first is to assure that electrolytes and minerals are either within acceptable ranges or easily corrected before the initiation of parenteral nutrition solutions. The second safeguard is not to use parenteral nutrition solutions to correct fluid or electrolyte problems. If abnormalities arise, they can be corrected through a second intravenous line.

Aside from transient abnormalities in serum glucose, triglycerides, or electrolytes, a few metabolic problems are common and vexing. The delivery of adequate calcium and phosphorus to prematurely born infants who require parenteral nutrition as a sole nutrient source for prolonged periods of time can be difficult. Table 30.23 lists advisable parenteral intakes for minerals. Infants and children who are malnourished often experience a drop in serum phosphorus and potassium when intravenous carbohydrate is administered. It is essential that adequate minerals be supplied; otherwise fat, rather than lean body mass, will accumulate.[102] In addition, very low serum phosphorus or potassium levels are associated with cardiac arrhythmias.[103]

Cholestasis is common when long-term parenteral nutrition support is necessary. Despite careful studies over more than a decade, no identifiable cause has been found.

Infections. With children on long-term parenteral nutrition support, sepsis is guaranteed. Meticulous attention to detail is important in maintaining as low a rate of infection as possible.

Table 30.23 Advisable Parenteral Intake for Calcium, Magnesium, Phosphorus, and Vitamin D

	Calcium	Magnesium	Phosphorus	Vitamin D
mmol/L	12.5–15.0	1.5–2.0	12.5–15.0	250–1000 IU/L
(mg/L)	(500–600)	(36–48)	(390–470)	
mmol/kg·day^{-1a}	1.5–2.25	0.18–0.3	1.5–2.25	
(mg/kg·day^{-1})	(60–90)	(4.3–7.2)	(47–70)	40–160 IU/kg/day+

Source: Reproduced with permission from Koo and Tsang.[106]

Mechanical. Mechanical problems can be related to the catheter and to the solutions. The literature is replete with case reports of line complications such as pneumothorax, hemothorax, air embolism, thrombosis, and catheter malposition. Solutions have been inadvertently infused into the pericardium, peritoneum, thorax, etc.

CONCLUSIONS

Nutrition support is a powerful medical tool to support children through critical illnesses or to sustain long-term growth and development. To appropriately and safely administer nutrition support, careful attention must be paid to the medical condition of the child, the child's specific nutrient requirements, and the child's developmental stage. Many possible complications can be anticipated and prevented with constant monitoring of the child and the nutrition support.

REFERENCES

1. Prichard JA. Fetal swallowing and amniotic fluid volume. *Obstet Gynecol* 1966; 28:606–610.
2. Goluber EL, Shuleikina KV, Vainshsteinii V. Development of reflex and spontaneous activity of the human fetus in the process of embryogenesis. *Obstet Gynecol (USSR)* 1959;3:59–62.
3. Treloar DM. The effect of nonnutritive sucking on oxygenation in healthy, crying full-term infants. *Appl Nutr Res* 1994;7:52–58.
4. Bernbaum JC, Pereira CR, Watkins JB, Peckham GJ. Nonnutritive sucking during gavage feeding enhances growth and maturation in premature infants. *Pediatrics* 1983;71:41–45.
5. Measel CP, Anderson GC. Nonnutritive sucking during tube feedings: effect on clinical course in premature infants. *J Obstet Gynecol Neonatal Nurs* 1979;8:265–272.
6. Field T, Ignatoff E, Stringer S, et al. Nonnutritive sucking during tube feedings: effects on preterm neonates in an intensive care unit. *Pediatrics* 1972;70:381–384.
7. Deren JS. Development of structure and function in the fetal and newborn stomach. *Am J Clin Nutr* 1971;24:144–159.
8. McLain CR. Amniography studies of the gastrointestinal motility of the human fetus. *Am J Obstet Gynecol* 1965;86:1079–1087.
9. Morris FH, Moore M, Weisbroodt NW, et al. Ontogenic development of gastrointestinal motility. IV: duodenal contractions in preterm infants. *Pediatrics* 1986; 78:1106–1113.

10. Slagle TA, Gross SJ. Effect of early low-volume enteral substrate on subsequent feeding tolerance in very low birth weight infants. *J Pediatr* 1988;113:526–531.
11. Morris FH. Neonatal gastrointestinal motility and enteral feeding. *Sem Perinatol* 1991;15:478–481.
12. Watkins JB, Ingall D, Szczepanik P, et al. Bile-salt metabolism in the newborn. Measurement of pool size and synthesis by stable isotope technic. *N Engl J Med* 1973;288:431–434.
13. Fredrikzon B, Hernell O, Blackberg L. Lingual lipase. Its role in lipid digestion in infants with low birth weight and/or pancreatic insufficiency. *Acta Paediatr Scand* 1982;296(suppl):75–80.
14. Hamosh M, Scanlon JW, Ganot D, et al. Fat digestion in the newborn. Characterization of lipase in gastric aspirates of premature and term infants. *J Clin Invest* 1981;67:838–846.
15. Sarles J, Maori H, Verger R. Human gastric lipase: ontogeny and variations in children. *Acta Petitur* 1992;81:511–513.
16. Areca S, Rubin A, Murset G. Intestinal glycosidase activities in the human embryo, fetus and newborn. *Pediatrics* 1965;35:944–954.
17. Bond JK, Levitt MD. Fate of soluble carbohydrate in the colon of rats and man. *J Clin Invest* 1976;57:1158–1162.
18. Grand RJ, Watkins JB, Torti FM. Development of the human gastrointestinal tract. A review. *Gastroenterology* 1976;70:790–810.
19. Kein CL, Sumners JE, Stetina JS, et al. A method for assessing carbohydrate energy absorption and its application to premature infants. *Am J Clin Nutr* 1982;36:910–915.
20. Mobashaleh M, Montgomery RK, Biller JA, et al. Development of carbohydrate absorption in the fetus and neonate. *Pediatrics* 1985;75(suppl):160–165.
21. Kelly EJ, Brownlee KG. When is the fetus first capable of gastric acid, intrinsic factor and gastrin secretion. *Biol Neonate* 1993;63:153–156.
22. Berfenstam R, Jagenburg R, Mellander O. Protein hydrolysis in the stomach of premature and full term infants. *Acta Pediatr* 1955;44:348–341.
23. Lebenthal E, Lee PC. Development of functional response in human exocrine pancreas. *Pediatrics* 1980;66:556–561.
24. Hirata Y, Matsur P, Kobubu H. Digestion and absorption of milk proteins in infants' intestine. *Kobe J Med Sci* 1965;11:103–106.
25. Lebenthal E, Lee PC, Heitlinger LA. Impact of development of the gastrointestinal tract on infant feeding. *J Pediatr* 1983;102:1–5.
26. Raiha NCR. Nutritional proteins in milk and the protein requirement of normal infants. *Pediatrics* 1985;75(suppl):136–142.
27. Fordtran JS, Rector FC, Maynard FE, et al. Permeability characteristics of the human small intestine. *J Clin Invest* 1965;44:1935–1944.
28. Jackson D, Walker-Smith JA, Phillips AD. Macromolecular absorption by histologically normal and abnormal small intestinal mucosa in childhood: an in vitro study using organ culture. *J Pediatr Gastroenterol Nutr* 1983;2:235–247.

29. Ziegler TR, Smith RJ, O'Dwyer ST, et al. Increased intestinal permeability associated with infection in burn patients. *Arch Surg* 1988;123:1313–1319.
30. Weaver LT, Laker MF, Nelson R. Intestinal permeability in the newborn. *Arch Dis Child* 1984;59:236–241.
31. Beach RC, Menzies IS, Clayden GS, et al. Gastrointestinal permeability changes in the preterm neonate. *Arch Dis Child* 1982;57:141–145.
32. Tuchman DN. Oropharyngeal and esophageal complications of enteral tube feeding. In Baker SS, Baker RD, Davis A, eds. *Pediatric Enteral Nutrition.* New York: Chapman and Hall, 1994:179–192.
33. Illingworth RS, Lister J. The critical or sensitive period, with special reference to certain feeding problems in infants and children. *J Pediatr* 1964;65:839–848.
34. Widdowson EM. Growth and body composition in childhood. Brunser O, Carrazza FR, Gracey M, et al. *Clinical Nutrition of the Young Child.* New York: Raven Press, 1991:1–14.
35. Pollitt E, Leibel R. Iron deficiency and behavior. *J Pediatr* 1976;88:372–381.
36. Walter R, De Andraca I, Chadud P, Perales CG. Iron deficiency anemia: adverse effects on infant psychomotor development. *Pediatrics* 1989;84:7–17.
37. Lozoff B, Jimenez E, Wolf AW. Long term developmental outcome of infants with iron deficiency. *N Engl J Med* 1991;325:687–694.
38. Dibley MJ, Staehling N, Nieburg P, et al. Interpretation of Z score anthropometric indicators derived from the International Growth Reference. *Am J Clin Nutr* 1987;20:503–510.
39. Solomon SM, Kirby DF. The refeeding syndrome: a review. *J Parent Ent Nutr* 1990;14:90–97.
40. Walravens PS, Hambridge KM, Koeffer DM. Zinc supplementation in infants with a nutritional pattern of failure to thrive: a double-blind, controlled study. *Pediatrics* 1989;83:532–538.
41. Costillo-Duran C, Heresi G, Fisberg M, Uauy R. Controlled trial of zinc supplementation during recovery from malnutrition: effects on growth and immune function. *Am J Clin Nutr* 1987;45:602–608.
42. Castillo-Duran C, Garcia H, Venegas P, et al. Zinc supplementation increases growth velocity of male children and adolescents with short stature. *Acta Paediatr* 1994;83:833–837.
43. Castillo-Duran C, Rodriguez A, Venegas G. Zinc supplementation and growth of infants born small for gestational age. *J Pediatr* 1995;127:206–211.
44. Seidman EG. Gastrointestinal benefits of enteral feeds. In Baker SS, Baker RD, Davis A, eds. *Pediatric Enteral Nutrition.* Chapman and Hall, 1994:46–67.
45. Jenkins M, Gottschlich M, Alexander JW, et al. Effect of immediate enteral feeding on the hyper metabolic response following severe burn injury. *J Parent Ent Nutr* 1989;13(suppl 1):12S–15S.
46. Berseth CL. Effect of early feeding on the maturation of the preterm infant's small intestine. *J Pediatr* 1992;120:947–953.

47. Slagle TA, Gross SJ. Effect of early low-volume enteral substrate on subsequent feeding tolerance in very low birth weight infants. *J Pediatr* 1988;113:526–531.
48. Davis A. Indications and techniques for enteral feeds. In Baker SS, Baker RD, Davis A, eds. *Pediatric Enteral Nutrition.* New York: Chapman and Hall, 1994:67–94.
49. Kulkarni PB, Hall RT, Rhodes PG, et al. Rickets in very-low-birth weight infants. *J Pediatr* 1980;96:249–253.
50. Hambridge K. Zinc deficiency in the premature infant. *Pediatr Rev* 1985;6:209–215.
51. Stocks RJ, Davies DP, Allen F, et al. Loss of breast milk nutrients during tube feeding. *Arch Dis Child* 1985;60:164–167.
52. Lavine M, Clark RM. The effect of short term refrigeration and addition of breast milk fortifier on the delivery of lipids during tube feeding. *J Pediatr Gastroenterol Nutr* 1989;8:496–500.
53. Narayanan I, Singh B, Harvey D. Fat loss during feedings of human milk. *Arch Dis Child* 1984;59:745–748.
54. Geer FR, McCormick A, Laker J. Changes in fat concentration of human milk during delivery by intermittent bolus and continuous mechanical pump infusion. *J Pediatr* 1984;105:745–749.
55. American Academy of Pediatrics, Committee on Nutrition. Hypoallergenic infant formulas. *Pediatrics* 1994;(In Press).
56. Eastham EJ, Lichauco T, Grady MI, Walker WE. Antigenicity of infant formulas: role of immature intestine on protein permeability. *J Pediatr* 1978;93:561–564.
57. Crawford LV, Grogan FT. Allergenicity of cow's milk protein: II. Studies with serum-agar precipitation technique. *Pediatrics* 1961;28:362–366.
58. Kaufmann S, Murray ND, Wood RP, et al. Nutritional support for the infant with extrahepatic biliary atresia. *J Pediatr* 1987;110:679–681.
59. Kaufman SS, Scrivner DJ, Murray ND, et al. Influence of portagen and pregestamil on essential fatty acid status in infantile liver disease. *Pediatrics* 1992;89:151–154.
60. Khoshoo V, Pillai BV, Cowan G, et al. Use of energy-dense formula for treating infants with non-organic failure to thrive. *J Pediatr Gastroenterol Nutr* 1994;(In Press).
61. Ziegler EE, Ruy JE. Renal solute load and diet in growing premature infants. *J Pediatr* 1976;89:609–615.
62. American Academy of Pediatrics, Committee on Nutrition. Commentary on breast feeding and infant formulas, including proposed standards for formulas. *Pediatrics* 1979;63:52–54.
63. Davis A, Baker SS. The use of modular nutrients in pediatrics. *Pediatrics* 1994;(In Press).
64. Listernick R, Sidransky E. Hypernatremic dehydration in children with severe psychomotor retardation. *Clin Pediatr* 1985;24:440–445.
65. Chernoff R, Block AS. Liquid feedings: considerations and alternatives. *J Am Diet Assoc* 1977;70:4–6.

66. Leider A, Sullivan L, Mullen MA. Intermittent tube feeding: pros and cons. *Nutr Supp Serv* 1984;4:59–61.
67. Lavine JE, Hattner RS, Heyman MB. Dumping in infancy diagnosed by radio nuclide gastric emptying technique. *J Pediatr Gastroenterol Nutr* 1988;7:614–617.
68. Parathyras AJ, Kassak LA. Tolerance, nutritional adequacy, and cost-effectiveness in continuous drip versus bolus and/or intermittent feeding techniques. *Nutr Supp Serv* 1983;3(5):56–62.
69. Parker P, Stroop S, Greene H. A controlled comparison of continuous versus intermittent feeding in the treatment of infants with intestinal disease. *J Pediatr* 1981;99:360–364.
70. Coben RM, Weintraub A, DiMarino AJ, et al. Gastroesophageal reflux during gastrostomy feeding. *Gastroenterology* 1994;106:13–18.
71. Davis A. Transitional and combination feeds. In Baker SS, Baker RD, Davis A. eds. *Pediatric Enteral Nutrition.* New York: Chapman and Hall, 1994:139–156.
72. Kurkchubasche AG, Rowe MI, Smith SD. Adaptation in short-bowel syndrome: reassessing old limits. *J Pediatr Surg* 1993;28:1069–1071.
73. Leider Z, Sullivan L, Mullen MA. Intermittent tube feeding: pros and cons. *Nutr Supp Serv* 1984;4:59.
74. Heitkemper ME, Martin DC, Hansen BC, et al. Rate and volume of intermittent enteral feeding. *J Parent Ent Nutr* 1981;5:125.
75. Satter EM. The feeding relationship. *J Am Diet Assoc* 1986;86:352–356.
76. Harris NB. Oral motor management of the high risk neonate. *Phys Occup Ther Pediatr* 1986;6:231–235.
77. Byzyk S. Factors associated with the transition to oral feeding in infants fed by nasogastric tubes. *Am J Occup Ther* 1990;44:1070–1072.
78. White WE, Acuff TE, Sykes TR, et al. Bacterial contamination of enteral solution: a preliminary report. *J Parent Ent Nutr* 1979;3:459–461.
79. Hostetler C, Lipman T, Gearaghty M, et al. Bacterial safety of reconstituted continuous drip tube feedings. *J Parent Ent Nutr* 1983;3:232–375.
80. Macuard SP, Stegall KL, Trogdon S, et al. Clearing obstructed feeding tubes. *J Parent Ent Nutr* 1989;13:81.
81. Haynes-Johnson V. Tube feeding complications: causes, prevention, and therapy. *Nutr Supp Serv* 1986;6:17.
82. Cataldi-Betcher EL, Seltzer MH, Slocum BA, et al. Complications occurring during enteral nutrition support. A prospective study. *J Parent Ent Nutr* 1983;7:546.
83. Marcuard SP, Perkins AM. Clogging of feeding tubes. *J Parent Ent Nutr* 1988;12:403.
84. Nicholson LJ. Declogging small-bore feeding tubes. *J Parent Ent Nutr* 1987;11:594.
85. Marcuard SP, Stegall KS. Unclogging feeding tubes with pancreatic enzyme. *J Parent Ent Nutr* 1990;14:198.
86. Brammer EM. Shortcomings of current formula for long-term enteral feeding in pediatrics. *Nutr Clin Pract* 1990;5:160.

87. Bowen PE, Mobarhan S, Henderson S, et al. Hypocarotenemia in patients fed enterally with commercial liquid diets. *J Parent Ent Nutr* 1988;12:484–489.

88. Kenny F, Sriram K, Hammond JB, et al. Clinical zinc deficiency during adequate enteral nutrition. *J Am Coll Nutr* 1989;8:83–86.

89. Edes TE, Walk B, Austin JL. Diarrhea in tube-fed patients: feeding formula not necessarily the cause. *Am J Med* 1990;88:91–93.

90. Board of Directors. Guidelines for use of total parenteral nutrition in the hospitalized adult patient. *J Parent Ent Nutr* 1986;10:441–445.

91. Driscol DF. Total nutrient admixtures; theory and practice. *Nutr Clin Pract* 1995; 10:115–119.

92. Warshawsky KY. Intravenous fat emulsions in clinical practice. *Nutr Clin Pract* 1995;7:187–196.

93. Pencharz P, Stefee WP, Cochran W, et al. Protein metabolism in human neonates: nitrogen-balance studies, estimated obligatory losses of nitrogen and whole-body turnover of nitrogen. *Clin Sci Mol Med* 1977;52:485–498.

94. Heird WD, Gomez MR. Parenteral nutrition. In Tsang RC, Lucas A, Uauy R, Zlotkin S, eds. *Nutritional needs of the preterm infant.* Baltimore: Williams and Wilkins, 1993:267–280.

95. Zlotkin S, Bryan MH, Anderson GH. Intravenous nitrogen and energy intakes required to duplicate in utero nitrogen accretion in prematurely born human infants. *J Pediatr* 1981;99:115–120.

96. Rivera Jr. A, Bell EF, Stegnik LD, et al. Plasma amino acid profiles during the first three days of life in infants with respiratory distress syndrome: effect of parenteral amino acid supplementation. *J Pediatr* 1989;115:465–468.

97. Sani J, MacMahon P, Morgan JB. Early parenteral feeding of amino acids. *Arch Dis Child* 1989;64:1362–1366.

98. Kashyap S. Nutritional management of the extremely low-birth-weight infant. In Cowett RM, Hay WW, eds. *The Micropremie: The Next Frontier.* Report of the 99th Ross Conference on Pediatric Research. Columbus, OH: Ross Laboratories, 1990:115–122.

99. Helms RA, Christensen ML, Mauer EC, et al. Comparison of a pediatric versus standard amino acid formulation in preterm neonates requiring parenteral nutrition. *J Pediatr* 1987;110:466–472.

100. Uauy R, Greene HL, Heird WC. Conditionally essential nutrients: cysteine, taurine, tyrosine, arginine, glutamine, choline, inositol and nucleotides. In Tsang RC, Lucas A, Uauy R, Zlotkin S, eds. *Nutritional needs of the preterm infant.* Baltimore: Williams and Wilkins, 1993:267–280.

101. Davies PA. Low birth weight infants: immediate feeding recalled. *Arch Dis Child* 1991;66:551–553.

102. Rudman D, Millikan WJ, Richardson TJ, et al. Elemental balances during intravenous hyper alimentation of underweight adult subjects. *J Clin Invest* 1975;55:94–104.

103. Weinseir RL, Krumdieck CL. Death resulting from overzealous total parenteral nutrition: the refeeding syndrome revisited. *Am J Clin Nutr* 1980;34:393–399.

104. Altman PL, Dittmer DS. *Metabolism.* Bethesda, MD: Federation of American Societies for Experimental Biology, 1968:344.

105. Page CP, Hardin TC, Melnik G. *Nutritional Assessment and Support. A Primer.* 2nd ed. Baltimore: William and Wilkins, 1989:32.

106. Koo WSK, Tsang RC. Calcium, magnesium, phosphorus, and vitamin D. In Tsang RC, Lucas A, Uauy R, Zlotkin S, eds. *Nutritional needs of the preterm infant.* Baltimore: Williams and Wilkins, 1993:135–155.

107. Levy JS, Winters RW, Heird WC. Total parenteral nutrition in pediatric patients. *Pediatr Rev* 1980;2:99.

108. Kerner JA. Monitoring the patient on parenteral nutrition. In Kenner JA, ed. *Manual of Pediatric Parenteral Nutrition.* New York: Wiley, 1983:228.

109. Forlaw L, Wong M, Little CA. Recommended daily allowances of maintenance parenteral nutrition in infants and children. *Am J Health Syst Pharm* 1995;52:651–653.

110. Greene HL, Hambridge KM, Scheanler R, Tsang RC. Guidelines for the use of vitamins, trace elements, calcium, magnesium, and phosphorus in infants and children receiving total parenteral nutrition. *Am J Clin Nutr* 1988;48:1324–1342.

Management of Patients on Home Parenteral and Enteral Nutrition

CHAPTER 31

Lyn Howard, M.D.,
Margaret Malone, Ph.D.,
Suzanne Murray, R.N., M.S.N., and
Laura Ellis, Ph.D.

INTRODUCTION

Current economic pressures in medicine dictate ever-earlier hospital discharges and the expansion of traditionally hospital-based treatments into the home. Patients who cannot adequately consume and/or absorb sufficient oral nutrients to maintain weight and strength can often be successfully managed at home. This approach may allow the patient to return to a relatively normal lifestyle and reduce many of the costs associated with institutional care. In the era of increasing emphasis on medical cost containment, this field has seen tremendous growth. However, for nutritional support to be provided safely in the home setting, a working knowledge of the pertinent concerns including patient selection criteria, careful monitoring, and close follow up are essential.

This chapter discusses the current use of home parenteral and enteral nutrition in the United States and their clinical outcome, factors that affect patient selection, and the management of these patients in the home setting.

CURRENT USAGE AND COST OF HOME PARENTERAL AND ENTERAL NUTRITION IN THE UNITED STATES

There is no mandatory reporting of patients starting on home parenteral and enteral nutrition (HPEN), and hence there is no central source for measuring HPEN use and clinical outcome in a random sample of these patients from the general population. Despite this deficit, there are two large sources of HPEN patient information from which estimates of use and outcome can be made.[1]

Medicare is the largest single payer for HPEN therapies and hence Medicare utilization provides useful information about national use, growth, and costs but does not assess HPEN therapy outcome. Between 1985 and 1992 the North American HPEN Patient Registry collected yearly outcome data on over 9000 newly initiated HPEN patients managed by 217 nutrition support programs across the United States. The experience of these patients provides useful information about HPEN clinical outcome in different diagnoses and different age groups.

As shown in Figure 31.1, over a 4-year period (1989 to 1992 inclusive), usage

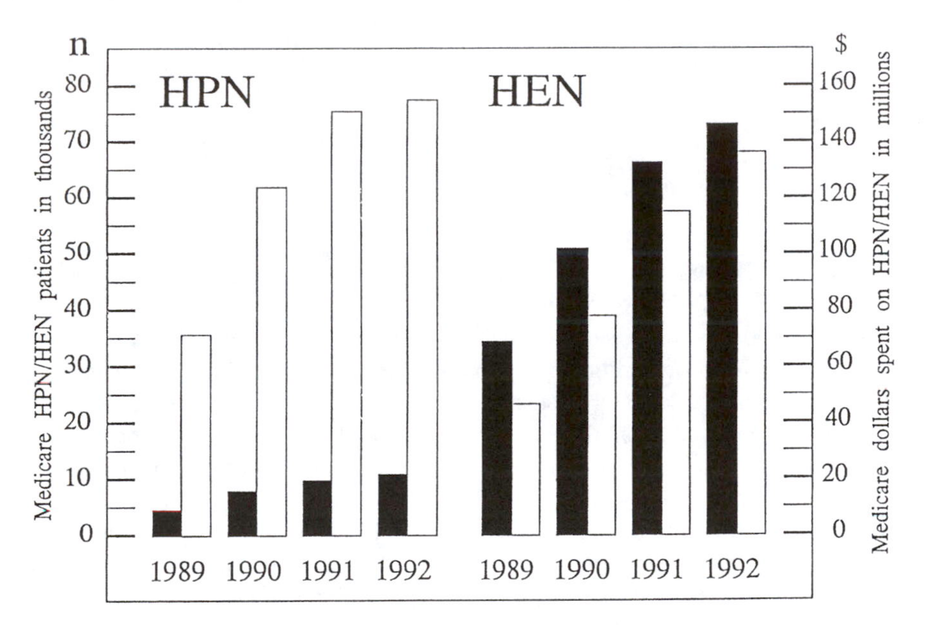

Figure 31.1 An estimate of the number of home parenteral and enteral nutrition Medicare patients in thousands (black bars) and the dollars paid in millions (white bars) between 1989 and 1992: These estimates were derived from Medicare Part B parenteral and enteral nutrition workload statistics compiled by Blue Cross/Blue Shield of South Carolina (personal communication). This carrier processed approximately 75% of all Medicare parenteral and enteral nutrition claims. Their workload statistics have been increased to provide an estimate of national Medicare activity.

of HPEN in Medicare beneficiaries doubled. Medicare beneficiaries on home parenteral nutrition (HPN) increased from 4500 to 10,000 and dollars paid from 73 million to 156 million. Medicare beneficiaries on home enteral nutrition (HEN) increased from 34,000 to 73,000 and dollars paid from 47 million to 137 million. These dollars paid were 80% of the allowable charges shown in Table 31.1. These patient numbers and dollars do not include Medicare beneficiaries receiving parenteral and enteral nutrition (PEN) in nursing homes or renal dialysis centers.

Between 1989 and 1992, the average prevalence of HPN in the Medicare population was 238 patients per million of the Medicare population and the Registry sampled 5% of these patients. The average prevalence of HEN in the Medicare population was 1660 per million of the Medicare population and the registry sampled 0.8% of these patients. Although Medicare beneficiaries accounted for 13% of the general population between 1989 and 1992, they accounted for 27% of all HPN patients and 46% of all HEN patients in the large Registry sample. This implies that Medicare use overestimates general population use by a factor of two for HPN and a factor of four for HEN. Thus prevalence in the general population is closer to 120 per million for HPN and 400 per million for HEN. This estimate of HEN prevalence is similar to that found in a population-based study of feeding enterostomies in nonhospitalized patients in rural Minnesota.[2]

The United States estimates of yearly prevalence indicate a U.S. usage of HPN ten times higher, and of HEN four times higher than in other medically advanced countries.[3] The greater use of HPEN in the United States probably reflects health care entrepreneurial trends characteristic of the United States[4–7] but also an increasing shift towards greater use of HPEN in short-term patients. Medicare data show that in 1992 the average treatment time was 60 days for an HPN beneficiary and 70 days for an HEN beneficiary.[1]

In the early years of HPN, Shils reported the collective experience of 29 centers treating 168 patients; 63% of these patients had short-bowel syndrome due to Crohn's disease or ischemic bowel disease, and only 17% had a resected malignancy.[8] In these early years unresectable cancer was felt to be a contraindication at least to HPN and this is still the prevailing opinion in Canada[9] and Britain.[10,11] In the United States the use of HPN in active cancer patients has grown dramatically. In a subset of 37 programs that consistently reported to the Registry, 35% of all new patients had active cancer in 1985, 43% in 1987, and 46% in 1989.[12] The percentage of Registry patients with acquired immune deficiency syndrome (AIDS) has also increased.[13] Recent reports from the United Kingdom[3] and Europe[14] show that HEN use is rapidly expanding, and, although HPN therapy is far more constrained, the shift toward wider use of HPN in cancer patients has already happened in many European countries.

By knowing the percentage of Medicare patients in the large Registry sample and the absolute number of Medicare patients on HPEN, it is possible to estimate

Table 31.1 Medicare Allowable Charges[a] for HPEN Therapy in 1992 ($/day)

	Parenteral	Tube Enteral
Nutrient solution		
Glucose	158–298	10–35
Amino acids		
Lipids	30–40	
Additives	7	
Dressing kit	7	0.5–2.0
Administration set	22	11
Pump loan (15 months only)	12	3.6
Mean[b] (range)	280 (238–390)	33 (25–50)

[a]Medicare paid 80% of these allowable charges.

[b]The mean daily Medicare allowable charge was calculated from actual Medicare reimbursement of 1000 Medicare HPN and 1000 Medicare HEN patient-days to five large nutrition support programs. This mean daily reimbursement was increased from 80% to 100% to derive the mean daily Medicare allowable charge.

the number of patients in the general population on HPEN. Thus in 1992 Medicare accounted for 25% of the HPN Registry sample and there were 10,000 Medicare HPN beneficiaries that year. Therefore in 1992 there were approximately 40,000 HPN patients in the U.S. population. That same year Medicare accounted for 48% of the HEN Registry sample and there were 73,000 Medicare HEN beneficiaries that year. Therefore in 1992 there was approximately 150,000 HEN patients in the U.S. population.

If an assumption is made that the Medicare allowable charge was the average amount paid for patients with all different types of insurance coverage, then the national cost of HPEN was approximately $800 million in 1992 and all nonhospitalized nutrition support was about twice this amount.[1] Nutrition support accounts for 25% to 30% of all nonhospital infusion therapies, which had revenues of $5.1 billion in 1993.[15]

With the growing move toward managed care and capitated agreements, it is likely that reimbursement for these therapies will decline. It is not known how much cost restraint can occur before the quality and safety of these therapies are jeopardized. It should be noted that HEN costs one tenth of HPN (Table 31.1).

CLINICAL OUTCOME ON HPEN

Between 1985 and 1992, the North American HPEN Patient Registry collected yearly outcome data on over 12 thousand patients from 217 nutrition support programs across the United States.[13] However, the outcome analysis was confined to 9288 patients who were reported to the Registry during their first year on therapy. Of these patients, 5357 were on HPN and 3931 were on HEN. The diagnostic distribution of these patients is shown in Figure 31.2. It is likely that the outcome data reported are biased toward programs with greater-than-average experience, because 59% of these HPEN patients were in large treatment programs (managing more than 21 patients/year), 19% were treated in medium-sized programs (6 to 20 patients/year), and 22% were treated in small programs (1 to 5 patients/year).

For the purpose of the Registry report, the quality of the HPEN outcome was assessed by the primary physician or a clinical designee and was based on four parameters: (a) Survival rates while receiving HPEN were calculated by the Life Table method and derived from 36 months of follow-up data except where otherwise stated. If the patient died or discontinued therapy to resume oral nutrition, this change was noted at year end. Thereafter, the surviving patient off HPEN was not tracked unless HPEN was resumed. (b) Therapy status assessed the patients clinical status one year after starting HPEN therapy. It described the patients who resumed oral nutrition, who stayed on HPEN or who died. (c)

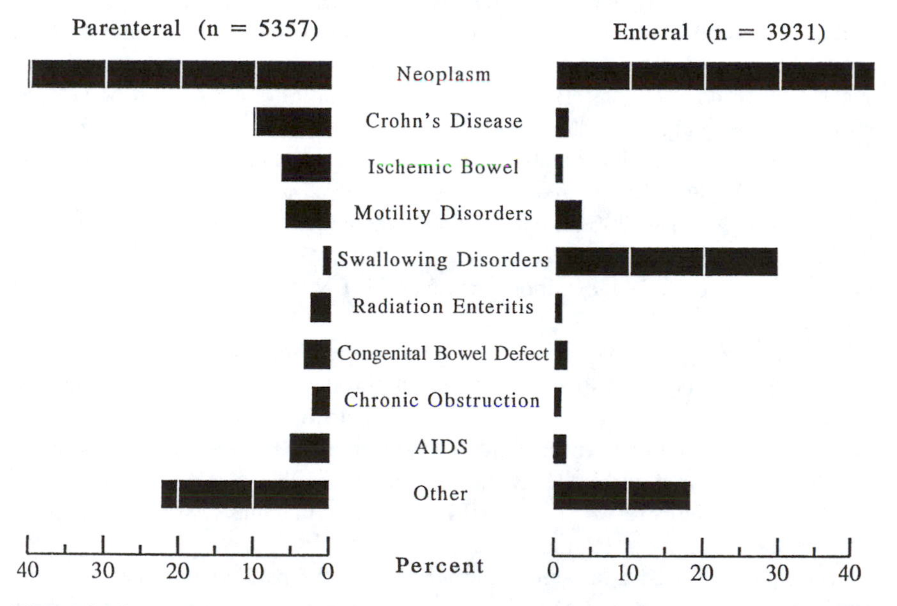

Figure 31.2 The distribution of diagnoses in new patients on home parenteral and enteral nutrition reported to the North American HPEN Registry from 1985–1992. The values shown were percents of the total for each therapy category.

Rehabilitation status on HPEN was assessed over the first treatment year and not at a particular time point. Rehabilitation was defined as complete, partial, or minimal. Complete rehabilitation meant "normal" activity, partial rehabilitation meant some limitations in activity or function, and minimal rehabilitation meant requiring extensive support or being primarily bedridden. (d) Complication rates per year were recorded; only those complications were included that required hospital admission. The complication rates were further stratified into those related to the HPEN and those related to the medical diagnosis.

Table 31.2 summarizes the clinical outcome for 11 diagnostic groups treated with HPN and 2 groups treated with HEN.[1] The number of patients within each group and their average age at HPEN onset are given. Overall there were approximately equal numbers of males and females, however, within particular groups this ratio was skewed reflecting the gender distribution prevalent within certain disease categories. For example, there was a predominance of women (85%) with radiation enteritis, presumably due to the use of radiotherapy for gynecological malignancy. The majority of patients (90%) in the AIDS population were male.

The survival on HPN for some of the major diagnostic groups obviously varied. HPN patients with Crohn's disease, ischemic bowel disease, motility disorders, congenital bowel defects, hyperemesis gravidarum, and chronic pancreatitis all had relatively good outcomes, with 87% or better annual survival rates and a 50% to 70% likelihood of complete rehabilitation. This outcome appears to be primarily determined by the relatively benign nature of these particular diagnoses. In comparison, HPN patients with radiation enteritis and bowel obstruction due to chronic adhesions had similar survival rates, but only 20% to 40% experienced complete rehabilitation. In contrast, patients with cystic fibrosis, cancer, or AIDS had the poorest outcome in terms of survival. After 1-year only 50% of cystic fibrosis, 20% of cancer, and 10% of AIDS patients were still alive. These data may not be as bleak as they seem since 32% of cystic fibrosis, 25% of cancer, and 10% of AIDS patients came off HPEN in the first year and resumed oral nutrition.

Only two groups of enteral nutrition patients were represented in the Registry in sufficient numbers to allow meaningful interpretation of their data. The largest group (n = 1644) were patients with cancer. Their mean (s.d.) age was 61 ± 17 years. The survival rate of these patients is presented in Figure 31.3. The percentage survival on therapy for HEN cancer patients (30%) was similar to that of HPN cancer patients (20%). Of the HEN patients with cancer, 80% experienced only partial or minimal rehabilitation.

The second largest group receiving HEN were the patients with neurological swallowing disorders (n = 1134). This group included primarily older patients who had suffered a cerebrovascular accident (CVA). Whereas the overall analysis for this group creates a rather discouraging outcome picture with median survival

Table 31.2 Summary of Outcome on HPEN. Derived from North American HPEN Patient Registry

Diagnosis	Number of Patients	Age in Years (S.D.)	% Survival on Therapy[a]	Therapy Status: 1% at 1 year[b,c]			Rehabilitation[a,d] Status % in 1st year			Complications[e] (per patient-year)	
				Full Oral Nutrition	Continued on HPEN Rx	Died	C	P	M	HPEN[c]	NonHPEN[c]
HOME PARENTERAL NUTRITION											
Crohn's disease	562	36 (17)	96 [31/2.9]	70 (2)	25 (2)	2 (1)	60 (5)	38 (5)	2 (n/a)[f]	0.9	1. 1
Ischemic bowel disease	331	49 (24)	87 [81/7.5]	27 (3)	48 (4)	19 (3)	53 (4)	41 (4)	6 (2)	1.4	1 .1
Motility disorder	299	45 (22)	87 [81/3.1]	31 (3)	44 (4)	21 (3)	49 (4)	39 (4)	12 (3)	1.3	1 .1
Congenital bowel defect	172	5 (14)	94 [20/1.6]	42 (6)	47 (6)	9 (3)	63 (6)	27 (5)	11 (4)	2.1	1.0
Hyperemesis gravidarum	112	28 (5)	100 [0/0.1]	100 (n/a)	0 (n/a)	0 (n/a)	83 (4)	16 (4)	1 (n/a)	1.5	3.5
Chronic pancreatitis	156	42 (17)	90 [9/0.6]	82 (3)	10 (3)	5 (2)	60 (5)	38 (5)	2 (n/a)	1.2	2.5
Radiation enteritis	145	58 (15)	87 [47/3.2]	28 (5)	49 (5)	22 (4)	42 (6)	49 (6)	9 (3)	0.8	1.1
Chronic adhesive obstructions	120	53 (17)	83 [30/1.1]	47 (6)	34 (5)	13 (4)	23 (7)	68 (7)	10 (n/a)	1.7	1.4
Cystic fibrosis	51	17 (10)	50 [254/0.05]	38 (7)	13 (5)	36 (7)	24 (6)	66 (7)	16 (5)	0.8	3. 7

continued on next page

Table 31.2 *Continued*

Diagnosis	Number of Patients	Age in Years (S.D.)	% Survival on Therapy[a]	Therapy Status: 1% at 1 year[b,c]			Rehabilitation[a,d] Status % in 1st year			Complications[e] (per patient-year)	
				Full Oral Nutrition	Continued on HPEN Rx	Died	C	P	M	HPEN[c]	NonHPEN[c]
Cancer	2122	44	20	26	8	63	29	57	14	1.1	3.3
		(24)	[1336/8.7]	(1)	(1)	(1)	(3)	(3)	(2)		
AIDS	280	33	10	13	6	73	8	63	29	1.6	3.3
		(12)	[182/0.8]	(3)	(2)	(4)	(n/a)	(7)	(6)		
HOME ENTERAL NUTRITION											
Neurologic disorders of swallowing	1134	65	55	19	25	48	5	24	71	0.3	0.9
		(26)	[447/31.5]	(2)	(2)	(2)	(1)	(2)	(3)		
Cancer	1644	61	30	30	6	59	21	59	21	0.4	2.7
		(17)	[885/13.6]	(2)	(1)	(2)	(3)	(3)	(3)		

[a]Survival rates on therapy are values at one year, calculated by the life table method. This will differ from the percentage listed as died under Therapy Status, since all patients with known end points are considered in this latter measure. The ratio of observed versus expected deaths is equivalent to a standard mortality ratio and is shown in brackets.

[b]Not shown are those patients who were back in hospital or who had changed therapy type by 12 months. Standard error of the mean is shown in parentheses.

[c]Chi-square test, $p < 0.05$.

[d]Rehabilitation is designated complete (c), partial (P), or minimal (M), relative to the patient's ability to sustain normal age-related activity. Standard error of the mean is shown in parentheses.

[e]Complications refer only to those complications that resulted in rehospitalization.

[f]n/a, not applicable because group was too small.

of 1.5 years, only 19% resuming full oral nutrition and 75% experiencing minimal rehabilitation, a separate look at those 65 years or over compared to those who are 25 years or younger shows younger HEN patients with neuromuscular disorders do quite well. Younger patients experienced complete or partial rehabilitation at a much higher rate (55%) than older patients (14%). Of younger patients, 89% survived 12 months compared to 54% of older patients. The differences in expected survival related to age account for only a small percentage of this difference, and, therefore, it is likely that the etiology of a neuromuscular swallowing disorder in younger patients is usually less devastating than the CVAs of the older subjects.

The influence of age on HPN therapy was analyzed for three diagnostic groups generally associated with a good clinical prognosis, namely, patients with Crohn's disease, ischemic bowel disorder or motility disorders (FIgure 31.3). In general, the outcome was better for younger than older patients in terms of the likelihood of resuming oral intake and rehabilitation status. However, pediatric patients had a greater rehospitalization rate for catheter-related sepsis than the older patients. Interestingly, the geriatric patients did not experience any greater number of therapy-related complications than the middle-aged patients.

PATIENT SELECTION FOR HPEN

A number of requirements must be addressed prior to selecting a patient for HPEN. Firstly, a complete history must be taken and a physical examination must be done to assess whether the patient truly requires invasive nutritional intervention. Once it is established that the patient cannot voluntarily consume adequate nutrition, issues pertaining to choosing the most appropriate feeding modality (i.e., enteral versus parenteral) and access option must be addressed. In addition, a thorough assessment of the patient, the family, and the home setting is undertaken to determine whether the patient and/or family are willing and able to follow instructions, whether the patient lives in a safe home environment, and whether the patient's medical condition is stable. Even reimbursement options must be analyzed.

Home nutrition support places a great deal of responsibility on patients and their families. Therefore, appropriate selection of the HPEN patient has to involve both a clinical and social decision; this is best done by an experienced HPEN professional often with the input of a seasoned HPEN patient, describing to the potential new HPEN patient what is involved and what risks have to be taken. If the home treatment is acceptable to the patient and their family, the professional still has to ensure that the patient is medically stable enough and the treatment can be administered at home without undue hazard. If HEN is feasible, this is a more physiologic, generally safer, and the more cost effective approach. If HPN

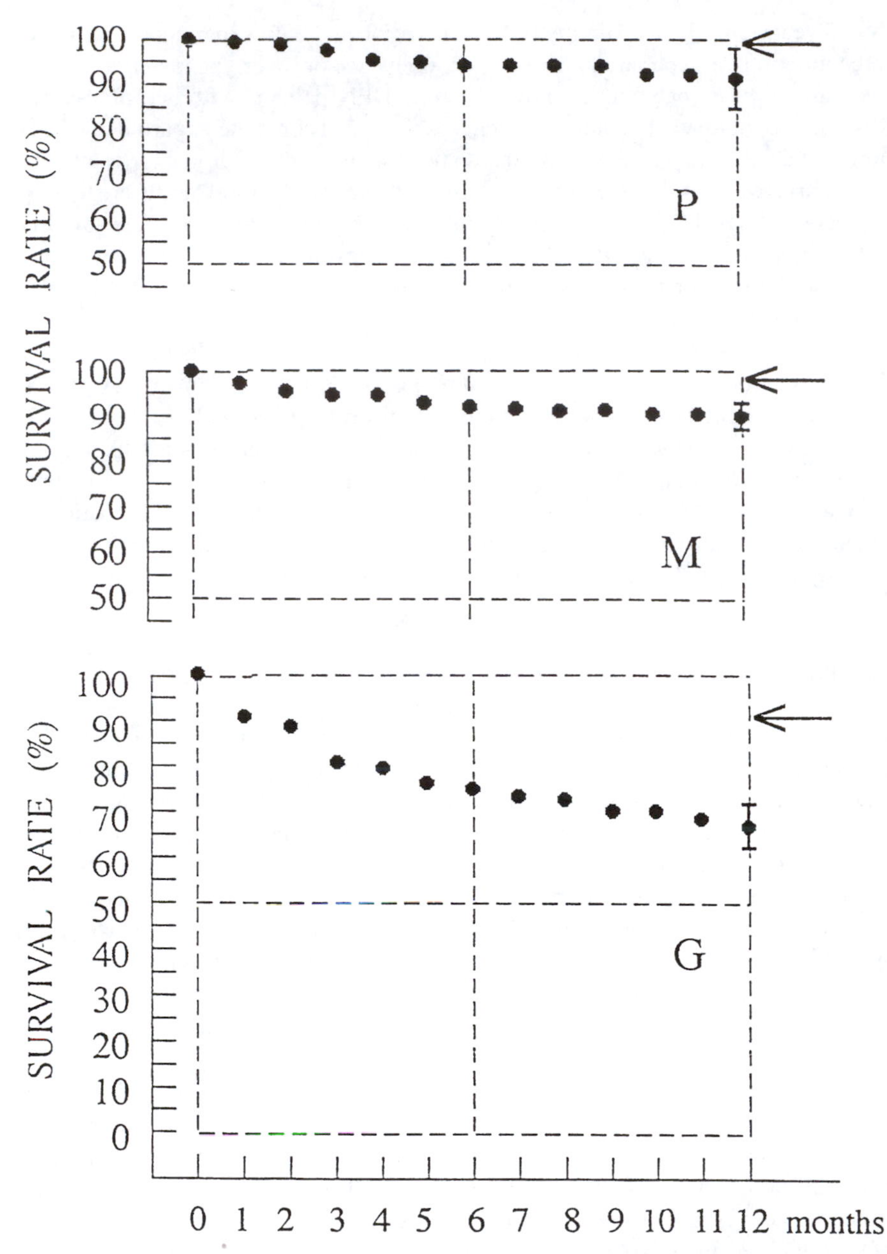

Figure 31.3 Survival rates on home parenteral nutrition for pediatric (P, 0–18 yr), middle age (M, 35–55 yr), and geriatric (G, 65+ yr) patients with Crohn's disease, ischemic bowel disorder, and motility disorder, showing the 95% confidence interval for patients surviving 12 months and indicating expected mortality rates (arrow) for the age- and sex-matched individuals in the general U.S. population.

is necessary, Registry data show that age per se should not be a reason for denying a patient parenteral support.

Both therapies are relatively safe. Registry data shows that only 5% of patient deaths were ascribed to the HPEN. The majority of patients died from the disease that led to HPEN or another medical condition. Although formal cost–benefit HPEN studies have not been done in the United States, Detsky et al. performed a cost–utility analysis on Canadian HPN patients[9] and determined that long-term patients with severe bowel impairment are less expensive and have a better quality of life when supported on parenteral nutrition at home compared with frequent rehospitalization for nutritional rehabilitation. Nutritional therapy at home even when it requires some professional nursing support is likely to be a significant cost saving compared with nutritional therapy delivered in hospital, particularly if it is delivered enterally rather than parenterally. In fact the low frequency of therapy-related complications in home patients suggests the possibility that nutritional treatment at home may be safer than nutritional treatment in the nosocomial hospital environment. This means the major cost-benefit issue is not where to provide long-term nutritional support but when to provide it.

The clinical-outcome profiles already presented show that the underlying diagnosis is the chief factor influencing HPEN outcome. However, even in the disorders where the overall prognosis is poor, there is a small percentage of patients who have a reasonable outcome. This means that diagnosis alone cannot determine the appropriateness of initiating HPEN and predicted good quality of life at home for several months seems to be the soundest justification.

The increasing use of HPEN in short-term clinical situations raises a question about the appropriateness of these therapies in terminal patients with bowel obstruction. *Terminal* is usually defined as a patient surviving 3 months or less. In a recent study, the role of nutrition and hydration was evaluated in 32 terminal patients. The majority of these patients never experienced thirst or hunger and there was no evidence that food or fluid administration beyond that specifically requested by the patient, contributed to patient comfort.[16] However, these dying patients were without specific gastrointestinal disorders leading to persistent vomiting or other significant gastrointestinal fluid losses where rapid dehydration and electrolyte imbalance is more likely and therefore comfort issues might be different. Studies are needed to determine whether HPEN is better than simple hydration or even antiemetic or narcotic therapy as advocated in Britain.[17]

The appropriateness of HPEN in short-term patients who are not dying is also a quality-of-life and cost–efficacy issue. This relates particularly to the preoperative home buildup of seriously depleted patients undergoing major elective surgery, such as a transplant. While the place for HPN preoperative buildup is not yet established, one study has shown a reduction of postoperative complications and length of hospital stay in malnourished patients with head and neck cancer who received preoperative HEN.[18]

The appropriateness of HPEN in patients who have no primary gastrointestinal disease but who suffer from severe nutritional depletion because of an inadequate oral intake is probably the most unresolved issue. This can occur because of psychological or cognitive disorders such as anorexia nervosa or dementia. It can occur with metabolic disorders such as chronic renal and hepatic failure, which induce profound anorexia. It can occur if patients have to choose between eating and breathing as in severe pulmonary or cardiac dyspnea. In all these clinical situations, randomized prospective controlled trials are needed to determine if nutritional support achieves nutritional repletion in these patients and if this repletion improved their quality of life and the overall cost efficiency of their care. In disorders where gastrointestinal dysfunction is not the primary disorder, HEN rather than HPN is likely to be more appropriate. Registry information suggests these types of patients account for about 20% of what is currently being prescribed.[13]

TEACHING PATIENTS TO ADMINISTER HPEN

Going home with tubes, pumps, and solutions for home nutrition infusion can be a frightening and overwhelming experience for patients and their families. A systematic approach to coordinating the home care and educating the patient is essential for safe home nutrition therapy.[19] This process generally begins when the patient is still in the hospital, but it may also be initiated for an outpatient. Getting to know the patients and their families and assessing factors influencing the psychomotor and cognitive abilities of these individuals is the first step in preparing for home nutritional support. Usually the patient or a parent is the primary person responsible for delivery of the therapy. Ideally a second person will be instructed as a back-up support for the primary care giver.

Instructional sessions are most effective when provided without interruptions or distractions and in an environment that simulates the home setting. The Patient Learning Center, a concept initiated at the University of Minnesota, provides a separate area for patient and family teaching and is like a home setting. It utilizes the same equipment provided to the patient at home and has nurse-teachers knowledgable in home nutrition therapy and educational principles (Figure 31.4).[20–22] An evaluation of such a designated teaching center found lower readmission rates for catheter reinsertion and fewer catheter related complications.[21,23] Patient learning centers have been instituted by many tertiary hospitals discharging large numbers of patients on an array of different self-administered home therapies.[24] At the Albany Medical Center Patient Learning Center, the Center staff in collaboration with the patient's physician, the patient, and the specialized discharge planner, coordinate the multiple aspects of the home care. This approach includes developing criteria for selection of a home infusion company, clarifica-

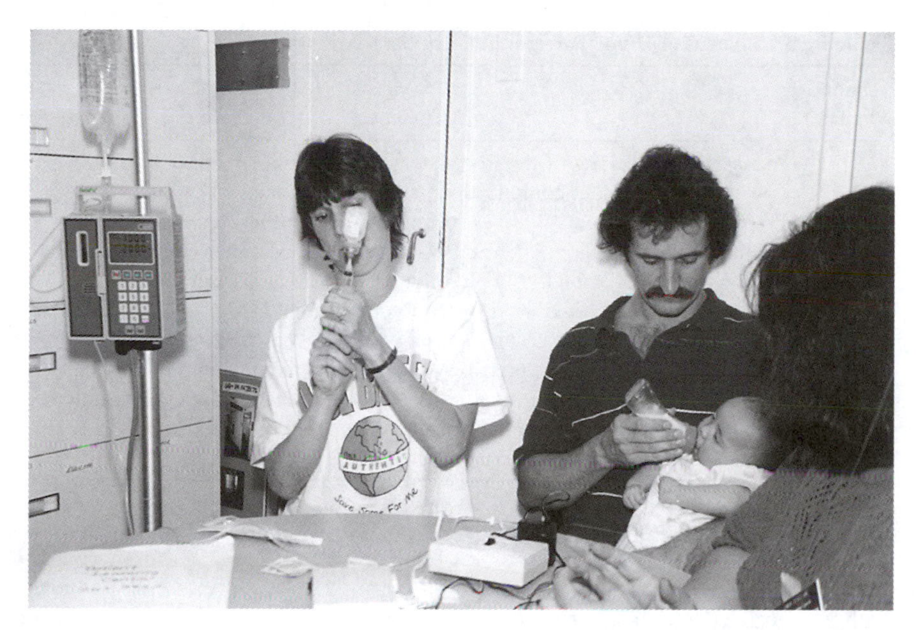

Figure 31.4 The parents of a small child with extreme short bowel syndrome secondary to gastroschisis being instructed by a nurse on how to administer home parenteral nutrition in the Patient Learning Center in Albany, NY.

tion of the physician's home orders, deciding the type of infusion pump (portable or stationary) and other equipment, assessing the learner's abilities and details of their home environment, education of the patient and the family, and communication of the discharge requirements to the companies providing services at home.[25]

Assessment of the learner's abilities includes such factors as their vision, hearing, manual dexterity, ability to read, primary language, any previous utilization of needles, syringes or medical equipment, motivation to learn, and understanding of the reason for their therapy. Factors in the physical and social environment of the home influence the specific teaching, determine what equipment will be used, and the health professional support needed to provide safe nutritional therapy for the patient at home.

The patient education occurs in one or sometimes 2 hr teaching sessions. Longer sessions are too exhausting and are not conducive to retention of information. The room is arranged specifically for teaching with the equipment that the patient will be using at home. Usually four to six sessions are required to present the information and techniques necessary for HPN. An outline of the content of these teaching sessions for HPN and HEN patients is presented in Tables 31.3 and 31.4.

Table 31.3 Content of Home Parenteral Nutrition Teaching Session

Session Number	Content
1	Assessment; overview of the PN system; central venous line (CVL) care with demonstration; complications and action to take if these occur.
2	Preparing PN solution, additives, priming tubing, pump function, alarms; initiating PN infusion; cycling PN. Learner demonstrates techniques.
3	Discontinuing PN; monitoring/recording activities, intake and output, weight, glucose checking, review of complications.
4	Pump programming, CVL cap, dressing change, complications. Accessing implanted port. Learner prepares solution; uses home pump to administer.
5	Discontinuing PN solution. Learner demonstrates emergency cycling down (pump/manual). Infection control measures are reinforced. Supply management/ordering is reviewed.

PN, parenteral nutrition.

Most of the teaching is done with simulation of the procedures, using a model with a central venous line, or an enteral feeding tube. The last sessions occur with the patient, with supervision by the nurse-teacher, preparing, infusing, and discontinuing their own infusion therapy as if they were at home. Examples of parenteral nutrition infusion devices and enteral feeding pumps are presented in Tables 31.5 and 31.6, respectively. Pumps vary in their complexity and cost. Ideally the patient should be trained using the simplest pump that will meet their needs. Some pumps are designed to be attached to an intravenous pole, whereas others are portable and can be carried in a backpack, which is excellent for the patient that requires a prolonged infusion time.

All patients are given step-by-step instructions individualized for them. Patients are encouraged to highlight and edit these instructions in their own words. Many nutrition programs and home infusion companies have written instructions

Table 31.4 Content of Home Enteral Nutrition Teaching Session

Session Number	Content
1	Assessment; overview of HEN system; preparing and administering feeding; pump functions, alarms. Complications and actions to take if these occur. Learner demonstrates techniques.
2	Demonstration of nasogastric (NG) insertion. Learner inserts and removes NG tube on self or on child. Review of session 1 content.
3	Learner inserts NG tube on self and removes NG tube. Review of other content as needed.

Table 31.5 Examples of Parenteral Infusion Devices

Name	Manufacturer	Cost	Comments
	VOLUMETRIC PUMPS		
Flo gard 8100	Baxter Healthcare	$	Alarms for flow rate error, complete infusion, occlusion, low battery, cassette problems
IMED 980	IMED	$	Alarms for air in line, door open, low battery, malfunction, occlusion, automatic priming.
	PROGRAMMABLE PUMPS—MULTIPLE RATES		
CADD-TPN	Pharmacia Deltec	$$	Alarms for completed infusion, low reservoir volume, back pressure, invalid rate. Portable in backpack.
Quest 521 Intelligent	McGaw	$$	Alarms for same features as above pumps. Nine programmable cycles of time and rate.
	PROGRAMMABLE PUMPS—MULTIPLE SOLUTIONS		
Gemini PC 2	IMED	$$$	Two channels permit delivery of two different fluids at their own rate, alarm features as above.
Omni Flow 4000	Abbott	$$$	Can deliver up to four fluids at any one time either simultaneously or intermittently. Alarm features as above.

available for the HPEN consumer.[26] These can be used if appropriate. All techniques are demonstrated by the nurse-teacher, followed by a return demonstration by the patient or caregiver. If multiple persons are learning, each person performs the technique.

Evaluation of the learner's competence involves observation of their technique while administering their therapy. The learner is also verbally quizzed on aspects of line care and infection preventing measures. The knowledge and understanding

Table 31.6 Examples of Enteral Feeding Pumps

Name	Flow rate	Manufacturer	Comments
Clintec 2200 volume	1–295 mL/hr	Clintec Nutrition	Memory for flow rate, dose, delivered, alarm for low battery, battery on, dose complete, occlusion/empty
Corflo 300	1–300 mL/hr	Corpak Inc.	Tracks volume infused on each shift, rates and amounts stored in memory for 24 hr, alarms for free flow, dose limit, low battery, occlusion/empty
EP85	1–295 mL/hr	Elan Pharma	Records and displays accumulated amount of formula delivered over several feedings, alarms for free flow, occlusion/empty, dose complete, low battery
Rate Saver Plus	0–300 mL/hr	EntraCare Corp.	Has set detection and set security features, alarms for no flow, set out, dose complete, system error, low battery
Flexiflo 111	1–300 mL/hr	Ross Products Division, Abbott	Easy to use touch controls. Flexiflo Companion available for portable use. Flexiflo Quantum has automatic flushing feature
Compat	1–295 mL/hr	Sandoz Nutrition	Model #199235: memory for infusion rate, dose limit, volume delivered and accumulated, alarms for low battery, occlusion/empty, dose complete, free flow, rate change Model #199225: Infusion rate and volume constantly displayed. Alarm features as above.

of the HPEN consumer is continually enhanced by home care professionals and by their periodic clinic visits with their physician.

MEDICAL SUPERVISION OF PATIENTS ON HOME PARENTERAL AND ENTERAL NUTRITION

This medical supervision commonly involves a physician knowledgeable about HPEN therapies, a nutrition support pharmacist, and a nurse; in some centers, a dietician, a social worker, and/or a psychologist are also part of the team. Some of these professionals may be part of the hospital's nutrition support service who manage the patient in hospital and then at home, others may be employees of the homecare service that supplies the patient with their infusion equipment and solutions. Complex homebound HPEN patients may also require involvement of a certified health care agency to provide more extensive home-nursing support.

To keep this professional team adequately informed, the responsible physician, or his clinical designee, must stay in frequent phone contact with the patient and family and with these other co-professionals. This telephone supervision although often time consuming is seldom reimbursed. Medicare recently introduced a reimbursement code for physician oversight of a home patient (CPT Code 99375) health care plan. Such a charge requires written documentation of conferencing about the particular patient with other health care professionals for 30 minutes or more per month.

The physician also sees the patient in follow-up outpatient clinical visits and may occasionally visit the patient at home. However, most home visits are made by nurses.

An outpatient visit usually involves an assessment of three aspects of the HPEN patient. First is the evaluation of the patient's nutrition delivery mechanism; the catheter or feeding tube, dressing, pump, and ostomy equipment if present. Second is the metabolic assessment, evaluating the patient's fluid and electrolyte balance, and their weight gain, while looking for evidence of infection or nutrient imbalance. Third is checking the patient and family psychosocial stress; finding out if the patient is able to assume most of their own care; asking if parents are getting enough sleep and respite time, and addressing financial stresses.

The frequency of telephone and outpatient follow-up visits may be weekly or biweekly in the first few months or during a complication. Except in unusual circumstances, all patients should be seen by the physician every 3 to 4 months, as long as they are on the infusion therapy. For most stable patients, the laboratory testing interval should depend on the particular patient and their clinical findings. Routine weekly or biweekly testing is not usually necessary or desirable.

POTENTIAL COMPLICATIONS

Complications associated with HPEN do not necessarily result in rehospitalization but nevertheless require monitoring and intervention by the supervising clinician. Complications associated with parenteral and enteral nutrition including micronutrient deficiencies are covered elsewhere in this textbook. In this section particular problems associated with long-term nutrition support are discussed.

Complications of HPEN are often divided into infectious, mechanical, and metabolic.[27] Management of these complications is identical to the management in an acute-care setting. An alternative approach to classification of HPEN complications is by duration of therapy. A summary of this approach is presented in Table 31.7.

PSYCHOSOCIAL AND FINANCIAL PROBLEMS FOR PATIENTS ON HOME PARENTERAL AND ENTERAL NUTRITION

Psychosocial issues confronting new HPEN patients relate in large part to the underlying cause for the initiation of the home therapy and its anticipated duration. For long-term patients, major issues include employment and financial considerations, interpersonal relationships, and lifestyle concerns.

The Initial Reaction

For new HPEN patients, immediate psychosocial issues relate to the underlying cause for initiating the therapy. For example, a man with short-bowel syndrome who starts HPN after 15 years of debilitating Crohn's disease and finds his body weight and energy level restored has a much different initial reaction to the therapy than does a man who has sustained a bowel infarction and awakens from surgery to find that he now requires daily intravenous infusions for survival. In the first instance, HPN is often viewed positively and is welcomed; in the second, it may be resented as the daily reminder of a sudden and unexpected change in health status. For patients with a chronic underlying medical condition such as radiation enteritis, its exacerbation or recurrence is a persistent, overriding concern. In contrast, for those patients whose initiation of HPN resulted from an acute event such as severance of the mesenteric artery by an abdominal gun shot wound, the status of the underlying condition is usually stable and a much less salient issue.

Immediate psychosocial issues also relate to the anticipated duration of the home therapy. If therapy is needed for only a few weeks, self-sufficiency is usually not attained. In this instance, technical and psychosocial support at home

Table 31.7 Complications Associated with Home Parenteral and Enteral Nutrition

Duration of Therapy	Common Complications	Management
Short term (weeks)	Metabolic: fluid and electrolyte imbalance.	Appropriate monitoring and tailoring of regimen
	Catheter-related sepsis[28]	Antibiotic/urokinase locks Systemic antibiotics
Medium term (months)	Catheter-related sepsis	Antibiotic/urokinase locks Systemic antibiotics
	Catheter occlusion caused by lipids,[29] drugs,[30] calcium phosphate[31]	Dissolution techniques with appropriate solvent, e.g., ethanol,[32] hydrochloric acid, sodium bicarbonate.
Long term (years)	Thrombosis	Use of thrombolytic locks, e.g., urokinase, streptokinase, tissue plasminogen activator[33–35] Prevention using low dose warfarin (1 mg/day)[36]
	Hepatic dysfunction.[37–45] Incidence greater in infants than adults. Cholestasis is more common in adults, children are more likely to develop cirrhosis	Etiology unclear, possible therapeutic options include decreasing calorie load,[45] cycling of TPN,[46], choline,[47–50] taurine,[51] s-adenosyl-L-methionine,[52,53] metronidazole.[44]
	Bone disease is common. 42%–100% HPN patients reported to have bone disease or decreased bone mineral density	Cause is multifactorial and includes underlying disease, drug therapy, e.g., corticosteroids, heparin, furosemide, chronic malabsorption of calcium, magnesium, phosphorus, excess of aluminum, protein, vitamin D, chronic acid–base imbalance.[54,55]
	Renal dysfunction. Buchman reported a 3.5 ± 6.3% decline in creatinine clearance per year in 33 HPN patients[56]	Cause is multifactorial. 46% of decline in CrCl accounted for by frequency of infectious episodes and use of nephrotoxic drugs. No link with protein load

through visiting nurses, home care company professionals, home health aides, and hospice services (for terminal patients) are essential.

In cases where HPEN therapy is needed on a long-term basis, psychosocial adjustment is enhanced by encouraging independence and self-sufficiency. These patients are often overwhelmed with the thought of needing to learn the highly technical aspects of the therapy at a time when they are recuperating from a medical crisis. This is where quiet, focused teaching sessions in the hospital's patient learning center can be very reassuring. Patient and family concerns about their ability to do self-infusion safely are universal, extending even to those patients with prior work experience in a health care environment.

Financial Considerations

Financial considerations become a major concern for many long-term nutrition support patients, especially for those on HPN. With HPN costs of $100,000 to $150,000 or more a year, an insurance policy's lifetime maximum coverage can be reached quickly. Thus strategies to maximize coverage and to explore all funding sources should be addressed early. With HEN, the costs are usually one-tenth of parenteral therapy costs.

Patients who receive health insurance through managed care organizations (health maintenance organizations and preferred provider organizations) should consider a few important issues: Does their plan include physicians with specialization in home nutrition support? Are patients allowed to consult with medical specialists outside their network? What is the cost associated with this "point of service" option? Can they choose the type of durable medical equipment that best fits into their lifestyle? For example, are they permitted to use an ambulatory pump, which is usually more convenient, but may be more expensive than a stationary, pole-mounted pump? Finally, patients and families need to know what recourse they have if dissatisfied with their medical care, home care services, or medical equipment.

Insurance coverage is frequently linked to employment issues. Although the therapy may permit patients to maintain full-time employment, employers are often reluctant to hire a person with an expensive "prior medical condition." For this reason, rehabilitated HPEN patients sometimes stay in an unsatisfying job for fear that they will lose their health insurance coverage if they take a new position elsewhere.

If rehabilitated HPEN patients cannot obtain employment or if their private insurance options run out, those who are eligible for Social Security benefits may have to claim total disability to receive Social Security support and coverage of their home therapy through Medicare. In this situation, patients are often forced to confront issues of diminished income and lost self-identity. These pressures often have a profound impact on the spouses and children of patients. For example, when a foreman who had spent his entire career with the same manufacturing company, stopped working to receive Medicare and Social Security disability benefits, the job loss was more disabling than the Crohn's disease and managing his home nutrition therapy. In this case, the pressure contributed to alcohol abuse and eventually to the break up of his marriage.

Relationship and Lifestyle Concerns

Home parenteral and enteral nutrition therapy influences interpersonal relationships throughout the initial adjustment period and beyond. At the beginning, all energies focus on stabilizing the patient's medical condition and educating him or her on the technical aspects of the therapy. Throughout this process, the psychosocial adjustment of family members is often overlooked. The mother of a young boy on home nutrition support recounts how she felt when her son was coming home from the hospital the first time on HPN, following a traumatic accident: "Although I was always there for my son, it was hard for me to accept that I would have to be his primary caregiver. I was sent home to do procedures I didn't even understand. It was hard for me to see my son go through so much emotional and physical pain. And I felt I didn't have enough time for other family members." For longer term patients, the family's needs become paramount, particularly when a family member functions as the primary caregiver. In this instance, consideration of the caregiver's psychological and social needs become highly relevant. Respite care for that person can be crucial in enabling the family and patient to sustain long-term HPEN.

Long-term patients also confront the need to adjust to changes in their body image. For young, single patients, body-image issues complicate their ability to establish intimate relationships. Adjustments to changes in body image are best dealt with by giving patients an opportunity to voice concerns and ask questions. The consensus of experienced patients is summarized in one patient's analysis: "You don't have to like your new battered and torn body, but you do have to accept it, as it is, scars and all. Find some goodness in it and build from there. Coping skills will grow, and you can enjoy life again, and life includes sexuality." Support and encouragement from other home patients may help ease the adjustment.

For long-term HPEN patients, incorporating the therapy into a patient's lifestyle is of paramount importance. Over time, the significant issues and priorities of patients and their families often change. For example, after a young child began on HPN therapy, his parents were most concerned with preserving family traditions and maintaining normal relationships with this child's siblings. After several years had passed, the family's concerns centered around their college-bound son's new lifestyle away from home.

Support Organizations

Because psychosocial stresses are complex, health professionals managing patients on HPEN must set aside time to talk about these issues and should arrange for patients to get together for mutual support and problem solving. Organizations devoted to HPEN patients and their families include the Oley Foundation, which sponsors regional support groups and national conference activities for patients on HPEN, the Crohn's and Colitis Foundation, the American Society of Adults with Pseudo-Obstruction, and the United Ostomy Association. These organizations often play a vital role in the successful adjustment of long-term HPEN patients and their family members.

SUMMARY

Over the past 20 years HPEN therapies have become widely available and are relatively user friendly and safe. The chief questions that remain are the appropriateness or inappropriateness of HPEN in terminal patients and in malnourished patients without primary severe gastrointestinal disease. In an era of growing managed care, these issues will probably be addressed in terms of quality of life and the patient's overall medical cost, with or without nutrition support.

Since reimbursement, for HPN especially, is sliding, the question of when cost restraint jeopardizes patient care is critical. This requires ongoing large-scale monitoring of HPEN clinical outcome.

REFERENCES

1. Howard L, Ament M, Fleming CR, et al. Current use and clinical outcome of home parenteral and enteral nutrition therapies in the United States. *Gastroenterology* 1995;109:355–365.
2. Bergstrom LR, Larson DE, Zinsmeister AR, et al. Utilization and outcomes of surgical gastrostomies and jejunostomies in an era of percutaneous gastrostomy. A population based study of outcome. *J Parent Ent Nutr* 1995;19:32S.
3. Elia M. An international perspective on artificial nutrition support in the community. *Lancet* 1995;345:1345–1349.
4. Relman AS. The health care industry; where is it taking us? Shattuck Lecture. *N Engl J Med* 1991;325:854–859.
5. Kusserow KP. *Fraud alert, joint venture arrangement.* Baltimore: Office of Inspector General, (OIG-89=04; US GRP:1989;0-235-622).
6. Burrows WP, Fernandez H. Patient referrals in health law. Update. Bond, Schoeneck, King. Oct 1, 1992.

7. Lord M. A high priced hook up. The convenience of home infusion is a god send, but abuses abound. *US News World Rept* 1994;63–69.
8. Shils ME. *Home TPN Registry Annual reports.* New York: New York Academy of Medicine, 1978–1983.
9. Detsky AS, McLaughlin JR, Abrams HB, et al. A cost utility analysis of the home parenteral nutrition program at Toronto General Hospital. *J Parent Ent Nutr* 1986;10:49–57.
10. Mughal M, Irving MH. Home parenteral nutrition in the United Kingdom and Ireland. *Lancet* 1986;2:383–387.
11. O'Hanrahan T, Irving MH. The role of home parenteral nutrition in the management of intestinal failure—report of 400 cases. *Clin Nutr* 1992;11:331–336.
12. Howard L. Home parenteral and enteral nutrition in cancer patients. *Cancer* 1993; 72:3531–3541.
13. *North American Home Parenteral and Enteral Nutrition Patient Registry Annual Reports 1985 to 1992.* Albany, NY: Oley Foundation, 1987–1994.
14. Bakker H, De Francesco A, Ladefoged K, et al. Current practice in home parenteral nutrition in adults: results of a large survey in Europe. *Clin Nutr* 1995 (in press).
15. Scott L. Home care revenues soar to $5.1 billion via mergers. *Modern Health Care* 1994;85–88.
16. McCann RM, Hall WJ, Groth-Juncker A. Comfort care for terminally ill patients. *JAMA* 1994;272:1263–1266.
17. Bains M, Oliver DJ, Carter RL. Medical management of intestinal obstruction in patients with advanced malignant disease. *Lancet* 1985;2:990–993.
18. Flynn MB, Leighty FF. Preoperative outpatient nutritional support of patients with squamous cancer of the upper aerodigestive tract. *Am J Surg* 1987;154:359–362.
19. Weinstein SM. *Plumer's principles and practice of intravenous therapy,* 5th ed. Philadelphia: J.B. Lippincott, 1993.
20. Sumpmann M. An education center for patient's high tech learning needs. *Patient Ed Counsel* 1989;13:309–323.
21. Sumpmann M, Goldstein N. Patients learning center: a model for discharge skills training. *Pickler/Commonwealth Rept.* Spring 1991.
22. Rifas E, Morris R, Grady R. Innovative approach to patient education. *Nurs Outlook* 1994;42(5):214–216.
23. Goldstein NL. Patient learning center reduces patient readmissions. *Patient Ed Counsel* 1991;17:177–190.
24. Johnson SC. Learning centers help patients and caregivers master treatment skills for home care. *Patient Ed Management* 1994;1(2):17–21.
25. Murray SA. *Patient Learning Center Unit Structure Manual.* Albany, NY: Albany Medical Center Hospital.
26. Evans MA, Czopek S. Home nutrition support materials. *Nutr Clin Pract* 1995;10:37–39.

27. Malone M, Howard L. Long term hyperalimentation. *Curr Op Gastroenterol* 1994;10:227–234.
28. O Keefe, Burnes JU, Thompson RL. Recurrent sepsis in home parenteral nutrition patients: an analysis of risk factors. *J Parent Ent Nutr* 1994;18:256–263.
29. Erdman SH, McElwee CL, Kramer JM, et al. Central line occlusion with three in one nutrition admixtures administered at home. *J Parent Ent Nutr* 1994;18:177–181.
30. Orr ME, Ryder MA. Vascular access devices: perspectives on designs, complications and management. *Nutr Clin Pract* 1993;8:145–152.
31. FDA Safety Alert. *Hazards of precipitation associated with parenteral nutrition.* Washington, DC: Department of Health and Human Services, April 18, 1994.
32. Pennington CR, Pithie AD. Ethanol lock in the management of catheter occlusion. *J Parent Ent Nutr* 1987;11:507–508.
33. Monturo CA, Dickerson RN, Mullen JL. Efficacy of thrombolytic therapy for occlusion of long term catheters. *J Parent Ent Nutr* 1990;14:312–314.
34. Dollery CM, Sullivan ID, Bauraind O, et al. Thrombosis and embolism in long term central venous access for parenteral nutrition. *Lancet* 1994;344:1043–1045.
35. Mailloux RJ, Delegge MH, Kirby DF. Pulmonary embolism as a complication of long term total parenteral nutrition. *J Parent Ent Nutr* 1993;17:578–582.
36. Bern MM, Lokich JJ, Wallach SR, et al. Very low doses of warfarin can prevent thrombosis in central venous catheters: a randomized prospective trial. *Ann Intern Med* 1990;112:423–428.
37. Quigley EMM, Marsh MN, Shaffer JL, Markin RS. Hepatobiliary complications of total parenteral nutrition. *Gastroenterology* 1993;104:286–301.
38. Fein BI, Holt PR. Hepatobiliary complications of total parenteral nutrition. *J Clin Gastroenterol* 1994;18(1):62–66.
39. Nanji AA, Anderson FH. Sensitivity and specificity of liver associated tests in the detection of parenteral nutrition associated cholestasis. *J Parent Ent Nutr* 1985;9:307–308.
40. Lindor KD, Fleming CR, Abrams A, Hirschkorn M. Liver function values in adults receiving total parenteral nutrition. *JAMA* 1979;241:2398–2400.
41. Payne James JJ, Silk DBA. Hepatobiliary dysfunction associated with total parenteral nutrition. *Digest Dis* 1991;9:106–124.
42. Nanji AA, Anderson FH. Cholestasis associated with parenteral nutrition develops more commonly with hematologic malignancies than with inflammatory bowel disease. *J Parent Ent Nutr* 1984;8:325.
43. Leaseburge LA, Winn NJ, Scloerb PR. Liver test alterations with total parenteral nutrition and nutritional status. *J Parent Ent Nutr* 1992;16:348–352.
44. Elleby H, Solhang JH. Metronidazole, cholestasis and total parenteral nutrition. *Lancet* 1983;1:1161.
45. Buchmiller CE, Kleiman-Wexler RL, Ephgrave KS, et al. Liver dysfunction and energy source: results of a randomized clinical trial. *J Parent Ent Nutr* 1993;17:301–306.

46. Matuchansky C, Messing B, Jeejeebhoy KN, et al. Cyclical parenteral nutrition. *Lancet* 1992;340:588–592.
47. Chawla RK, Wolf DC, Kutner MH, Bonkovsky HL. Choline may be an essential nutrient in malnourished patients with cirrhosis. *Gastroenterology* 1993;97:1514–1520.
48. Buchman AL, Moukarzel A, Jenden DJ. Low plasma choline is prevalent in patients receiving long term parenteral nutrition and is associated with hepatic aminotransferase abnormalities. *Clin Nutr* 1993;12:33–37.
49. Buchman AL, Dubin M, Moukarzel AA. Choline deficiency causes TPN associated hepatic steatosis in man and is reversed by choline supplemented TPN. *Gastroenterology* 1993;104:A881.
50. Buchman AL, Dubin M, Jenden D, et al. Lecithin increase plasma free choline and decreases hepatic steatosis in long term parenteral nutrition patients. *Gastroenterology* 1992;102:1363–1370.
51. Desai TK, Reddy J, Kinzie JL, Ehrinpreis MN. Taurine supplementation and cholestasis during bone marrow transplant. *Gastroenterology* 1993;104:A616.
52. Chawla RK, Bonkovsky HL, Galambos JT. Biochemistry and pharmacology of s-adenosyl L-methionine and rationale for its use in liver disease. *Drugs* 1990;40(3):98–110.
53. Osman E, Owen JS, Burroughs AK. Review article: s-adenosyl L-methionine—a new therapeutic agent in liver disease? *Aliment Pharmacol Ther* 1993;7:21–28.
54. Hurley DL, McMahon MM. Long term parenteral nutrition and metabolic bone disease. *Endocrin Metab Clin North Am* 1990;19:113–131.
55. Koo WWK. Parenteral nutrition related bone disease. *J Parent Ent Nutr* 1992;16:386–394.
56. Buchman AL, Moukarzel A, Ament ME, et al. Serious renal impairment is associated with long term parenteral nutrition. *J Parent Ent Nutr* 1993;17:438–444.

Index

O

P

R

S